MAGNETISM

Materials and Applications

MAGNETISM

Materials and Applications

Edited by

Étienne du TRÉMOLET de LACHEISSERIE
Damien GIGNOUX
Michel SCHLENKER

 Springer

Library of Congress Cataloging-in-Publication Data

A C.I.P. Catalogue record for this book is available
from the Library of Congress.

ISBN 1-4020-7223-6 (hardcover)   Printed on acid-free paper.
© 2003 Kluwer Academic Publishers

First Springer Science+Business Media, Inc. softcover printing, 2005
ISBN 0-387-23000-9   E-book ISBN 0-387-23063-7

Printed in the United States of America.

9 8 7 6 5 4 3 2 1   SPIN 11313922

springeronline.com

Authors

Michel CYROT - Professor at the Joseph Fourier University of Grenoble, France

Michel DÉCORPS - Senior Researcher, INSERM (French Institute of Health and Medical Research), Bioclinical Magnetic Nuclear Resonance Unit, Grenoble

Bernard DIÉNY - Researcher and group leader at the CEA (French Atomic Energy Center), Grenoble

Etienne du TRÉMOLET de LACHEISSERIE - Senior Researcher, CNRS, Laboratoire Louis Néel, Grenoble

Olivier GEOFFROY - Assistant Professor at the Joseph Fourier University of Grenoble

Damien GIGNOUX - Professor at the Joseph Fourier University of Grenoble

Ian HEDLEY - Researcher at the University of Geneva, Switzerland

Claudine LACROIX - Senior Researcher, CNRS, Laboratoire Louis Néel, Grenoble

Jean LAFOREST - Research Engineer, CNRS, Laboratoire Louis Néel, Grenoble

Philippe LETHUILLIER - Engineer at the Joseph Fourier University of Grenoble

Pierre MOLHO - Researcher, CNRS, Laboratoire Louis Néel, Grenoble

Jean-Claude PEUZIN - Senior Researcher, CNRS, Laboratoire Louis Néel, Grenoble

Jacques PIERRE - Senior Researcher, CNRS, Laboratoire Louis Néel, Grenoble

Jean-Louis PORTESEIL - Professor at the Joseph Fourier University of Grenoble

Pierre ROCHETTE - Professor at the University of Aix-Marseille 3, France

Michel-François ROSSIGNOL - Professor at the Institut National Polytechnique de Grenoble (Technical University)

Yves SAMSON - Researcher and group leader at the CEA (French Atomic Energy Center) in Grenoble

Michel SCHLENKER - Professor at the Institut National Polytechnique de Grenoble (Technical University)

Christoph SEGEBARTH - Senior Researcher, INSERM (French Institute of Health and Medical Research), Bioclinical Magnetic Nuclear Resonance Unit, Grenoble

Yves SOUCHE - Research Engineer, CNRS, Laboratoire Louis Néel, Grenoble

Jean-Paul YONNET - Senior Researcher, CNRS, Electrical Engineering Laboratory, Institut National Polytechnique de Grenoble (Technical University)

Grenoble Sciences

"Grenoble Sciences" was created ten years ago by the Joseph Fourier University of Grenoble, France (Science, Technology and Medicine) to select and publish original projects. Anonymous referees choose the best projects and then a Reading Committee interacts with the authors as long as necessary to improve the quality of the manuscript.

(Contact : Tél. : (33)4 76 51 46 95 - E-mail : Grenoble.Sciences@ujf-grenoble.fr)

"Magnetism" (2 volumes) is an improved version of the French book published by "Grenoble Sciences" in partnership with EDP Sciences with support from the French Ministry of Higher Education and Research and the "Région Rhône-Alpes".

BRIEF CONTENTS

I - FUNDAMENTALS

Foreword

Phenomenological approach to magnetism

Theoretical approach to magnetism

Coupling phenomena

Appendices

General references

Index by material and by subject

II - MATERIALS AND APPLICATIONS

Foreword

Magnetic materials and their applications

Other aspects of magnetism

Appendices

General references

Index by material and by subject

TABLE OF CONTENTS
II - MATERIALS AND APPLICATIONS

17 - Soft materials for high frequency electronics 155

XIII

XV

OTHER ASPECTS OF MAGNETISM

FOREWORD

Thousands of years before our time, our ancestors already knew about the amazing properties of lodestone, or magnetite. Ever since, man has been fascinated by magnetic phenomena, especially because of their action at a distance. They are found everywhere in our daily lives: in refrigerator doors, cars, cellphones, suspension systems for high speed trains etc. In pure science they are present at all scales, from elementary particles through to galaxy clusters, not forgetting their role in the structure and history of our Earth.

The last thirty years have seen considerable progress in most of these fields, whether fundamental or technological. The purpose of this book is to present this progress. It is the collective work of faculty members and researchers, most of whom work in laboratories in Grenoble (Universities, CNRS, CEA), often in close cooperation with local industry, and the large international organizations established in the Grenoble area: Institut Laue-Langevin, ESRF (large European synchrotron), etc. This is no surprise, since activities concerning Magnetism have consistently been supported in Grenoble ever since the beginning of the 20th century.

Most of the chapters are accessible to the University graduate in science. Those notions which require a little more maturity do not need to be fully mastered to be able to understand what comes next. This treatise should be read by all who intend to work in the field of magnetism, such an open-ended field, rich in potential for further development.

New magnets, with higher performance and lower cost, will surely be found. The magnetic properties of materials containing unfilled electronic shells are not yet fully understood. Hysteresis plays a key role in irreversible effects. While its behavior is fairly well understood both in magnetic fields which are small with respect to the coercive field, and in very strong fields near saturation, the processes occurring within the major loop have not yet been very well described. When hysteresis depends on the combined action of two variables, such as magnetic field and very high pressure, we know nothing. How are we, for instance, to predict the magnetic state of a submarine cruising at great depth, depending on its diving course?

French scientists, with Pierre Curie, Paul Langevin and Pierre Weiss, played a pioneering role in magnetism. They will certainly have worthy successors, notably in biomagnetism in a broad sense.

This work includes interesting features: exercises with solutions, references fortunately restricted to the best papers and books, and various appendices: lists of symbols, special functions, properties of various materials, economic aspects, and, last but not least, a very necessary summary of units, which the dual coulombic-amperian presentation made so unnecessarily complicated and unpalatable in the past.

I believe this book should satisfy a broad readership, and be a valuable document to students, researchers, and engineers. I wish it a lot of success.

Louis NEEL
Nobel Laureate in Physics,
Member of the French Academy of Science

PREFACE

Magnetic materials are all around us, and understanding their properties underlies much of today's engineering efforts. The range of applications in which they are centrally involved includes audio, video and computer technology, telecommunications, automotive sensors, electric motors at all scales, medical imaging, energy supply and transportation, as well as the design of stealthy airplanes.

This book deals with the basic phenomena that govern the magnetic properties of matter, with magnetic materials, and with the applications of magnetism in science, technology and medicine.

It is the collective work of twenty one scientists, most of them from Laboratoire Louis Néel in Grenoble, France. The original version, in French, was edited by Etienne du Trémolet de Lacheisserie, and published in 1999. The present version involves, beyond the translation, many corrections and complements.

This book is meant for students at the undergraduate and graduate levels in physics and engineering, and for practicing engineers and scientists. Most chapters include exercises with solutions.

Although an in-depth understanding of magnetism requires a quantum mechanical approach, a phenomenological description of the mechanisms involved has been deliberately chosen in most chapters in order for the book to be useful to a wide readership. The emphasis is placed, in the part devoted to the atomic aspects of magnetism, on explaining, rather than attempting to calculate, the mechanisms underlying the exchange interaction and magnetocrystalline anisotropy, which lead to magnetic order, hence to useful materials. This theoretical part is placed, in volume I, between a phenomenological part, introducing magnetic effects at the atomic, mesoscopic and macroscopic levels, and a presentation of magneto-caloric, magneto-elastic, magneto-optical and magneto-transport coupling effects. Volume II, dedicated to magnetic materials and applications of magnetism, deals with permanent magnet (hard) materials, magnetically soft materials for low-frequency applications, then for high-frequency electronics, magnetostrictive materials, superconductors, magnetic thin films and multilayers, and ferrofluids. A chapter is dedicated to magnetic recording. The role of magnetism in magnetic resonance imaging (MRI), and in the earth and the life sciences, is discussed. Finally, a chapter deals with instrumentation for magnetic measurements. Appendices provide tables of magnetic properties, unit conversions, useful formulas, and some figures on the economic place of magnetic materials.

We will appreciate constructive comments and indications on errors from readers, via the web site *http://lab-neel.grenoble.cnrs.fr/magnetism-book*

ACKNOWLEDGMENTS

We are grateful for their helpful suggestions to the members of the Reading Committee who worked on the original French edition: V. ARCHAMBAULT (Rhodia-Recherche), E. BURZO (University of Cluj-Napoca, Rumania), I. CAMPBELL (Laboratoire de Physique des Solides, Orsay), F. CLAEYSSEN (CEDRAT, Grenoble), J.M.D. COEY (Trinity College, Dublin), G. COUDERCHON (Imphy Ugine Précision, Imphy), A. FERT (INSA, Toulouse), D. GIVORD (Laboratoire Louis Néel), L. NEEL, Nobel Laureate in Physics (who passed away at the end of 2000), B. RAQUET (INSA, Toulouse), A. RUDI (ECIA, Audincourt), and P. TENAUD (UGIMAG, St-Pierre d'Allevard). The input of many colleagues in Laboratoire Louis Néel or Laboratoire d'Electrotechnique de Grenoble was also invaluable: we are in particular grateful to R. BALLOU, B. CANALS, J. CLEDIERE, O. CUGAT, W. WERNSDORFER. Critical reading of various chapters by A. FONTAINE, R.M. GALERA, P.O. JUBERT, K. MACKAY, C. MEYER, P. MOLLARD, J.P. REBOUILLAT, D. SCHMITT and J. VOIRON helped considerably. Zhang FENG-YUN kindly translated a document from the Chinese, J. TROCCAZ gave helpful advice in biomagnetism, D. FRUCHART, M. HASSLER and P. WOLFERS provided figures, and P. AVERBUCH gave useful advice on the appendix dealing with the economic aspects.

We also would like to thank all our fellow authors for their flawless cooperation in checking the translated version, and often making substantial improvements with respect to the original edition. We are happy to acknowledge the colleagues who, along with the two of us, took part in the translation work: Elisabeth ANNE, Nora DEMPSEY, Ian HEDLEY, Trefor ROBERTS, Ahmet TARI, and Andrew WILLS.

We enjoyed cooperating with Jean BORNAREL, Nicole SAUVAL, Sylvie BORDAGE and Julie RIDARD at Grenoble Sciences, who published the French edition and prepared the present version.

Damien GIGNOUX - Michel SCHLENKER

MAGNETIC MATERIALS
AND THEIR APPLICATIONS

CHAPTER 15

PERMANENT MAGNETS

The number of materials systems used to make magnets on an industrial scale is limited. They are, in chronological order of appearance, AlNiCo (late 1930s), hard ferrites (1950s), the family of Sm-Co magnets which gave rise to two different magnet groups, $SmCo_5$ (late 1960s) and $Sm(CoFeCuZr)_{7\text{-}8}$, often referred to as the "2-17" type magnets (late 1970s), and the neodymium-iron-boron or NdFeB family (mid-1980s).

The magnetism of the first two families is due only to the 3-d elements (Fe and Co for the AlNiCo magnets, Fe^{3+} for the hard ferrites) whereas the 4-f magnetism of the rare earth elements Sm and Nd is also involved in the new magnet families.

Two principal magnet types may be made with these materials: sintered magnets (dense and oriented) which show the highest performance, and bonded magnets (magnetic powders injected into a non-magnetic matrix, often a polymer) which are simpler and cheaper to manufacture, but show poorer magnetic performance.

The first part of this chapter deals with the use of magnets, and introduces criteria for selecting the most suitable material and magnet type for a given application. The working principles of a magnet as well as the characterization of its performance are tackled in § 1; we distinguish between oriented and isotropic magnets in § 2; § 3 deals with the materials used while in § 4 we present various areas of application, and go on to review the principal electromagnetic and electromechanical systems which use magnets.

The second part of this chapter is concerned with the fabrication of magnets: in § 5 we discuss microstructure and preparation techniques, while in § 6 we deal with the choice of composition and basic alloy properties. Coercivity mechanisms are discussed in § 7.

1. *IMPLEMENTATION OF MAGNETS*

1.1. *THE TWO HYSTERESIS LOOPS OF A MATERIAL: MAGNETIZATION LOOP M(H) AND INDUCTION LOOP B(H)*

A typical M(H) magnetization loop of a hard material (i.e. a permanent magnet material) was presented in figure 6.2-b. This loop is characterised by its width, which indicates the value of the coercive field. A hard material with sufficient coercivity stores magnetic energy, and the amount stored depends both on the value of remanent magnetization and on the *squareness* of the loop. High squareness indicates that the magnetization varies little under the application of a reverse magnetic field until it reverses. Another type of hysteresis loop may be plotted, with the magnetic induction field B in the material replacing the magnetization M on the y-axis. This type of loop, B(H), is sometimes more appropriate for describing the working properties of the material. It is called an *induction loop* or simply a B-H loop. It is deduced from the magnetization loop by applying the classical equation (2.36) $B = \mu_0 (H + M)$. For $H = 0$, we have: $B = \pm B_r$ (remanent induction), and for $B = 0$, $H = \pm H_{cB}$ the coercive field for induction.

To within the factor μ_0, the M(H) and B(H) loops differ by $\mu_0 H$. Their values of remanence coincide, i.e. $B_r = \pm \mu_0 M_r$, while for high performance materials, the value of coercivity of the B(H) loop, H_{cB}, differs from that of the M(H) loop, denoted H_{cM}.

Remark - *Rather than plotting M(H) and B(H), equation (2.36) prompts one to plot $\mu_0 M (\mu_0 H)$ for the magnetization loop, and $B (\mu_0 H)$ for the induction loop. The same unit (the Tesla) is then used on both axes for both loop types. The quantity $\mu_0 M$ is called the polarization J. Consequently the coercive field for magnetization is often denoted H_{cJ}.*

1.2. *OPERATION OF AN IDEAL PERMANENT MAGNET WITHIN A SYSTEM*

Consider a hard material, with remanence M_r, and intrinsic coercivity $H_c = H_{cM}$, assuming it has an ideal M(H) or $\mu_0 M (\mu_0 H)$ loop as shown in figure 15.1-a. The corresponding induction loop, a parallelogram inclined at 45°, intersects the axes at $\pm B_r$ and $\pm H_{cB}$ (fig. 15.1-b). If this ideal material is coercive enough ($\mu_0 H_{cM} > B_r$), we have: $|H_{cM}| > |H_{cB}|$, and the part of the B loop situated in the second quadrant (between points B_r and H_{cB}) reduces to a straight line with slope 1, given by the equation:

$$B = B_r + \mu_0 H \qquad (15.1)$$

This part of the loop, called the *demagnetization curve*, is the basis of the properties characterising the use of the magnet, and defines its operation.

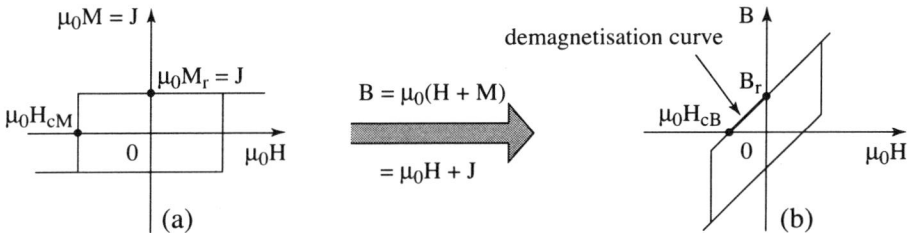

Figure 15.1 - The two hysteresis loops of an ideal material:
(a) magnetization loop - (b) induction loop

1.2.1. Load line and working point of a magnet

The most common use of a permanent magnet is as an integral part of a magnetic circuit (as already discussed in a previous chapter: see fig. 2.22-b). Schematically, such a circuit consists, apart from the permanent magnet itself, of two parts:

♦ one part, called the *pole pieces*, made from a soft ferromagnetic material (generally Fe-based), which channels the flux, and

♦ a section between the pole pieces, which may remain empty or alternatively consist of water or any other non-magnetic substance, known as the *air-gap*.

The principal air-gap is the space in which usable flux is generated. The conditions under which the magnet works are defined by the geometry and dimensions of the circuit as well as the geometry and dimensions of the magnet itself.

Consider an ideal magnet placed in a simple magnetic circuit as shown in figure 15.2. The average magnetic field line (denoted Γ) is divided into three parts. Its total length ℓ is given by: $\ell = \ell_m + \ell_p + \ell_g$, where ℓ_m is the length of the magnet, ℓ_p, the length of the pole pieces, and ℓ_g, the length of the air-gap.

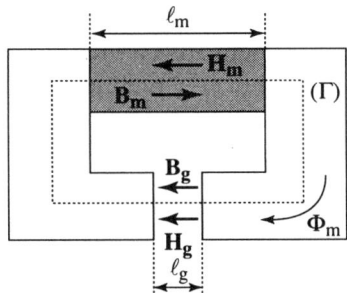

Figure 15.2 - Simplified magnetic circuit

In normal use, the magnet generates a magnetic induction of flux Φ_m, which is channelled by the circuit. The flux is constant along the entire loop (conservation of flux). For each element of the circuit one can write:

$\Phi_m = B_m S_m$ (in the magnet of cross-sectional area S_m),

$\Phi_m = B_p S_p$ (in the pole pieces of cross-sectional area S_p, considered constant),

$\Phi_m = B_g S_g$ (in the air-gap, of cross-sectional area S_g).

According to Ampère's law, the line integral of the field **H** around Γ is zero, i.e.
$\oint_{(\Gamma)} \mathbf{H} . d\ell = \sum$ (ampere turns) $= 0$, because the magnetomotive force is due only
to the magnet. Supposing that **H** is constant within a given element, this can be written
as: $H_m \ell_m + H_p \ell_p + H_g \ell_g = 0$.

If the magnetic circuit is properly designed, the pole pieces are not saturated by the
flux Φ_m (otherwise, there is no flux channelling). Thus B_p is below the saturation
value for the soft material used. Accordingly, the soft material has a very high relative
permeability μ_r (see eq. 2.52), so that the field H_p is negligible within the pole pieces.
We deduce that $H_m \ell_m + H_g \ell_g \approx 0$, or: $H_g \ell_g \approx -H_m \ell_m$.

The negative sign means that the excitation field is negative inside the magnet: it is a
demagnetising field (only the $H \leq 0$ section of the hysteresis loop of a hard material is
useful).

For this model circuit, the working equations are: $B_m S_m = B_g S_g$ and $H_g \ell_g = -H_m \ell_m$.
In addition, B and H in the air-gap are given by: $B_g = \mu_0 H_g$. This provides a linear
relationship between the magnetic induction and the magnetic field in the magnet:

$$B_m = -\mu_0 H_m (S_g \ell_m / S_m \ell_g) \qquad (15.2)$$

This is the equation for the *load line*, the slope of which is:

$$\tan \alpha = B_m / \mu_0 H_m = -(S_g \ell_m / S_m \ell_g) \qquad (15.3)$$

which is a function of the relative dimensions of the magnet and the air-gap.

The point of intersection of this straight line with the demagnetization curve of a
magnetic material defines the *working point* for the conditions considered (point P in
figure 15.3).

In practice, the useful flux in the principal air-gap represents only a fraction of the total
flux created by the magnet. The remainder passes outside the pole pieces, as flux
leakage. This flux leakage is taken into consideration in the equation for the
conservation of flux: $B_m S_m = \sigma B_g S_g$, where σ is the *flux leakage coefficient of the
magnetic circuit*. This coefficient is often of the order of 2 to 5, which means that the
useful flux is between just one half and one fifth of the total flux created by the
magnet.

To account for the defects of the magnetic circuit (residual air-gaps, finite permeability
of the iron of the pole pieces, etc.), we introduce another coefficient r, the
magnetomotive force loss coefficient, such that: $H_m \ell_m = -r H_g \ell_g$. The value of r is
generally of the order of 1.05 to 1.2.

With these two corrective coefficients, equation (15.2), which relates the magnetic
induction to the magnetic field in the magnet, becomes:

$$B_m = \mu_0 H_m \frac{\sigma}{r} \frac{S_g \ell_m}{S_m \ell_g} \qquad (15.4)$$

The slope of the load line is then given by:

$$\tan \alpha = \frac{B_m}{\mu_0 H_m} = -\frac{\sigma}{r} \frac{S_g \ell_m}{S_m \ell_g}$$ (15.5)

The leakage coefficient of a magnetic circuit σ plays an important role when choosing the dimensions of a magnet. Though sometimes experimentally estimated, it can now be determined with much higher precision using electromagnetic calculation software.

1.2.2. Static and dynamic operation of a permanent magnet

As shown in equations (15.2) and (15.3), or (15.4) and (15.5), the position of the working point P along the demagnetization curve of a material is defined by the choice of the dimensions of the magnet and the air-gap.

In systems without moving parts, the fixed point P defines the static operation point of the permanent magnet (see fig. 15.3).

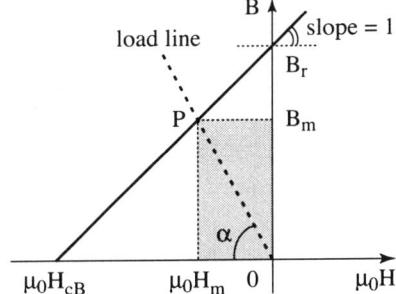

Figure 15.3 - Demagnetization curve
and load line, with slope α,
of an ideal material;
their intersection defines
the working point P

In motors, actuators and other systems where the geometry and/or the dimensions of the air-gap change, the load line rotates about the origin O (see fig. 15.4-a), or possibly (under the influence of additional coils, for example) it moves along the field axis (see fig. 15.4-b): the magnet is then said to operate in a dynamic mode.

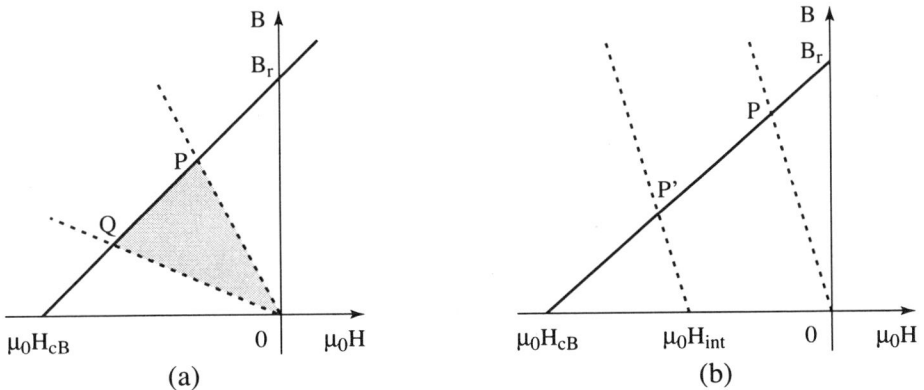

Figure 15.4 - Possible dynamic operation modes of a magnet
Displacement of the load line (a) in the case of a circuit with variable geometry
(b) under the influence of an intermittent field H_{int}.

1.2.3. The maximum energy product (static operation)

Within a magnet of given geometry, inserted in a given magnetic circuit, there exist a magnetic flux density **B** and a demagnetising field **H**. The product (**BH**) is proportional to the potential energy of the induction field created by the magnet in the air-gap. For this reason, it is a good measure of the performance of a given magnet in a given circuit.

This correlation can be established using Maxwell's equations for the situation in which there are no electric currents.

H and **B** satisfy equations (2.66) and (2.67): **curl H** = 0 (Ampère's law), and div **B** = 0, which stipulates the conservation of flux. For a permanent magnet of finite dimensions, it follows [1] that the product (**BH**) integrated over all space is zero:

$$\int_{\text{all space}} (\mathbf{B}.\mathbf{H})\, dV = 0$$

Written as the sum of an integral over the volume of the magnet, and an integral over the rest of space, the above expression becomes:

$$\int_{\text{magnet}} (\mathbf{B}.\mathbf{H})\, dV = -\int_{\text{remaining space}} (\mathbf{B}.\mathbf{H})\, dV.$$

Let us assume that the space outside the magnet does not contain magnetic substances, i.e. that outside the magnet $\mathbf{B} = \mu_0 \mathbf{H}$. Then the right-hand side of the equation is necessarily negative, implying that the left hand side is also negative. We find once again that inside the magnet **B** and **H** are oppositely oriented, or at least, are oriented at an obtuse angle to each other: the field **H** in the magnet is a demagnetising field. This result is still true if the space outside the material contains soft magnetic materials, because in this case **B**, and **H** are oriented in the same direction.

Assuming, as above, that the magnetic field **H** in the soft material is negligible, we can write:

$$\int_{\text{magnet}} (\mathbf{B}.\mathbf{H})\, dV = -\mu_0 \int_{\text{space outside the magnet}} H^2\, dV \qquad (15.6)$$

The right hand side of this equation represents twice the potential energy of the magnetic field outside the magnet, see equation (2.68). Now, in the ideal system considered here (fig. 15.2), the magnetic field **H** outside the magnet has a non-zero value only in the air gap: the potential energy of the field in the air gap is thus proportional to the product (**BH**) in the magnet.

Remark - *The above rigorous approach, based on Maxwell's equations, demonstrates the general character of the result previously stated. This result can be obtained in a less rigorous way, for the case of an ideal circuit, simply by multiplying the equations given above: $B_m S_m = B_g S_g$ and $H_m \ell_m = -H_g \ell_g$ from which: $B_m H_m V_m = -B_g H_g V_g = -\mu_0 H_g^2 V_g$, where V_m, and V_g are the volumes of the magnet and the air gap, respectively.*

The product (BH) of the moduli of the fields **B** and **H** in the magnet is called the *energy product*. Its value varies, depending on the position of the working point P. It is represented, to within a factor μ_0, by the area of the hatched region in figure 15.3. In the case where $H = H_d = 0$ (see chap. 2, § 1.3.1), i.e. for a magnet in the form of an infinitely long needle or a magnet inserted in a closed circuit, the energy product (BH) is zero. It is also zero in the case of a magnet in the form of a plate of infinite surface area in which **M** is perpendicular to the surface since, in this case, $H_d = -M$, and $B = 0$ (see chap. 2, § 1.2.6). The energy product takes its maximum value for a shape between these two extreme cases (see, further, fig. 15.7). The value of $(BH)_{max}$ characterises the magnet material.

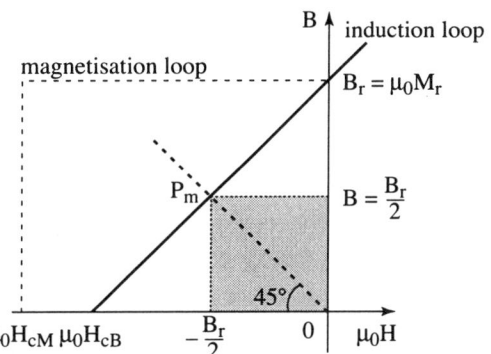

Figure 15.5 - Working point corresponding to $(BH)_{max}$ in an ideal magnet material

For the case of an ideal magnet, with a straight line demagnetising curve between B_r and $\mu_0 H_{cB}$, given by the equation $B = B_r + \mu_0 H$ (see eq. 15.1), the working point P_m for which $(BH) = (BH)_{max}$, corresponds to the centre point of this line, i.e.:

$$B = -\mu_0 H = B_r / 2 \qquad (15.7)$$

The value of the *ideal maximum energy product*, $(BH)_{max}^{ideal}$, depends only on the remanence B_r:

$$(BH)_{max}^{ideal} = \frac{B_r^2}{4\mu_0} \qquad (15.8)$$

This is the area of the square drawn from P_m (see fig. 15.5 above).

The material with the highest spontaneous magnetization at room temperature is an iron-cobalt alloy ($\approx 30\%$ Co) for which $\mu_0 M_{sat} = 2.4$ T. This value represents the highest possible value of remanent magnetization. From this we can deduce that the highest possible value of $(BH)_{max}$ attainable is 1150 kJ.m^{-3} (i.e. 144 MGOe). This is a highly idealised limit, as the coercivity of the Fe-Co alloys is very low. In 2000, the highest performance magnets have a $(BH)_{max}$ of the order of 440 kJ.m^{-3} (or 55 MGOe), as can be seen in figure 1.7 in chapter 1.

1.2.4. Free energy involved in the dynamic operation of a magnet

During an isothermal process, all variations in the free energy of a system correspond
to work provided by the external system. It can be shown [2] that the free energy
involved in the dynamic operation of a magnet equals the area of the sector swept out
by the load line during operation (grey area in fig. 15.4-a).

1.3. PERFORMANCE PARAMETERS OF REAL MAGNET MATERIALS

The concepts developed above for ideal materials are equally applicable to real
material.

1.3.1. Magnetization and induction loops
of various hard magnetic materials

Some hysteresis loops typical of various materials actually used to make magnets are
presented in figures 15.6-b-d. They should be compared with the loops which
characterise the ideal operation of a magnet, redrawn in figure 15.6-a. Real
demagnetization curves are always below the straight lines which characterise the ideal
operation of a magnet, and consequently real magnets show a poorer performance.

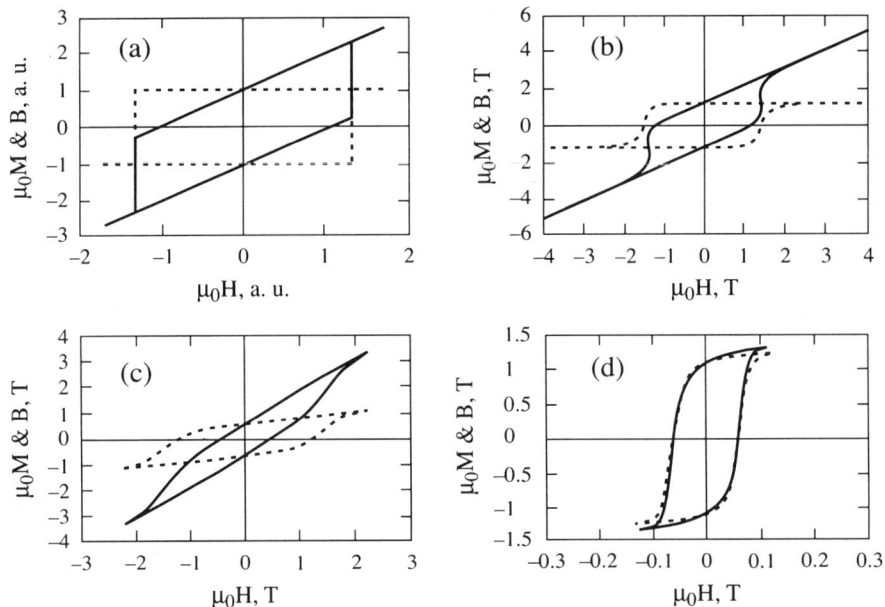

Figure 15.6 - Comparison of the behavior of magnets of varying performance
*(a) ideal magnet - (b) grain oriented magnet (sintered NdFeB) - (c) rapidly quenched
NdFeB (isotropic) magnet - (d) early design magnet (AlNiCo)*

The majority of grain oriented materials, presented in § 2 of this chapter (see
fig. 15.6-b) have hysteresis loops which are very close to ideal. For *isotropic*

materials, which are also treated in § 2, the differences between the real and ideal hysteresis loops are more pronounced (see fig. 15.6-c), especially concerning the linearity of the curves; however, H_{cM} is generally much higher than H_{cB}. For low coercivity magnets, based on early magnet design (e.g. AlNiCo), the M and B loops are almost the same (fig. 15.6-d); in particular: $H_{cB} \approx H_{cM}$.

1.3.2. The maximum energy product (BH)$_{max}$

Equation (15.6) is applicable to real materials for the same reasons that it is applicable to ideal materials. Thus the maximum potential energy which can be harnessed from a magnet material is proportional to the maximum energy product, $(BH)_{max}$. However, the working point P_m corresponding to $(BH)_{max}$ is not necessarily the middle of the demagnetising curve.

For any magnet, the position of P_m, and the value of $(BH)_{max}$ are deduced from the plot BH(H) for all points of the demagnetising curve (fig. 15.7-a). We can also remark that

♦ P_m is never far from the point where the diagonal of the rectangle constructed from the points B_r and H_{cB} intersects the demagnetising curve (fig. 15.7-a),

♦ among all the hyperbolae giving a constant value of (BH), that which corresponds to BH $= (BH)_{max}$ is tangent to the demagnetization curve at point P_m (fig. 15.7-b),

♦ near remanence, the demagnetization curves are reversible over a variable field range; they can be locally approximated by their tangents at B_r with slopes corresponding to the relative permeability μ_{rev} (see eq. 2.52).

This last remark often leads to the demagnetization curve in the neighbourhood of B_r being described by the linear equation: $B = B_r + \mu_0\mu_{rev}H$, and the value of $(BH)_{max}$ by $B^2/4\mu_0\mu_{rev}$. For an ideal magnet, $\mu_{rev} = 1$, while for a real magnet, $\mu_{rev} > 1$.

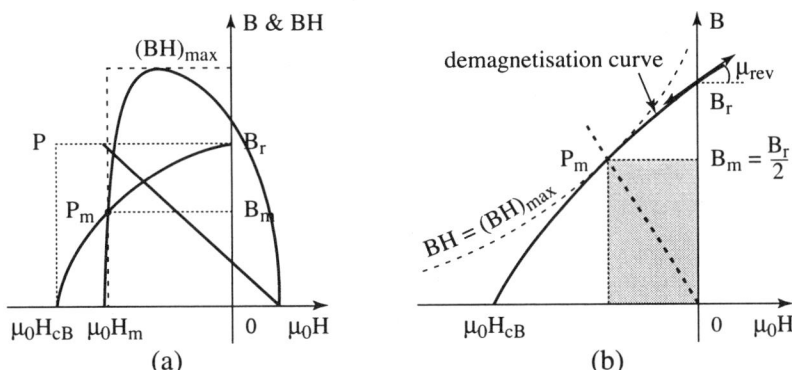

Figure 15.7 - Determination of (BH)$_{max}$ in a real magnet

(a) determination of the point P_m corresponding to the maximum energy product in any real magnet - (b) the hyperbola BH $= (BH)_{max}$ is tangent to the demagnetization curve at point P_m, and the tangent to the demagnetization curve at the point $B = B_r$ has a slope equal to μ_{rev} (if it is plotted as a function of $\mu_0 H$).

1.3.3. Performance parameters, and their range of variation

The parameters used to characterise the performance of a permanent magnet material are taken from one or other of the hysteresis loops of the material.

Principal parameters of permanent magnet materials

Parameters from the M(H) loop	Parameters from the B(H) loop
♦ Remanent magnetization M_r (remanent polarization J_r), and saturation magnetization M_{sat} (saturation polarization J_{sat})	♦ Remanent magnetic induction B_r
♦ Coercivity of the magnetization H_{cM} also called the intrinsic coercivity (H_C)	♦ Coercivity of the magnetic induction H_{cB}
♦ Squareness coefficient SQ of the loop	♦ Maximum energy product $(BH)_{max}$, and corresponding ideal maximum energy product: $(BH)_{max}^{ideal} = B_r^2 / 4\mu_0$
♦ Orientation coefficient of the material	

- ♦ The remanent magnetization and remanent magnetic induction are obtained in zero internal field (remember that $B_r = J_r = \mu_0 M_r$). Very large applied magnetic fields, larger than the anisotropy field H_A defined by equation (3.8), may be needed to measure the saturation magnetization.

- ♦ The coercivity of the magnetization, and the coercivity of the magnetic induction are the field values at which the magnetization M (the polarization J) and the magnetic induction B are zero, respectively.

- ♦ The squareness coefficient(≤ 1) is given by the ratio $S / (M_r \cdot H_{cM})$, where S is the area of the second quadrant of the M(H) loop. Therefore, the closer the operation of the magnet is to ideal, the closer the value of SQ to 1.

- ♦ The orientation coefficient of a material gives an indication on the distribution of the local directions of magnetization about the mean direction (see § 2). This coefficient is often defined as the ratio of the remanent magnetization to the saturation magnetization, M_r / M_{sat}. This value varies between 50% for isotropic materials (uniform distribution of the magnetization direction in the grains), and 100%, for a material in which the grains are ideally oriented along one direction. The distribution in the local orientation of moments can also be described by the susceptibility χ, which is the slope of the curve M(H) between M_r and M_{sat}.

- ♦ The ideal maximum energy product is $B_r^2 / 4\mu_0$. This sets the theoretical upper limit of the maximum energy product (approaching $B_r^2 / 4\mu_{magnet}$, where $\mu_{magnet} = \mu_0 \mu_{rev}$, see previous remark) for the case of ideal operation.

Other magnetic parameters

Other magnetic parameters of interest include the slope of the recoil curves (recoil permeability), maximum working field H_K, the Curie temperature T_C, the temperature coefficients of the principal parameters: B_r, J_r, M_r, H_{cM}, etc.

♦ The recoil curves (numbered 1 to 3 in fig. 15.8) are the portions of the curves along which the magnetic induction evolves when the applied magnetic field is reduced to zero from a point on the demagnetization curve which is sufficiently far from B_r (e.g. Q_1). These portions of the curves can be considered to be straight and reversible. They have a constant slope which is, in terms of dimensions, a permeability μ. In AlNiCo magnets (fig. 15.8 and 15.9), the slope μ is generally higher than μ_{magnet} ($= \mu_0\mu_{rev}$).

When the reverse field is increased once again, the magnetic induction B follows the same recoil curve in the opposite direction, and returns to point Q_1 for a field μ_0H_1; for $|\mu_0H| > |\mu_0H_1|$, the induction is again described by the principal demagnetization curve. Reducing the field once again, from Q_2, the magnetic induction traces a new recoil line (numbered 2 in fig. 15.8), of similar slope, situated lower down on the principal demagnetization curve, and so on.

During the dynamic operation of an AlNiCo magnet, we are concerned not with the demagnetization curve but with the recoil curve starting from the lowest working point corresponding to the highest reverse field applied (see fig. 15.9).

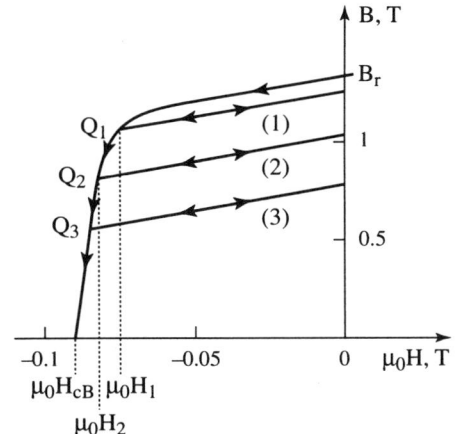
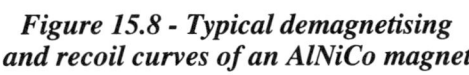
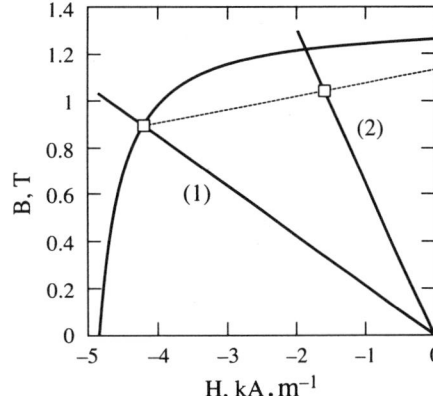

Figure 15.8 - Typical demagnetising and recoil curves of an AlNiCo magnet

Figure 15.9 - Demagnetising curve of an AlNiCo 5 magnet, and the recoil curve along which the magnet "works" between load lines (1) and (2)

♦ The maximum working field, $\mu_0 H_K$, is defined as the reverse field for which the magnetization is reduced by 10%; thus it corresponds to the point on the magnetization loop for which $\mu_0 M = 0.9 \mu_0 M_r$ ($J = 0.9 J_r$) (fig. 15.10). The value of this demagnetising field is the upper limiting value of the external field which may be applied to the magnet,. Beyond this value the magnetic properties of the magnet will irreversibly deteriorate.

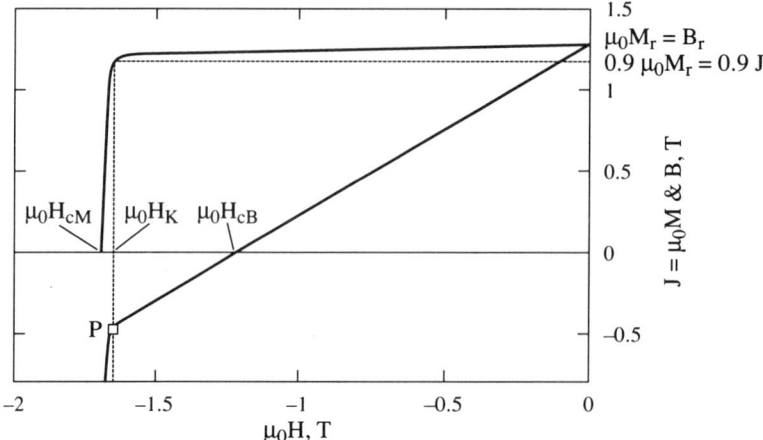

Figure 15.10 - Maximum working field, and corresponding working point (P) on the demagnetization curve

♦ The Curie temperature T_c defines the upper temperature limit for the ferromagnetic behavior of the material considered, and thus gives an indication of the temperature range over which the material may be considered for use.

♦ The temperature coefficients (valid over certain temperature ranges only) are used to evaluate the stability of the important magnetic parameters when the working temperature is different from ambient temperature or when the temperature may vary during operation of the magnet.

2. ORIENTED (TEXTURED) AND ISOTROPIC MAGNETS

The hysteresis loop (and consequently, the performance) of a magnet made of a given hard magnetic material depends on whether the magnet is *oriented* (or *textured*) or isotropic. We have already alluded to this characteristic, which we will now precisely define. In presenting these two magnet types, we will make a brief comparison of their magnetic behavior.

2.1. MAGNET CLASSIFICATION

We will see later (§ 5), that all hard magnet materials consist of an assembly of magnetic grains. For certain magnet types, each grain is a tiny single crystal. For other types (including most of the bonded magnets), each grain consists of a large number of entangled, randomly oriented crystallites.

If the grains are single crystalline, their easy axes may –before assembly– be oriented along a unique direction by an external magnetic field. The easy axes are then distributed within a cone, which is more or less open, around the field direction. This characterises the oriented (textured) magnets (case I, fig. 15.11).

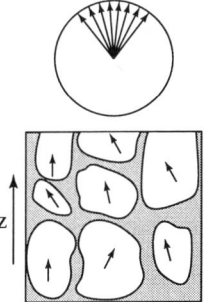

Figure 15.11 - Textured magnets

The hard magnetic phase is white while the non-magnetic phase is grey. The magnetic crystallites are separated, and the individual magnetic moments are more or less oriented along the same direction z.

Because of the orientation of the crystallites, the value of B_r is close to the maximum attainable value for B, i.e. $J_s = \mu_0 M_s$, the saturation polarization which characterises the material. This situation is very favourable for the magnetic performance. When the grains are polycrystalline (or single crystalline but non-oriented), the magnetic moments point in all directions. These are known as "isotropic" magnets (case II, fig. 15.12).

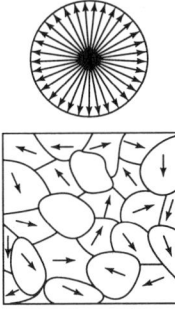

Figure 15.12 - Isotropic magnet

Entangled nanocrystallites with their magnetic moments oriented randomly.

2.2. COMPARISON OF THE MAGNETIC BEHAVIOUR OF ORIENTED AND ISOTROPIC MAGNETS

Although the application of an external field activates the same magnetization processes in both oriented and isotropic magnets, their hysteresis loops M(H) are quite different due to the difference in the orientation distribution of their easy magnetization directions (see figures 15.13, and 15.14).

♦ *Starting state: complete saturation* ①: in oriented magnets, the external magnetic field used to saturate the magnet is applied along the axis of the cone containing the easy magnetization axes of the individual crystallites. In isotropic magnets, the field is applied in any direction. In either case, saturation is achieved by applying a sufficiently high field so as to remove all domain walls, and then to align the moments of the crystallites along the field. For the maximum field H_{max}, we then have: $M \approx M_{sat}$.

♦ ***From saturation to remanence***: from M_{sat} to M_r. When the magnetic field is reduced from H_{max}, the magnetic moments of the crystallites progressively return to their respective closest easy directions of magnetization. A cone formed from the directions of the magnetic moments opens up ②. In oriented magnets, this is identical to the cone formed by the easy axes when the internal field is zero ③; the resulting magnetization is the remanent magnetization M_r. The value of M_r is typically 88% to 97% of M_{sat}. The process is similar for an isotropic magnet, though the value of M_r is much lower; it is exactly equal to 0.5 M_{sat}, because the cone of easy axes is a half-sphere, the axis of which is along the applied field.

♦ ***Increasing negative field***: opening of the cones, and reversal of the magnetic moments. As the reverse field progessively increases (④ and ⑤), the cone further opens up due to the torque the field exerts on the moments, while at the same time the moments of individual crystallites reverse, one after the other. Each moment orients itself along its easy axis which is closest to the applied field direction. Both phenomena, rotation and reversal of the moments, occur successively and/or simultaneously. The decrease in M under the influence of the applied reverse field is much stronger in isotropic than oriented magnets; in the isotropic magnets the moments are more spread out around the field direction, and thus experience a stronger torque. Consequently, oriented magnets have squarer hysteresis loops: this is another property which improves their performance.

The coercive field H_{cM} –for which magnetization is zero– does not depend much on whether the magnet is isotropic or oriented.

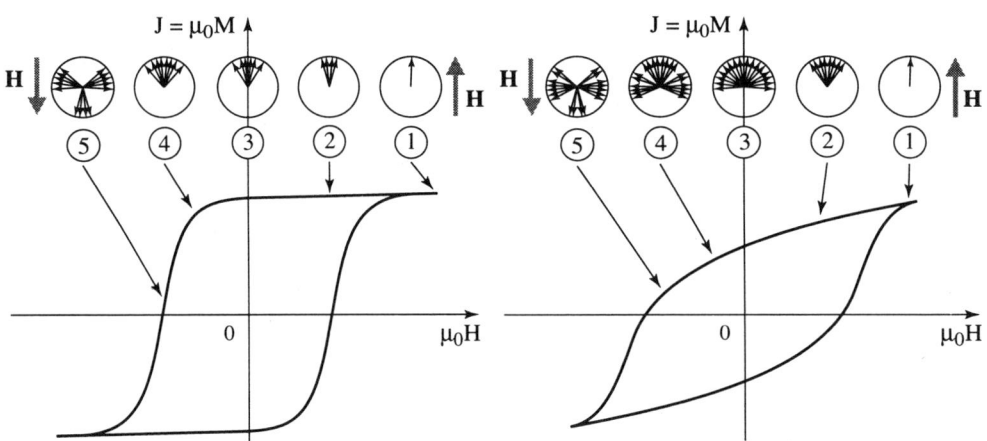

Figure 15.13
Hysteresis loop of an oriented magnet

Figure 15.14
Hysteresis loop of an isotropic magnet

3. *PRINCIPAL INDUSTRIAL MAGNET MATERIALS*

These consist of the four families already mentioned in the introduction to this chapter: *AlNiCo*, hard *Ferrites*, *SmCo* and *NdFeB*. The industrial exploitation of such diverse materials demonstrates that the magnetic performance criteria, especially the maximum energy product, are not the only criteria. Technical constraints –working temperature, thermal stability, miniaturization possibilities, etc.–, and cost constraints, weigh heavily on the choice of magnet type.

3.1. *THE VARIOUS TYPES OF SINTERED AND ORIENTED MAGNETS*

The four families of permanent magnets (sintered and oriented) industrially produced today are represented in a plot of the cost per joule *vs* the maximum energy available (figure 15.15). It is obvious from this plot that two distinct groups exist.

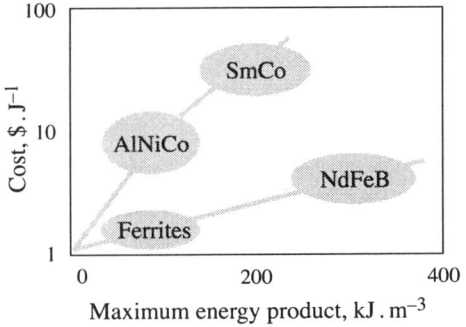

Figure 15.15 - The principal magnet families

The first group consists of the ferrites and NdFeB magnets which provide energy at a lower price, and, as a result, represent the greater share of the market. The choice between them depends on whether cost or performance is the priority:

♦ The ferrites are still commercially successful today, even though they were discovered in the early 1950's. They are sufficiently good to fulfil most classical industrial needs at low cost. They cover the majority of traditional and mass production applications: motors, car accessories...

♦ NdFeB magnets are preferred for high performance applications, including the ever increasing number of new applications which require miniaturization: actuators, such as for read/write heads for magnetic recording,...

The second group, situated in the upper part of the diagram, consists of AlNiCo and SmCo magnets. These magnets have a higher cost for a given available energy. They are used only where they are irreplaceable, i.e. where the cheaper NdFeB or ferrite magnets cannot be used.

The advantages and disadvantages of each magnet family are presented below:

3.1.1. AlNiCo magnets

This family consists of a wide range of magnets [3], the principal disadvantage of which is their low coercivity: $\mu_0 H_C$ achieves a maximum value of about 0.2 T only at the cost of a loss in remanence by nearly 50% [4]. Their rigidity is thus low. Nevertheless, due to their unequalled thermal stability ($\Delta B_r / B_r = -0.01$ to $-0.04\% / °C$), and the possibility of using them at high temperatures (of the order of 500°C), the AlNiCo magnets retain their significant share of the market for measurement applications (especially counters) and some sensors.

3.1.2. Ferrite magnets

This family of magnets includes a number of ferric oxides of the type $XO\text{-}6Fe_2O_3$, where X is a heavy element Ba, Sr..., and O is oxygen [5]. Their remanent magnetization (≈ 0.4 T) is the lowest of all magnets now used, so that their maximum energy product is not very high. Nevertheless, they dominate the magnet market (90% in weight, 65% in turnover). They are prepared in sintered oriented form as well as bonded form, the latter usually being isotropic. As well as their low cost, they show good long term stability and corrosion resistance (they are oxides), and are easily prepared in orientable coercive powder form.

3.1.3. Samarium-cobalt magnets

The properties of $Sm(CoFeCuZn)_{7\text{-}8}$, also called Sm_2Co_{17} and $SmCo_5$ type permanent magnets have been described in many reviews, notably by Strnat, the discoverer of the $SmCo_5$ magnets [6]. These are high performance magnets but with some disadvantages:

- samarium is the most expensive of the rare earth elements which can be used to prepare magnets (see § 6),
- cobalt, which has a high Curie temperature, and can thus be used to improve the temperature performance of magnets, is also an expensive material, and its magnetization is lower than that of iron. Above all, it is a strategic element, the reserves of which are concentrated in one country, Zaire (Democratic Republic of Congo); the political instability which characterises this country (most notably in 1978 and 1997) results in irregular supply and fluctuating prices.

Nevertheless, due to their high temperature behavior and their usability in potentially corrosive environments, 2-17 type SmCo magnets have kept their market share for applications where performance and reliability are a priority over cost: aeronautic and military applications, magnetic couplings in hot environments, etc. The use of $SmCo_5$ type magnets has moved to special applications where partial substitution of Gd for Sm results in materials with constant remanent induction over a wide temperature range.

3.1.4. Neodymium-iron-boron magnets

In terms of market share, the family of NdFeB magnets is the first challenger to the dominance of the ferrites. NdFeB magnets can be prepared by various processes (see § 5), with widely varying costs. In sintered form, they possess the highest energy densities, and their remanent magnetizations equal those of AlNiCo magnets. Due to their very square magnetization loops, they display quasi-ideal operation at ambient temperature. The major disadvantage of these materials is their low Curie temperature ($\approx 300°C$), which restricts their working temperatures. They are also easily oxidised on exposure to air, with the result that grains which are in direct contact with air are non-coercive. Consequently, the magnets must be protected at the surface. To date, this has made it impossible to prepare coercive, orientable NdFeB powders. Many review articles and monographs have been written on this magnet family, including an article by Herbst and Croat [7], the discoverers of the magnetic properties of sintered NdFeB, as well as a collective work prepared within the framework of a European research programme, edited by Coey [8].

3.2. PARAMETERS AND TYPICAL CURVES

3.2.1. Sintered and oriented magnets

Typical values of the principal magnetic parameters characterising sintered and oriented magnets of each material type are shown in table 15.1. The corresponding demagnetization curves, expressed in magnetization and in magnetic induction, are compared in figure 15.16.

Table 15.1 - Typical magnetic parameters for permanent magnet materials

	B_r (T)	$\mu_0 H_{cM}$ (T)	$\mu_0 H_{cB}$ (T)	$(BH)_{max}$ (kJ.m^{-3})	$B_r^2/4\mu_0$ (kJ.m^{-3})	T_C (°C)/(K)
AlNiCo	1.3	0.06	0.06	50	336	857/1130
Ferrites	0.4	0.4	0.37	30	31.8	447/720
SmCo$_5$	0.9	2.5	0.87	160	161	727/1000
SmCo (2-17)	1.1	1.3	0.97	220	241	827/1100
NdFeB	1.3	1.5	1.25	320	336	313/586

The M(H) loops of ferrite, SmCo and NdFeB magnets have similar shapes (in particular for the ferrite and NdFeB magnets). These materials show a near-ideal behavior: the variation of B as a function of H is linear in the second quadrant, which indicates that the slope of the load line, which is characteristic of the conditions under which these materials are used, may be very small. These materials can therefore be used to make magnets of any shape, even very thin plates.

3.2.2. Bonded magnets

Most hard magnetic materials are also used to make composite magnets, using a metallic or polymer matrix. These magnets are called *bonded magnets*, and they are often *isotropic*.

In these magnets, the magnetic powder is bonded by a polymer, a resin or even a low melting point metal, such as zinc. Due to the dilution of the magnetic material, and the fact that they are most often isotropic, some of the magnetic properties are inferior to those of sintered magnets. However, the loss in magnetic performance may be acceptable if accompanied by advantages such as:

♦ simplicity and low production cost,

♦ ease of forming, especially for magnetised parts, and

♦ mechanical resistance.

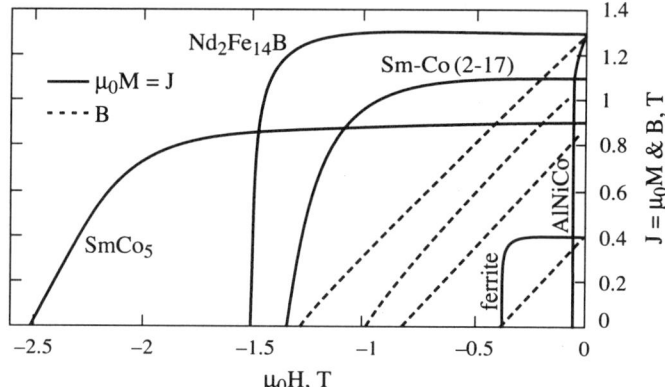

Figure 15.16 - Comparison of the performances of commercial magnets

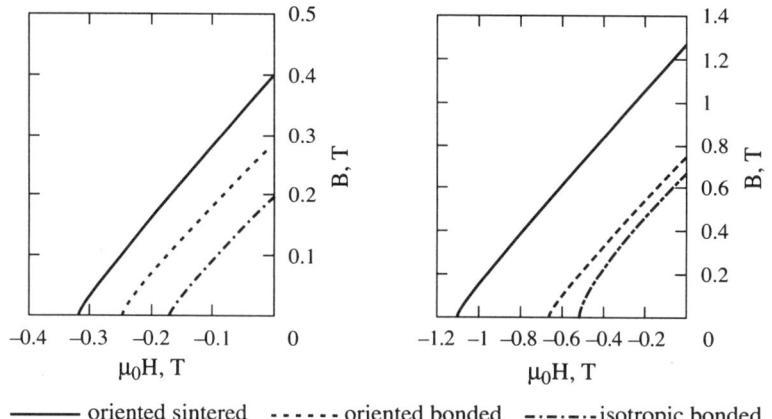

***Figure 15.17 - Comparison of the performances of oriented sintered magnets,
and isotropic and oriented bonded magnets***

Hard ferrites (left) and NdFeB based magnets (right). Note the different scales used.

Whether laminated, injected, pressed or extruded, bonded magnets are of considerable commercial interest. Their share of the market, which in 1999 was of the order of 30% of that of oriented magnets, is rapidly increasing. Demagnetization curves typical of isotropic bonded ferrite and NdFeB magnets are shown in figures 15.16 and 15.17, respectively, with –for comparison– the characteristic curves of oriented sintered magnets of the same magnet families. To a first approximation, for the same magnetic material, oriented bonded magnets have a magnetization which is two-thirds that of an equivalent sintered magnet.

4. USES OF PERMANENT MAGNETS

Permanent magnets are an integral part of a wide variety of systems where they are used to supply a permanent magnetic flux, and to generate a static magnetic field. The general public is aware that magnets are used for holding objects, and that they are an essential element in small DC motors; generally they are unaware of the multitude of other systems used in everyday life which depend on magnets.

It is possible to classify these systems in three categories, depending on how the magnet is used:

- *Electromechanical systems,* which produce forces or torques by interaction between the magnetic field created by the magnet and electric currents: motors, actuators, etc. These forces or torques are directly proportional to the product $I.J$ where I is the electric current, and J is the polarization of the magnet.

- *Magneto-mechanical systems*, which produce forces by direct interaction between magnets, or between a magnet and a piece of the assembly. The forces are proportional to J^2, where J is the polarization of the magnet. These include magnetic gears, bearings, shock absorbers, etc.

- Systems which use magnets as a *permanent source of magnetic field*. The effects are directly proportional to the polarization J of the magnet: field sources, magnetic sensors, etc.

4.1. THE PRINCIPAL FIELDS
OF PERMANENT MAGNET APPLICATIONS

4.1.1 Miniaturization

When magnets are used to create a magnetic field in an airgap, their principal competitors are solenoids. Comparison between these two field sources shows that magnets are better adapted for small systems. This comparison can be carried out with a simple magnetic circuit in which the field source may be a magnet or a solenoid (see fig. 15.18). The magnet and the solenoid create the same magnetic field H_g in the airgap g. The comparaison is made for a 2D system, i.e. the effect of depth is assumed to be negligible. The iron of the pole pieces is assumed to have infinite permeability. Calculation of the magnetic field in the airgap gives:

♦ for the circuit with a magnet (fig. 15.18-a):

$$H_g^{(a)} = -(l_m/g)H_m$$

where H_g and H_m are the fields in the airgap and the magnet.

♦ and for the circuit with a solenoid (fig. 15.18-b):

$$H_g^{(b)} = (S/g)j$$

where S is the cross-section of the solenoid in which a current density j flows.

If we now reduce all the cross-sectional dimensions of the magnetic circuit by the same factor k, the airgap will be reduced to: g' = g/k. We can then state that:

♦ for the circuit with a magnet (fig. 15.18-c), the field in the airgap remains unchanged:

$$H'_g^{(a)} = -(l_m/k)(k/g)H_m = H_g^{(a)}$$

♦ while for the circuit with a solenoid (fig. 15.18-d), the magnetic field in the airgap is reduced by k:

$$H'_g^{(b)} = (S/k^2)(k/g)j = H_g^{(b)}/k$$

This comparison indicates that it is difficult for a solenoid to create a constant high magnetic field in a circuit of small size. In practice, the ampere-turns needed to create a constant high field cannot be achieved. On the other hand, a system with permanent magnets can be used to create constant high magnetic fields, often of the order of 1 tesla, even in miniaturised systems.

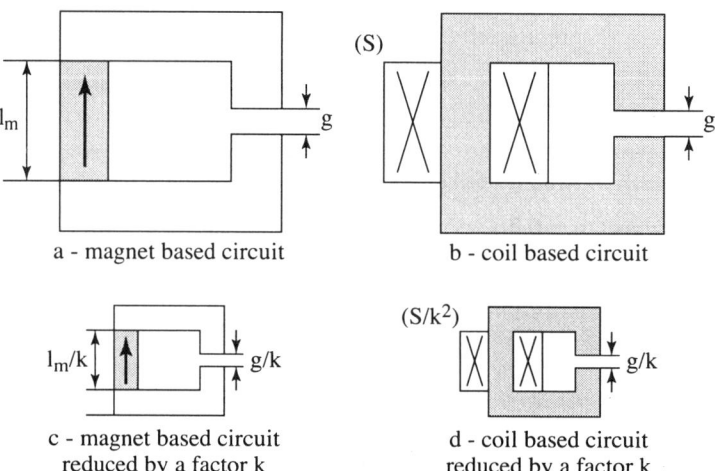

a - magnet based circuit

b - coil based circuit

c - magnet based circuit
reduced by a factor k

d - coil based circuit
reduced by a factor k

Figure 15.18 - Comparison of a magnet-based magnetic circuit and a solenoid

In the area of electric motors for example, all small sized motors use permanent magnets: watches, toys, rasors, ventilators, etc. In automobiles, all the electric motors except the ignition are made with permanent magnets, and now even some ignition motors are being made with permanent magnets. In miniaturised systems (sensors, micromotors, etc.), magnets are irreplaceable.

4.1.2. Permanent field sources

Another obvious application is to use permanent magnets to make constant flux sources. They can be used to make passive systems, which operate without external power sources.

Permanent flux sources are used for very diverse applications. Here we name just a few:

◆ Constant field sources for nuclear magnetic resonance imaging (MRI, see chap. 23). These magnet sources compete with superconducting coils. More than one tonne of neodymium-iron-boron (NdFeB) magnets are needed to make an MRI machine.

◆ Magnetic holding systems: latches, shutters, sign fixtures, etc. These systems use cheap magnets.

◆ Loudspeaker-type actuators, where a coil is placed in the permanent flux. These are also used in the magnetic recording industry to position the read/write heads over the hard disk: as high performance is demanded, the most powerful NdFeB magnets are used.

◆ Certain magnetic sensors, such as ABS systems, which operate thanks to the permanent flux produced by a magnet, often an AlNiCo magnet. The sensor signal comes from the variation of this flux within a detection coil.

4.1.3 Repelling magnets

There are not many systems which can create a repulsive force. Repulsive forces involving coils are usually small. Only superconducting systems produce very strong forces, but these require complex cryostats, and are thus used only for some large scale systems.

Permanent magnets can produce repulsive forces which are relatively large with respect to their own weight. However, magnets used in such a way are submitted to large reverse fields, and thus they must have high coercivity. Their magnetization should be as rigid as possible because they are submitted to an applied magnetic field.

Ferrite magnets were the first to be used in repulsion, in particular to levitate the rotating disks in electric counters. Friction forces, which prevent metering when power consumption is very low, are then reduced to a minimum. Although popular in the 1960's, these systems are hardly used nowadays.

Samarium-cobalt magnets are particularily well suited to repulsion applications, due to their good loop squareness and very high coercivites. Their advent has resulted in the development of magnetic bearings used in gas centrifuge systems, turbomolecular pumps and flywheels used to stabilise satellites, such as those of SPOT.

Another example of the use of repulsive forces is in permanent magnet based magnetic coupling. The magnets are attracted when there is zero torque, but part of the magnets work in repulsion when the torque is increased. These systems are widely used in industry to transmit torque across impenetrable walls; they are most often made with NdFeB magnets.

These different domains of use show that magnets form components for a wide variety of systems.

4.2. PROPERTIES OF INDUSTRIAL MAGNETS

We have seen above that the only magnets of commercial importance are AlNiCo magnets, ferrite magnets and rare earth based magnets. Of course there are other magnet materials, but they either are no longer produced commercially, or never went beyond the research stage. We will discuss only the three commercially produced magnet families.

Each magnet type should be considered as a magnet family. For each, there exists a number of compositional variants and ways in which to assemble the magnetic grains (sintering or bonding). Rare earth magnets is a generic designation for the "1-5" and "2-17"-type samarium-cobalt magnets, and for the "Nd-Fe-B" magnets.

There are a number of parameters which should be considered when comparing different magnet types:
- the magnetic characteristics, i.e. remanent magnetization, coercivity, as well as the $\mu_0 M(H)$ or $B(H)$ curves,
- the variation of these characteristics with temperature,
- the maximum working temperature,
- the mechanical properties of the magnet, mechanical strength, thermal expansion…
- the smallest size realisable, or the minimum thickness with respect to the other dimensions,
- fabrication tolerances,
- chemical and corrosion resistance, in particular for the Nd-Fe-B,
- magnetisability,
- material and production costs.

4.2.1. Principal properties of the AlNiCo magnets

AlNiCo magnets have a very high magnetization (B_r of the order of 0.8 to 1.3 T) but a rather low coercivity (50 to 160 kA.m^{-1}). Their maximum energy products $(BH)_{max}$ are generally of the order of 40 to 50 kJ.m^{-3}, but may even be greater than 80 kJ.m^{-3} for top grade magnets.

The AlNiCos are relatively expensive materials. This is mostly because of the high cost of cobalt, which is one of the constituent elements (from 20 to 40%).

These materials are still used, despite their high cost and low coercivity, because of their thermal stability and the fact that they can be used up to 450°C-500°C. Their magnetization varies little: $(\Delta J_r / J_r \Delta T) = -0.02\% / K$, and $(\Delta H_{CJ} / H_{CJ} \Delta T) = +0.03\%$ to $-0.07\% / K$.

AlNiCo magnets are prepared by casting, and are mechanically very hard. Only the surfaces which make contact with other components are machined, the other surfaces being left in their as-cast state. The main areas of application of AlNiCo magnets are directly related to their particular properties:

♦ *High temperature use*: couplings, fluid level detection...

♦ *Thermal stability*: galvanometer, electric meter braking, tachymetric dynamo, sensors (ABS), automobile speedometer...

4.2.2. Main characteristics of ferrite magnets

Ferrite magnets are widely used, their low cost compensating for their relatively low remanent magnetization, which is of the order of 0.4 T. In addition, their relatively high coercivity, 200 to 300 kA . m^{-1}, allows them to be used in a number of applications.

Ferrite magnets are quite temperature sensitive, and they cannot be used above 200°C. A particular phenomenon occurs at very low temperature, where the coercivity notably decreases, which can result in the demagnetization of the magnet. Around room temperature, the remanent magnetization and the coercivity vary appreciably: $(\Delta J_r / J_r \Delta T) = -0.2\% / K$ and $(\Delta H_{cJ} / H_{cJ} \Delta T) = +0.3\%$ to $+0.5\% / K$.

These values mean that a temperature variation of 100 K, in a motor for example, will result in a 20% variation in the remanent magnetization of the magnet.

Ferrites are used in many forms:

♦ sintered magnets; these are ceramics, machined by grinding ($J_r \approx 0.4$ T for the anisotropic ferrites and 0.25 T for the isotropic ferrites),

♦ magnetic rubber, where ferrite particles are oriented through calendering in a sheet of rubber ($J_r \approx 0.25$ T),

♦ plastic bonded magnets. These can be formed into complex shapes by injection moulding, but their remanent magnetization is further reduced (0.16 T to 0.25 T).

Sintered ferrite magnets are mechanically quite hard, and their working surfaces are machined at the end of the production cycle. The edges of the magnets are relatively fragile, and are easily damaged by shock. These magnets are easy to magnetise: a magnetic field which is of the order of 3 to 4 times their coercivity, corresponding to an induction of 0.6 to 1.2 T, is required.

The low cost of ferrites explains their wide range of use:

♦ *Magnetic rubber*: fixation, refrigerator door gaskets, magnetic flywheels, small dc motor stators, keyboards...

♦ *Bonded ferrites*: sensors, small synchronous motor rotors.

♦ *Sintered ferrites*: DC motor arc magnets, loudspeakers, stepping motor rotors, synchronous motor rotors (example: water pump), couplings, fixation, water meters.

In practice, ferrites are found wherever their magnetic characteristics are sufficient, and where their low cost is an important parameter.

4.2.3. *Main properties of rare earth magnets*

Samarium-cobalt type magnets

As we have already seen, this family contains two groups, the $SmCo_5$- and Sm_2Co_{17}-type magnets. These materials are relatively expensive, but they feature remarkable magnetic properties.

$SmCo_5$ magnets have an enormous coercivity, of the order of 2,000 kA.m^{-1}, and, as a result, this material is very difficult to demagnetise. Its magnetization is particularily rigid and practically insensitive to external magnetic fields. It is the ideal magnet for systems operating in repulsion, e.g. magnetic bearings. Its remanent induction is of the order of 0.9 T, and $(BH)_{max}$ is about 160 kJ.m^{-3}. $SmCo_5$ magnets can be used up to 250°C. The remanent induction is not very temperature sensitive:

$$(\Delta J_r / J_r \Delta T) = -0.04\% / K, \text{ and } (\Delta H_{cJ} / H_{cJ} \Delta T) = -0.2\% \text{ to } +0.5\% / K.$$

These values are slightly temperature dependent.

Despite their high coercivities, $SmCo_5$ magnets are relatively easy to magnetise, which is not the case for Sm_2Co_{17}-type magnets; this difference is due to the different coercivity mechanisms involved.

Sm_2Co_{17}-type magnets have a magnetization of the order of 1.15 T (higher than that of $SmCo_5$ magnets), a significant coercivity (greater than 1,000 kA.m^{-1}), and their $(BH)_{max}$ values can exceed 200 kJ.m^{-3}. These magnets can operate at temperatures up to 300 to 350°C, and their properties show weak temperature dependence: $(\Delta J_r / J_r \Delta T) = -0.03\% / K$, and $(\Delta H_{cJ} / H_{cJ} \Delta T) = -0.2\%$ to $-0.5\% / K$.

SmCo magnets are still used, despite their high costs and the recent development of NdFeB magnets. They fill the gaps in the high performance market where NdFeB magnets are not suitable:

♦ *High operating temperatures*: servo motors, magnetic couplings.
♦ *Miniature systems*: sensors, heart valves, micromotors…

NdFeB type magnets

The majority of NdFeB magnets exist in a sintered form (Sumitomo process) or in the form of grains prepared by rapid quenching (General Motors process), and then assembled by bonding or by mechanical compaction. The bonded magnets can be used up to about 100°C only while the sintered magnets may be used up to

150°C-200°C. Their magnetic properties are more temperature sensitive than those of the SmCo magnets: $(\Delta J_r / J_r \Delta T) = -0.1\% / K$, and $(\Delta H_{cJ} / H_{cJ} \Delta T) \approx -0.5\% / K$.

Sintered NdFeB magnets have some distinctive advantages: firstly, they have high magnetization, above 1.4 T in the best grades, while their $(BH)_{max}$ values may exceed 400 kJ / m^3. Secondly, they are cheaper than Sm-Co magnets, because they contain little or no cobalt and Nd is cheaper than Sm, as it is much more abundant.

High coercivity magnet grades are produced for use in motors (H_C greater than 1,000 kA . m^{-1} at 20°C), where their low remanent induction (inevitable in high coercivity materials) is tolerable.

A total of 10,000 tonnes of rare earth magnets were produced in 1998. Two countries dominate the market: China and Japan, each producing 4,000 tonnes while the remainder (2,000 tonnes) was produced in the United States and Europe.

The Chinese dominance comes from their mineral resources: more than 80% of the world's rare earth reserves are in China, in particular in Inner Mongolia, where the minerals are extracted from iron ores. China is the main supplier of the cheaper magnets, used in low performance applications such as acoustic systems.

The principal uses of NdFeB magnets depend on the magnet grade:
◆ *High J$_r$ sintered NdFeB:* reading head actuators for hard disks, loudspeakers,
◆ *High H$_C$ sintered NdFeB*: DC motors, synchronous motors, magnetic couplings, sensors (ABS),
◆ *Bonded NdFeB:* hard disk drive motors, stepper motors.

NdFeB magnets are quite susceptible to corrosion. The surfaces must be covered with a protective layer after machining, the nature of the protective layer depending on the operation temperature of the magnet. Magnet manufacturers propose various coatings, both organic and metallic.

4.3. ELECTROMAGNETIC SYSTEMS

Electromagnetic systems are systems in which magnetic fields interact with electric currents to create forces or torques, the strength of which are generally proportional to the product I . J (electric current.polarization). A variety of magnets are needed for the wide range of motor types and shapes which exist. DC motors, which use arc shaped magnets are produced in large numbers for a variety of applications; e.g. in the automobile industry. Original structures have been developed, most notably in synchronous machines.

Magnets used in actuators for magnetic recording read/write heads account for a significant fraction of the NdFeB magnet market.

4.3.1. The evolution of motors

Most of the magnets used in motors are either ferrite or rare earth magnets. The choice of magnet depends on the application and the performance demands, e.g.:

♦ SmCo magnets for servomotors (robotics, machine tools, servo control motors) as they meet the performance demands, and can operate at elevated temperatures,

♦ bonded NdFeB magnets for hard disk drives, because they can be prepared as very thin parts,

♦ ferrite magnets for all electric motors in the automobile industry, and small motors for the consumer market because of their low cost.

Figure 15.19 is a schematic diagram of the classic structure of a DC motor. The flux of the stator is created by two arc-shaped magnets. This is the type of motor used for many automobile accessories (fan motors, windscreen wipers, etc.), and is equipped with sintered ferrite magnets. These magnets are machined only on the surface which is in contact with the magnetic circuit, and on the airgap surface, i.e. on the interior, and exterior surfaces of the magnet. They are then fixed on the stator by glue or more frequently by inserts which keep them in position.

Synchronous machines using permanent magnets are being developed more and more. The magnets can be glued to the surface of the rotor, or inserted in an ortho-radial position (fig. 15.20). In this structure, the magnet flux is concentrated by the pole pieces of the rotor, which allows a high induction to be achieved in the airgap. What is more, by controlling the current supply of the motor, it is possible to vary the total flux of the machine. These motors, which allow variable drive speeds, are well adapted to certain applications such as electric vehicle propulsion. In the range of power required (30 kW peak value), these machines reach an efficiency of 90 to 95% (motor + undulator) while DC motors under the same conditions reach an efficiency of 75 to 80%.

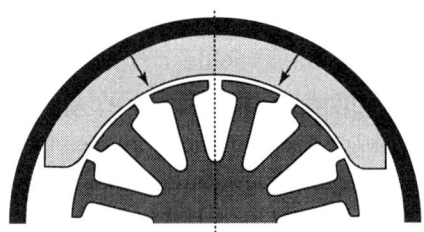

Figure 15.19
Classic structure
of a bipolar DC motor

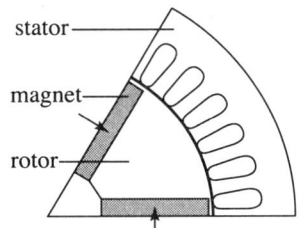

Figure 15.20 -
Permanent magnet synchronous machines
with orthoradial magnets (6 poles)

Permanent magnet synchronous machines have other advantages also. In particular, all the coils are placed on the stator, where the Joule heat is easily dissipated. Due to the progress being made in electronics, the cost of undulators is decreasing, and permanent magnet synchronous motors are beginning to replace DC motors, e.g. in the automobile industry.

The motor shown in figure 15.21 is also a synchronous motor, but its structure is a lot less conventional. The rotor consists of a multipole magnetised magnet, mounted on a shaft. The stator is made by glueing elementary coils on a flat, toothless magnetic support. Some of these motors use Hall effect sensors to determine the position of the rotor, which allows commutation of the stator phases. However, more often, these motors work without sensors, with integrated circuits determining the position of the rotor by measuring the electromotive force.

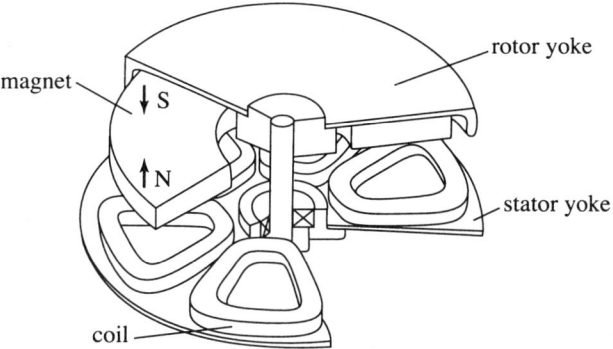

Figure 15.21 - Small brushless dc motor

This type of motor can produce a few watts of mechanical power. Its simple construction, compared to conventional motors, results in lower production costs. The magnet is axially magnetised and is multipolar.

4.3.2. Permanent magnet actuators

The loudspeaker is the best known actuator (fig. 15.22). The magnet is used to create a magnetic field in a cylindrical airgap. A coil fixed to the mobile membrane of the loudspeaker is placed in the airgap. When a current passes through the coil, the coil experiences a force (the Laplace force) which moves the membrane of the loudspeaker. Ferrite magnets are widely used for this application.

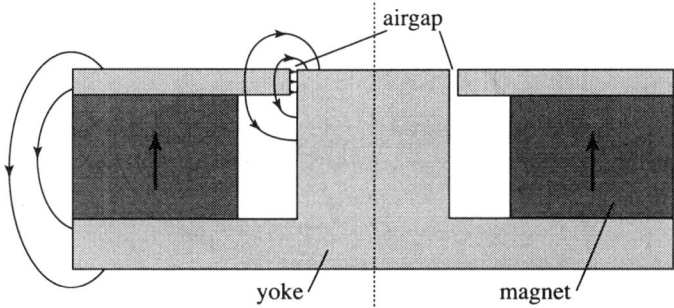

Figure 15.22 - Magnetic circuit of a loudspeaker

The same principle is exploited in actuators used to move read/write heads in magnetic disk recording systems (fig. 15.23). To achieve the lowest access time, the actuator must produce the highest possible force; for this reason they are made with sintered NdFeB magnets. The mobile part carries a coil which moves in the airgap of the magnetic circuit.

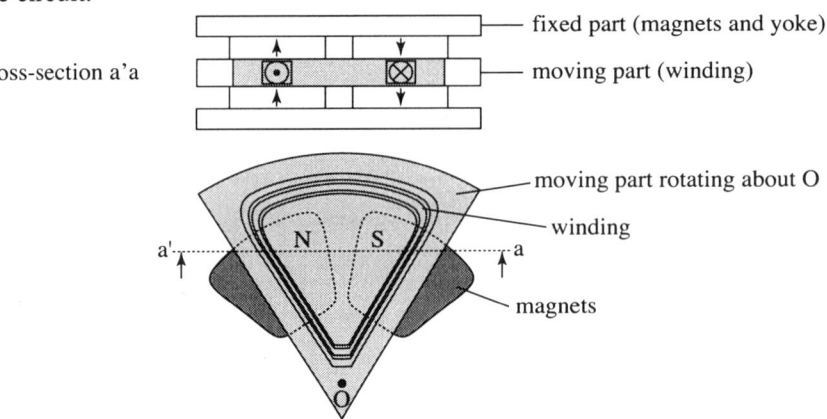

Figure 15.23 - Read head actuator for hard disk

4.4. MAGNETOMECHANICAL SYSTEMS

These are systems which operate due to the direct forces between permanent magnets. The forces obtained are proportional to the square of the polarization of the magnets. There is an obvious interest in using rare earth magnets: their high polarization results in significant forces. The magnets often operate in repulsion in these applications, therefore magnet grades with sufficiently high coercivities are needed.

The principal magnetomechanical systems are magnetic bearings and magnetic couplings. There are a number of other application including locks, fixation systems...

4.4.1. Magnetic bearings

Magnetic bearings are used for systems which rotate at high speed: flywheels turbines... Figure 15.24 shows a simple basic structure which is widely used, consisting of two annular magnets in repulsion.

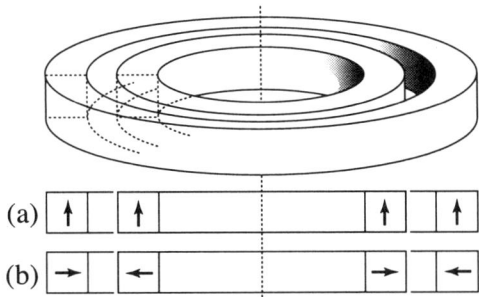

Figure 15.24 - Two types of magnetic bearings
(a) axial magnetization - (b) radial magnetization

The centring forces position one of the rings with respect to the other. The system is stable in the radial direction but unstable in the axial direction. This axial instability must be compensated by a mechanical thrust or servo control. It should be noted that the two configurations shown, A and B, create exactly the same forces. Configuration A, in which the magnetization is axial, is easier to make (ring with axial anisotropy) and to magnetise.

The magnets must have a very homogeneous magnetization, to limit the induction of currents in the facing ring, and they must be able to withstand large magnetic fields. Rare earth magnets are used because of their very high values of remanent induction, and their high coercivities.

4.4.2. Magnetic couplings

Magnetic couplings are used to transmit torque and motion through barriers (e.g. vacuum chamber walls, storage tank walls) when direct contact with the magnet is not possible. They are widely used in certain types of industrial systems, e.g. drying ovens. Magnetic couplings may also be used to limit torque and to damp vibrations in transmissions.

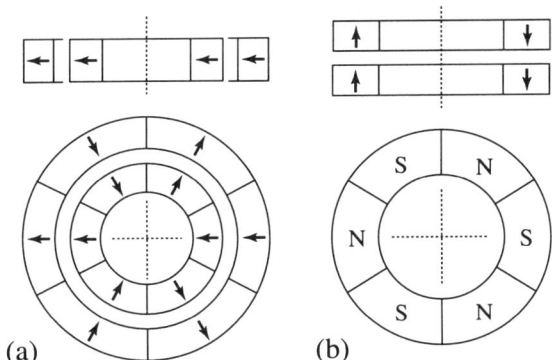

Figure 15.25 - Two examples of magnetic coupling
(a) coaxial coupling - (b) frontal coupling

Two basic configurations are used: a face-to-face system, where the separation wall is flat, and a coaxial system, where the separaration wall is cylindrical.

In the face-to-face system the maximum torque is adjusted by varying the spacing between the two coupled parts. The coaxial system is used to transmit large torques. As for magnetic bearings, magnetic coupling systems are made with rare earth magnets. If the temperature remains low, it is possible to use NdFeB; otherwise SmCo magnets are the most suitable.

Another example of the application of magnetic coupling is in the transmission systems in water meters. The external shaft is magnetically coupled to a turbine which rotates as water passes through the meter. The magnetic transmission operates with a relatively low transmission torque.

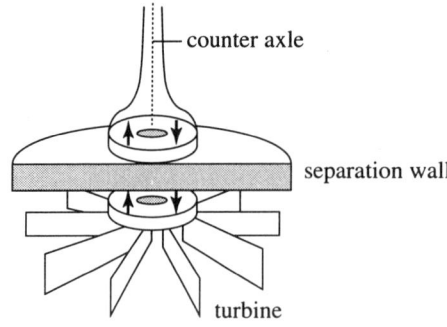

Figure 15.26
Magnetic transmission
in a water meter

The magnets attached to the turbine and the external shafts are usually ferrites, but some large counters are equiped with rare earth magnets. The magnetization of the magnets is quadrupolar to avoid the influence of external magnetic fields.

4.5. FIELD SOURCES

There are various applications which use magnets as magnetic field sources. These include sensors, systems which use magnets to bias a magnetic circuit, and systems which use magnets to create a permanent magnetic field. We cite just a few examples below.

4.5.1. Sensors

ABS sensors, which are installed in certain automobiles, have the most widespread use. The most common system design is shown in figure 15.27. The gear wheel turns with the wheel of the vehicle. At the passing of each tooth, the flux in the pole piece varies, and thus a voltage is induced in the coil. The permanent flux is created by a magnet, an AlNiCo in the structure shown.

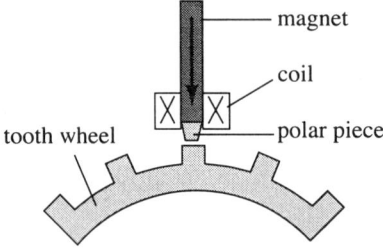

Figure 15.27
Example of an ABS sensor

In these ABS sensors, the magnets experience quite difficult working conditions, in particular with respect to temperature. AlNiCo magnets fulfill the working criteria, but now must compete with sintered ferrite magnets.

Magnets are also used in certain position sensors. The system shown in figure 15.28 operates in a differential mode: the sum of two measurements of induction, B_1 and B_2, is constant. In each airgap, the magnetic induction is measured by a Hall probe, and the measurement varies linearly with the angular position of the magnet. The output signal is given by $S = (B_1 - B_2)/(B_1 + B_2)$. The differential operation compensates for variations due to spurious effects, e.g. temperature changes. For this application, it is important to use magnets with a very homogeneous magnetization, though the actual value of magnetization itself is not as important.

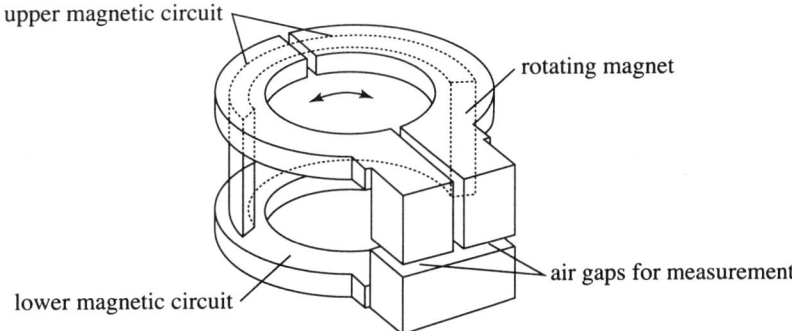

Figure 15.28 - Magnetic angular position sensor

Sensors using permanent magnets are being developed more and more, in particular with the increase in the number of electrically controlled systems, e.g. in the automobile industry. Magnet based sensors are robust, reliable, wear resistant, and cheap. They are also found in lifts, and are used to measure the speed of high speed trains (e.g. the TGV).

4.5.2. Eddy current systems

Less well known applications include the use of the magnetic field created by a magnet in braking and mechanical damping systems. Magnets are used in electric counters (fig. 15.29) where the rotating disk experiences a braking force when in close proximity to a system of permanent magnets.

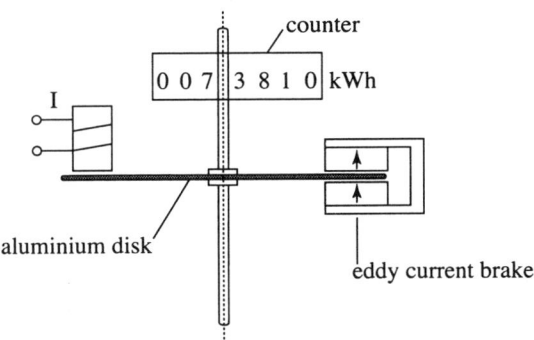

Figure 15.29 - Schematic of the working principle of an electric counter

It is an asynchronous linear motor with strong damping. In a simplified picture, the motor torque, proportional to the current I to measured, is balanced by that due to the eddy current brake which is proportional to the angular velocity. As a result the latter is proportional to the current and its temporal integration allows a measurement of the current consumption.

Another example is the needle drive in the speedometer of an automobile. A revolving cable rotates a magnet inside an aluminium bell which is stuck to the needle. The driving torque created is proportional to the angular velocity of the wheels. A spring develops a restoring torque which is proportional to the angular deviation (fig. 15.30). The counter needle thus turns by an angle proportional to the angular velocity of the wheels.

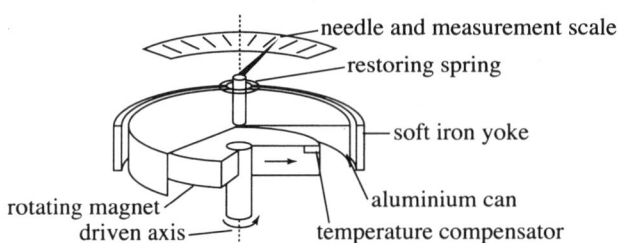

Figure 15.30 - Schematic of the working principle of a speedometer

In all these systems, the braking must be as insensitive as possible to changes in temperature. This is why AlNiCo magnets are almost always used in these measurement devices. We also mention damping created by eddy currents in other magnet-based systems: shock absorption in magnetic suspension systems, vibration damping in skis...

4.5.3. Field source

A magnet is an ideal constant field source, requiring no external power supply. Magnets now compete with superconducting coils for the fabrication of Magnetic Resonance Imaging (MRI) machines. A quantity of the order of 1 tonne of NdFeB magnets is needed for a *full-body* MRI system. Magnets are also suitable for smaller MRI systems, used to image limbs.

A constant field which is homogeneous over a large volume is needed for these imagers. Two structures are currently used:

♦ the first structure type uses magnets with pole pieces shaped to produce a constant field, and a ferromagnetic yoke (fig. 15.31)

♦ the second structure type is called a «Halbach cylinder» or *magic cylinder*. The direction of magnetization changes continuously around the perimeter of the cylinder, in such a way that the field inside the cylinder is constant. (fig. 15.32).

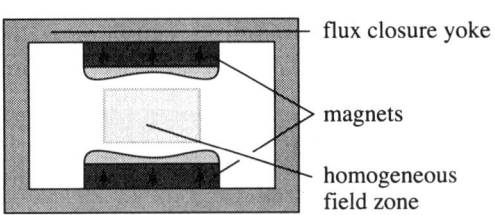

Figure 15.31
System which creates a homogeneous magnetic field

Figure 15.32
Halbach cylinder

The magnetic induction inside the cylinder is given by:

$$B = J_r \ln (r_2 / r_1) \tag{15.9}$$

If the ratio (r_2 / r_1) is large enough, the magnetic induction obtained can be much larger than J_r. The upper limit to the magnetic induction is determined by the coercivity of the magnets used.

This type of field source is used, not only for large scale MRI machines, but also for small measuring devices, on the centimeter and decimeter scales. To obtain a high magnetic induction, most systems are made of NdFeB magnets.

An intense flux source, producing a field greater than 4 T, was made on the basis of such a structure. It consists of a roughly spherical array of NdFeB magnets, containing a core of iron-cobalt, and with an external diameter of 100 mm. The grade of NdFeB magnets used is chosen in terms of the demagnetising field which the individual blocks experience locally. The prototype produces a flux density of 4.6 T in a usable volume of 6 mm (diameter) × 0.5 mm (height). The airgap can be extended from 0.5 mm to 6 mm, but the field is then reduced to 3 T. The device is presently used for magneto-optical experiments on a beam line at the ESRF [9].

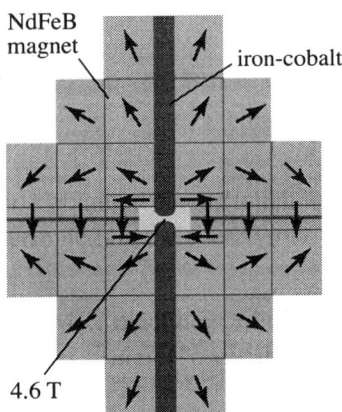

Figure 15.33 - 4 T magnet

The permanent field is created with magnets and FeCo.

Magnets can also be used to create staggered fields as in the *wigglers* and *undulators* used in the Grenoble Synchrotron (ESRF). NdFeB magnets generate a vertical staggered alternating field that curves the path of particles which turn in the ring. It is during this motion that the particles emit synchrotron radiation.

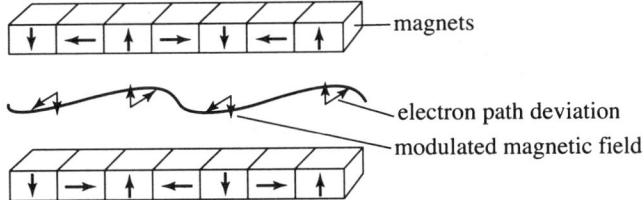

Figure 15.34 - Schematic diagram of the working principles of a wiggler and an undulator

4.6. *MODELLING PERMANENT MAGNET SYSTEMS*

Calculation methods available to model permanent magnet systems have greatly evolved in the last few years, due to developments in analytical methods as well as in numerical calculation approaches.

4.6.1. *Closed circuit calculations*

Closed circuit calculations, i.e. the calculation of the magnetic flux of a magnet channelled by pole pieces towards an air-gap, are often carried out with the method outlined at the beginning of this chapter (§ 1.2.1). The difficulty is to estimate σ, the flux leakage coefficient of the magnetic circuit, and r, the loss coefficient of the magnetomotive force. Equations (15.4) and (15.5) can be used to determine the working point of the magnet, defined as the intersection of the straight line $-B_a / \mu_0 H_a$, and the demagnetising curve of the magnet B(H), as shown in figure 15.3 [10].

The flux leakage coefficient σ depends greatly on the geometric shape of the airgap. The value of this coefficient is given for pole pieces of different shape in the works of Parker [11] and Hadfield [12].

The magnetic induction in the airgap can be determined quite rapidly with these methods, but the precision is limited because flux leakage outside the useful zone is not well known. The results are reasonable for circuits with little leakage, e.g. DC motors (fig. 15.19), but the error is quite large for circuits with significant leakage, e.g. loudspeakers (fig. 15.22).

4.6.2. *Open circuit calculations*

In magnetomechanical applications such as magnetic bearings, magnetic couplings, and certain fixation systems, magnets work with large airgaps. This is also the case when magnets operate in repulsion. The calculation methods discussed above are not applicable to such systems. To model these systems, we can replace the magnets by distributions of equivalent magnetic charges at the poles. The problem can then be treated like an electrostatic problem, and the forces can be calculated by integrating over the elementary forces between the distributions of equivalent magnetic charges.

The presence of near-by pole pieces or ferromagnetic sheets can be treated as a mirror effect if their size and their permeability are sufficiently big. The "magnet-sheet" system can be replaced by a system of two magnets where the second magnet is the image of the first by symmetry with respect to the surface of the sheet. The calculation of forces or torques then reduces to summing the interactions among elementary magnets.

The magnetic field created around a bar of rectangular cross-section, or a magnetised parallelepiped, can be calculated analytically. So too can the interactions between two magnets, in 2D or 3D: interaction energy, forces, torques, stiffness…

An entire chapter ("Magnetomechanical devices") in a recent book by J.M.D. Coey [8] deals with analytical calculations. To give just one example, the forces between two parallelepiped magnets with parallel magnetizations (fig. 15.35) are given

by: $F = \dfrac{J\,J'}{4\pi\mu_0} \displaystyle\sum_{i,j,k,l,p,q=0,1}(-1)^{i+j+k+l+p+q}\,\Phi\big(u_{ij},v_{kl},w_{pq},r\big)$

where:

for F_x, $\Phi_x = \dfrac{1}{2}\big(v^2-w^2\big)\ln(r-u)+u\,v\,\ln(r-v)+u\,w\,\tan^{-1}\dfrac{u\,v}{r\,w}+\dfrac{1}{2}r\,u$

for F_y, $\Phi_y = \dfrac{1}{2}\big(u^2-w^2\big)\ln(r-v)+u\,v\,\ln(r-u)+u\,w\,\tan^{-1}\dfrac{u\,v}{r\,w}+\dfrac{1}{2}r\,v$

for F_z, $\Phi_z = -u\,w\,\ln(r-u)-v\,w\,\ln(r-v)+u\,w\,\tan^{-1}\dfrac{u\,v}{r\,w}-r\,w$

with: $u_{ij} = \alpha + (-1)^j A - (-1)^i a$; $v_{kl} = \beta + (-1)^l B - (-1)^k b$;

$w_{pq} = \gamma + (-1)^q C - (-1)^p c$; and $r = (u_{ij}^2 + v_{kl}^2 + w_{pq}^2)^{1/2}$.

These expressions may seem complex, but they are easily incorporated into quite simple programmes, alloying a rapid optimization of the system studied.

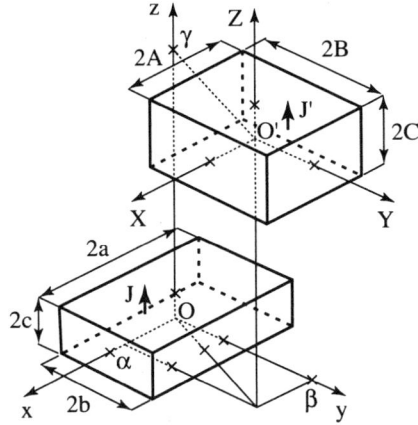

Figure 15.35 - Interaction between two parallelepiped magnets

4.6.3. Numerical methods

Numerical methods are needed for calculations involving magnets which have complex shapes, or if the soft magnetic material of the magnetic circuit is saturated. Of the many numerical methods used, the *finite element method* is the most popular. 2D and 3D calculation software exist, with the latter needing high computing power, and the precision of the results obtained is determined by the number of elements. It is preferable to use 2D software whenever possible, and 3D software for the more difficult cases: absence of symmetry, end effects, etc.

We illustrate the use of this type of software for the case of a synchronous motor made with magnets, shown in figure 15.20. The software used, called *FLUX2D*, operates by finite elements, and is based on work carried out at the Laboratoire d'Electrotechnique de Grenoble (LEG). The area studied is divided into elements (figure 15.36), with the density of elements much higher in areas where the magnetic energy is significant, i.e. around the airgap.

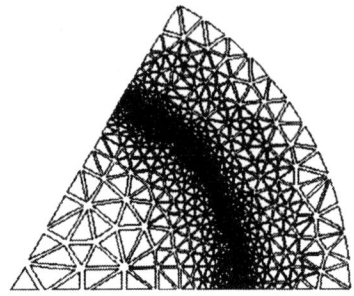

***Figure 15.36 - Dividing the studied
area into elements (FLUX2D)***

An example of results where the software was used to calculate the distribution of field lines is shown in figure 15.37. In the absence of current, figure 15.37-a, the distribution of field lines is perfectly symmetric. With a very small current I_n, figure 15.37-b, the field line map is slightly changed. We can see that the principal flux in the machine is created by the magnets, and that the flux due to stator currents only slightly modifies the flux distribution.

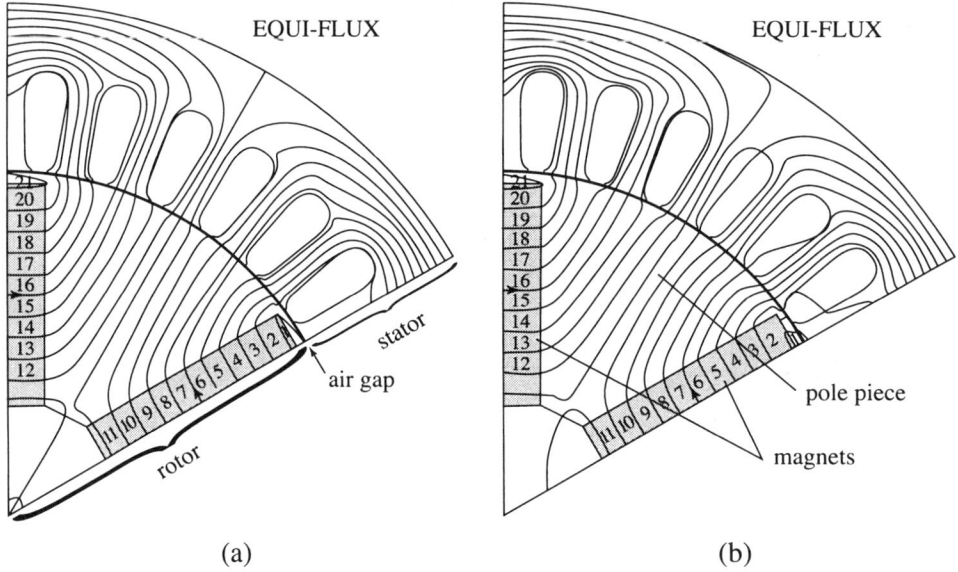

(a)   (b)

Figure 15.37 - Example of a magnetic field map
(a) at zero current - (b) at a given current

4.6.4. Magnet characteristics

To correctly use the magnet grades which are commercially available, it is necessary to know their magnetic and physical properties. This information is available from the magnet producers, which include:

- Ugimag, 38 St Pierre d'Allevard (France)
- Vac*, Hanau (Germany)
- Widia, Essen (Germany
- Sumitomo Special Metals, Osaka (Japan)
- Magnequench*, Anderson (USA)

This list is far from exhaustive.

5. MAGNET MATERIALS: MICROSTRUCTURE AND PREPARATION TECHNIQUES

5.1. COERCIVITY AND HOW TO ENHANCE IT

Coercivity, the fundamental property of hard magnetic materials, defines a material's capability to resist reversal of its magnetization **M** when it experiences a magnetic field applied antiparallel to **M**. It is quantified by measuring the magnetization reversal field (denoted H_R), the field at which irreversible susceptibility is maximum (in the half-plane $H < 0$).

Note - *In practice, H_R is generally not distinguished from the intrinsic coercive field $H_c \equiv H_{cM}$ (the reverse field for which magnetization reaches zero).*

Before magnetization reversal, the magnetization may change in a reversible manner (due to rotation of local moments), and, though this change may be significant, the magnet conserves its power. However, once magnetization reversal has occurred, the magnet properties are lost.

Magnetization reversal has already been referred to and in part studied in the two preceeding chapters: chapter 5, § 8.2 and chapter 6, § 3 and § 4.3. We thus know that two conditions must be fulfilled if a magnetic material is to resist reversal in significant reverse magnetic fields. The first of these conditions involves an intrinsic property of the material, its anisotropy: this must be uniaxial and as large as possible. The second condition concerns the microstructure: a specific microstructure, in which defects are controlled, must be developed. Due to this second condition, coercivity is considered as an extrinsic property of the material.

---

* Producers of rare earth magnets only.

5.1.1. Strong uniaxial anisotropy

The Stoner-Wohlfarth model (§ 8.2 of chap. 5) shows that nothing can prevent magnetization reversal once the applied reverse field reaches the value of the total anisotropy field $H_A^{tot} = H_A + H_A^f$ (H_A = magnetocrystalline anisotropy field and H_A^f = shape anisotropy field). Conversely, it is theoretically impossible to reverse the magnetization of a perfectly homogeneous material in a reverse magnetic field which is less than the magnetocrystalline anisotropy field H_A, an idea formulated in equation (5.64) and known as Brown's inequality. Thus, the reversal field of an ideally homogeneous substance lies between the magnetocrystalline anisotropy field and the total anisotropy field:

$$H_A \leq H_R \leq H_A^{tot} \qquad\qquad (15.10)$$

For this reason, the anisotropy fields, especially the magnetocrystalline anisotropy field H_A, are the absolute references for the reversal field.

5.1.2. The role of defects and the need for a microstructure

In reality, we observe that $H_R \ll H_A$, a result known as *Brown's paradox* (chap. 6, § 3.1). This indicates that the magnetization reversal process begins at magnetic defects in the material (compositional defects, crystal defects, stresses, etc.), and is not a collective phenomenon as it involves only a small area of the material at a given time. The first phase of the process (the possible mechanisms of which will be discussed later in § 7.2.1 of this chapter) is localised at and in the vicinity of triggering defects: this is the *nucleation* phase of magnetization reversal. After this phase comes the *propagation* phase, in the form of a wave: the wall created around an initial nucleus grows, and, driven on by the reverse field, it crosses the entire sample.

An increase in the reversal field H_R thus requires:
♦ materials with a strong uniaxial anisotropy, and
♦ a reduction in the number and influence of defects in the material so as to retard reversal nucleation.

When nucleation intervenes at low fields, the domain walls created may be pinned by suitable defects, thus retarding domain wall propagation, and consequently retarding magnetization reversal.

The nature and distribution of defects entirely determines the conditions in which coercivity can be developed. These conditions are determined by the microstructure, which is itself determined by the process used to manufacture the magnet.

5.2. THE MAGNET MATERIAL MUST BE REDUCED TO FINE PARTICLES

The microstructure to be developed should allow the following objectives to be reached: maximum coercivity (maximum reversal field), maximum remanent induction, and a rectangular magnetization loop. Each of these objectives is easier to reach if a granular microstructure is developed. Note that particle size refers to the size of the physical entity while grain size refers to the crystallite size; one particle may contain many grains, or alternatively each particle may be a single crystal.

5.2.1. Reduction into grains to suppress nucleation

In the nucleation phase, the magnetization reversal process begins at those particular defects which are most detrimental, and the magnetization then reverses throughout the magnetic volume containing the defect once the applied magnetic field allows the propagation of the domain wall. Major defects must thus be isolated to reduce their influence. This is the first reason why the material is divided into independent magnetic particles.

Figure 15.38 shows the influence of particle milling on the coercivity of SmCo5 magnets. The coercivity increases with milling time, then levels off, and finally it decreases when milling introduces new defects into the particles [13].

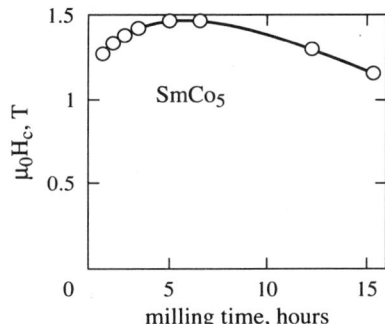

Figure 15.38 - Influence of milling time on the coercivity of SmCo5 magnets, after [13]

The optimum particle size is between 0.5 and 10 microns for *SmCo5* and *Nd2Fe14B*, while it is of the order of 1 μm for *ferrites*. After milling, the particles are usually single crystalline. Milling directly leads to coercivity in ferrite and SmCo5 powders. In the case of NdFeB, grains of Nd2Fe14B are not coercive: the surface of the grains is oxidised on contact with air. Only at the end of the process, during a post-sintering anneal, do NdFeB magnets become coercive. During this step, an intergranular phase appears, which plays an essential role in protecting the grains, and allowing coercivity to develop in these magnets.

5.2.2. Reduction into grains to increase remanence and the squareness of the M(H) loop

When particles produced by milling (or another suitable treatment) are single crystals, they can be oriented in a magnetic field. The advantages resulting from this orientation have already been explained (see § 2.2).

This is the sole objective of the milling stage in the processing of 2-17 magnets, where the starting material is a Sm(CoFeCuZn)$_{7-8}$ alloy, composed of two phases. Coercivity in these materials is not due to the suppression of nucleation but to the pinning of domain walls which hinders magnetization reversal: this is known as *pinning* type coercivity (see § 7.2.1), and it already occurs in the bulk material.

A schematic diagram of a typical microstructure of an oriented sintered magnet was shown in figure 15.11: the grains have their magnetization vectors pointing in the same direction, but they are separated from each other by an intergranular phase which serves to magnetically isolate the individual grains (i.e. the exchange interactions between neighbouring grains are suppressed).

5.3. GENERAL PRINCIPLES OF THE PROCESSES USED TO PREPARE MAGNET MICROSTRUCTURES

Each magnet has an optimum microstructure, resulting from a very carefully controlled preparation process. A number of characteristics defines the microstructure, notably: the particle size, the grain size, the arrangement of the particles and grains in the presence or absence of intergranular phases the precise nature of intergranular phases, and that of other secondary phases, the optimization of these phases by the addition of chosen additives or particular heat treatments, the alignment of crystallites in a magnetic field where possible, corrosion protection if necessary, etc.

5.3.1. Oriented sintered magnets

The metallurgical processes used to prepare oriented sintered magnets aim at the following effects:
- ◆ reduction into grains by simple milling, or by phase segregation followed by milling of the starting alloy,
- ◆ protection and magnetic isolation of the grains through a secondary phase produced by the alloy itself during heat treatment,
- ◆ grain orientation, through a strong applied magnetic field, before or during the compression stage, or during the heat treatment in the case of AlNiCo magnets,
- ◆ densification of the magnet, usually by sintering.

The various processes used for each of the principal magnet types are given in table 15.2. Only the main steps are listed. Note that the powder metallurgy route is the reference process [5,14].

The quality of crystallite alignment is a function of the strength of the magnetic field applied and the compression method used. It is possible to
- ◆ align and compress along the same axis (axial compression),
- ◆ align perpendicular to the compression direction (transverse compression), or
- ◆ apply an isostatic pressure on a prealigned powder (isostatic compression)
(as schematically shown in figures 15.39-a, 15.39-b and 15.39-c, respectively). The degree of alignment of *SmCo* and *NdFeB* magnets increases on going from axial compression ($M_r/M_{sat} \approx 87\%$) to transverse compression ($M_r/M_{sat} \approx 92\%$), and finally to isostatic compression (($M_r/M_{sat} \geq 95\%$), as do the production costs.

Table 15.2 - Magnet fabrication processes

	Sintered Ferrite	Sintered NdFeB	Sintered SmCo$_5$	SmCo 2-17	AlNiCo
Principal phase	BaO-6Fe$_2$O$_3$ SrO-6Fe$_2$O$_3$	Nd$_2$Fe$_{14}$B	SmCo$_5$	Sm$_2$(CoFe)$_{17}$	FeCo
Anisotropy	------------------ magnetocrystalline ------------------				shape
Reduction into grains	--------------- by milling ---------------			by thermal phase segregation +milling	by thermal phase segregation
Inter-granular phase	BaO-nFe$_2$O$_3$ SrO-nFe$_2$O$_3$	Nd rich eutectic	Sm eutectic	Sm(CoCu)$_5$	NiAl
Orientation of grains	-------------- by compression under field --------------				by heat treatment under field
Densification	-------------------- during sintering --------------------				during melting or sintering
Nature of generic process used	-------------------- Powder metallurgy --------------------				casting or powder metallurgy

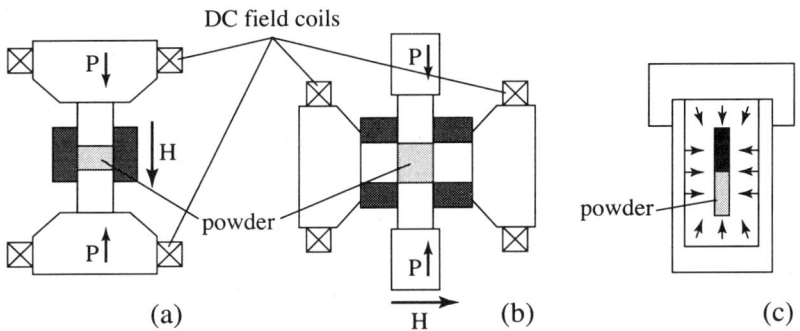

Figure 15.39 - Various modes of powder compression under a magnetic field
(a) axial compression - (b) transverse compression - (c) isostatic compression
Alignment under magnetic field is carried out just before the compression step.

An interesting mode of compression may be used for sintered NdFeB magnets [15] (see fig. 15.40): a rubber mold turns a uniaxial compression (easy to apply) into a quasi-isostatic compression (the most efficient compression mode). This is known as the RIP (rubber isostatic pressing) mode. Compression is carried out directly after alignment (in a pulsed magnetic field of about 4 T), without handling the sample in between the two steps. The degree of alignment achieved is usually of the order of 95% but may even reach 97%.

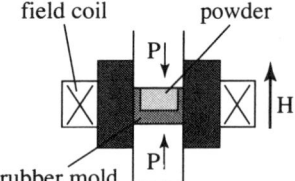

Figure 15.40 - Quasi-isostatic pressure obtained using a flexible rubber mold

In AlNiCo magnets, uniaxial anisotropy is provided by the elongated shape of the magnetic grains (here the anisotropy itself is determined by the microstructure). The desired microstructure is obtained during a thermal phase segregation treatment under magnetic field. The presence of cobalt in the starting alloy increases the Curie temperature of the principal phase, and the phase segregation begins below T_C. Phase segregation is carried out in a field of the order of 0.3 T, and the new particles thus created (of cubic structure) grow longer in the <100> direction which is closest to the field direction. All of these axes can be aligned in the same direction if this alignment favours crystal growth along the preferential direction (columnar crystal growth). An almost perfect microstructure is obtained (see fig. 15.41), where the grains, which are elongated and well separated (the volume fraction of the magnetic phase is about 65%) have a length which is 10 to 50 times greater than their diameter [16].

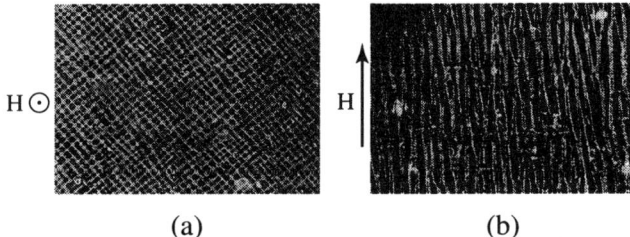

(a)   (b)

Figure 15.41 - Texture of AlNiCo magnets obtained by phase segregation under field, *after [16]: (a) plane parallel to H - (b) plane perpendicular to H*

5.3.2. Coercive powders and bonded magnets

Magnetic powders having permanent magnet properties and a typical particle size between 10 and 200 microns, are needed to prepare bonded magnets. The most widely produced powders are ferrite powders and NdFeB powders.

In the industrial preparation of ferrite powders, the starting materials (Fe_2O_3, and $SrCO_3$ or $BaCO_3$) are calcinated, after which the resulting hexaferrites react with the remaining Fe_2O_3 to form coarse powders of $SrFe_{12}O_{19}$ or $Ba\ Fe_{12}O_{19}$ which are then mechanically milled. A chemical preparation route is used to prepare ferrites on the laboratory scale. Fine magnetic powder is produced in one step with this route, i.e. a final milling step is not needed.

NdFeB powders are produced by rapid quenching. This original technique was proposed by Croat and Herbst of General Motors in 1983 [17]. A molten jet of alloy is projected onto the cold rim of a water cooled wheel (made of copper or copper-beryllium) which rotates at very high speed (fig. 15.42). Cooling rates of up to 10^6 K.s^{-1} can be achieved with this technique. The quenching takes place in a controlled atmosphere chamber, containing helium or argon. The resulting ribbon, which has a thickness of the order of 30 μm, is very brittle, and spontaneously breaks up into short pieces. The magnetic properties of the ribbon depend on the mechanical and thermal parameters of the quench: temperature of the molten alloy, ejection speed, wheel speed, cooling rate, etc. The coercivity depends very much on the wheel speed; it is most easily controlled by overquenching the material and then adjusting the grain size by subsequent annealing.

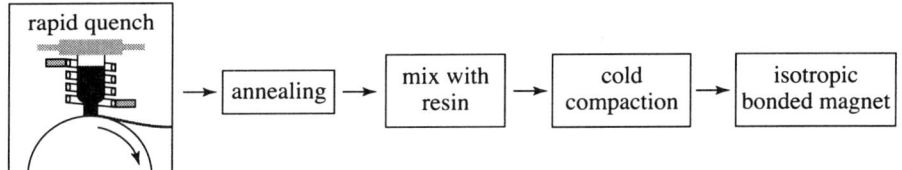

Figure 15.42 - Fabrication process for a bonded isotropic magnet

The pieces of ribbon thus processed consist of crystallites with an average grain size of about 50 nm (about a hundred times smaller than the crystallites of sintered magnets). The anisotropy axes of these crystallites are randomly oriented: the material is thus isotropic. At the time of writing (2,000), it is still not possible to obtain NdFeB powders which are both coercive and orientable in a magnetic field, and thus it is not possible to prepare oriented bonded magnets, which would have much higher energy products than their isotropic counterparts.

A number of magnet grades, denoted as MagneQuench (MQ) magnets, are produced from rapidly quenched NdFeB ribbon.

6. THE BASIC MATERIALS FOR PERMANENT MAGNETS

6.1. MAGNETIC CHARACTERISTICS OF 3d AND 4f ELEMENTS VS THE PROPERTIES REQUIRED TO OBTAIN A HARD MAGNETIC MATERIAL

Permanent magnet materials must have very high magnetization and strong anisotropy from room temperature up to 100 or 150°C. The first requirement dictates that they must be ferro- or ferrimagnets up to high temperatures (above 500 K) while the second dictates the need for uniaxial anisotropy favoring a high symmetry axis.

With these requirements in mind, it is instructive to recall specific properties of magnetic elements used to make magnets:

♦ transition elements (Fe, Co, Ni…), in which magnetism is due to the 3d electronic shell,

♦ rare earth elements (Nd, Sm…), in which magnetism is due to the 4f electronic shell.

A detailed review of the intrinsic magnetic properties of all these elements, and the origin of these properties was given in chapters 7, 8 and 9.

6.1.1. Magnetic moments and exchange interactions in rare earth and transition metals

Magnetism in rare earth elements is due to the 4f electron shell which is very localised (see fig. 7.8). The electron-electron interactions are strong (of the order of 10 eV), and the atomic moment is built up according to Hund's rules: $S = S_{max}$, then $L = L_{max}$ (see chap. 7 § 1.2.2). The L and S moments are strongly coupled by the spin-orbit coupling term $\lambda L S$ (of the order of 1 eV) between the orbital moment and the spin moment. The value of the magnetic moment (up to $10 \mu_B$ / atom) is well defined (atomic moment $= g_J J \mu_B$).

Due to the local character and depth of the magnetic shell within the atoms, there are no direct interactions between 4f electrons of different rare earth atoms. However, the 5d and 6s electrons of a rare earth atom, which have an itinerant character, interact with the 4f electrons of the atom. The 5d and 6s states are polarised, and give rise to an indirect coupling between the localised 4f electrons (see chap. 9, § 3.1: RKKY interaction). This coupling is relatively weak in rare earth metals (0.01 eV), and leads to ordering temperatures below room temperature (the highest, 20°C, is found for gadolinium). Moreover, the polarization of conduction electrons oscillates with distance, alternately taking positive and negative values (see fig. 9.4). This can lead to various types of magnetic coupling, not just ferromagnetic coupling.

In transition metals, such as iron and cobalt, the 3d electrons, which are responsible for the magnetism of these metals, are itinerant, and form a band of typical width 5 eV. The exchange interactions, of the order of 1 eV, tend to favour the occupation of states in the same half-band (i.e. with the same spin state: a phenomenon which is equivalent to the one giving rise to Hund's first rule), but this obviously gives rise to an increase in kinetic energy. Competition between exchange and kinetic energy can result in a splitting of the two half-bands, that of spin ↑ (in the direction of the resultant moment), and that of spin ↓ (in the opposite direction), thus giving rise to a permanent magnetic moment (see chap. 8). This occurs only in few cases, principally in systems based on Mn, Fe, Co or Ni. The maximum magnetic moment of a simple metal is obtained for iron: $2.2 \mu_B$ per atom of Fe. The experimentally determined magnetic moments of 3d metals and alloys, plotted as a function of the number of 3d electrons per atom, follow a curve known as the *Slater-Pauling* curve, the significance of which was briefly explained in chapter 8.

Magnetic interactions in transition metals are generally stronger than those in rare earth metals. Qualitatively, ferromagnetism dominates when the 3d band is nearly full. Cobalt orders ferromagnetically at 1,392 K, and nickel, which has a magnetic moment of just 0.6 μ_B / atom, orders at 631 K. In the middle of the band, competition exists between ferromagnetic and antiferromagnetic interactions. Iron orders at 1,043 K, but many iron based alloys order at low temperature, due to competition between opposing interactions. Manganese and its alloys are most often antiferromagnetic.

In conclusion, iron, cobalt and nickel are the only elements which, in the metallic state, have a Curie temperature above room temperature. All magnetic materials to be used for applications must contain one of these elements, and/or manganese.

6.1.2. Magnetocrystalline anisotropy of 3d and 4f elements

Magnetocrystalline anisotropy is due to attractive-repulsive interactions exerted on electrons of the magnetic shell of an atom by the electric field created by all nearby ions (the crystal field). If neither the electronic orbitals of the magnetic shell considered, nor the equipotential surfaces of the crystal field have spherical symmetry, this coupling leads to a preferential orientation (corresponding to a minimum energy) of the electronic distribution and the associated orbital moments. In addition, the spin moment is induced to follow the orbital moment by spin-orbit coupling. Therefore, the global magnetic moment of the atom orients along a particular crystallographic direction.

A more quantitative discussion of magnetocrystalline anisotropy should take into account all the energy terms involved, notably exchange (which determines the spin moment), spin-orbit coupling and coupling between the magnetic shell, and the crystal field. The relative importance of these contributions is different for transition metals, and rare earth metals.

Transition elements Fe, Co, Ni

The 3d shell of a transition metal ion is strongly coupled to the local crystal field because it is an external shell. This coupling is the dominant energy term. It competes with the intra-atomic electrostatic repulsion which leads to Hund's rules in insulating systems. The electronic distribution adopts a configuration which minimises its interaction with the crystal field, and the values of the orbital and spin moments are strongly modified. In particular, the orbital moment vanishes or is weak for cubic symmetry. The energy terms to be considered are, in order of decreasing intensity, crystal field, exchange coupling between spins and spin-orbit coupling ($\lambda L S$).

The intensity of this last term is of the order of 10^{-2} eV / atom. The spin moment S_z can freely orient itself along any crystallographic direction without notably affecting the energy of the system. Thus the magnetic anisotropy is weak (of the order of 10^{-5} eV / atom): this is the case for metallic iron and nickel.

The orbital moment is non-zero when the symmetry of the crystal field is uniaxial, and it is maximum along a particular crystallographic direction. Due to the spin-orbit coupling $\lambda L S$, the spin moment follows the orbital moment, and this particular direction becomes the easy magnetization direction. The anisotropy energy thus obtained in hexagonal cobalt alloys has a value of the order of 10^{-4} eV / atom.

Rare earth metals (see chap. 7, § 4.4)

The deep 4f electron distribution is much less influenced by the crystal field as it is shielded by more external electronic shells. Thus, spin-orbit coupling is the dominant term in the rare earths. For the systems of interest here, exchange interactions are next; the crystal field comes third.

The electronic distribution of the 4f shell is entirely determined by exchange and spin-orbit coupling, S_z and L_z taking their maximum values. This distribution orients itself with respect to the crystallographic directions so as to minimise the energy of interaction with the crystal field, and this defines the easy magnetization direction. The magnetic anisotropy represents the variation in coupling energy between the 4f electronic distribution and the crystal field depending on whether the orientation of the orbital is along an easy direction or a hard direction. In a uniaxial system, the crystal field is strongly aspheric, and the anisotropy can be very high (10^{-3} eV / atom).

6.2. TRANSITION METAL BASED MAGNET MATERIALS

Until 1940, all known magnet materials were transition metal based with cubic crystal symmetry. Thus their magnetocrystalline anisotropy was often much smaller than their shape anisotropy. This is why shape anisotropy was exploited in AlNiCo type magnets (as well as in other similar systems, FeNiAl or FeCoCr).

In the 1950's, new families of compositions having hexagonal structures appeared, allowing the development of significant magnetocrystalline anisotropy: MnBi, MnAl, PtCo. The hard ferrites –barium ($BaO\text{-}6Fe_2O_3$) and strontium ($SrO\text{-}6Fe_2O_3$) hexaferrites– are the most interesting materials of this type. The crystal structure, of hexagonal symmetry, is based on a compact stacking of layers of O^{2-} ions, some of which have been substituted by heavy ions of similar dimensions, Ba, Sr, Pb… The M phase contains only Fe^{3+} ions in the interstitial sites.

A description of the structure of this phase was given by Braun [18] starting from two types of simple structure elements, called R and S blocks (three other block types –T, Q and HBT– must also be considered to describe more complex ferrites). The S or spinel block (fig. 15.43-a) has a conventional cubic spinel structure with the [111] axis vertical. Its chemical formula (Fe_6O_8) contains two Fe^{3+} ions in tetrahedral sites, and four Fe^{3+} ions in octahedral sites. The R or hexagonal block, with chemical formula ($BaFe_6O_{11}$), also contains six Fe^{3+} ions, five in octahedral sites, and one which is surrounded by five oxygen ions located at the apices of a trigonal bi-pyramid (fig. 15.43-a).

The M phase can be represented by a stacking along the c-axis of R and S blocks according to the sequence RSR*S* (where * denotes a rotation through 180° around the c-axis): see fig. 15.43-b. The unit cell (fig. 15.43-c) contains two complete $BaFe_{12}O_{19}$ formula units, and has cell parameters: a = 0.589 nm and c = 2.32 nm.

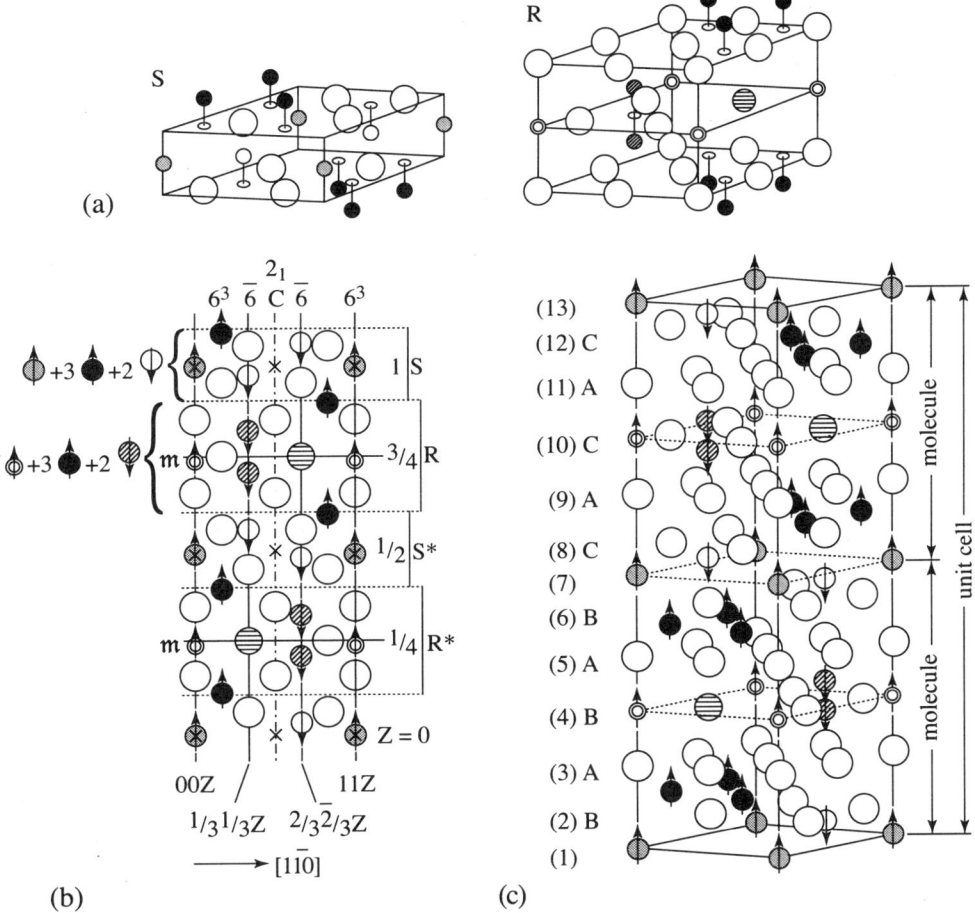

Figure 15.43 - Crystallographic structure of hexaferrites *after [5]:*
(a) R and S building blocks - (b) M phase - (c) hexagonal unit cell of barium ferrite

The positions of the O^{2-} ions are represented by large white circles, those of Ba^{2+} (or Sr^{2+}) correspond to the horizontally hatched big circles. The small circles indicate the positions of the Fe^{3+} ions in their various sites: the bipyramidal sites occupied by Fe^{3+} ions, to which the magnetocrystalline anisotropy is attributed, are represented by two concentric circles.

The general analysis of the magnetic properties of oxides led Néel to deduce that the magnetic moments are not all oriented in the same direction, and to propose the theory of ferrimagnetism. Based on Anderson's theory of indirect exchange [19], Gorter [20] proposed a colinear model of the arrangement of the Fe^{3+} magnetic moments, along the c-axis of the structure, in the hexaferrites. Figure 15.44 shows how the moments interact to form five sub-lattices corresponding to the five interstitial sites.

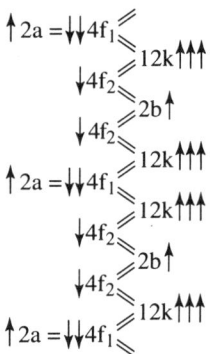

Figure 15.44 Model of interactions between sub-lattices

Since each Fe^{3+} ion has a moment of 5 μ_B at zero Kelvin, the saturation moment per formula unit is $\mu_{sat} = 5 \times (6 - 2 - 2 + 1 + 1) = 20 \, \mu_B$, which corresponds to a moment per unit mass $\sigma_{sat} = \dfrac{6.02 \times 10^{23}.20.0.927 \times 10^{-23}}{1114.48 \times 10^{-3}} = 100.4 \, A.m^2.kg^{-1}$, as shown in figure 15.45 [21]; the magnetization is then $M_{sat} = \sigma_{sat} \times d = 52.5 \times 10^3 \, A.m^{-1}$.

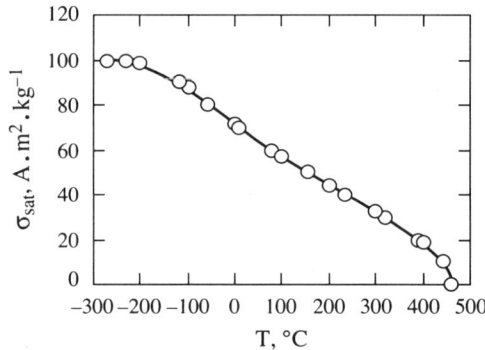

Figure 15.45 - Thermal variation of the magnetization of hexagonal barium ferrite
after [21]

The uniaxial anisotropy of ferrites is generally attributed to the Fe^{3+} ions in the bipyramidal site [22]. It is strong and may be described by a single second order constant K_1. At 300 K, K_1 is of the order of $3.2 \times 10^5 \, J.m^{-3}$ for $BaFe_{12}O_{19}$, and of the order of $3.5 \times 10^5 \, J.m^{-3}$ for $SrFe_{12}O_{19}$.

The thermal variation of K_1 is shown in figure 15.46 for $BaFe_{12}O_{19}$ [23, 24], and that of the anisotropy field H_A is given for $BaFe_{12}O_{19}$ in figure 15.47, and for $SrFe_{12}O_{19}$ [25] in figure 15.48.

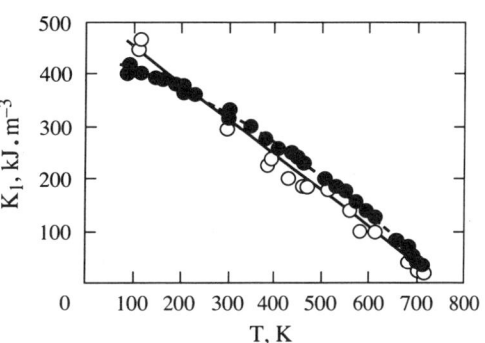

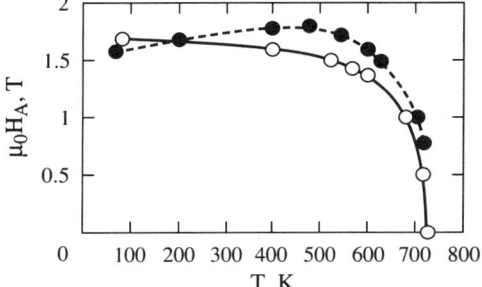

Figure 15.46 - Thermal variation of the magnetocrystalline anisotropy constant K_1 of BaO-6Fe$_2$O$_3$ ferrite: open circles, after [23], and full circles, after [24]

Figure 15.47 - Thermal variation of the anisotropy field H_A of BaO-6Fe$_2$O$_3$ ferrite: open circles, after [23], and full circles, after [24]

Figure 15.48 - Thermal variation of the anisotropy field H_A of SrO-6Fe$_2$O$_3$ ferrite, after [25]

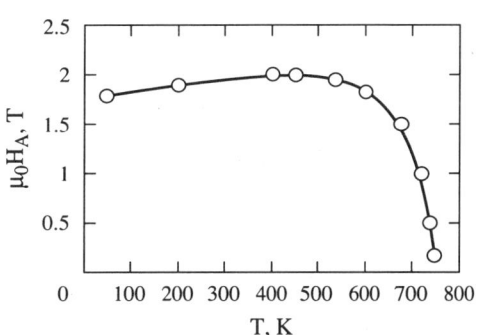

6.3. R-M INTERMETALLIC ALLOYS (R = RARE EARTH, M = TRANSITION METAL)

The special magnetic properties of transition metals M (ferromagnetic up to high temperature) and rare earth metals R (very strong magnetocrystalline anisotropy) are complementary with respect to permanent magnet properties. In certain R-M compositions, strong coupling between the M moments and the R moments may produce the required permanent magnet properties in a single material. This is the case for the SmCo$_5$ and Nd$_2$Fe$_{14}$B compositions which are the basic components of the two main families of high performance magnets.

6.3.1. R-M exchange coupling, via the d electrons

Starting from a matrix of iron (or cobalt), suppose that we substitute x rare earth atoms for x iron atoms, to obtain the metallic composition R$_x$Fe$_{1-x}$. The potential due to the nucleus is stronger for the 3d shell of iron or cobalt (elements at the end of the transition series) than for the 5d shell of a rare earth element (elements at the beginning of the transition series). Therefore the 3d bands of iron and cobalt are lower in energy than the 5d bands of the rare earths: see figure 15.49. Each time a rare earth ion is introduced into the matrix to replace a 3d ion (iron or cobalt), the z' d electrons which it brings find themselves positioned above the Fermi level, while the 10 3d states of the M ion, and the z electrons they contained, disappear (see chap. 8).

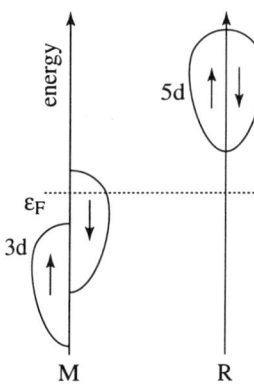

**Figure 15.49 - Band structure
in R-M alloys**

The distribution of the total number of electrons over the remaining 3d states determines the d spin magnetic moment of the system. If we suppose that the ↑ spin half bands of the M atoms are full (this is the case for systems known as strong ferromagnets, e.g. cobalt), the 3d magnetic moment, per atom of the alloy, is, in Bohr magnetons (μ_B): $m_d = 5(1-x) - (z-5)(1-x) - z'x$.

The total d magnetic moment decreases as x increases. The theoretical values of moments, shown in figure 15.50, were calculated using values of z and z' deduced from band structure calculations: $z = 7.9$ for Co, $z = 6.6$ for Fe, and $z' = 1.5$ for R. There is good agreement between calculated and experimentally determined values for cobalt-rich RCo compositions. This agreement is poor for RFe compositions, for which the assumption of a full spin ↑ half-band does not hold (iron is a weak ferromagnet). Agreement is also bad for the RCo alloys richest in R, for which the assumption of strong ferromagnetism does not hold either.

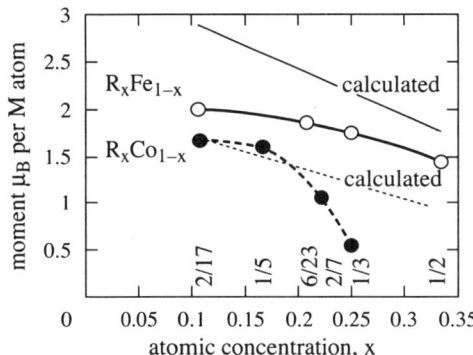

**Figure 15.50 - Theoretical, and
experimental values of the
magnetic moment as a function
of chemical composition
in R_xM_{1-x} alloys (M = Fe, Co)**

Beyond this overall result, how does the redistribution of d electrons occur on the atomic scale? We could be tempted to deduce from the previous discussion that the 5d electrons of R are simply transferred to the 3d band of Fe or Co. However, the appearance of charge on the atoms would cost too much energy, and thus this is a weak effect.

In fact, instead of the 5d electrons (coming from the substituting R ions) transferring to the 3d states of surrounding M atoms, a part of the 3d states ($\uparrow$ and $\downarrow$) of the M ions are transferred to the R ions (fig. 15.51). This phenomenon is known as *hybridization*. It involves an interaction between 3d and 5d electrons of neighbouring atoms. This hybridization is the stronger, the closer the hybridising states are in energy. The 3d $\uparrow$ band is further in energy from the 5d $\uparrow$ band than the 3d $\downarrow$ band is from the 5d $\downarrow$ band. The resulting hybridization is thus stronger between the d $\downarrow$ states, so that, on the R site, the number of available d $\downarrow$ states is greater than the number of d $\uparrow$ states. The d magnetic moment which appears on the R sites is thus antiparallel to the moment on the M sites.

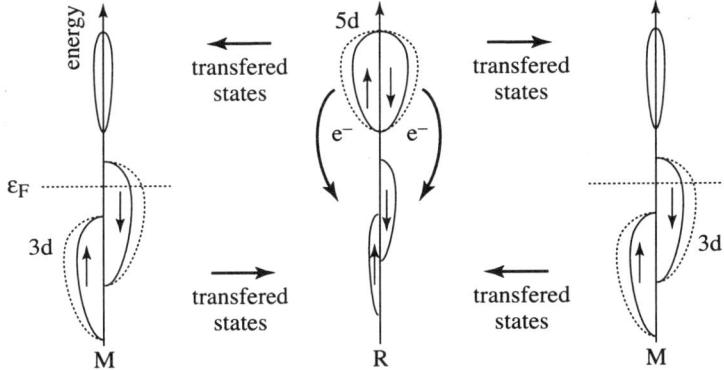

Figure 15.51 - Hybridization of d states in R-M compounds

The original properties of R-M alloys is due to this antiparallel coupling between the spin magnetic moments of 3d electrons of the M ions and the spin magnetic moments of 5d electrons of the R ions. On each R atom, the spins of the 4f electrons are ferromagnetically coupled with the spins of the 5d electrons. The coupling chain 3d-5d (interatomic, negative) and 5d-4f (intra-atomic, positive) results in strong antiparallel exchange coupling between the 3d spins (of the M ions) and the 4f spins (of the R ions). This coupling keeps the magnetic moments of the rare earth atoms ordered up to high temperatures, sometimes far above room temperature, allowing the transfer to the M atoms of the intrinsic anisotropy of the rare earth atoms. This is why R-M systems are the best candidates for the preparation of permanent magnets.

In the framework of a simple molecular field model, the energy associated with R-M coupling is given by: $E_{RM} = -n_{RM} M_R^{spin} M_M^{spin}$, where n_{RM} is a molecular field coefficient, and M_R^{spin} and M_M^{spin} are the spin moments of the rare earth and 3d metal, respectively [26].

The contribution, T_{RM}, of this energy term to the Curie temperature of the compound is given by: $T_{RM} = 2\dfrac{g_J - 1}{g_J} n_{RM} \sqrt{C_R C_M}$ where C_R and C_M are the Curie constants deduced from susceptibility measurements in the paramagnetic state. The determination of T_{RM} thus makes it possible to evaluate n_{RM}, the coefficient which measures the strength of the exchange interaction experienced by the rare earth.

Knowing that the Curie temperature is given by $T_C = \dfrac{1}{2}\left(T_M + \sqrt{T_M{}^2 + 4\,T_{RM}{}^2} \right)$ and supposing that T_M, the contribution of the dominant M-M interactions to T_C, is almost constant (in fact it slightly increases from lanthanum to lutecium), we can deduce the value of T_{RM}, and thus n_{RM}, from the measurement of T_C.

For example, n_{RFe} was deduced from the Curie temperatures of $R_2Fe_{14}B$ compounds (fig. 15.52). n_{RFe} decreases by a factor 3 from Pr compounds to Tm compounds (see fig. 15.53-a). A similar behavior is observed throughout the series of R-M compounds. Considering the similarity of the band structures across a given composition series, the 5d-3d interactions must be nearly constant. The variation of n_{RFe} thus reveals a decrease of the 4f-5d interactions. This is related to the fact that the distance between the 4f and the 5d shells increases notably across the rare earth series, from La to Lu (fig. 15.53-b).

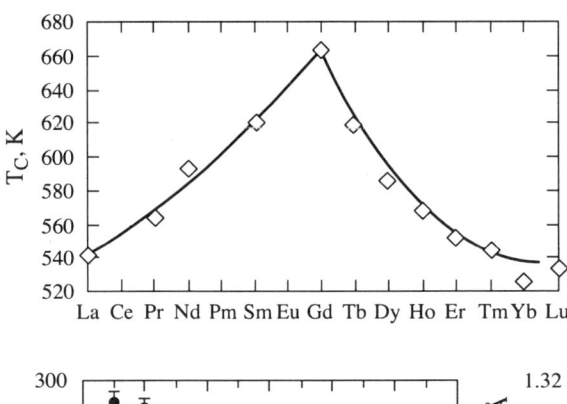

Figure 15.52 - Evolution in the Curie temperature of $R_2Fe_{14}B$ compounds as a function of the rare earth element involved

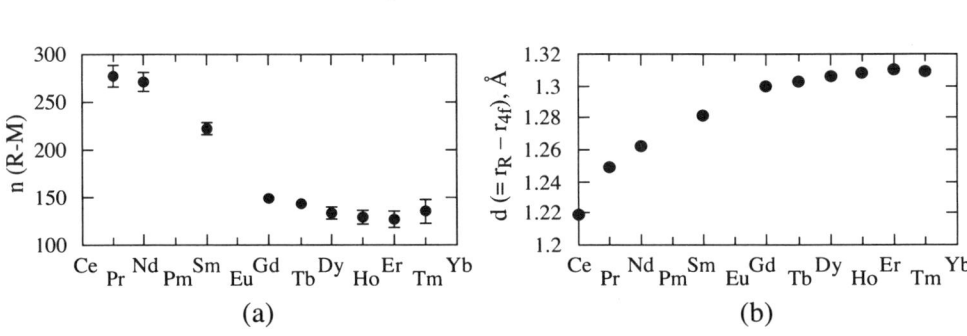

(a)   (b)

Figure 15.53 - Evolution in (a) the n_{RM} coefficient, and (b) the distance between the 4f and 5d shells in $R_2Fe_{14}B$ compounds as a function of the rare earth element

6.3.2. R-M coupling = ferromagnetism or ferrimagnetism

The antiparallel coupling between spin moments of M and R atoms combines with spin-orbit coupling at the R site (see fig. 15.54). S_M is coupled to S_R by exchange via a d electron, and S_R is coupled to L_R via spin-orbit coupling.

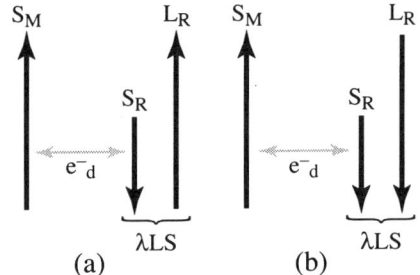

Figure 15.54 - Relative arrangement of magnetic moments in R-M compounds
 (a) R = light rare earth (1st series)
 (b) R = heavy rare earth (2nd series)
 The coupling is indicated in grey.

For heavy rare earth elements (from Tb to Tm), L and S are parallel, and the total moments of M and R are antiparallel. For the light rare earth elements (Pr, Nd and Sm), L and S are antiparallel, and $L\mu_B > 2S\mu_B$, so that the total moments of M and R are parallel. This ferromagnetic coupling is the basis for the fact that, within a given series of compounds, the magnetization is stronger in those compounds involving light rare earths (see fig. 15.55, and, as an example, the magnetization of RCo_5 or $R_2Fe_{14}B$ compounds in fig. 15.56).

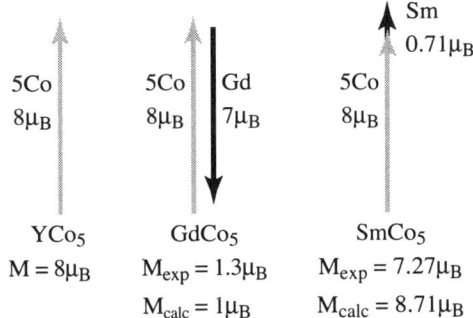

Figure 15.55 - Resultant magnetic moment for three RCo_5 compounds

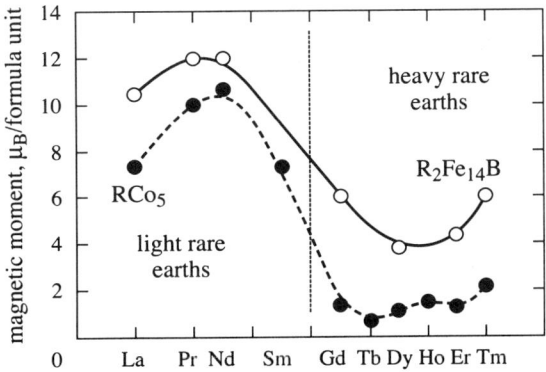

Figure 15.56 - Magnetic moments of RCo_5, and $R_2Fe_{14}B$ compounds

6.3.3. *Magnetocrystalline anisotropy in R-M compounds with uniaxial crystallographic structure: easy axis or easy plane*

In uniaxial symmetry the electrostatic potential favours the electronic distribution with the most aspherical shape consistent with the filling of the 4f shell. The asphericity of the latter is largely described by the sign of a coefficient α_J, called the Stevens coefficient, which characterises the ion. In uniaxial symmetry, α_J is positive when the electron distribution that leads to the moment being along the high symmetry axis is elongated along this axis, i.e. takes a prolate shape (e.g. samarium). In this case ($\alpha_J > 0$), the magnetic moment being perpendicular to the high symmetry axis corresponds to an oblate electron distribution, flattened perpendicular to the axis (see fig. 15.57). $\alpha_J < 0$, as for neodymium, praseodymium and terbium, corresponds to the complementary situations [27].

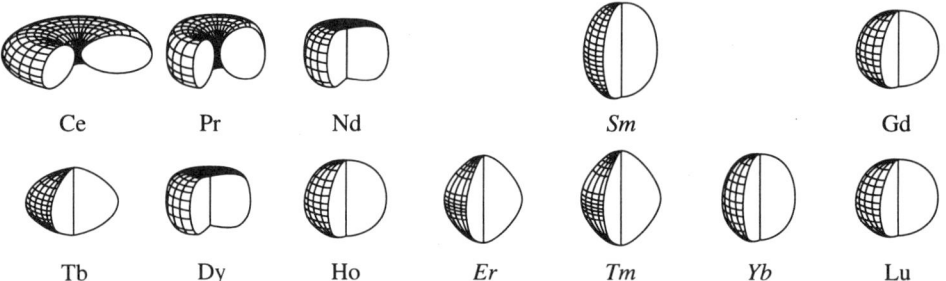

Figure 15.57 - Shape of the 4f shell of R^{3+} ions, when the magnetic moment is along the symmetry axis (vertical), *after [27]:*
$\alpha_J < 0$ for the rare earths shown on the left (oblate distribution)
$\alpha_J > 0$ for the rare earths shown on the right (prolate distribution)

The electronic distribution thus determined experiences the crystalline field of the environment, which is also non-spherical, notably in the uniaxial structures which we are now considering. These uniaxial structures define two different types of environments, characterised by the term A_2^0 (referred to in chap. 7, § 4.4). These favor different shapes of the 4f orbital.

- $A_2^0 > 0$ corresponds to the case where negative charges (repulsive for electrons) are located on the c-axis. A 4f magnetic orbital of the R ion flattened perpendicular to the c-axis (fig. 15.58-a) is then privileged.

- $A_2^0 < 0$ corresponds to the case where the charges on the c-axis are positive (thus attracting electrons), and the preferred shape for the 4f orbital is elongated along the c-axis (fig. 15.58-b).

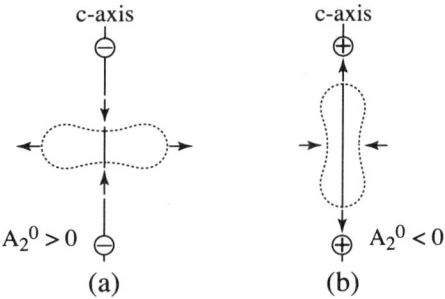

Figure 15.58 - Preferred shape for the 4f orbitals, in a situation with symmetry of revolution around c, depending on the sign of the A_2^0 crystal field term:
(a) $A_2^0 > 0$, oblate with axis c
(b) $A_2^0 < 0$, prolate along c

In a given series of R-M compounds (M fixed, R variable), the crystallographic structure stays the same, and, in principle, A_2^0 does not vary. Thus, whatever the element R, the same orientation of the 4f electron distribution is favoured. Consequently, the rare earth moments are oriented along the c-axis for R elements with Stevens coefficient α_J of the appropriate sign, while the others have their moments perpendicular to the c-axis. Thus, for the c-axis of the uniaxial structure to be the easy magnetization axis, we need either $A_2^0 > 0$ (e.g. $R_2Fe_{14}B$) and $\alpha_J < 0$ (Nd or Pr): see figure 15.59-a, or $A_2^0 < 0$ (RCo_5) and $\alpha_J > 0$ (e.g. Sm): see figure 15.59-b. For a more detailed discussion, the reader should refer to chapter 7.

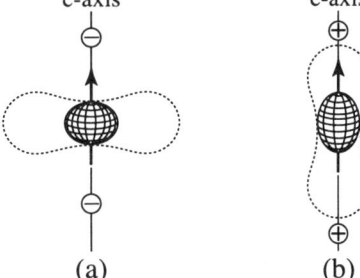

Figure 15.59 Necessary conditions for easy c-axis magnetization:
(a) $A_2^0 < 0$ and $\alpha_J > 0$
(b) $A_2^0 > 0$ and $\alpha_J < 0$

6.4. REVIEW OF INTERMETALLIC COMPOUNDS

In the series of R-M compounds, two requirements must be fulfilled when choosing elements to be used to prepare permanent magnets:

♦ there must exist a stable compound which is magnetic up to high temperature,
♦ the criteria concerning magnetization and anisotropy must be satisfied.

In general, these requirements are best satisfied in 3d transition metal rich R-M compounds.

The Curie temperature is particularly high in Co rich compounds but the magnetization of Co is lower than that of Fe and it is very expensive. This is why Fe based compounds are preferred. We present the principal R-M compounds in the following sections, and underline their value as permanent magnet materials.

6.4.1. Binary $R_x M_{1-x}$ compounds

There are substantial differences in the electronegativities, and atomic radii of rare earth and 3d elements (tab. 15.3). As a result of these differences, R and M elements form a number of stable intermetallic compounds.

Table 15.3 Electronegativities and atomic radii of some 3d- and 4f- series atoms

3d	Cr	Mn	Fe	Co	Ni	Cu	Zn
electronegativity, eV	1.66	1.55	1.83	1.88	1.91	1.90	1.65
atomic radius, nm	0.185	0.179	0.172	0.167	0.162	0.157	0.153
4f	Ce	Pr	Nd	Pm	Sm	Eu	Gd
electronegativity, eV	1.12	1.13	1.14	1.13	1.17	1.20	1.20
atomic radius, nm	0.270	0.267	0.264	0.262	0.259	0.256	0.254
4f	Tb	Dy	Ho	Er	Tm	Yb	Lu
electronegativity, eV	1.20	1.22	1.23	1.24	1.25	1.10	1.27
atomic radius, nm	0.251	0.248	0.247	0.245	0.242	0.240	0.225

As an example we show the phase diagram of samarium-cobalt (figure 15.60). The crystallographic structure of transition metal rich compounds, from RM_2 to RM_{12}, are based on the simple $CaCu_5$-type hexagonal structure, and they often have a uniaxial character. In cobalt alloys, the Curie temperature evolves approximatively as the square of the cobalt moment (fig. 15.61), which proves that the exchange interactions are not significantly modified from one alloy to another.

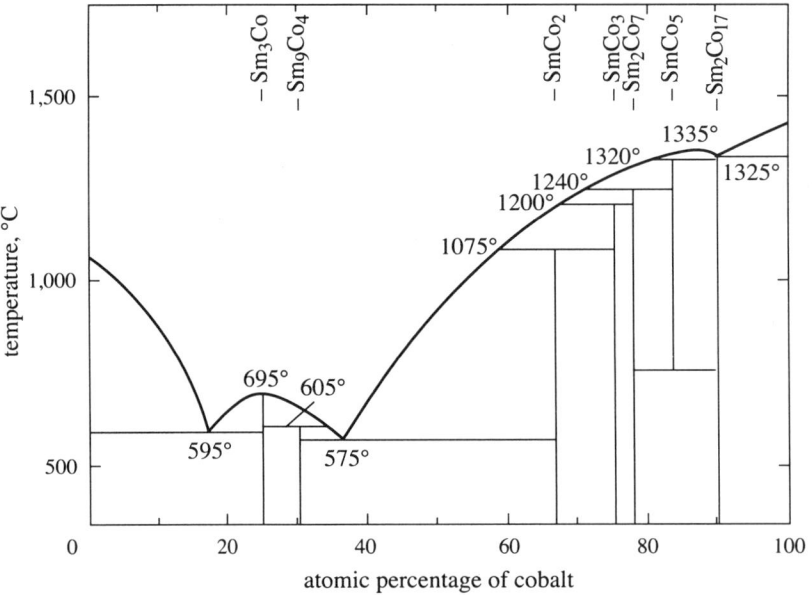

Figure 15.60 - Sm-Co phase diagram

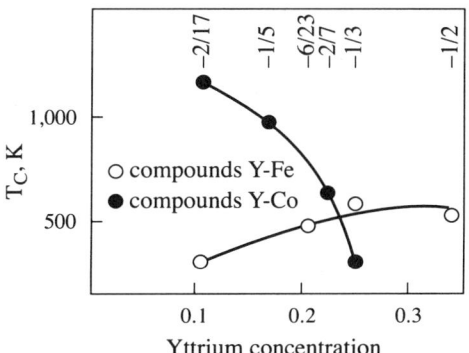

Figure 15.61 - Variation in the Curie temperature of Y-Fe and Y-Co compounds as a function of yttrium content

The R_2Co_{17} and RCo_5 type compounds are richest in 3d metal. Their hexagonal (or sometimes rhombohedral for the first category) crystallographic structures make them good candidates for permanent magnet materials.

RCo_5 compounds

These compounds have a $CaCu_5$-type hexagonal structure (space group P6/mmm), consisting of a compact alternate stacking of two different atomic layers (unit cell shown in figure 15.62). One of the layers (z = 0) consists of a compact arrangement of R and Co atoms (of different dimensions), in the ratio of 1 R for 2 Co atoms. The R atom is located at the centre of a hexagon of Co atoms (Co_I atoms on 2c site). The other layer (z = 1/2) consists of a non-compact hexagonal arrangement of Co atoms (Co_{II} on 3g site).

Figure 15.62 Hexagonal CaCu₅ structure

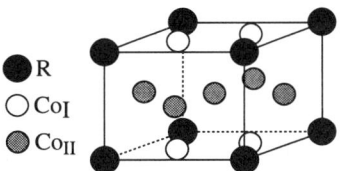

The R atoms form chains parallel to the six-fold axis, with short inter-atomic spacing. The ionic environment of each R^{3+} ion leads to a crystal potential characterised by $A_2^0 < 0$. The rare earth elements which lead to an easy direction of magnetization parallel to the c-axis are thus Sm, Er, Tm –for which the Stevens coefficient α_J is positive.

The anisotropy of cobalt, measured on a single crystal of the isotypic material YCo_5 where Y is non-magnetic (fig. 15.63), is very high for a 3d metal (see § 5.1.2.1). H_A is of the order of 12 T, and it favours the orientation of the Co moments parallel to the c-axis. For the R atoms cited (Sm, Er, Tm), the rare earth and Co anisotropies thus combine to make the c-axis the easy magnetization axis, with a maximum anisotropy field.

Conversely, with the rare earth elements of the other group (Pr, Nd, Tb, Dy, Ho), for which $\alpha_J < 0$, there is a competition between the anisotropies of Co (which dominates at high temperature), and the rare earth (stronger at low temperature). As the R-Co exchange interaction is sufficiently strong to impose a parallel or anti-parallel coupling of M_R and M_{Co}, the resultant magnetization M_{tot} will be perpendicular to the c-axis at low temperature, and parallel to it at high temperature.

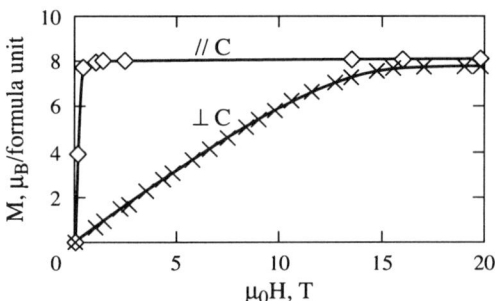

Figure 15.63 - Magnetization of a YCo₅ single crystal

In a certain temperature range, the balance between the anisotropies of the two moment sub-lattices may give rise to an intermediate orientation of M_{tot}. These phenomena are characteristic of R-M systems. Rotation of M_{tot}, as a function of field or temperature, can be observed, as well as orientations intermediate between the c-axis and the plane perpendicular to the c-axis. This is the case for TbCo₅ (fig. 15.64).

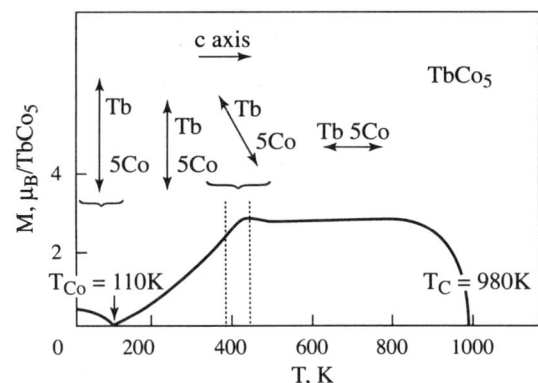

**Figure 15.64 - Evolution in the orientation
of the magnetic moment of TbCo₅ with temperature**

In conclusion, from the point of view of the properties required for permanent magnet materials, SmCo₅, ErCo₅ and TmCo₅ could be suitable as far as anisotropy is concerned. However, due to the anti-parallel coupling of moments in R-Co compounds when R is a heavy rare earth, the magnetization of ErCo₅ and TmCo₅ is low. SmCo₅ is the only compound which combines all the desired properties. Its high anisotropy field, which is above 50 T, its polarization $\mu_0 M$ in the region of 1.15 T, and its Curie temperature close to 1,000 K, make it one of the best permanent magnet materials known.

R_2Co_{17} compounds

The R_2Co_{17} compounds also cristallise in a uniaxial structure. They are richer in cobalt than the RCo_5 compounds, and have a higher magnetization and Curie temperature.

The crystallographic structure of the R_2Co_{17} system [6] is derived from that of the RCo_5 system. It is obtained by replacing one third of the rare earth atoms by pairs of cobalt atoms. For steric reasons, the number of substitutions in a chain of R atoms cannot exceed 50%. Two substitution configurations are possible (fig. 15.65) leading to either a hexagonal or a rhombohedral unit cell.

Figure 15.65
The two 2-17 unit cells

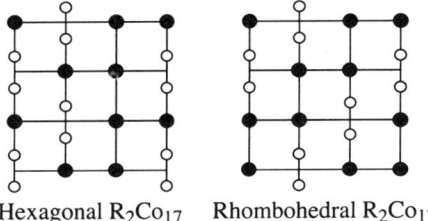

Hexagonal R_2Co_{17} Rhombohedral R_2Co_{17}

The anisotropy of the rare earth sublattice in the R_2Co_{17} system is similar to that in the RCo_5 system. However, the substituted pairs of cobalt atoms strongly influence the anisotropy of the cobalt sublattice, which now favours the plane perpendicular to the c-axis.

Thus, as regards anisotropy, Sm_2Co_{17} cannot compete with $SmCo_5$. However, in $Sm(Co,Cu,Fe,Zr)_{7.8}$ alloys, the $SmCo_5$ and Sm_2Co_{17} phases are intimately mixed, and benefit from their respective strong points, namely high anisotropy and high magnetization. The very large value of spontaneous magnetization of $Sm(Co,Cu,Fe,Zr)_{7.8}$ alloys gives them even better properties than $SmCo_5$.

R-Fe compounds

The Nd-Fe phase diagram is shown in figure 15.66.

Figure 15.66
The Nd-Fe phase diagram

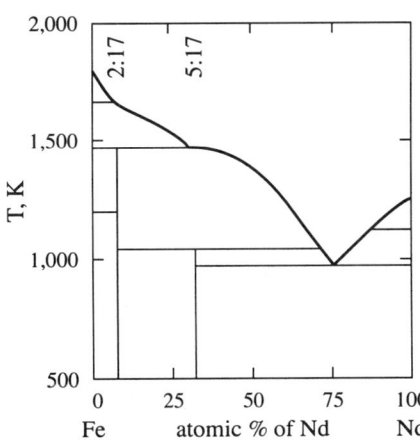

The RM_5 phase does not exist with Fe. In the R_2Fe_{17} compounds, the ordering temperature is particularly low, close to 300 K (the low value of T_C in many iron alloys was already mentioned in § 5.1.1). In these compounds, Fe has a planar-type anisotropy, and the only compound in which the overall anisotropy favours an easy c-axis magnetization is Tm_2Fe_{17}, which has a very low magnetization (rare earth of the second series) and a Curie temperature close to 300 K.

In conclusion, no binary Fe-rich R-Fe compound has the properties required to make permanent magnets.

6.4.2. Ternary compounds

The low Curie temperatures of Fe-rich binary R-Fe compounds, a phenonenon which is not well understood from a fundamental point of view, can be empirically explained by the fact that the distances between the iron atoms are very short. The idea of adding a third element to stabilize a less compact uniaxial structure, where the larger Fe-Fe distances would lead to stronger positive interactions (which would persist to sufficiently high Curie temperatures), is at the origin of the development of ternary and interstitial compounds.

In particular, the $R_2Fe_{14}B$ compounds order magnetically at about 600 K. They have a relatively complex tetragonal crystallographic structure (see fig. 15.67). The crystal field due to the environment of the R ions, in the $R_2Fe_{14}B$ structure, privileges a distribution of the 4f electrons which is flattened perpendicular to the c-axis ($A_2^0 > 0$). Thus, it is the rare earth elements with 4f shells having an oblate angular distribution ($\alpha_J < 0$: Pr, Nd, Tb, Dy or Ho) which will have the c-axis as the easy magnetization direction. As we have seen, the heavy rare earths are unsuitable because of the anti-parallel coupling between the rare earth and iron moments, and thus only the Pr and Nd compounds are suitable candidates. In these compounds, the anisotropy of iron favours the c-axis, and thus reinforces the rare earth anisotropy, e.g. iron contributes 1/3 of the room temperature anisotropy of $Nd_2Fe_{14}B$.

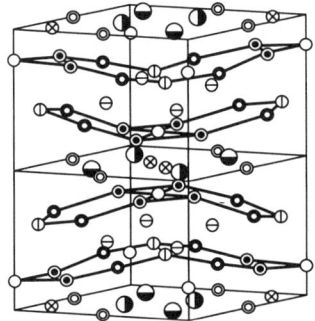

Nd	Fe	B
◑4f	○4e ⬭8j₁ ◉16k₁	⊗4f
●4g	◎4c ⊖8j₂ ●16k₂	

Figure 15.67
The crystallographic structure
of $R_2Fe_{14}B$ compounds

To summarise, the characteristic properties of $Nd_2Fe_{14}B$ are: $T_C = 587$ K, and, at 300 K, $\mu_0M_s = 1.62$ T and $\mu_0H_A = 8$ T. This favourable combination of properties, discovered simultaneously by Sagawa *et al.* [28], and Croat *et al.* [29], led to the development of the highest performance permanent magnets of today.

6.4.3. Interstitial ternary compounds

The discovery, in 1990, of $R_2Fe_{17}N_y$-type interstitial ternary compounds by Coey *et al.* [30] was the result of efforts carried out within the framework of a Concerted European Action on Magnets (CEAM). Stimulated by the sucess of NdFeB magnets, the idea was, not to obtain a new compound, but to introduce interstitial atoms into the R_2Fe_{17} compounds, with the hope that the resulting expansion of the unit cell would lead to a significant increase in the Curie temperature, due to the increase in the Fe-Fe distances. Rarely in physics do experimental results verify a starting hypothesis, but here it happened. In the $R_2Fe_{17}N_y$ compounds, atoms of nitrogen ($y \approx 2.7$) [or also carbon ($y \approx 1.7$)] can be introduced into an octahedral site where they are surrounded by two R atoms and four iron atoms. This results in a significant increase in the cell volume (by 6-8%), which leads to an increase in both the iron moment and the Curie temperature. It turns out that the change in crystal field caused by the introduction of interstitial atoms also results in a strong increase in anisotropy, which remains uniaxial for Sm, Er or Tm ($\alpha_J > 0$) compounds. The magnetic properties are comparable to those of $Nd_2Fe_{14}B$ (tab. 15.4). The fact that their magnetic properties are not remarkably better than those of $Nd_2Fe_{14}B$, coupled with the fact that there are more restrictions on how these materials may be processed (they thermally decompose above 450°C) explains why these materials are not yet developed on a large industrial scale.

Table 15.4 - Intrinsic properties of the principal interstitial compounds, compared with those of the starting compounds

Material	T_C (K)	μ_0M_s (T) (at 300 K)	μ_0H_A (T) (at 300 K)
$SmCo_5$	1,000	1.14	25.0
Sm_2Co_{17}	1,193	1.25	6.5
$Nd_2Fe_{14}B$	588	1.60	8.0
$Sm_2Fe_{17}C_{1.1}$	552	1.24	5.3
$Sm_2Fe_{17}N_{2.3}$	749	1.54	14.0
$SmFe_{11}Ti$	584	1.14	10.5
$NdFe_{11}TiN_{1.5}$	729	1.32	7.0

The $ThMn_{12}$-type tetragonal structure (derived from RCo_5) does not exist for M = Fe. However, it may be stabilized by substituting some Fe with other elements such as Mo, Ti, V, Cr, W or Si. In compounds such as $RFe_{10.5}Mo_{1.5}$, $RFe_{11}Ti$ or $RFe_{10}V_2$, the c-axis is the easy magnetization direction if the rare earth is characterised by a positive Stevens coefficient α_J (Sm, Er and Tm). On the contrary, after introduction of nitrogen or carbon atoms, leading to cell expansion, the electric potential surrounding the R ions changes shape, and the c-axis becomes the easy axis for R ions characterised by $\alpha_J < 0$, i.e. Pr or Nd. The increase in the cell volume also leads to a strong increase in the Curie temperature, as happens in the $R_2Fe_{17}X_y$ compounds mentioned above. Compounds with Nd or Pr, interstitially modified with nitrogen or carbon (N is more efficient than C because it enters in larger quantities), appear to be potential candidates for magnet materials (tab. 15.4).

7. MAGNETIZATION REVERSAL MECHANISMS

The results referred to here have been established in chapter 5.

The anisotropy in energy which creates the barrier to magnetization reversal (fig. 15.68) has two contributions:

♦ One contribution acts on the atomic scale, and is due to local magnetocrystalline interactions, electrostatic in origin (see chap. 7, § 4). This anisotropy acts on each individual magnetic moment; it gives rise to the anisotropy field H_A (see eq. 5.50).

♦ One contribution acts on the scale of the grains themselves, and is related to the shape of the grains. This contribution is due to the anisotropy of the demagnetising field, and is a function of the resultant magnetization only; it gives rise to the shape anisotropy field (see eq. 5.41) denoted $H_A{}^f$ and equal to $(N_a - N_c) M_s$.

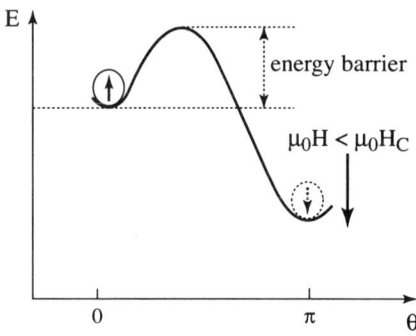

Figure 15.68 - The energy barrier and the metastable state of a magnet

7.1. MAGNETIZATION REVERSAL IN MAGNETIC SYSTEMS DEVOID OF MAGNETOCRYSTALLINE ANISOTROPY: APPLICATION TO ALNICO MAGNETS

In systems without magnetocrystalline anisotropy (such as AlNiCo), where the uniaxial anisotropy is due to the microstructure itself (through the elongated shape of the grains, see § 5.3.1), three modes of magnetization reversal are possible:

♦ the uniform rotation of moments (Stoner-Wohlfarh model), for which: $H_C = v = H_A{}^{tot}$ (remember that in this case the reversal field, H_R, is taken to be identical to the intrinsic coercivity H_C),

♦ a *non-uniform collective* mode, among those already discussed in chapter 5, § 8.3.2,

♦ the non-collective mode involving nucleation and then propagation of a domain wall.

Whatever the mode in question, we have considered until now that the grains are magnetically independent of each other. In a real assembly, there are always magnetostatic interactions between particles. We can estimate the importance of these interaction effects, starting from the following principle: the potential energy of a particle in the field from other particles is proportional to $p\,M_s^2$, where p is the volume fraction of magnetic particles in the system considered (p can be estimated from the ratio $\overline{M}_s / M_s$ of the global saturation magnetization of the magnet $\overline{M}_s$. to the saturation magnetization of a grain M_s).

Thus, for an assembly of cylindrical particles in a field parallel to the axes of the cylinders, the magnetization reversal field for a uniform collective mode is a function of p, and is given by [31]:

$$H_c(p) = H_c(0)(1 - p) \qquad (15.11)$$

For the same cylindrical particles, the most favourable non-uniform collective mode is the curling mode. In this case, the reversal field is independent of p, because the reversal does not produce any interaction field [32]. This mode occurs only if it costs less energy than the uniform rotation mode, and this can occur only if the radius of the cylinder R is greater than a critical value, R_c, given by [33]:

$$R_c(p) = R_c(0)(1 - p)^{-1/2} \qquad (15.12)$$

with:
$$R_c(0) = 3,69\sqrt{A_{exch} / \mu_0 M_s^{\,2}} \qquad (15.13)$$

According to this equation, the critical radius increases with the volume fraction of particles, p. Thus, for a given particle radius R_1, such that $R_c(0) < R_1$, we expect the coercive field $H_C(p)$ to remain independent of p so long as $R_c(0) < R_c(p) < R_1$ (magnetization reversal by curling without a magnetostatic interaction field). As p increases, $H_C(p)$ decreases linearly (according to eq. 15.11) as soon as $R_1 \leq R_c(p)$, because reversal then occurs by uniform rotation.

Experimental data for AlNiCo magnets do not make it possible to assign reversal to one of these mechanisms. We observe that the coercive field is 2 to 5 times smaller than the theoretical reversal field for uniform rotation, and that the radii of the grains, from 15 nm for AlNiCo 8 [34] to 30 nm for AlNiCo 5 [35], are similar in value to the critical radii R_c predicted by the models ($R_c(0) = 8$ nm and $R_c(p = 0.65) = 15$ nm). Considering that the theoretical values come from approximate models, and that the experimental values (notably measurements of R) are inevitably of a statistical nature, the assumption of reversal by curling –for example– cannot be identified with certainty over that of reversal by coherent rotation in particles with dimensions distributed about the average values measured ([2], p. 63-64).

7.2. *MAGNETIZATION REVERSAL IN SYSTEMS WITH STRONG MAGNETOCRYSTALLINE ANISOTROPY: NON-COLLECTIVE REVERSAL IN STAGES*

Magnetization reversal in materials with strong magnetocrystalline anisotropy is dominated by the influence of defects (see § 5 and chap. 6). It occurs within the material by a *non-collective* process, so called because at successive moments it involves different small volumes of the material. One after the other, small volumes of the material play host to particular physical processes which are linked. A specific energy barrier and consequently a critical field (active within the small volume concerned) is associated with each of these mechanisms. The crossing of a given energy barrier leads on to the next stage of magnetization reversal.

7.2.1. *Stages of the process*

Magnetization reversal of a grain involves four identifiable successive mechanisms:

1. *Nucleation:* the creation of a reverse domain and associated domain wall at the position where the anisotropy barrier is lowest (i.e. at a defect); the critical field is the nucleation field, denoted H_N.

2. *Passage* of the domain wall from the defect into the principal phase (where exchange and anisotropy have their normal values). The barrier to overcome depends on the surface energy of the wall ($\gamma \sim \sqrt{A_{ex} K}$); the characteristic field is denoted H_p (passage field).

3. *Expansion* of the domain wall in the principal phase; there is a big increase in the surface area of the wall associated with this expansion and thus in its energy cost; the expansion field is denoted H_{exp}.

4. *Pinning*: pinning of the domain wall at magnetic heterogeneities within the bulk; as for the passage mechanism, unpinning requires an energy barrier, due to the difference in wall energy inside and outside the defect, to be overcome. Unpinning re-triggers wall propagation; the propagation field is denoted H_{pr}.

These successive events happen at definite places in the grains. They involve volumes of non-arbitrary dimensions. During the nucleation process itself, magnetization reversal causes a local deviation from saturation in the material. In a ferromagnetic material, the characteristic dimension over which magnetization can reverse at least cost is the domain wall width δ. Thus we deduce that the critical volume is of the order of δ^3.

The other mechanisms which occur to complete magnetization reversal bring about the displacement of the domain wall across local energy barriers. The critical volume involved is again of the order of δ^3.

The largest of the four critical fields defined above determines the effective reversal field. The mechanisms involved in stages 1-3 are believed to determine the coercivity of ferrites, $SmCo_5$ and NdFeB magnets, though the exact mechanisms have not been clearly identified. It is generally accepted that the coercivity of SmCo:2-17 magnets is determined by the pinning-unpinning of domain walls during the fourth stage of reversal.

7.2.2. The driving forces of reversal: magnetic fields and thermal effects

Two energy terms contribute to overcoming the local energy barriers which are the source of coercivity: coupling of the magnetic moments of the material with local magnetic fields, and local thermal activation.

The internal field of the material, which is what the magnetic moments of the material experience, differs from the applied field, principally because of local dipolar interactions. If we suppose that the macroscopic demagnetising field is negligible (or if we subtract it from the applied field), the effective critical field is the sum of the internal field, equal to H_R, and a local dipolar contribution $H_{loc\ dip}$:

$$H_{eff\ crit} = H_R + H_{loc\ dip} \qquad (15.14)$$

Thermal activation assists the crossing of energy barriers, contributing an energy proportional to $k_B T$ (k_B = Boltzmann's constant) which, randomly in time, can be concentrated into volumes of different sizes, the activation volumes (denoted V_a). Thermal activation allows some local energy barriers to be overcome when the activation volume V_a equals the critical volume associated with the energy barrier. Thermal activation can be evidenced by magnetic viscosity measurements performed near the coercive field (see chap. 6, § 5.2). The barriers to magnetization reversal are then sufficiently reduced under the effect of the applied field, and can thus be crossed in a reasonable time through thermal activation.

7.3. THE MAGNETIZATION REVERSAL FIELD H_R: RELATIONSHIP WITH THE INTRINSIC MAGNETIC PROPERTIES OF THE PRINCIPAL PHASE

We have emphasised that the magnetization reversal field strongly depends on the microstructure. We could thus think that H_R does not depend on the magnetic properties of the principal phase. However, the ratio H_C / H_A (H_A is characteristic of the principal phase) is similar in value for all categories of magnets (see fig. 15.69). Thus, even if only indirectly, the coercivity must depend on the intrinsic magnetic properties of the material used.

This observation has led to two approaches aiming at evidencing a behavior law for the reversal field H_R.

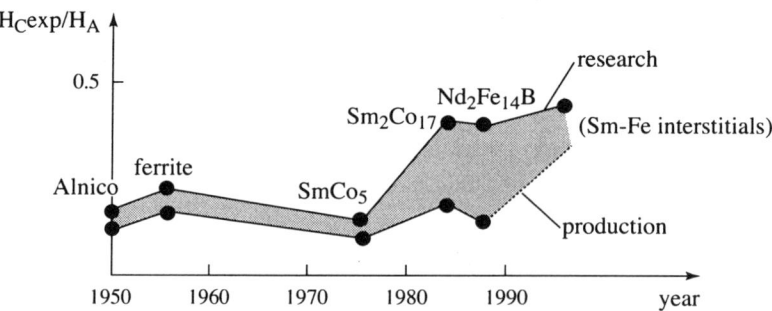

Figure 15.69 - Evolution of the ratio H_C/H_A over the years

7.3.1. H_R as a function of H_A and local dipolar effects

Starting from equation (15.14), we make two simplifying assumptions:

♦   that $H_{eff\,crit}$ is proportional to the anisotropy field of the principal phase, H_A

♦   that $H_{loc\,dip}$ is proportional to M_s (the spontaneous magnetization of the grains: $M_s = M_{sat}$ of the magnet), since the dipolar fields are a function of the magnetization of the material.

We thus obtain:

$$H_R(T) = \alpha\,H_A(T) - N_{eff}\,M_s(T) \qquad (15.15)$$

where α and N_{eff} are parameters which are experimentally determined in the temperature range where the law is obeyed.

F. Kools was the first to develop this type of analysis that he applied to anisotropic ferrites [36]. Subsequently this analysis was applied to PrFeB magnets [37], for which H_R/M_s was plotted as a function of H_A/M_s at all temperatures between 4.2 K, and T_C (see fig. 15.70).

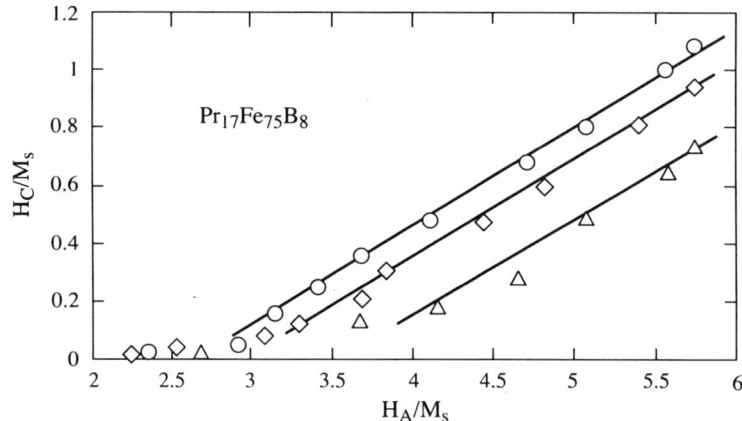

Figure 15.70 - Analysis of coercivity in PrFeB magnets
using an energy barrier proportional to the anisotropy energy
Samples annealed at 1353 K for 1.5 h (circles), 1373 K for 1.5 h (diamonds), and as sintered (triangles).

This result confirms the existence of an essentially linear relationship between H_R and H_A over a wide temperature range. The value of α deduced is of the order of 0.4 (the significance of this value is not understood). N_{eff} has a value of $+1$, indicating that the local dipolar field helps reversal. The value of the local dipolar field is significant at 300 K ($\mu_0 H_{loc\ dip} \approx 1.6$ T) as it is comparable to the value of the applied field H_R.

7.3.2. H_R as a function of the energy barrier for the critical mechanism

This approach is justified by the fact that the successive mechanisms which intervene during the reversal process –these mechanisms were presented above (§ 7.2.1)– have a common characteristic: they all involve a local deviation from saturation, which can be likened to elements of a domain wall. The energy necessary to create or expand the corresponding nucleus constitutes an energy barrier which must be overcome. Consequently, whatever the mechanism determining the process, the effective reversal field must be proportional to:

♦ the local value of the domain wall energy γ_ℓ, and

♦ the surface area of the domain wall element involved.

This approach may be modelled, without introducing variable parameters, by making two assumptions:

1. The critical volume V of the nucleus involved is taken to be the activation volume V_a involved in thermal activation occurring during measurement); V can thus be evaluated from measurements of the magnetic after-effect,

2. The surface area s of the wall may be related to V through simple dimensional analysis: $s \sim V^{2/3}$.

The energy barrier Δ which blocks magnetization reversal is thus given by: $\Delta = a\gamma_\ell V^{2/3} = a'\gamma V^{2/3}$, if γ_ℓ (local value of the wall energy) is proportional to γ (value of the wall energy in the principal phase) in accordance with the starting assumption.

The energy barrier thus defined may be overcome under the influence of:

♦ an applied field H_R, which supplies an energy equal to $\mu_0 M_s H_R V$

♦ a local dipolar field, which contributes an energy $\mu_0 N_{eff} M_s^2 V$

♦ thermal activation, $k_B T \ln(t/\tau_0)$ which, with $\tau_0 \approx 10^{-11}$ s (see chap. 6, § 5.2) and taking a measurement time t of a few seconds, approximately equals $25 k_B T$.

Thus, the equation which defines H_R is:

$$H_R = \alpha \frac{\gamma}{\mu_0 M_s V^{1/3}} - N_{eff} M_s - 25 S_V \qquad (15.16)$$

where:

$$S_V = k_B T / (\mu_0 M_s V) \qquad (15.17)$$

The physical quantities which appear on the right hand side of this equation are determined by experiment at each temperature T: γ, M_s, V and S_V (V and S_V are deduced from magnetic viscosity measurements). α and N_{eff} are determined by comparison of the experimental values of $H_R(T)$ with the behavior law given above, when applicable.

The plot of H_0/M_s as a function of $\gamma/\mu_0 M_s^2 V^{1/3}$, with $H_0 = H_R + 25S_V$, for a NdFeB magnet, is shown in figure 15.71-a, after [38]. A linear relation is obtained over the entire temperature range, between 4.2 K and the Curie temperature. This gives: $\alpha = 1$ and $N_{eff} = 0.8$. This analysis was also successfully applied to ferrite [39] and SmCo [40] magnets.

Two principal pieces of information have been extracted from this analysis:

- N_{eff} is always of the order of 1, in accordance with the result of the previous analysis. This estimation has been confirmed by a more direct experimental determination of $H_{loc\ dip}$ [38, 41].

- The critical volume, deduced from thermal activation measurements, is close to the cube of the domain wall width (V is of the order of $10\,\delta^3$) for all the magnets studied. The ratio V/δ^3 (fig. 15.71-b) diverges only in the vicinity of the Curie temperature. This result validates, *a posteriori*, the approach of relating the coercivity to intrinsic physical properties because it establishes the connection to δ, the characteristic length for any deviation from saturation. However, the divergence observed near T_C indicates that the magnetic parameters which describe the critical nucleus are not exactly the same as those of the principal phase.

In rare earth based magnets, the critical nucleus corresponds to a cube with sides a few nanometers in length (5 to 8 nm), which is much smaller than the critical size for a single domain particle (of the order of 200 nm a side).

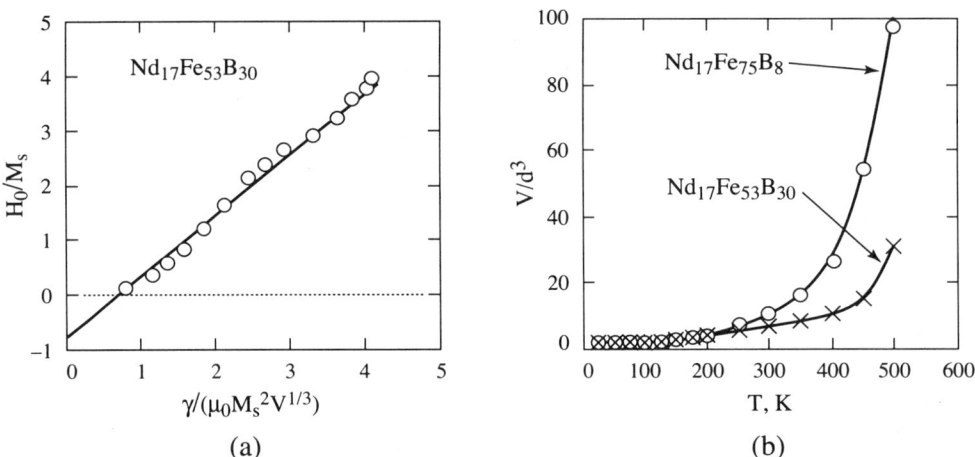

Figure 15.71 - Coercivity analysis in NdFeB magnets using an energy barrier proportional to the energy of formation of a domain wall

7.4. WHICH MECHANISM DETERMINES MAGNETIZATION REVERSAL?

The approaches which have been presented establish the connection between the reversal field H_R and certain fundamental physical parameters, but they say nothing about the actual reversal mechanism at work. No conclusive experimental means are available to characterise these mechanisms. Direct observation is impossible because of the small size of the critical nucleus involved as well as the suddenness and brevity of the phenomenon. The connection between the magnetization reversal mechanism and perceptible elements of the microstructure are difficult to establish, because of the size of the active nucleus, and also the complexity of the microstructure. Traditional magnetization characterization measurements, although they access only macroscopic statistical effects, can provide important information.

7.4.1. Analysis of the initial magnetization curve, particularly the initial susceptibility

The analysis of the initial magnetization curve, measured on a thermally demagnetised sample, is used to judge the mobility of domain walls within the magnetic grains.

For certain magnet categories, most notably for sintered $SmCo_5$ (fig. 15.72) or NdFeB magnets, the initial susceptibility is equal to the reciprocal of the demagnetising field slope, and it is reversible. This behavior characterises the free motion of domain walls. It confirms that the coercivity cannot be attributed to a domain wall pinning-unpinning mechanism (step 4).

$\times$: initial magnetisation

$\bigcirc$: measurement in decreasing field
  starting from saturation

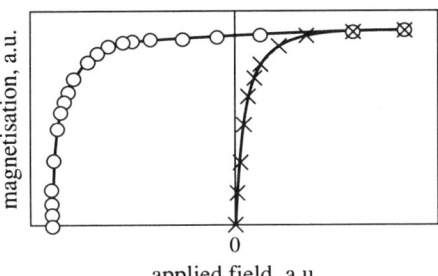

Figure 15.72 - Typical M (H) curve of a $SmCo_5$ magnet

Conversely, low initial magnetization, accompanied with an irreversible variation in magnetization, is evidence that the domain walls do not move easily. This may be understood as either domain wall pinning, or the absence of domain walls (single domain grains). The increase in susceptibility observed when the applied field reaches a value close to H_R then indicates that the magnetization mechanism up to saturation is identical to the mechanism involved during magnetization reversal from the saturated state (coercivity mechanism). However, this is not enough to identify the mechanism. The initial magnetization curves obtained for a ferrite magnet (fig. 15.73-a) and a $CeCo_{3.8}Cu_{0.9}Fe_{0.5}$ magnet (fig. 15.73-b) illustrate this fact: they are very similar, although the coercivity of the ferrite is considered to be nucleation controlled while the coercivity of the $CeCo_{3.8}Cu_{0.9}Fe_{0.5}$ magnet is believed to be pinning controlled.

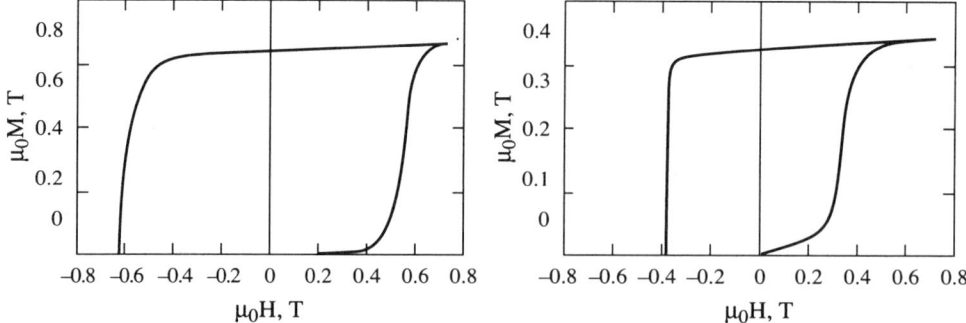

Figure 15.73 - Initial magnetization curves:
(a) Ferrite magnet (nucleation controlled coercivity)
(b) Ce-Co-Cu-Fe magnet (pinning controlled coercivity)

7.4.2. Observation of domains and domain wall motion in the thermally demagnetised state

Domain and domain wall observations, made with an optical microscope and polarised light (Kerr effect microscopy) or an electron microscope (Lorentz microscopy), complement initial susceptibility measurements.

A domain structure exists within the grains of sintered NdFeB and $SmCo_5$ magnets in the thermally demagnetised state. The domain configuration evolves under the influence of an applied magnetic field which easily moves domain walls in these materials. In modern ferrite magnets and NdFeB magnets made from melt spun ribbons, the grains are generally single domain.

Observations of Sm_2Co_{17} magnets lead to quite different conclusions. Their microstructure is obtained by precipitation of a $Sm(CoCu)_5$ high coercivity phase within a $Sm_2(FeCo)_{17}$ phase of lower coercivity (see § 5). It consists (fig. 15.74) of an assembly of cells which are more or less bi-pyramidal, of length 100 to 300 nm, containing the principal 2-17 phase. The grains are separated by boundaries consisting of the 1-5 phase, of thickness 5 to 10 nm. The magnetic domains of a thermally demagnetised sample, observed by Lorentz microscopy [42], seem to follow the cell boundaries, with the domain wall taking on a zig-zag shape (fig. 15.75-a). The direct visualization of Bloch walls forming saw-tooth lines which follow the cell boundaries exactly (fig. 15.75-b), and the fact that the application of a magnetic field does not alter the observations, confirms that the domain walls are pinned at the intergranular phase boundaries.

(a)   (b)

Figure 15.74 - Microstructure of a SmCo 2-17 type magnet

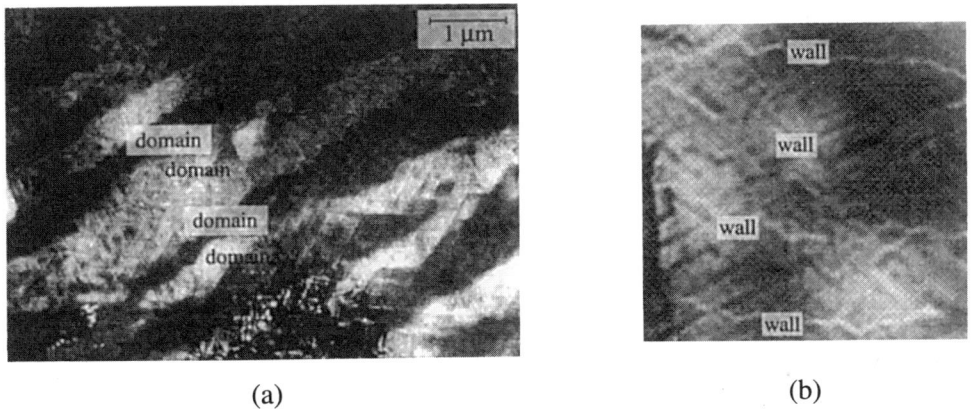

(a)   (b)

Figure 15.75 - Visualization of domains (a) and domain walls (b)
in a SmCo 2-17 type magnet

7.4.3. Variation of the reversal field with the direction of the applied field: $H_R(\theta_c)$

According to the Stoner-Wohlfarth model, the value of the reversal field in a homogeneous particle varies as a function of the angle at which the magnetic field is applied. This variation is given by:

$$H_R^{\text{local}}(\theta_c) = \frac{H_A^{\text{local}}}{(\cos^{2/3}\theta_c + \sin^{2/3}\theta_c)^{3/2}} \qquad \text{(see eq. 5.64)}$$

This gives rise to the curve shown in figure 15.76-a, the angle θ_c being the angle that the field makes with the anisotropy axis of the grain (see fig. 15.76-b).

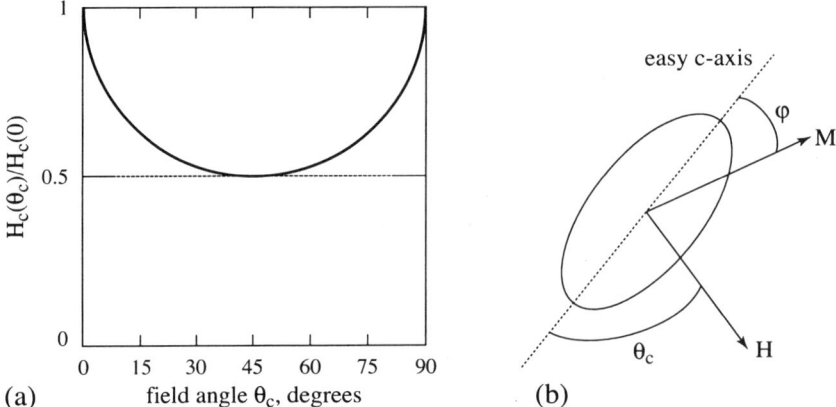

(a)

(b)

Figure 15.76 - Variation of the magnetization reversal field as a function of the direction of the applied field, according to the Stoner-Wohlfarth model

This law is obeyed in nanoparticles (in particular in ferrites [43]: see fig. 15.77), which led to believe that magnetization reversal in these particles occurs by collective uniform rotation of the moments. However, detailed experimental studies of certain systems, and numerical simulations of magnetization reversal in nanoparticles have shown that this type of behavior rather characterises the fact that the reversal field H_R is close to the magnetocrystalline anisotropy field H_A (i.e. the magnetic system is non-rigid close to the coercive field), without pointing to a particular reversal mechanism: non-collective reversal is often more probable than uniform rotation.

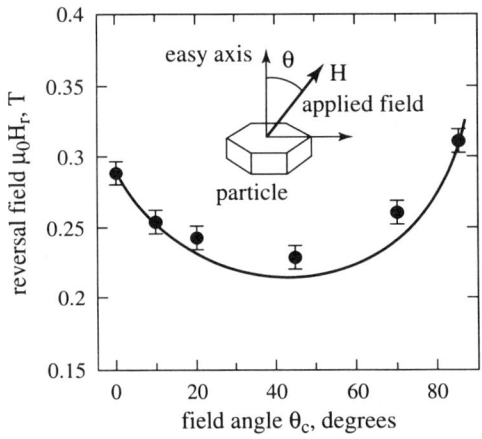

Figure 15.77 - Experimental analysis of $H_R(\theta_c)$ for the case of an isolated particle (ferrite)

In fact, equation (5.64), recalled above, does not describe the behavior observed in the majority of magnet materials. Analysis of the angular dependence of the magnetization reversal field measured in these materials shows that, in almost all cases, the magnetization of each grain reverses in a field which varies with the angle of the applied field θ_c according to a $1/\cos\theta_c$ law (fig. 15.78), as long as $0 \leq \theta_c \leq \pi/3$ (fig. 15.79).

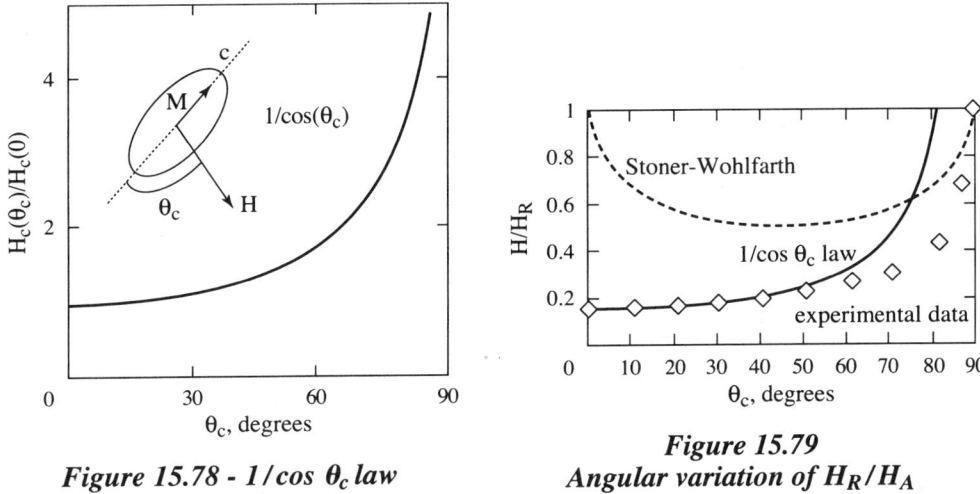

Figure 15.78 - $1/\cos\theta_c$ law

Figure 15.79
Angular variation of H_R/H_A

To understand this behavior, we must distinguish the roles played by the applied field projections. The $H_{//}$ component, along **c** (the easy magnetization axis), tends to reduce the barriers to magnetization reversal, whereas the $H_\perp$ component (perpendicular to **c**) tends to rotate the magnetization. For this rotation to be significant, the value of the perpendicular component must be of the order of the anisotropy field H_A. When the direction of the applied field is close to the c-axis, the $H_\perp$ component is only a small fraction of the applied field. Besides, if the field which triggers magnetization reversal (roughly speaking, the coercive field) is small compared with H_A, $H_\perp$ is itself much smaller than the anisotropy field, and the moments do not rotate before magnetization reversal. The $H_{//}$ component, which is the only active component, acts on a rigid magnetization distribution. Its value, $H_{//} = H \cos\theta_c$, naturally leads to the observed behavior.

For higher values of θ_c, typically between $\pi/3$ and $\pi/2$, the field needed for reversal grows quickly ($1/\cos\theta_c$ law), and the $H_\perp$ component, which then becomes predominant, approaches H_A. The system is no longer rigid, and its behavior progressively approaches that described by equation (5.64), (fig. 15.79).

7.4.4. Modelling of the different mechanisms involved in non-collective magnetization reversal

The mechanisms involved in each stage of the non-collective process have been modelled so as to evaluate the theoretical critical field associated with each mechanism. From the evaluation of different critical fields resulting from these models: H_N [44], H_p [44, 45], H_{exp} [41] and H_{pr} [46, 47, 48, 2, 49], it turns out that all the mechanisms invoked can give rise to critical fields of the right size. However, these calculations, based on simplifying approximations, most notably concerning the dimensionality of the problem, do not allow us to unambiguously identify the involvement of certain mechanisms, nor, conversely, to rule out the involvement of others.

EXERCISES

E.1 - Design of a magnetic "suction cup" fixation device

While exercice E13 at the end of chapter 2 is centered on the comparison between a coil and a permanent magnet, the present exercise is mainly intended to demonstrate how spectacularly the attraction force is increased when a ferromagnetic circuit is associated to the magnet.

E.1.1 - Attraction between a magnet and an iron plate

The magnet has a cylindrical shape, with diameter D = 10 mm and height h = 5 mm. It is made of a ferrite, and its polarization is assumed to be constant and uniform ($J = B_r = 0.4$ T). Calculate the attraction force between the magnet and the iron plate.

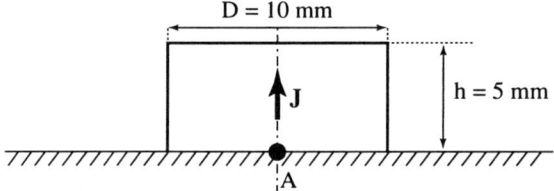

Figure 15.80 - Dimensions of the magnet in attraction with an iron plate

To perform this calculation, assume that the induction at any point on the contact surface is identical to the induction that can be calculated at point A, on the symmetry axis.

E.1.2 - The same magnet inside a magnetic circuit closing the flux

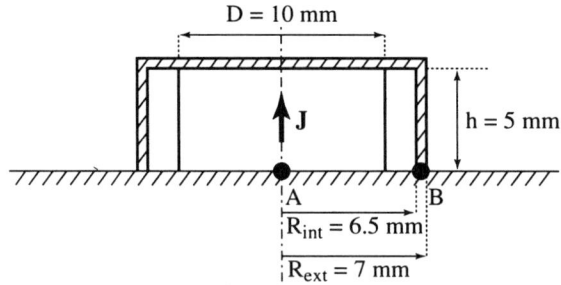

Figure 15.81 - The magnet inside its iron cup

This magnetic circuit has the shape of an iron cup, 0.5 mm thick.

E.1.2.1 - What is the operating point of the magnet? Deduce the magnetic flux it sends into the magnetic circuit.

E.1.2.2 - Calculate the cross-section for flux transmission at the contact between the cup and the iron plate.

E.1.2.3 - Calculate the induction at point B. Check that the cup is not saturated.

E.1.2.4 - Calculate the two contributions of the force (F_m for the magnet and F_i for the cup) and the total attraction force.

E.1.2.5 - Compare with the result of E.1.1.

E.2 - Design of a loudspeaker

We consider a loudspeaker in which the magnetic circuit involves a hard ferrite magnet. The magnetization curve of this ferrite is trapezoidal in the second quadrant, with $B_r = J_r = 0.4$ T, $J = 0.38$ T for $\mu_0 H = -0.2$ T, $\mu_0 H_{cJ} = 0.2$ T.

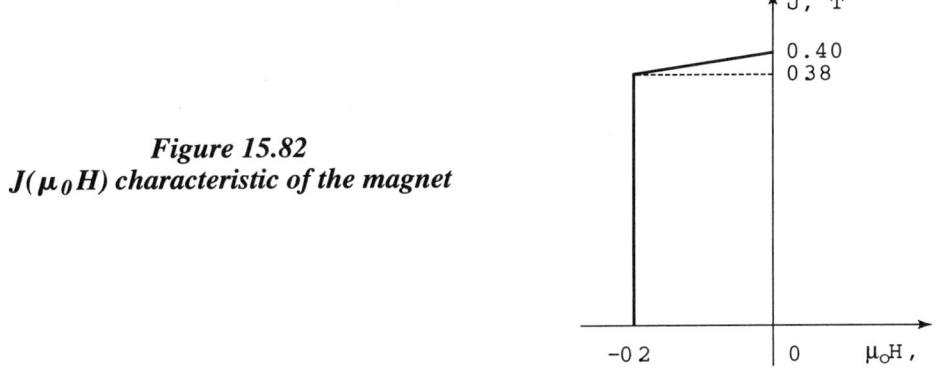

Figure 15.82
$J(\mu_0 H)$ characteristic of the magnet

The design of the circuit is shown below.

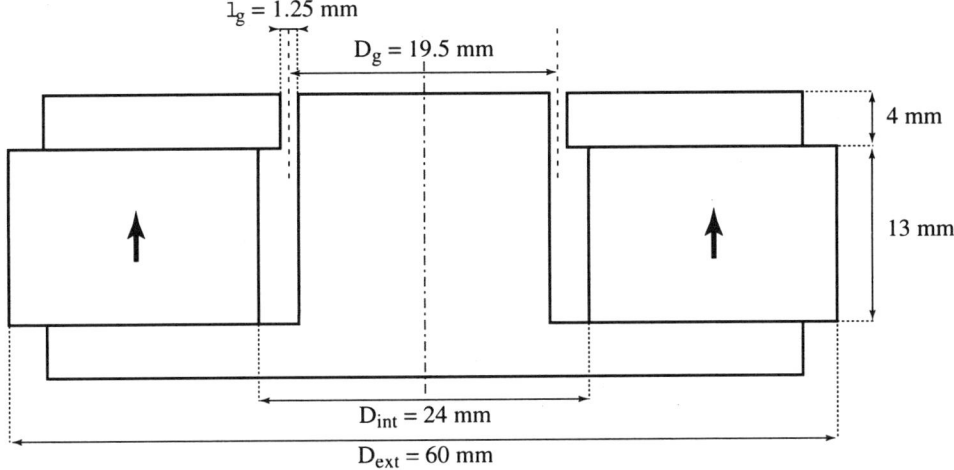

Figure 15.83 - Loudspeaker dimensions

E.2.1 - Draw the B(H) characteristic in the second quadrant

E.2.2 - What is the value of BH_{max}?

E.2.3 - The leakage coefficient for this magnetic circuit is $\sigma = 2.1$, and the coefficient of magnetomotive force loss is $r = 1.2$. Determine the operation point (B_m, H_m) of the magnet.

E.2.4 - Calculate the induction B_g in the gap.

E.2.5 - Determine the Laplace force acting on the coil, which consists of 50 turns on an average diameter $D_g = 19.5$ mm, made of copper wire with diameter $d = 0.14$ mm, when the current density is 10 A . mm^{-2}.

E3 - Permanent magnet based linear position sensor

This is the linear version of the angular position sensor presented on figure 15.28.

As shown on the figure below, it is designed to measure a linear position along Ox. The central ferrite magnet ($B_r = J = 0.4$ T, $\mu_{rev} = 1.1$) moves across a fixed magnetic circuit. The latter includes two identical, oppositely located, air gaps, where the inductions B_1 and B_2 are measured using Hall probes.

We assume that the flux coming out of the magnet is very efficiently channeled by the pole pieces (μ_r infinite), and that no flux closes via gap 3. The flux is thus entirely channeled toward gaps 1 and 2. Also neglect the gaps between the magnet and the pole pieces ($\varepsilon = 0$).

Figure 15.84 is a cross-sectional view of the device. We assume it is infinitely long in the direction perpendicular to the drawing (Oz).

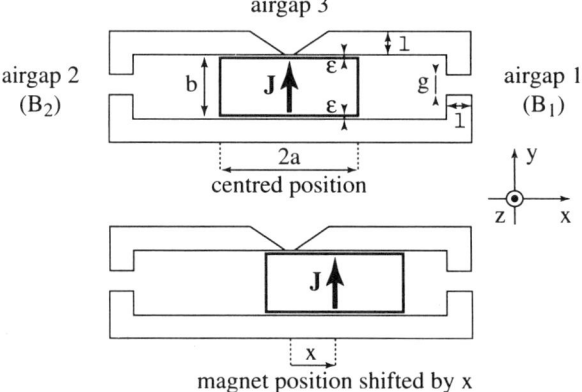

Figure 15.84 - Linear position sensor

E.3.1 - A simplified study of the effect of a shift of the magnet

We assume the gaps g to be very small with respect to the height b of the magnet. This is equivalent to assuming H_m to be zero and $B_m = J$.

The gaps 1 and 2 have a leakage coefficient σ equal to 2. The magnetomotive force loss coefficient is taken to be equal to 1.

E.3.1.1 - Calculate the inductions B_1 and B_2 when the magnet is centred.

E.3.1.2 - Determine the numerical values for $g = 1$ mm; $a = 10$ mm; $b = 10$ mm; $l = 4$ mm.

E.3.1.3 - Determine how the inductions B_1 and B_2 vary as a function of x, and plot their variations for x between $-a$ and $+a$.

E.3.1.4 - Calculate $B_1 + B_2$ and $B_1 - B_2$.

E.3.1.5 - Determine the maximum value of the induction in the magnetic circuit

E.3.1.6 - What value should the thickness l of the magnetic circuit have so that B never exceeds 1.2 T in the soft magnetic material?

E.3.2 - *The real operating point of the magnet*

We now assume that g is no more negligible with respect to b and that the other parameters of the circuit are unchanged ($\sigma = 2$; $r = 1$; $J_r = 0.4$ T; $\mu_{rev} = 1.1$). The magnet is in the center position.

E.3.2.1 - Determine the operating point of the magnet (B_m, $\mu_0 H_m$)

E.3.2.2 - Calculate the inductions B_1 and B_2.

E.3.2.3 - Numerical values for $g = 1$ mm; $a = 10$ mm; $b = 10$ mm; $l = 4$ mm.

E.3.2.4 - Do you think the approximation in part E.3.1 is justified?

E.3.3 - *Sensitivity of the sensor*

The ferrite magnet has a temperature dependence $(\Delta J / J) = -2 \times 10^{-4} . °C^{-1}$

The output signal of the sensor is proportional to the ratio $S = (B_1 - B_2) / (B_1 + B_2)$. Calculate the temperature sensitivity of the sensor.

E.4 - Biased magnetic bearing

Such bearings are used for axial stabilization of the flywheels in some satellites, such as SPOT-1.

We consider the actuator part of an axial magnetic bearing, with vertical rotation axis zz' and mean diameter $d = 10$ cm. Since the radius of curvature is large with respect to the other dimensions, the system will be treated as if it was two-dimensional. Figure 15.85 shows the device. All the air-gaps have the same cross-section $S_g = 10^{-3}$ m². The windings all have the same number of turns $N = 1,000$.

We make the following simplifying assumptions:
- the mass of the moving part is zero;
- the materials used for flux closure are infinitely permeable;
- there is no leakage flux.

For zero current, the gaps are equal: $g_1 = g_2 = 10^{-3}$ m (initial position).

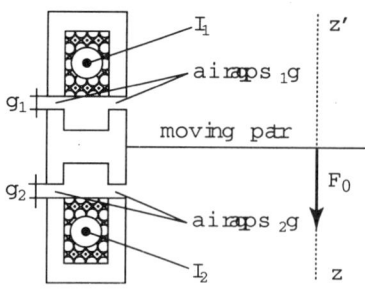

Figure 15.85 - Half of the cross-sectional view of the unbiased magnetic bearing with rotation symmetry around z' z

E.4.1 - Unbiased moving part

E.4.1.1 - What happens when I_1 and I_2 are equal and non zero?

E.4.1.2 - Give as a function of I_1 and I_2 the expression of the magnetic fields H_1 and H_2 in the air gaps, the bearing being maintained in its initial position.

E.4.1.3 - Give as a function of I_1 and I_2 the expression of the force F that makes it possible to hold the bearing in its initial position.

E.4.1.4 - Assume $I_2 = 0$. The current I_1 is adjusted so that it balances the bearing when it is submitted to a vertical, downward oriented force F_0 of 100 N. Calculate I_1.

E.4.1.5 - The force F_0 is varied by a small amount ΔF. Determine the sensitivity $\Delta F / \Delta I_1$ that makes it possible to maintain the device in its equilibrium position through a variation ΔI_1 of current I_1.

E.4.2 - Moving part biased by a magnet

The moving part in the above system is replaced by another moving part, including a magnet with constant and uniform polarization J_m ($J_m = 1$ T), with cross-section S_m ($S_m = 10^{-3}$ m²), and length ℓ_m ($\ell_m = 1$ cm), and the windings of the fixed part are connected in series. The current in these windings is the same, with the same direction ($I_1 = I_2 = I$). This bearing is represented in figure 15.86.

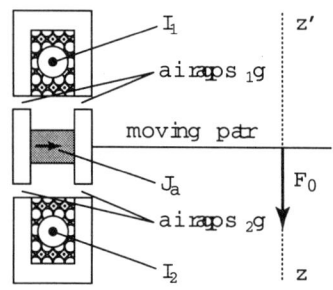

Figure 15.86 - Half of the cross-sectional view of the biased magnetic bearing

E.4.2.1 - Determine the expression of the magnetic fields H_1 and H_2 in each air gap when $g_1 = g_2 = g$ and $I = 0$.

E.4.2.2 - Determine, then calculate the current I needed to balance the bearing when it is submitted to a downward directed vertical force $F_0 = 100$ N.

E.4.2.3 - Determine the sensitivity $\Delta F / \Delta I$ of this device around this equilibrium situation.

E.4.2.4 - Compare the sensitivities obtained in E.4.1.5 and in E.4.2.3.

SOLUTION TO THE EXERCISES

S.1 *S.1.1*   The ferromagnetic surface is a symmetry plane. The system is then equivalent to two identical magnets in contact and of height h. The induction at A is identical to that at the center of a magnet with height 2h.

Figure 15.87

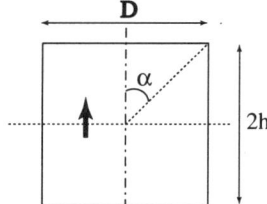

This was calculated in exercise E1 of chapter 2. We obtain the operation point of the magnet at point A:

$$B_A = \mu_0 M \cos\alpha = J \cos\alpha$$

$$H_A = -M(1 - \cos\alpha) = -\frac{J}{\mu_0}(1 - \cos\alpha)$$

For our dimensions, $\alpha = 45°$, $B_A = 0.71$ T, and $\mu_0 H_a = -0.29$ T.

Note - *another approximation method can be used. Assume that the induction on the contact surface is the same as inside an ellipsoid that would just fit in the volume of the magnet. In this special case (D = 10 mm, 2 l = 10 mm), the ellipsoid is a sphere, its demagnetising field coeficient is 1/3 (see chap. 2, § 1.2.7), whence*

$$B_A = 0.67 J = 0.67 \mu_0 M$$

$$\mu_0 H_A = -0.33 J.$$

The force turns out to be $F = B_A{}^2 \dfrac{S_a}{2\mu_0} = J^2 \cos^2\alpha \dfrac{\pi D^2}{8\mu_0}$

with $\cos\alpha = \dfrac{2h}{\sqrt{D^2 + 4h^2}}$.

Numerically, $B_A = 0.28$ T; $S_m = 0.79$ cm²; $F = 2.5$ N.

S.1.2.1 The magnet is in a closed circuit. Its demagnetising field is zero. The operating point is then: $H_m = 0$ and $B_m = J = \mu_0 M = B_A$. If S_m is its cross-section, the magnetic flux ϕ_m it sends into the magnetic circuit is:

$$S_m = \pi \frac{D^2}{4} \; ; \; B_m = J = \mu_0 M = B_A$$

$$\phi_m = B_m . S_m = \mu_0 M S_m = \mu_0 M \pi \frac{D^2}{4}$$

Numerically, $\phi_m = 0.31 \times 10^{-4}$ Wb.

S.1.2.2 Call S_i this cross-section: $S_i = \pi (R_{ext}^2 - R_{int}^2)$.

Numerically, $S_i = 0.21$ cm^2.

S.1.2.3 The whole flux created by the magnet passes through the contact area between the cup and the plate: $\Phi_m = B_B S_i$, whence

$$B_B = J \frac{S_m}{S_i} = \mu_0 M \frac{D^2}{4 \left(R_{ext}^2 - R_{int}^2 \right)}$$

Numerically, $B_B = 1.48$ T.

This value is smaller than the saturation induction of the iron in the cup. The magnetic circuit thus operates correctly, without saturation.

S.1.2.4 *Contribution of the magnet*

$$F_m = B_m^2 \frac{S_m}{2\mu_0}$$

Numerically, $F_m = 5.0$ N.

Contribution of the cup

$$F_i = B_B^2 \frac{S_m}{2\mu_0}$$

Numerically, $F_i = 18.6$ N.

Note - *at constant flux Φ_m, restricting the cross-section from S_m to S_i leads to multiplying the induction by the ratio S_m/S_i (numerically $S_m/S_i = 3.7$). The attraction force is multiplied by the same factor:*

$$F_i = \frac{1}{2\mu_0} B_B^2 S_i = \frac{1}{2\mu_0} \frac{\phi_m^2}{S_i} = \frac{1}{2\mu_0} B_m^2 S_m \frac{S_m}{S_i} = F_m \frac{S_m}{S_i}$$

Total force: $F_t = F_m + F_i = F_m \left(1 + \frac{S_m}{S_i} \right)$

Numerically, $F_t = 23.6$ N.

S.1.2.5 Using the same magnet, flux closure in the cup makes it possible to multiply the attraction force at contact by a large factor.

S.2 *S.2.1*   *B(H) curves*

See figure 15.1, which shows the $\mu_0 M (\mu_0 H) = J (\mu_0 H)$ and $B (\mu_0 H)$ loops. Going over from one to the other is performed through the relations $B = \mu_0 H + J = \mu_0 H (H + M)$.

Coordinates of the kink corresponding to $\mu_0 H = -0.2$ T:
- On the $J (\mu_0 H)$ loop: $J = \mu_0 M = 0.33$ T.
- On the $B (\mu_0 H)$ loop: $B = \mu_0 H + J = -0.2 + 0.38 = 0.18$ T.

Figure 15.88

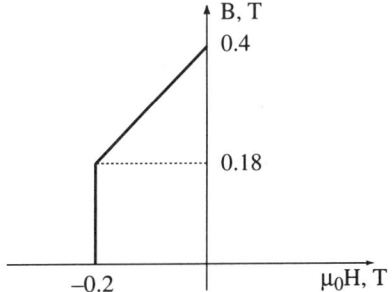

This kink is just below the point corresponding to $(BH)_{max}$ for the magnet.

S.2.2   *$(BH)_{max}$ value for the magnet*

The oblique part of the B(H) curve can be expressed as:
$B = 1.1 \mu_0 H + 0.4$.

Multiplying by $\mu_0 H$, we get the expression:
$(\mu_0 BH) = 1.1 (\mu_0 H)^2 + 0.4 (\mu_0 H)$.

The maximum is obtained from the condition $\dfrac{d(\mu_0 BH)}{d(\mu_0 H)} = 0$

We obtain: $(\mu_0 BH)_{max} = 0.036$ T^2
$(BH)_{max} = 29$ kJ.m^{-3}

This point with $(BH)_{max}$ corresponds to B = 0.2 T, $\mu_0 H = 0.182$ T, H = 146 kA.m^{-1}.

Note - *For an ideal magnet with $B_r = 0.4$ T, we get (eq. 15.8):*
$(BH)_{max} = B_r^2 / 4 \mu_0 = 32$ kJ.m^{-3}.

S.2.3   *Determination of the operating point*

Method 1: permeance line

This is a straight line with slope p (eq. 15.3 and 15.4)

$$p = \frac{B_m}{\mu_0 H_m} = \frac{\sigma}{r} \frac{S_g}{S_m} \frac{\ell_m}{\ell_g}$$

Numerically, $\sigma = 2.1$ $r = 1.2$.

$$\ell_m = 13 \text{ mm} \qquad \ell_g = 1.25 \text{ m}$$

$$S_m = \frac{\pi}{4}\left(D_{ext}^2 - D_{int}^2\right) \approx 23.7 \text{ cm}^2$$

$$S_g = \pi D_g h = 2.45 \text{ cm}^2$$

$$p = 1.88$$

Graphical solution $\rightarrow B_m = 0.25 \text{ T}, \mu_0 H_m = -0.13 \text{ T}.$

Method 2: setting up the equations and solving them

$$\left.\begin{array}{l} B_m S_m = \sigma B_g S_g \\ H_m \ell_m + r H_g \ell_g = 0 \\ B_g = \mu_0 H_g \\ B_m = 0.4 + 1.1 \mu_0 H_m \end{array}\right\} \quad \text{4 equations and 4 unknowns}$$

This yields the same result.

S.2.4   Calculation of B_g

$$B_g = \frac{B_m S_m}{\sigma S_g} = 1.15 \text{ T}.$$

S.2.5   Laplace force

$$F = I \ell B_g$$
$$= j.S_{wire}.n.\pi.D_g.B_g$$

Numerically, $j = 10 \text{ A.mm}^2$

$$\left.\begin{array}{l} S_{wire} = \frac{\pi}{4}d^2 = 154 \times 10^{-4} \text{ mm}^2 \end{array}\right\} \quad I = 0.154 \text{ A}$$

$$\left.\begin{array}{l} n = 50 \\ D_g = 19.5 \end{array}\right\} \quad \ell = 3.06 \text{ m}$$

$$F = 0.54 \text{ N}$$

Note - *Resistance of the loudspeaker:* $R = \rho \dfrac{\ell}{S_{wire}} = 3.6 \text{ } \Omega$

S.3 *Linear position sensor*

S.3.1.1 $B_1 = B_2 = J \dfrac{a}{\sigma \ell}$

S.3.1.2 Numerically, $B_1 = B_2 = 0.5 \text{ T}$

S.3.1.3 $B_1 = J \dfrac{a + x}{\sigma \ell} \qquad B_2 = J \dfrac{a - x}{\sigma \ell}$

S.3.1.4 $B_1 + B_2 = J \dfrac{2a}{\sigma \ell} \qquad B_1 - B_2 = J \dfrac{2x}{\sigma \ell}$

S.3.1.5 $(B_i)_{max} = J\dfrac{2\,a}{\ell}$, numerically $(B_i)_{max} = 2$ T

S.3.1.6 $\ell \geq J\dfrac{2\,a}{1.2} = 6.7$ mm

S.3.2.1 $B_m = J\dfrac{\sigma\,b\,\ell}{\sigma\,b\,\ell + \mu_{rev}\,g\,a}$ $\mu_0\,H_m = -J\dfrac{a\,b}{\sigma\,b\,\ell + \mu_{rev}\,g\,a}$

S.3.2.2 $B_1 = B_2 = B_g = J\dfrac{a\,b}{\sigma\,b\,\ell + \mu_{rev}\,g\,a}$

S.3.2.3 Numerically, $B_m = 0.35$ T, $\mu_0\,H_m = -0.044$ T, $B_g = 0.44$ T

S.3.2.4 The actual induction is 14% lower than the result obtained through the simplified method. However the latter provides a much simpler approach.

S.3.3   *Temperature sensitivity*

The ratio $(B_1 - B_2)/(B_1 + B_2)$ is independent of any variation in the magnet polarization.

S.4 *S.4.1.1* Symmetry of the attraction forces $\rightarrow$ (unstable) balance.

S.4.1.2 It can be easily shown, in particular using the reluctance formalism, that the flux due to I_1 (I_2) closes through the moving part, and that it does not go through the air gaps g_2 (g_1). The sysem is thus equivalent to two independent magnetic systems: one around I_1 with the air gaps g_1, and the other around I_2 with the air gaps g_2.

Ampère's law $2\,H_1 = N\,I_1$

$H_1 = \dfrac{N\,I_1}{2\,g_1}$ $H_2 = \dfrac{N\,I_2}{2\,g_2}$

S.4.1.3 Attraction force in each air gap $F = \dfrac{\mu_0\,H^2}{2}\,S_g$

Resulting force $F = \dfrac{\mu_0\,S_g\,N^2}{4\,g^2}\left(I_1{}^2 - I_2{}^2\right)$

S.4.1.4 $I_2 = 0 \Rightarrow H_2 = 0 \Rightarrow F = \dfrac{\mu_0\,S_g\,N^2}{4\,g^2}\,I_1{}^2$

At equilibrium $I_1 = \dfrac{2\,g}{N}\sqrt{\dfrac{F_0}{\mu_0\,S}}$, numerically $I_1 = 0.56$ A.

S.4.1.5 $\dfrac{\Delta F}{\Delta I} = \dfrac{\mu_0\,S\,N^2\,I_1}{2\,g}$, numerically $\dfrac{\Delta F}{\Delta I} = 350$ N $\cdot$ A^{-1}

S.4.2.1 The fields in the air gaps are the sum of those due to the system without the magnet and with the currents, and of those due to the system with the magner and without current.

$$\left.\begin{array}{l} B_m S_m = 2 B_g S_g \\ H_m \ell_m + 2 H_g g = 0 \\ B_g = \mu_0 H_g \\ B_m = \mu_0 H_m + J_m \end{array}\right\} \text{ 4 equations and 4 unknowns}$$

$$H_1 = \frac{J_m}{\mu_0} \frac{\ell_m/2}{g + \ell_m \left(S_g/S_m\right)}$$

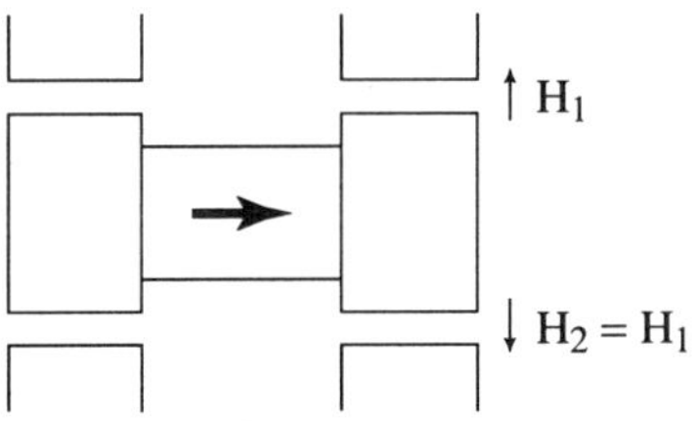

Figure 15.89

S.4.2.2 $H_1 = \dfrac{NI}{2g} + \dfrac{J_m \ell_m/2}{\mu_0 \left[g + \ell_m \left(S_g/S_m\right)\right]}$

$H_2 = \dfrac{NI}{2g} - \dfrac{J_m \ell_m/2}{\mu_0 \left[g + \ell_m \left(S_g/S_m\right)\right]}$

whence $I = \dfrac{F_0 \, g \left[g + \ell_m \left(S_g/S_m\right)\right]}{N S_g J_m \ell_m}$, numerically $I = 110$ mA.

S.4.2.3 Sensitivity $\dfrac{\Delta F}{\Delta I} = \dfrac{F}{I}$, numerically $\dfrac{\Delta F}{\Delta I} = 900 \, \text{N} \cdot \text{A}^{-1}$.

S.4.2.4 The current is 5,000 times smaller and the biased device is more than 2,500 times more sensitive.

REFERENCES

[1]   W.F. BROWN Jr, Magnetostatic Principles in Ferromagnetism (1962) Amsterdam, North Holland.

[2]   H. ZIJLSTRA, *in* Ferromagnetic Materials **3** (1982) 37, E.P. Wohlfarth & K.H.J. Buschow Eds, North-Holland, Amsterdam.

[3]   R.A. MCCURRIE, *in* Ferromagnetic Materials **3** (1982) 107, E.P. Wohlfarth & K.H.J. Buschow Eds, North-Holland, Amsterdam.

[4]   C. BRONNER, E. PLANCHARD, J. SAUZE, *Cobalt* **32** (1966) 124.

[5]   H. STÃBLEIN, *in* Ferromagnetic Materials **3** (1982) 441, E.P. Wohlfarth & K.H.J. Buschow Eds,North-Holland, Amsterdam.

[6] K.J. STRNAT, *in* Ferromagnetic Materials **4** (1988) 131, E.P. Wohlfarth & K.H.J. Buschow Eds, North-Holland, Amsterdam.K.J. STRNAT, R.M.W. STRNAT, *J. Magn. Magn. Mater.* **100** (1991) 38.

[7] J.F. HERBST, J.J. CROAT, *J. Magn. Magn. Mater.* **100** (1991) 57.

[8] J.M.D. COEY, Rare Earth Iron permanent Magnets (1996) Clarendon Press, Oxford.

[9] O. CUGAT, F. BLOCH, J.C TOUSSAINT, Rare Earth Magnets and their Applications, REM XV, Vol. 2 (1998) 853, L. Schulz & K.H. Müller Eds, Werkstoffinformationsgesellschaft, Dresden.

[10] P. BRISSONNEAU, Magnétisme et matériaux magnétiques pour l'électrotechnique (1997) Hermès, Paris.

[11] R.J. PARKER, R.J. STUDDERS, Permanent magnets and their applications (1962) John Wiley & Sons, New York.

[12] D. HADFIELD, Permanent magnets and magnetism (1962) Illiffe Books Ltd., London.

[13] M. VELGE, K.H.J. BUSCHOW, *J. Appl. Phys.* **39** (1968) 1717.

[14] J. OMEROD, *J. Less-Common Met.* **111** (1985) 49;
D.K. DAS, *IEEE Trans. Magn.* **Mag-5** (1969) 214.

[15] M. SAGAWA, H. NAGATA, *IEEE Trans. Magn.* **Mag-29** (1993) 2747;
M. SAGAWA, H. NAGATA, O. ITATANI, T. WATANABE, 13th Int. Workshop on RE Magnets and their Applications (1994) 13, Birmingham, UK.

[16] K.J. DE VOS, Thesis (1966) Technical University Eindhoven, The Netherlands.

[17] J.J. CROAT, J.F. HERBST, R.W. LEE, F.E. PINKERTON, *Appl. Phys. Lett.* **44** (1984) 148;
J.J. CROAT, *J. Mat. Eng.* **10** (1) (1988) 7;
J.J. CROAT, *J. Less-Common Met.* **148** (1989) 7;
J.J. CROAT, 13th Int. Workshop on RE Magnets and their Applications (1994) 65, Birmingham, UK.

[18] P.B. BRAUN, *Philips Res. Rep.* **12** (1957) 491.

[19] P.W. ANDERSON, *Phys. Rev.* **79** (1950) 350.

[20] E.W. GORTER, *Proc. IEEE* **104B** (1957) 255 S.

[21] H.B.G. CASIMIR, J. SMIT, U. ENZ, J.F. FAST, H.P.J. WIJN, E.W. GORTER, A.J.W. DUYVESTEYN, J.D. FAST, J.J. DE JONG, *J. Phys. Rad.* **20** (1959) 360.

[22] J. SMIT, *J. Phys. Rad.* **20** (1959) 370.

[23] G.W. RATHENAU, J. SMIT, A.L. STUYTS, *Z. Phys.* **133** (1952) 250.

[24] B.T. SHIRK, W.R. BUESSEM, *J. Appl. Phys.* **40** (1969) 1294.

[25] L. JAHN, H.G. MüLLER, *Phys. Stat. Sol.* **35** (1969) 723.

[26] E. BELORIZKY, M.A. FREMY, J.P. GAVIGAN, D. GIVORD, H.S. LI, *J. Appl. Phys.* **61** (1987) 3971.

[27] R.L. COEHOORN, *in* Supermagnets - hard magnetic materials (1991) chap. 8, L. Schulz & K.H. Müller Eds, Kluwer Academic Publishers, Dordrecht.

[28] M. SAGAWA, S. FUJIMURA, N. TOGAWA, H. YAMAMOTO, Y. MATSUURA, *J. Appl. Phys.* **55** (1984) 2083.

[29] J.J. CROAT, J.F. HERBST, R.W. LEE, F.E. PINKERTON, *J. Appl. Phys.* **55** (1984) 2078.

[30]   J.M.D. COEY, J. SUN HONG, *J. Magn. Magn. Mater.* **87** (1990) L-251;
        J.M.D. COEY, J. SUN HONG, Y. OTANI, 11th Int. Workshop on RE Magnets and their
        Applications, Vol. 2 (1990) 36, Pittsburgh.

[31]   L. NéEL, *C.R. Ac. Sci.* **224** (1947) 1550;
        C. KITTEL, *Rev. Mod. Phys.* **21** (1949) 541;
        S. SHTRIKMAN, Thesis (1957) Haifa;
        W.F. BROWN, *J. Appl. Phys.* **33** (1962) 1308;
        K. COMPAAN, H. ZIJLSTRA, *Phys. Rev.* **126** (1962) 1722.

[32]   S. SHTRIKMAN, Thesis (1957) Haïfa;
        E. KNELLER, Fine particle theory *in* Magnetism and Metallurgy, Vol. 1 (1969) 365,
        A.E. Berkowitz & E. Kneller Eds, Academic Press, New York-London.

[33]   A. AHARONI, *J. Appl. Phys.* **30 suppl.** (1959) 70 S.

[34]   K.J. DE VOS, Magnetism and Metallurgy (1969) 473, A.E. Berkowitz & E. Kneller Eds,
        Academic Press, New York-London.

[35]   J.J. DE JONG, J.M.G. SMEETS, H.B. HAANSTRA, *J. Appl. Phys.* **29** (1958) 297.

[36]   F. KOOLS, *J. Phys.* **C6** (1985) 349.

[37]   M. SAGAWA, S. HIROSAWA, H. YAMAMOTO, S. FUJIMURA, Y. MATSUURA, *Jap. J. of
        Appl. Phys.* **26** (1987) 785;
        M. SAGAWA, S. HIROSAWA, *J. Phys.* **49**, **C8** (1988) 617;
        S. HIROSAWA, Y. TSUBOKAWA, R. SHIMIZU, 10th Int. Workshop on RE Magnets and
        their Applications (1989) 465, Tokyo.

[38]   D.W. TAYLOR, V. VILLAS-BOAS, Q. LU, M.F. ROSSIGNOL, F.P. MISSELL, D. GIVORD,
        S. HIROSAWA, *J. Magn. Magn. Mater.* **130** (1994) 225.

[39]   D. GIVORD, Q. LU, M.F. ROSSIGNOL, P. TENAUD, T. VIADIEU, *J. Magn. Magn. Mater.*
        83 (1990) 183.

[40]   D. GIVORD, M.F. ROSSIGNOL, D.W. TAYLOR, A. RAY, *J. Magn. Magn. Mater.* **104-
        107** (1992) 1126.

[41]   D. GIVORD, M.F. ROSSIGNOL, *in* Rare Earth Iron permanent Magnets (1996) 272,
        J.M.D. Coey Ed., Clarendon Press, Oxford.

[42]   J. FIDLER, P. SKALICKY, Proc. 3rd Int. Symp. on Magnet. Anisotropy and Coercivity in
        RE Transition Metal Alloys (1982) 585, J. Fidler Ed., Baden, Austria;
        G.C. HADJIPANAYIS, *ibid.*, 609.

[43]   K.D. DURST, Thesis, Stuttgart (1986) figures 8 and 9, after unpublished results of Krupp
        Widia, GmbH, Essen.

[44]   G.C. HADJIPANAYIS, *in* Rare Earth Iron permanent Magnets (1996) 286, J.M.D. Coey Ed.,
        Clarendon Press, Oxford.

[45]   W. WERNSDORFER, E. BONET OROZCO, K. HASSELBACH, A. BENOIT, D. MAILLY,
        O. KUBO, H. NAKANO, B. BARBARA, *Phys. Rev. Lett.* **79** (1997) 4014.

[46]   A. AHARONI, *Phys. Rev.* **119** (1960) 127.

[47]   C. ABRAHAM, A. AHARONI, *Phys. Rev.* **120** (1960) 1576.

[48]   H.R. HILZINGER, H. KRONMüLLER, *Phys. Lett.* **51A** (1975) 59.

[49]   H.R. HILZINGER, H. KRONMüLLER, *Appl. Phys.* **12** (1977) 253.

[50]   H. KRONMüLLER, *J. Magn. Magn. Mater.* **7** (1978) 341.

[51]   P. GAUNT, *Phil. Mag.* **B48** (1983) 261.

CHAPTER 16

SOFT MATERIALS FOR ELECTRICAL ENGINEERING AND LOW FREQUENCY ELECTRONICS

The materials presented in this chapter are used at low frequencies, i.e. below 400 Hz for power electrical engineering, and 100 kHz for power electronics. After a general presentation which will include a detailed discussion on the origin of magnetic dissipation, we will discuss the various families of soft steels, of ferrites, and high or very high permeability alloys, from the oldest steels which date back to the nineteenth century to the newcomers, viz the amorphous and nanocrystallized materials. We will finally review the low frequency applications of these soft materials.

1. GENERAL PRESENTATION OF SOFT MATERIALS

A convenient classification consists in distinguishing among the soft magnetic materials for low frequencies:

- the materials for electrical engineering and power electronics, used in the production, the transportation, and the use of electrical energy (dynamos and alternators, transformers, motors), the iron-silicon alloys being by far the most important representatives of this category,

- the materials for low power applications: telecommunications and electronics, small transformers, sensors and security devices… This class has long included various metallic (iron-nickel, iron-cobalt), and insulating (ferrites) materials. More recently, materials obtained by fast quenching have appeared: amorphous and nanocrystallized alloys, the economical status of which is still uncertain [1].

This schematic classification is far from spanning the whole range of soft materials. Magnetic shielding can use materials belonging to one or another of these categories. Iron-nickel alloys can have quite different applications in bulk form or as thin films. The use of some materials is mainly determined by their magnetoelastic or magneto-optical coupling properties. Ferrites for radiofrequency and microwave applications, magnetostrictive soft materials, and soft films for magnetic recording will be presented in chapters 17, 18 and 21 respectively.

1.1. *THE PROPERTIES REQUIRED OF A SOFT MATERIAL*

A soft material is used to enhance or to channel the induction flux arising from electrical currents, magnets or external waves (in the case of shielding). This requires the following characteristics:

♦ as high as possible a *saturation polarization* $J_s = \mu_0 M_s$, the distinction between *magnetic polarization* ($J = \mu_0 M$), *and magnetic induction* $B = \mu_0 (H + M)$ being here of little importance, on account of the weak values of the magnetising fields.

♦ a large permeability, since small variations of the magnetising field have to lead to large variations in magnetization. This general notion must be specified depending on the actual use. One may look for a high initial or maximal permeability, a rapidly variable permeability or on the contrary as constant as possible a permeability along the hysteresis loop, and all this in the static or dynamic regime...

♦ as small as possible an energy dissipation within the material, if use in the dynamic regime is considered. The general trend towards an increase of operating frequency makes this requirement crucial. Furthermore, materials often work under non sinusoidal wave forms (in particular trapezoidal) with sizeable amounts of high order harmonics.

Obviously, the ideal material should also feature good mechanical characteristics (parts rotating at high speed), be easy to shape and to cut, devoid of aging effects, resistant to oxidation, and corrosion, cheap, and easily recyclable. Since this ideal material is of course inaccessible, the use of a soft material results from compromise between many often contradictory requirements.

1.2. *ROLE OF STRUCTURAL AND ELECTROMAGNETIC CHARACTERISTICS*

We now discuss the influence of the various characteristics of the material on the three main points we just defined.

1.2.1. *Polarization*

While physicists speak about magnetization, electrical engineers prefer to think in terms of polarization. The saturation polarization is an intrinsic characteristic of the material, mainly determined by its composition. Except in some very special applications, the polarization must remain as high as possible over a large temperature range. Accordingly a high Curie temperature is simultaneously required.

1.2.2. Permeability

Permeability results from many subtle influences. It reflects at our scale the phenomena occurring at the scale of magnetic domains.

- **Anisotropy**: in spite of their complexity, the bottom line of magnetization processes always involve the rotation of magnetic moments. It is thus necessary to minimize all phenomena which can oppose moment rotation, and in the first place magnetocrystalline anisotropy (see § 2.3 of chap. 3). High crystal symmetry is thus normally more favorable than uniaxial symmetry, but anisotropy also depends on composition.
   However, it can be worthwhile in some cases to induce a uniaxial anisotropy in order to tailor the hysteresis loop shape or to optimize a characteristic (permeability, energy losses).

- **Magnetostriction**: weak magnetostriction is also required in order to minimize the magnetic anisotropy of magnetoelastic origin arising from internal stresses in the material (chap. 12). Like magnetocrystalline anisotropy, this characteristic depends on the crystal symmetry and composition.

- **Wall area**: to within a few notable exceptions (alloys with skewed loops), the magnetization mechanisms involving wall motion are preponderant in soft materials for electrical engineering. It is thus necessary in the first place to maximize the wall area in the material (see § 2.4 below). This area essentially results from the compromise between wall surface energy and magnetostatic energy.
   It is therefore worthwhile to decrease the wall energy, which is a function of the exchange interactions and anisotropies. It is difficult to affect the exchange interactions, which also determine the Curie temperature. On the other hand, weak magnetocrystalline or induced anisotropy and weak magnetostriction will here again be beneficial.
   The control of magnetostatic effects is an efficient method for domain and wall multiplication. For this purpose, it is possible to play with the grain texture, which governs magnetostatic effects at the surfaces, and grain boundaries. Specific surface processing of the material (application of stress, scratching...) is also especially interesting. Transformer sheets made with "Hi-B" iron-silicon alloys are remarkable examples of applications of these principles.

- **Metallurgical characteristics of the material**: it is also necessary to offer wall motion as smooth an environment as possible. A soft material is obtained after careful metallurgy (purification, annealing...) aiming at reducing all defects responsible for wall pinning: dislocations and internal stresses, vacancies and impurities... The defects with sizes comparable to the wall thickness strongly pin them, and are the most detrimental. In particular, the elimination of metalloid precipitates is essential.

1.2.3. Energy dissipation

We will discuss later the subdivision of losses into (quasi)-static and dynamic losses. In conductors, energy is dissipated into heat by the induced currents. Three parameters are involved:

♦ resistivity: this is the main electromagnetic parameter from this point of view. It depends essentially on the material composition. One tries to maximize it.

♦ domain size: a crude argument shows the influence of this parameter. If a single wall has to move at speed v in order to give rise to an induction flux variation Φ, the same flux variation shared over n walls requires for each of them only a speed v/n. If the induced current overlap is neglected, the current densities are then changed by a factor $1/n$, and the total loss through Joule dissipation by a factor $n(1/n)^2 = 1/n$. A decrease of the domain width is thus beneficial, and it also allows a gain in permeability.

♦ thickness: a conductor must be divided into thin sheets in order to restrict the development of eddy currents.

1.3. ENERGY LOSS ANALYSIS

1.3.1. Macroscopic aspects

Energy dissipation comes from the irreversibility inherent to magnetization mechanisms. It shows up at the macroscopic scale through the area of the hysteresis loop, equal to the integral $\int \mathbf{H}\,d\mathbf{B}$ taken over the loop.

Losses increase very little with frequency in a material with high resistivity (ferrite). The situation is quite different in conductors. The loop area, and accordingly the losses, increase with frequency as shown in figures 16.1 and 16.2.

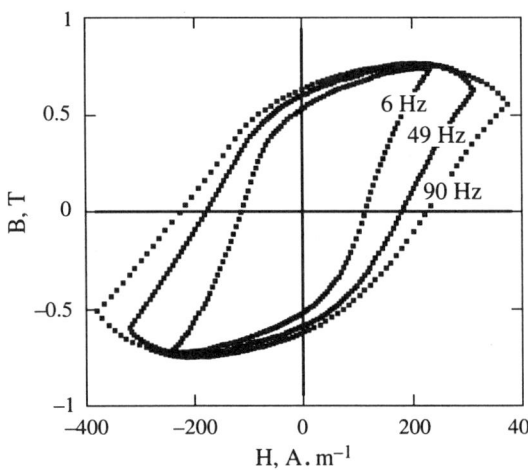

Figure 16.1 - Hysteresis loops of soft iron at different frequencies; maximum induction 0.75 T

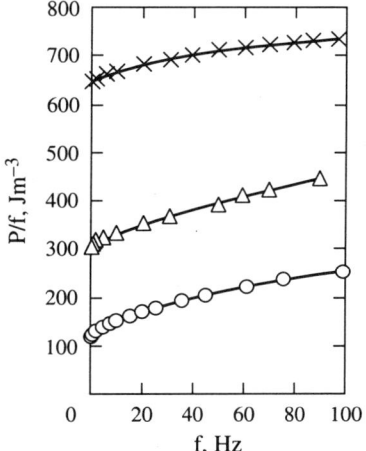

Figure 16.2 - Loss per cycle (P/f) in different samples of thickness e for a maximum induction B_0

Circles: Fe 3%Si single crystal (110)[001], e = 0.28 mm, B_0 = 1.7 T
Triangles: Fe 3%Si non oriented sheets, e = 0.33 m, B_0 = 1.5 T
Crosses: commercial iron sheet, e = 0.1 mm, B_0 = 1.3 T (after [2])

Note - *Two conventions are used in the literature to express losses. Theoretical approaches focus on the energy loss per cycle, whereas technologists rather consider the power loss at a frequency f. In this chapter we give numerical values of the dissipated power P. The energy loss per cycle is obviously equal to P/f.*

◆ In a slowly variable field, one measures hysteresis or quasi-static losses corresponding to the point $(P/f)_0$ where the curve P/f versus f cuts the vertical axis. Steinmetz's empirical formula states that, in the whole range of medium inductions, the loss over a loop with extreme inductions $\pm B_0$ is proportional to B_0^α. The exponent α generally ranges between 1.7 and 1.9.

◆ At a given frequency f, the difference $(P/f) - (P/f)_0$ represents the dynamical losses per cycle.

1.3.2. Calculation of losses in a conductor

We want to estimate the integral over a loop, for unit volume of the material, of the quantity $\rho\,|\mathbf{j}(\mathbf{r},t)|^2$ where $\mathbf{j}(\mathbf{r}, t)$ represents the eddy current density. The solution of the problem requires the knowledge of the space and time variations of the polarization vector $\mathbf{J}(\mathbf{r},t)$. The simplest approach consists in ignoring the domain distribution, and in considering the material as a homogeneous medium; this is the so called *classical loss model* approximation.

For loops with extreme inductions $\pm B_0$, performed at a frequency f, low enough that the skin effect is negligible, the classical loss per cycle in a sheet of thickness e is:

$$(P/f)_c = (\pi e B_0)^2 f / 6\rho \qquad (16.1)$$

This prediction disagrees with experiment: numerical values deduced from the above expression are too small, and the losses do not vary linearly with frequency. One calls *additional* or *excess losses* the difference between actual and classical losses. Of course, the disagreement originates from the fact that the existence of domains was ignored. In fact, magnetization rotation only occurs in the regions swept by the walls.

In a fictitious conducting circuit enclosing a wall in motion, there is a flux variation. The resulting eddy currents are not uniformly distributed within the material, but concentrated around walls in motion, which gives rise to additional losses [3].

The Pry and Bean model [4] idealizes this real situation both in space and time. The domains in a slab of thickness e are assumed to be of infinite length and of uniform width d (fig. 16.3).

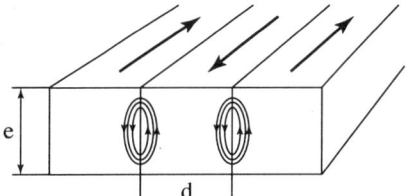

Figure 16.3 - Pry and Bean model: 180° domains of uniform width d in a sheet of thickness e

Eddy currents are localized in the vicinity of the walls.

Such a model is probably rather close to the situation in iron-silicon sheets with oriented grains. All the walls, assumed rigid, move in phase and sinusoidally as a function of time. Maxwell's equations then predict a loss per cycle given by:

$$(P/f)_{PB} = \frac{16 e d B_0^{\,2}}{\pi \rho} f \sum_{n\,odd} \frac{1}{n^3} \cot\left(\frac{n\pi d}{e}\right) \tag{16.2}$$

This expression gives the classical losses when the ratio $(d/e) \to 0$, i.e. if the flux variations can be considered as uniformly distributed. Otherwise, losses calculated in this way exceed the classical losses by a factor which can reach several units.

The Pry and Bean model confirms, at least qualitatively, the need to decrease both sheet thickness and domain width. However it is still too crude to accurately fit the experiments. On the one hand, as eddy currents expand more easily in the core of the sheet than at its surface, the central regions of the walls are submitted to more intense slowing down, and wall deformations appear at high frequencies. However this phenomenon remains quite insufficient to explain the difference between classical and actual losses.

On the other hand, magnetization processes are for sure more irregular in space and time than assumed. This point is essential because the periodicity assumptions of the Pry and Bean model give rise to destructive interference of eddy currents in the main part of the material. Accordingly, losses are strongly underestimated.

A more complete approach must take into account the random nature of wall motion [3]. The energy loss in a sheet of cross-section S, during a cycle of period T, is expressed in Fourier space $(\mathbf{k}, \omega)$ as:

$$(P/f) = \frac{4}{\rho S} \sum_{\mathbf{k}} \int_{-\infty}^{+\infty} \frac{d\omega}{2\pi} \frac{|\mathbf{k}|^2}{|\mathbf{k}|^4 + \left(\dfrac{\omega \mu}{\rho}\right)^2} \frac{|\mathbf{J}(\mathbf{k}, \omega)|^2}{ST} \tag{16.3}$$

μ being the reversible permeability of the material.

We will not go into the rather heavy mathematical details. The main result is that one can reduce the domain evolution to the dynamics of statistically independent "magnetic objects" (MO). Physically speaking, a MO represents a group of neighbouring walls coupled through magnetostatic interactions. The actual nature of a magnetic object depends on the material considered: one could for instance consider as a MO a single Bloch wall in a grain-oriented Fe-Si alloy with large domains, or the whole set of walls in a grain in the case of a disoriented polycrystal.

The physical meaning of this approach can be summarized as follows: each magnetic object experiences a random potential (obstacle of varied nature), the derivative of which represents an opposition field. The difference between the magnetising field and the opposition field is the *excess field* H_{exc}. An infinitely small excess field would only lead to quasi-static magnetization processes.

When a MO jumps over a local maximum of the opposition field, H_{exc} becomes positive, and the MO starts an irreversible motion at finite speed. The slowing down due to eddy currents being of viscous type, the time derivative of the flux during the jump is proportional to the excess field. The irreversible motion stops when the magnetic object meets an opposition field larger than H_{exc}.

Increasing the frequency of the magnetising field leads to an increase in the number of active objects at a given time: an increase of the induction variation dB/dt requires a larger instantaneous wall velocity, and accordingly a larger pressure of the driving field. This pressure is obtained by increasing the applied field, which leads to releasing new magnetic objects.

Figure 16.4 illustrates this behavior. At high frequency, magnetization processes tend to become homogeneous in space and time, because all magnetic objects simultaneously become active and "fly over" the obstacles at a high speed which mainly depends on the applied field.

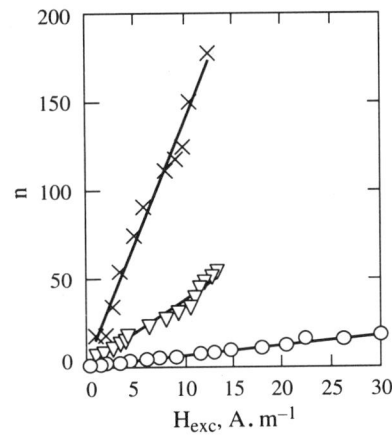

Figure 16.4 - Number n of magnetic objects (walls or correlated sets of walls) simultaneously active as a function of the excess field H_{exc}

The samples and the symbols are the same as those of figure 16.2 (after [3])

Loss calculation is simplified by reducing the number of degrees of freedom to be taken into account. The correlated motions of the walls which belong to a MO strongly reduce the eddy currents within this MO. The main losses take place outside.

This model leads to a significant unification of the loss concept. The medium where the MO moves (opposition field) determines both the statistics of the obstacles to be overcome in a quasi-static regime, and the wall motion dynamics at finite frequency. The distinction between hysteresis losses and dynamical losses thus tends to become less marked.

This is quite logical: in a conductor, dissipation phenomena eventually reduce to heating by eddy currents that are more or less localized in space and time.

This model applies quite well to a large number of materials, and predicts for the excess losses a variation roughly proportional to $\sqrt{f}$, in agreement with experimental results. It also leads to a dynamical extension of the usual quasi-static Preisach model [5].

1.4. LOSSES IN ROTATING OR TRAPEZOIDAL FIELD

It often occurs that a magnetic material is submitted to a field with changing direction, for instance in some regions of the stators in motors or of three-phase current transformers. The magnetization vector then also rotates, but with an angle shifted with respect to the field. This shift is the cause of energy dissipations different from those occurring in a field with fixed direction.

A small field (typically in the Rayleigh region) does not drastically modify the domain distribution. A reasonable approximation then consists in splitting the rotating vectors representing the field and the magnetization into two perpendicular contributions, out of phase by $\pi/2$. By superposing the phenomena along both directions, losses twice as large as the unidirectional ones are predicted. Experiments confirm the validity of this prediction.

In the range of intermediate inductions where a material usually operates, experiments show that losses pass through a maximum for an induction generally around 70% of the saturation induction. The superposition principle previously invoked is no more valid, and for the moment there is no quantitative model of losses. The situation is still more complex if the field contains a significant amount of harmonics.

Finally, in very high fields, the magnetization vector is alway collinear with the field. Only the reversible rotation processes remain, and losses tend toward zero, contrary to the case of the unidirectional field where they are maximum.

2. IRON BASED CRYSTALLINE MATERIALS

2.1. IRON AND SOFT STEELS

Iron should be an excellent material. It has a polarization of 2.16 T at room temperature, a large Curie temperature (1,043 K), and a cubic crystal structure. One uses pure iron or soft steels under quasi-static conditions in magnetic shieldings, in magnetic circuits and in cores for relays and electromechanical devices, in electromagnets, and in bottom-range motors. The main difficulty lies in removing impurities which lower the polarization, but even more importantly reduce the permeability when they form inclusions which hinder wall motion. Careful metallurgy (high-level refining, casting under vacuum…) allows to master these problems. After shaping, the material is generally improved by an anneal which relaxes manufacturing stresses; annealing also improves the time stability of magnetic characteristics. The temperature must not exceed 1,125 K, in order to avoid the $\alpha \to \gamma$ phase transition of iron.

When good mechanical stability is required, especially for machine rotors, iron is replaced by forged steels with high elastic limit (up to 700 MPa), containing 3d metals (vanadium, chromium, manganese, nickel), molybdenum, silicon… The concomitant improvement of mechanical properties allows the manufacturing of very big rotating parts, the weight of which can exceed 200 tons. These characteristics are obtained at the expense of saturation polarization, which however remains around 2 tesla.

By adding 17% of chromium in weight, one obtains stainless steels with mechanical properties similar to those of soft irons, which are used in the manufacturing of immersed devices (relays, electric valves, motors).

2.2. CLASSICAL IRON-SILICON ALLOYS

Note - *For Fe-Si alloys, we use the metallurgists' convention of giving the weight fraction (x). The atomic weights of iron and silicon being 56 and 28 g, respectively, the atomic fractions are $2x/(1 + 0.01x)$, i.e. approximately 2x for low silicon contents.*

- *The alloys for which the characteristics are given in this section are those produced by Ugine S.A.*

Pure or weakly alloyed iron is not the best material for power electrical engineering. It has long been known that the addition of silicon presents decisive advantages.

The first one is of metallurgical order. Pure iron passes from the α body centred cubic structure to the γ face centred cubic one at 910°C (1,183 K). This transition makes high temperature treatments very complicated. The addition of a small amount of silicon stabilizes the α phase at any temperature from 1.8%, and allows a large range of metallurgical treatments without phase transformation: rolling, refining, recrystallization… Moreover, silicon, and sometimes aluminium, additions have beneficial effects on the main electromagnetic characteristics (fig. 16.5):

♦ the magnetocrystalline anisotropy constant K_1, at room temperature, decreases from 4.8×10^4 J.m^{-3} for pure iron to 3.4×10^4 J.m^{-3} for the 3.5% Si alloy;

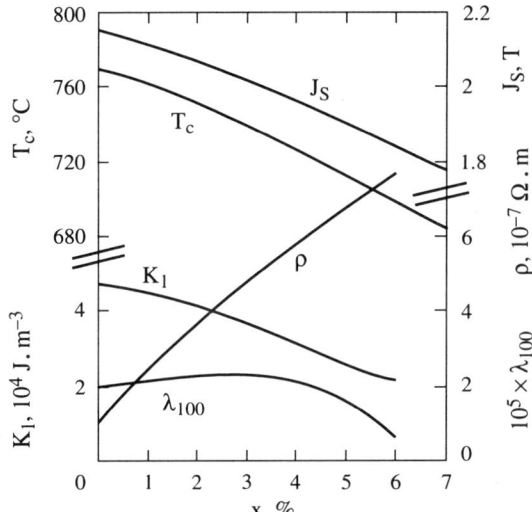

Figure 16.5 - Effects of silicon addition on the electromagnetic properties or iron at 20°C

x silicon weight content
K_1 magnetocrystalline anisotropy constant
λ_{100} magnetostriction
ρ resistivity
J_s saturation polarization
T_C Curie temperature

♦ simultaneously, the magnetostriction constant λ_{100} changes from -20×10^{-6} to about -5×10^{-6}, λ_{111} remaining close to 20×10^{-6};

Note - *Magnetostriction coefficients vary by ±20% depending on the authors.*

♦ the resistivity rapidly and almost linearly increases with silicon and aluminium contents. x and y being the weight contents of these latter elements, one can account for the resistivity with the approximate formula:

$$[1.36 + 1.10\,(x + y)] \times 10^{-7}\,\Omega.\,m.$$

A 3.5% Si alloy is already 3 or 4 times more resistive than pure iron;

♦ the saturation polarization and the Curie temperature are not too much lowered, as they decrease from 2.16 T down to around 2 T, and from 771°C down to 760°C, respectively.

Classical metallurgical techniques do not allow to exceed a silicon content x of 3.5 to 4%. Beyond this content, the alloy becomes too brittle for rolling. The specific weight decreases, for moderate values of x, as $(7,860 - 60\,x)\,kg.\,m^{-3}$. It is important to take into account this factor when comparing two materials, especially from the point of view of energy losses. Literature generally gives the weight loss whereas the important parameter is the volume loss.

The synthesis of Fe-Si alloys must also satisfy the general requirements already quoted:

♦ the impurity content, especially for metalloids, must be as low as possible. Care taken at all steps in the preparation allows to keep it below 60 ppm for C, 20 ppm for N, O, and S,

♦ the grains must be large enough in order to reduce Bloch wall pinning due to magnetostatic effects at the grain boundaries,

♦ one seeks to suppress residual stresses by suitable annealings,

♦ the reduction of energy losses in the AC regime requires that the material be produced in the form of sheets, as thin as possible, and electrically isolated from each other. One is limited in this direction by the increasing difficulty of rolling operations, and by the decrease in filling factor. Modern rolling techniques allow to get a well calibrated thickness and a high quality surface, whence filling factors which can reach 0.98. The manufacturer generally delivers the material under the form of coils several tons in weight.

Classical Fe-Si alloys are divided into two main categories depending on whether their grain texture is oriented or not [4].

2.3. FE-SI SHEETS WITH NON ORIENTED (NO) GRAINS

These sheets were for a long time hot rolled. Since the 1960s, one uses cold rolling which gives products of better quality for a smaller cost. The sheet thickness varies from about 1 mm for bottom range products to 0.35 mm for the best grades. The final product is coated with a very thin film of organic or mineral insulator. The silicon content of NO sheets varies from 0% for the lowest grades to around 3.2% for the best ones. The latter also usually contain an aluminium amount close to 0.3%.

2.3.1. Magnetic characteristics

The coercive field of NO sheets ranges from 80 to 25 $A.m^{-1}$ according to their grade. On account of grain misorientation, the quasistatic magnetization curve (fig. 16.6) exhibits a rather slow approach to saturation. They thus require a sizeable amount of excitation electric energy to reach high induction levels. Manufacturers indicate the minimal guaranteed inductions at 2,500, 5,000, and 10,000 $A.m^{-1}$, in general close to 1.5, 1.6, and 1.7 tesla.

Figure 16.6 - B(H) curve for a 0.35 mm thick NO Fe-Si sheet *(after Ugine S.A. documentation)*

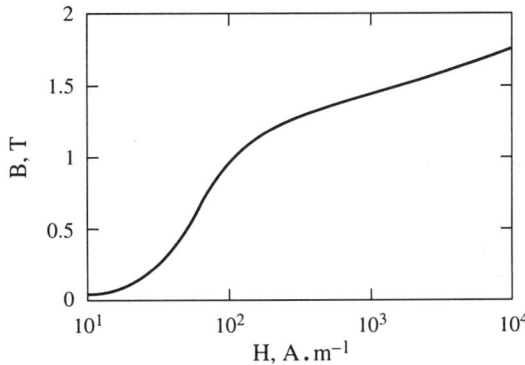

The energy losses in AC regime represent the most important sheet characteristic. They increase with frequency and induction level. Manufacturers usually indicate the losses at 50 hertz and 1.5 tesla, which are around 10 $W.kg^{-1}$ for the lowest qualities.

The table below gives the guaranteed maximal specific losses in $W.kg^{-1}$ for three varieties of normalized sheets.

Table 16.1 - Specific losses in $W.kg^{-1}$ of three NO Fe-Si sheets

Thickness (mm)	0.35	0.50	0.65
Losses P $(W.kg^{-1})$	2.50	2.70	3.50

Figure 16.7 shows the increase of losses with induction in NO sheets. It is worth noticing that the expression "non oriented sheet" is a slight misnomer. The successive metallurgical treatments and the unavoidable presence of impurities actually lead to a preferential grain orientation along the rolling direction. The result is an anisotropy of the losses which sometimes reaches 20%.

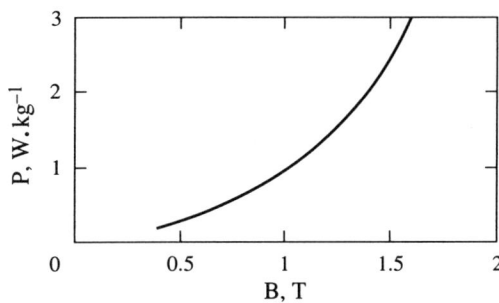

Figure 16.7 - Specific losses at 50 hertz in a 0.35 mm thick NO Fe-Si sheet (after Ugine S.A. documentation)

2.3.2. Applications

NO sheets are available under two varieties called "fully-processed", and "semi-processed", depending on whether the material already has its optimal magnetic characteristics or will get them only after a final thermal treatment performed by the user.

One uses the "fully-processed" sheets in three main categories of devices:
♦ rotating machines (motors, dynamos, alternators) which constitute the main use;
♦ low power transformers;
♦ some devices such as switches ou electricity meters, in which one takes advantage of their quasi-static characteristics.

Domestic electrical appliances and the automotive industry (motors, transformers, alternators) are the preferential application area of "semi-processed" sheets. The latter are also employed in rotating machines with powers smaller than about 20 kW.

2.3.3. Evolution and prospects of NO sheets

New varieties called "P quality", with improved permeability obtained thanks to improved texture control, have been on the market for a few years. For instance, the Ugine S.A. "P" sheets have minimal guaranteed inductions of 1.66 T, 1.75 T, and 1.87 T at 2.5, 5, and 10 $kA.m^{-1}$, respectively. These values, together with thermal conductivities twice larger than those of conventional NO sheets, make them especially suitable for electrical machine manufacturing. In this way one can reduce down to 65% the required Ampere-turn number in a machine of given size. Many advantages ensue: reduction of copper weight and of conductor insulation, decrease of Joule losses and of stray fields, thermal exchange improvement...

One often uses the NO alloys with an induction vector rotating within the sheet plane. It would thus be very interesting to industrially obtain textures involving easy magnetization directions close to the plane. Such would be, for instance, cubic or planar textures of (110)[001] (all grains have the same orientation) or (100)[0vw] (all grains have their axis [100] perpendicular to the sheet) type. Samples have already been obtained in laboratories after a very important purification effort since, from some ppm, impurities play a key role in recrystallization.

2.4. GRAIN-ORIENTED (GO) FE-SI SHEETS

The GO sheet metallurgy is based on a highly oriented grain texture: the *Goss texture*. Rolling induced shearing causes, in the body-centred-cubic structure, glide along the atomic planes of highest density. One thus obtains a (110)[001] texture called "cube on edge", characterized by an easy magnetization axis in the sheet plane and very close to the rolling direction (fig. 16.8).

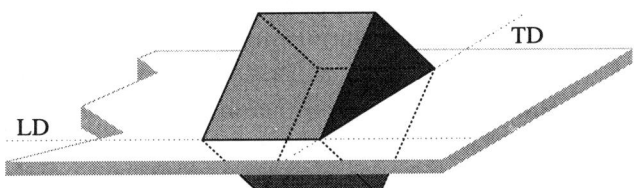

Figure 16.8 - "Cube on edge" crystal orientation in a GO sheet

In the ideal situation, the [001] and [1̄1̄0] axes are parallel (LD) and perpendicular (TD) to the rolling direction, respectively. Actually, misorientations of some degrees remain.

A secondary recrystallization anneal leads to grain size increase through the anomalous selective growth of some of them.

The classical GO sheets, 0.23 to 0.30 mm thick, are manufactured by continuous casting and a series of hot rolling and annealing steps. The final step is coating with a phosphatizing solution.

The high permeability sheets ("Hi-B"), 0.23 to 0.30 mm thick, are obtained by a single but very strong cold rolling process around 250°C, with a reduction rate larger than 80%. After secondary recrystallization, the grain size can reach 30 mm. Finally the sheet is coated with a phosphatizing colloid containing silicon. Beyond its protective and insulating function, we will see further on that this coating plays an important role in the magnetic properties.

2.4.1. Domain structure optimization

The quest for the Goss texture aims at an ideal domain structure presenting elongated 180° domains parallel to the rolling direction. The history of loss reduction in GO sheets follows closely that of texture optimization. The reduction of dynamic and hysteresis losses requires the elimination of hindrances of any nature which slow down the wall motion on the one hand, and the refinement of the domain width on the other hand. When impurities and cold-rolling stresses have been thoroughly eliminated, wall pinning mainly comes from magnetostatic effects.

Too large a discontinuity of the magnetization vector component perpendicular to an interface causes the onset of harmful *secondary* structures: closure domains at the grain boundaries, *spike* domains at the surface of the material (fig. 16.9). Metallurgists therefore worked at improving the key parameter, grain orientation. However experiment shows that it is good to retain a slight misorientation of the easy magnetization direction with respect to the sheet plane. As long as the misorientation angle does not exceed about 2°, the decrease in magnetostatic energy occurs through a width reduction of the 180° domains without creating spike domains. The corresponding increase of the total wall area also favours high quasi-static permeability. Furthermore, too large grains are undesirable: from a given grain size (around 0.5 mm), the decrease in hysteresis loss is smaller than the increase of the dynamic losses resulting from the domain widening.

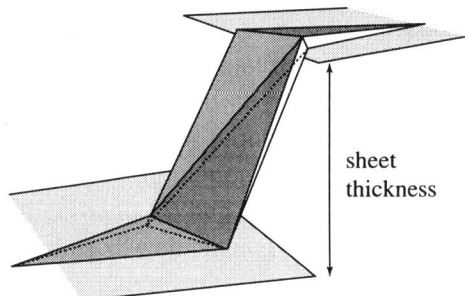

sheet
thickness

Figure 16.9 - Schematic representation of a spike domain appearing in a GO sheet on account of the crystal misorientation

The protective coating, only 2 or 3 μm thick, exerts on the sheet a longitudinal stress of several MPa. On account of magnetoelastic coupling, the easy magnetization axis tends to be parallel to the rolling direction. The result is a reduction of the secondary (closure domain) structure, and a refinement of the 180° domain main structure.

Finally, an especially efficient method of domain refinement consists in producing on the sheet surface a set of lines perpendicular to the lamination direction, and regularly spaced by a few mm to 30 mm (fig. 16.10) [7]. This treatment can be made using several processes based on two different principles. Mechanical scratching, spark-machining, continuous or pulsed laser irradiation, and plasma beams give rise to a system of tensile stresses parallel to the lamination direction. These treatments do not withstand a possible final anneal designed to relax machining stress.

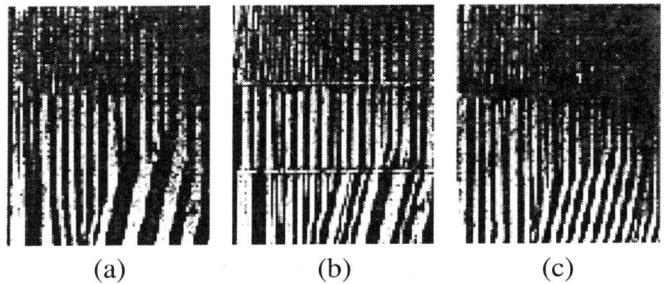

| (a) | (b) | (c) |

Figure 16.10 - Refinement of 180° domains in a GO sheet (after [7])

(a) non treated sheet - (b) electrolytically grooved sheet - (c) sheet irradiated with a plasma beam

On the other hand, the grooving processes based on a fluted wheel or electrolytic etching are mediated by magnetic poles which appear in the vicinity of grooves (fig. 16.11) [8]. They withstand further thermal treatments.

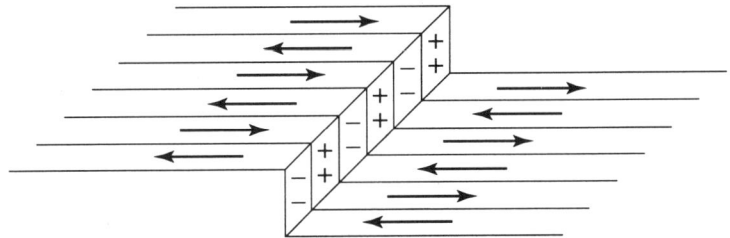

Figure 16.11 - 180° domain refinement in a GO sheet:
model of magnetic pole distribution appearing on the groove sides (after [8])

2.4.2. Magnetic characteristics

The coercive field of GO sheets is typically 5 A.m^{-1}. On account of the almost perfect domain orientation along the magnetising field direction, magnetization vector rotations practically do not occur, and the magnetization curve of GO alloys is much steeper than that of NO ones (fig. 16.12). In a 800 A.m^{-1} field, the induction is already close to 1.75 T for a conventional sheet, and 1.85 T for a high permeability sheet.

Figure 16.12 - Induction (B)
versus magnetising field (H)
at 50 hertz for a GO "Hi-B"
sheet 0.23 mm thick (after
Ugine S.A. documentation)

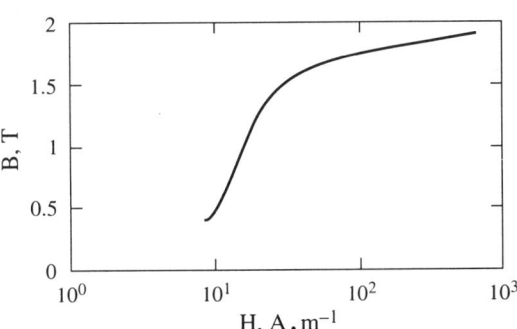

Table 16.2 gives the guaranteed maximal values of the specific losses P in $W.kg^{-1}$ for four different GO sheets ("Hi-B" sheets are not used below about 1.6 T).

Table 16.2 - Specific losses in $W.kg^{-1}$ for four GO sheets

Sheet type	Thickness (mm)	50 Hz 1.5 T	50 Hz 1.7 T	60 Hz 1.5 T	60 Hz 1.7 T
Conventional	0.35	1.05	1.50	1.38	1.98
	0.23	0.77	1.20	1.01	1.57
"Hi-B"	0.30	–	1.05	–	1.38
Laser treated	0.23	–	0.90	–	1.19

Figure 16.13 shows the increase of losses with induction in GO sheets. Of course, permeability decreases, and losses rapidly increase as soon as the field is no more applied parallel to the rolling direction. This is the reason why GO sheets are in practice used only for transformers.

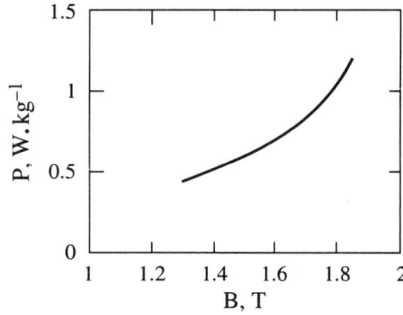

Figure 16.13 - Losses at 50 hertz (P) as a function of the induction (B) in a 0.23 mm thick "Hi-B" GO sheet (after Ugine S.A. documentation)

2.5. THIN IRON-SILICON SHEETS

When the working frequency increases, too thick a sheet is less and less efficiently used because only two thin sheets contribute to induction on account of the skin effect. A rigorous calculation of the latter is only possible in a material with linear response and constant permeability μ, which obviously is not the case.

However one obtains useful orders of magnitude by using as average permeability the slope of a hysteresis loop branch. With a relative permeability of 20,000, typical of a GO sheet, the skin thickness $\delta = \sqrt{\rho / \pi \mu f}$ reaches 0.35 mm at 50 Hz, and falls below 0.1 mm at 1 kHz. This is the reason why the general trend towards an increase in operating frequency requires thinner products. One speaks of *thin sheets* below 0.2 mm. As for classical thickness sheets, they are marketed under GO and NO forms.

As an example, a 0.1 mm thick GO sheet dissipates 15 $W.kg^{-1}$ at 400 hertz under 1.5 tesla, whereas a NO sheet of the same thickness dissipates at the same frequency 14 $W.kg^{-1}$ under 1 tesla.

As a general rule, thin sheets are used whenever a magnetic material works under a rapidly variable induction: machines rotating at high speed in order to increase the specific power (air-borne devices); special machines with a large number of poles; converters working at medium frequencies (400 Hz to 10 kHz) made with semiconductor components such as thyristors or power transistors.

2.6. *HIGH SILICON CONTENT ALLOYS*

It has long been known that a silicon content close to 6.5% optimizes the electromagnetic characteristics of the Fe-Si alloy. This composition corresponds to vanishing magnetostriction, whereas the magnetocrystalline anisotropy and the resistivity also evolve in the right direction, going from 4.8×10^4 to 2×10^4 J.m^{-3}, and from 10^{-7} to 7×10^{-7} Ω.m, respectively (fig. 16.5). Another advantage is the reduction of magnetic aging effects through the pinning of interstitial impurities, among others carbon. On the other hand, the huge increase in brittleness of these alloy forbids the use of rolling processes.

2.6.1. *Alloys obtained by fast solidification*

The melt-spinning technique, initially developed for amorphous alloys, produces continuous ribbons of crystalline Fe-Si [9].

The free flow or *melt-spinning* process consists in injecting the melted alloy through a circular nozzle on a spinning wheel located a few millimeters from the nozzle. Although very simple to implement, it has two major drawbacks: a limited ribbon width, and the lack of morphology control on account of the liquid flow instabilities.

This is why the *planar flow casting* process is preferentially used. The small gap (never more than 0.5 mm) between the nozzle and the spinning wheel virtually eliminates the instability problems. Furthermore the thinness of the liquid puddle improves thermal exchange with the wheel. The optimization of ribbon characteristics results from a complex compromise between the various casting parameters: molten alloy temperature, ejection pressure, nozzle-spinning wheel gap, nature and spinning velocity of the wheel. The purity grade of the starting alloy plays an important role as impurities noticeably modify the liquid viscosity. Besides, the composition and pressure of the gas inside the preparation zone play a role in the weldability and the liquid-wheel thermal exchange.

The ribbon thickness generally ranges from about 30 μm to 150 or 200 μm. An as cast ribbon exhibits a columnar structure approximately perpendicular to the ribbon surface. The average grain diameter is 5 to 10 μm.

Magnetic properties [10] are optimized after a recrystallization thermal treatment around 1,100°C. A high purity of the starting alloy favours the expansion of a (100)[0vw] cubic texture at the expense of (110) grains which contain hard magnetization directions within the ribbon plane. After recrystallization the grain size can reach 300 μm.

It is important that the thermal treatment does not exceed 1,100°C in order to avoid a noticeable silicon loss (the Si content falls down to about 5% after annealing at 1,300°C).

Unfortunately the thermal treatment degrades the mechanical characteristics. It causes, in the Fe-Si alloy, the growth of so called B_2 and DO_3 ordered crystallographic phases which make the ribbon more brittle. It is thus necessary to cool it down as fast as possible after annealing (more than $1,000°C . min^{-1}$) in order to restrict the development of structural order. Interstitial impurities also have a harmful effect on ductility, by blocking dislocation motion.

Depending on the purity grade and the quality of the recrystallization anneal, the quasi-static coercive field of ribbons can approximately range from 20 to 70 $A . m^{-1}$. This order of magnitude is quite comparable to that of classical NO sheets, but in the AC regime quenched ribbons are preferable.

Figure 16.14 shows that the losses per cycle, of the order of 10 $mJ . kg^{-1}$ at 1 tesla, increase only slowly with frequency. Also, the relative permeability weakly decreases, passing from 15,000 in the quasi-static regime to 10,000 at 10 kHz. The thinnest ribbons obviously have the smallest losses. A grain size of the order of 150 to 200 μm after recrystallization seems optimal. The need to wind the ribbon in order to make a magnetic core restricts the use of these rapidly quenched alloys to devices of rather reduced size: they are interesting in those technological applications where the dynamical properties are essential, such as transformers above 400 hertz, switching power supplies, and high frequency motors.

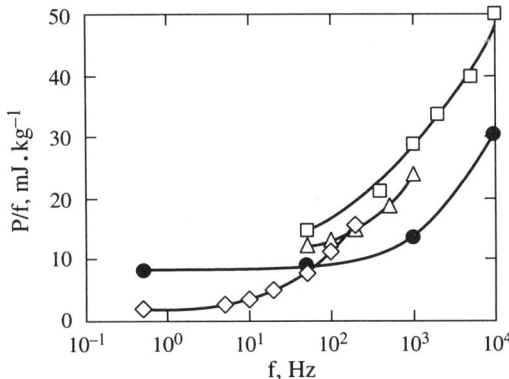

Figure 16.14 - Losses per cycle (P/f) as a function of frequency in a 6.5% Si microcrystalline Fe-Si ribbon (full circles)

These losses are compared to those in 0.05 mm (squares), 0.1 mm (triangles), and 0.3 mm (diamonds) thick 3.2% Si GO sheets

It is also worth noticing a minor, but interesting, application of the fast quenching technique to the preparation of Sendust. This alloy, with composition $Fe_{85}Si_{9.6}Al_{5.4}$, is well suited to magnetic recording head manufacturing, because it has a saturation polarization well above that of iron-nickel alloys, a high permeability, and a weak coercive field, together with an excellent wear resistance. Unfortunately it is difficult to use on account of its large brittleness. The ribbons obtained by fast quenching keep the qualities of the alloys obtained by the usual process, and withstand the cutting steps involved in the manufacturing of magnetic circuits for the recording heads.

2.6.2. *Diffusion enriched alloys*

Another route consists in performing all metallurgical treatments on pure iron or on a non-brittle classical Fe-Si alloy, and to silicon enrich it at the end. *Chemical vapour deposition* (CVD) provides an interesting solution. It consists in putting the material to be treated in contact with a volatile compound of the element to be deposited, at a temperature high enough so that it provokes a chemical reaction giving rise to one or several solid products. The static version of this process is easy to implement, in the laboratory as well as industrially. The treatment is performed in a semi-leakproof box, at a temperature usually ranging from 800 to 1,000°C. The volatile compound is supplied by a cement, i.e. a mixture of powders composed of silicon, a halogen compound (e.g. ammonium fluoride or chloride) which, after decomposition, provides the transport of silicon under the form of volatile halides, a moderator controlling the donor activity, and a thinner in order to prevent cement sintering. This technique allows simultaneous aluminium enrichment.

In the dynamic version of the process, a silicon based gaseous compound (halide, silane), diluted by hydrogen or argon, circulates around the part to be treated.

In both cases silicon settles at the surface, and then migrates by diffusion toward the interior of the material. The latter is then submitted to an homogeneization anneal around 900-1,000°C [11].

Dissipation mainly depends on sheet thickness. Table 16.3 indicates the specific losses and the maximal relative permeability for two 0.2 mm thick NO and GO samples.

Since 1993, NKK Corporation markets 0.3 to 0.05 mm thick 6.5% silicon NO sheets. The potential technological applications of this material are the same as those of quenched ribbons. In particular, the vanishing magnetostriction is favorable to reduce the acoustic noise of transformers working at audible frequencies. However we have to recognize that the relative contributions of magnetostriction and magnetic forces in the excitation of sheet vibrations are not clearly established.

Table 16.3 - Specific losses and maximal relative permeability
of two 0.2 mm thick NO and GO samples

	f (hertz)	1	50	400	1,000
NO	P (W. kg^{-1})	0.012	1.1	19	88
"	$\mu_r{}^{max}$	2,300	2,200	2,100	2,000
GO	P (W. kg^{-1})	0.006	0.55	10	40
"	$\mu_r{}^{max}$	8,500	8,000	7,000	5,000

3. IRON-NICKEL AND IRON-COBALT ALLOYS

Contrary to the mass-produced and low-cost iron-silicon alloys used for a few well defined applications, iron-nickel and iron-cobalt alloys are characterized by low production volumes, large added value, and an extreme versatility [12].

3.1. IRON-NICKEL FAMILY

Note - *In this section percentages are given in weight; numerical values are those of the alloys manufactured by Imphy Ugine Précision.*

Fe-Ni alloys crystallize in the face-centred-cubic phase when the nickel fraction is larger than 30%.

Figure 16.15 shows the Ni concentration dependence of their main electromagnetic characteristics. Three interesting situations, determining their main domains of applications, can be noticed right away:

♦ a Curie temperature close to room temperature around 30% Ni,

♦ a maximum saturation polarization around 50% Ni,

♦ and the almost simultaneous vanishing of anisotropy and magnetostriction around 80% Ni.

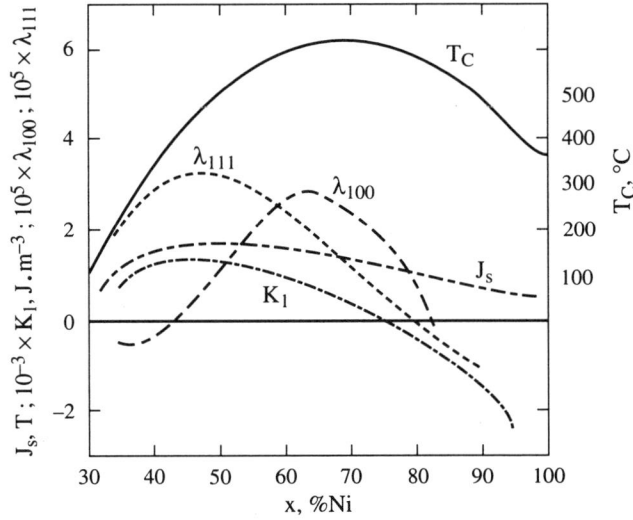

Figure 16.15 - Electromagnetic properties of the FeNi alloys

T_C: *Curie temperature -* J_s: *saturation polarization -* K_1: *magnetocrystalline anisotropy constant -* λ_{100} *and* λ_{111}: *magnetostriction constants*

These alloys have low resistivity except for a weight percentage of nickel around 30%. This is remedied by adding small amounts of chromium or molybdenum. For the 78.5% Ni alloy for instance, the resistivity increases with the molybdenum weight fraction x according to the approximate law: $\rho = (2 + 0.85\,x) \times 10^{-7}\ \Omega.m$.

The metallurgy of iron-nickel alloys allows a wide range of interesting treatments: cold rolling down to thicknesses of the order of 10 μm, annealing at any temperature.

A thermal treatment often applied to these material is *field annealing*. Magnetization being polarised by a magnetic field, atomic diffusion leads to a statistically anisotropic distribution of the atomic bonds in the alloy. An *induced anisotropy* of the order of 100 J.m^{-3} and with uniaxial symmetry thus superimposes on the magnetocrystalline anisotropy of cubic symmetry. As the latter does not exceed 1,000 J.m^{-3}, the magnetic properties can be noticeably modified.

FeNi preparation satisfies the general requirements of soft material synthesis. It is absolutely essential to eliminate the strongly electronegative (N, O, S) or electropositive (Al, Mg, Ca, Ti) impurities which tend to form precipitates, and on the other hand the impurities of small atomic radius (C, N) which could diffuse in solid solution and then precipate, leading to the aging of magnetic properties. Each step of FeNi metallurgy is extremely meticulous: raw material choice, melting under vacuum, thermal treatment under purifying atmospheres.

3.1.1. Alloys around 30% Ni

These alloys exhibit a fast and reversible variation of the saturation polarization in a temperature range adjustable through the composition. This feature is used in magnetic shunts used to compensate the temperature drift of permanent magnets in the magnetic circuits of devices such as electric meters, moving coil meters, tachymeters, loudspeakers. In fact, the soft material shunts off a fraction of the magnetic flux which decreases when temperature increases. The corresponding alloys contain between 28 and 32% nickel.

Some domestic electrical applications (rice cookers, saucepan bottoms with induction heating) require Curie temperature around 150-200°C. One then uses 38-40% nickel with 9 to 10% chromium alloys.

In television tubes, a grille containing hundreds of thousands of holes, the *shadow-mask*, is located near the internal face of the screen. It contributes to the magnetic shielding of the tube on its front face. Besides, as most electrons hit, and accordingly heat the mask, the latter must not expand for fear of image degradation. This is one of the many applications of the Invar alloy (36% Ni, registered trademark of Imphy S.A.) which is characterized by a very weak thermal expansion coefficient (see chap. 18). Invar also presents a relatively high resistivity (7.5×10^{-7} Ω.m) allowing high frequency applications, such as radar transformers.

3.1.2. Alloys around 50% Ni

The Anhyster type 48% Ni alloys have a high saturation polarization and interesting permeabilities (see tab. 16.4). They are used in bulk form in relays and safety devices. They are present as thin sheets in wound circuits and small size motors. For telephone earpieces, sheet stacks operate at low fields. The best linearity of the magnetization curve is then desirable in order to minimize signal distorsion. Secondary recrystallization above 1,100°C of the strongly cold-rolled 48% Ni sheet gives rise to a textured material with the easy magnetization axis predominantly along the rolling direction. The permeability in this direction is strongly increased (Supranhyster, see tab. 16.4).

Severe rolling of the 50% Ni alloy followed by an anneal gives it a large grain cubic texture. This yields a material with a virtually rectangular hysteresis loop: $J_r / J_s \approx 0.98$ (Rectimphy). Toroids of this material are used in magnetic amplifiers.

The 56% Ni alloy has an isotropic magnetostriction, $\lambda_{100} = \lambda_{111} = 25 \times 10^{-6}$. After annealing under field, this alloy gains an induced anisotropy which can virtually compensate for the magnetocrystalline anisotropy. The result is an almost isotropic material with a high permeability (Satimphy, see tab. 16.4) of the same order of magnitude as that of the medium-grade Permalloys.

3.1.3. Alloys around 80% Ni (Permalloys)

One benefits here from the weak anisotropy and magnetostriction levels. In fact, things are not so simple, because these two characteristics do not vanish quite simultaneously. The solution lies in adding molybdenum, copper or chromium, in combination with a treatment around 500°C. One then induces a short range order which counterbalances the residual magnetocrystalline anisotropy. The final adjustment of the anisotropies requires fine tuning of the processing parameters (temperature, cooling rate) to the individual composition of each casting. The best alloys have the $Fe_{15}Ni_{80}Mo_5$ composition (Permimphy, see tab. 16.4).

Table 16.4 - Coercive fields, and relative initial and maximal permeabilities at 50 hertz, of some FeNi based alloys

% Ni	48	48	50	56	80	80
Treatment	–	S	T	R	–	C
H_c (A/m)	8	2.5	9	1	0.4	0.7
μ_r^{in}	4,000	15,000	–	80,000	220,000	100,000
μ_r^{max}	35,000	75,000	100,000	150,000	360,000	130,000

Special treatments: S secondary recrystallization (high permeability), T cubic texture (squared loop), R annealing under field (high permeability), C annealing under transverse field (skewed loop)

Furthermore the additions have a beneficial effect in increasing the resistivity up to around 6×10^{-7} Ω.m. Hence, losses at 50 hertz and 0.5 tesla can fall down to 0.01 W.kg^{-1}.

The alloys with very high permeability are mainly used in safety devices (differential circuit-breakers). A toroid made of a thin strip winding (0.05 to 0.15 mm thick) surrounds the conductors feeding the electric installation. An insulation defect gives rise to a current unbalance which magnetises the toroid, and activates a relay.

Another typical use of these alloys is magnetic field shielding. The shielding efficiency being roughly proportional to the thickness-permeability product, the advantage of a very high permeability material in terms of weight and size are clear. The main applications concern cathodic tubes, magnetic recording devices, and amagnetic chambers for measurements in small fields (geomagnetism, and more recently magnetoencephalography and magnetocardiography).

On account of its small anisotropy, the above alloy has a reduced remanence J_r / J_s close to the theoretical value $2 / \pi = 0.64$. When it is slowly cooled from above 1,000°C, it exhibits a negative magnetocrystalline anisotropy constant K_1 leading to easy magnetization directions along <111>, and to a theoretical reduced remanence of 0.87. One thus has a material with a more rectangular loop (Pulsimphy).

On the contrary, an anneal under a field in the transverse direction gives a *skewed loop* with a reduced remanence of the order of 0.2 only (Permimphy C, see tab. 16.4). Such a material provides strong induction changes when the field varies without changing sign. It has many applications in the area of unipolar electronics.

3.2. IRON-COBALT ALLOYS

Their main advantages are a large polarization and a high Curie temperature. On the other hand, several drawbacks restrict them to special applications: an $\alpha \rightarrow \gamma$ phase transition occurs around 900-1,000°C over most of the phase diagram, which makes many thermal treatments difficult; their anisotropy and magnetostriction being markedly larger than those of FeNi alloys, their maximal permeabilities do not exceed about 20,000; their resistivity is smaller than that of FeNi alloys; and finally, cobalt is a rare and expensive metal.

The resistivity is increased to about 4×10^{-7} Ω.m by adding chromium and vanadium, the latter element also reducing the brittleness. The three main classes of Fe-Co soft alloys contain 25, 50 and 94% cobalt, respectively. In bulk form, their large polarization is an advantage in pole pieces for electromagnets. As sheets, these alloys are used in air borne medium frequency electrical engineering, for which the strong specific power is a key parameter. On the other hand, their large Curie temperature, together with weak aging, allows their use in devices working at high temperature. The 94% Co alloy is used up to around 950°C in electromagnetic pumps for molten metals.

4. *SOFT FERRITES*

A general presentation of ferrites can be found in the book of Smit and Wijn [13]. Ferrites for microwaves, generally with garnet structure, are treated in chapter 17 of the present book. We focus here on MM'_2O_4 ferrites with spinel structure, where M represents one or several divalent ions, and M' usually the Fe^{3+} trivalent ion. The most usual industrial compositions are $Ni_xZn_{1-x}Fe_2O_4$ and $Mn_xZn_{1-x}Fe_2O_4$. These compounds are cubic, as we will see in § 6.1 of chapter 17. Magnetism originates from metallic ions distributed on two antiparallel sublattices *a* and *d*, with magnetic moments of different magnitude, whence a non zero resulting magnetization.

The first stages in the preparation of polycrystalline ferrites are the milling and mixing of oxides of the metals involved in the composition. A first thermal treatment around 1,000°C triggers a solid phase reaction and the partial formation of ferrite. After additional milling and the addition of organic compounds, the powder is shaped under pressure. The final stage, which can last several tens of hours, is a sintering process above 1,200°C. During this stage, organic compounds disappear, and solid phase reactions put the finishing touches to ferrite synthesis, by welding the grains to each other and by leading to a large density increase (linear size reduction of 15 to 20%). The final size of the crystalline grains varies from some μm to some tens of μm.

For some applications, one synthezises spinel single crystals by pulling, between 1,400 and 1,000°C, from a molten bath containing *fluxes* (BaO, B_2O_3, PbO, PbF_2...), and *oxides of 3d metals* (Fe_2O_3, NiO, MnO...).

Ferrites, being mechanically hard and brittle, resist abrasion very well. Their thermal conductivity is small (about 100 times less than for copper), and they withstand thermal shocks poorly.

4.1. *ELECTROMAGNETIC PROPERTIES*

4.1.1. *Saturation polarization*

The large fraction of non magnetic oxygen ions and the partial compenzation of magnetic moments among the sublattices lead to a weak polarization, which in practice does not exceed 0.35 T at room temperature for nickel-zinc ferrites, and 0.6 T for manganese-zinc ferrites.

4.1.2. *Curie temperature*

Depending on composition it can range from 100 to around 600°C for Ni-Zn ferrites, and from 130 to 250°C for Mn-Zn ferrites.

4.1.3. Anisotropy

The anisotropy constant K_1 of soft ferrites is generally negative, the <111> axes being the easy magnetization directions, and its magnitude decreases when temperature increases. For instance it reaches $-5,100$ J.m^{-3} at room temperature for nickel ferrite $NiFe_2O_4$. The substitution of nickel by divalent ions such as Co^{2+} and Fe^{2+} brings a positive contribution. Hence it is possible to virtually cancel the overall anisotropy at a given temperature which can be adjusted with the substitution rate, and to optimize the permeability of the material at this temperature.

4.1.4. Magnetostriction

For the industrially relevant ferrites (those relatively rich in zinc), it ranges from 0 to -10×10^{-6} for Ni-Zn ferrites, and from 0 to -1×10^{-6} for Mn-Zn ferrites.

4.1.5. Resistivity

It ranges from 10^3 to more than about 10^4 Ω.m for polycrystalline Ni-Zn ferrites, and from 1 to 20 Ω.m for Mn-Zn ferrites, but it reaches 10^{10} Ω.m in some bismuth-lithium ferrites. Grain boundaries strongly contribute to the bulk resistivity of the material, and all the more so since they tend to gather impurities. The single crystal resistivity is markedly smaller (10^{-2} Ω.m) but still much larger than that of metallic alloys.

4.1.6. Permeability - cutoff frequency product

The initial relative permeability μ_i of a ferrite is virtually constant up to a cutoff frequency f_c. These two quantities vary in opposite directions with magnetocrystalline anisotropy (see chap. 14). One can consider the product $\mu_i f_c$ as a quality factor which is almost constant within a composition family. This product reaches about 8,000 MHz for Ni-Zn ferrites, and 4,000 MHz for Mn-Zn ones. Therefore it will be hardly possible to exceed 400 kHz with a Mn-Zn ferrite with an initial permeability of 10,000.

On account of these characteristics, ferrites form a category of their own among soft materials. They are weakly or not interesting for quasi-static applications since they carry little flux and their permeability is limited by the large density of grain boundaries. On the other hand they are very valuable above some hundreds of hertz, and unrivaled above one MHz. Ferrite characteristics being strongly composition dependent, it is not possible to give an exhaustive table. More details can be found in reference [14], and in chapter 17.

Table 16.5 gives some typical characteristics for the manganese-zinc ferrite, by far the most widespread in power applications (Ferrinox B 50, Thomson LCC). Specific losses (P) in W.kg^{-1} at 0.1, and 0.2 tesla are measured at 100 kHz.

Table 16.5 - Some typical characteristics of the B50 Ferrinox

J_s (T)	H_c (A . m^{-1})	μ_r^{in}	μ_r^{max}	ρ (Ω . m)	$P_{0.1T}$	$P_{0.2T}$
0.48	12	2,500	4,500	> 1	45	200

4.2. APPLICATIONS OF FERRITES

One can schematically distinguish two main classes of applications:

4.2.1. Power electronics

Flux transportation being basic for these applications, one uses Mn-Zn ferrites in transformers for frequencies larger than 400 hertz, DC converters, switching power supplies... Eddy current losses larger than those of Ni-Zn ferrites are at least partially counterbalanced by the weaker hysteresis losses. As the Curie temperature is never very high, it is necessary to account for the thermal variation of J_s, and to optimize the coupled parameters frequency-temperature-induction. For instance, the B 50 Ferrinox composition, used up to 500 kHz, is optimized at a working temperature of 50°C, between 16 and 150 kHz, in a maximum induction of 0.25 tesla.

4.2.2. Low power applications

These applications rather emphasize the permeability: inductives devices, filters, pulse transformers, sensors, recording heads where the resistance to abrasion is valuable, electron beam deviators in TV tubes... One uses Ni-Zn ferrites up to 300 MHz, and Mn-Zn ones up to some MHz only, but with larger permeabilities (see chap. 17). One has to avoid exposing Ni-Zn ferrites to mechanical shock and to fields above some mT for fear of degrading their properties. However the latter can be restored by heating above T_C.

5. AMORPHOUS ALLOYS

Note - *Contrary to Fe-Si, Fe-Ni, and Fe-Co alloys, the compositions of amorphous, and nanocrystalline alloys are given in atomic %.*
- *A detailed study of these materials can be found in reference [15].*

Theorists considered as early as 1960 the possibility for an amorphous alloy to be ferromagnetic in spite of atomic position disorder. After the first ferromagnetic amorphous alloy $Fe_{80}P_{13}C_7$ was synthesized [16], the industrial development of amorphous metals really began only in 1971; even then it was mainly based on their mechanical properties. The first commercial amorphous alloy, Metglas 2826 ($Fe_{40}Ni_{40}P_{14}B_6$), from the Allied Chemical Company, appeared in 1973. The high price of these new materials, initially sold per meter (300 $ per kilo in 1978), first limited their use to scientific applications. Soft amorphous alloys are obtained as ribbons by solidification of a melt on a fast rotating cooled wheel. A minimum quenching rate of 10^6 K.s^{-1} is required; as a result, the ribbon thickness does not exceed about forty μm. The techniques described above for rapidly quenched Fe-Si alloys are industrially used. In the so called *melt-spinning* technique, melt instabilities prevent from getting a ribbon width larger than a few cm. The *planar flow casting* technique can reach about 20 cm.

5.1. GENERAL CHARACTERISTICS

The total atomic disorder leads to the lack of magnetocrystalline anisotropy. Amorphous alloys are also, by definition, free from the structural defects of crystalline solids: grain boundaries, dislocations. Besides, chemical disorder gives them a resistivity of the order of 1.5×10^{-6} Ω.m, three times larger than that of crystalline alloys. The small thickness of ribbons also contributes to reducing dissipation by eddy currents. In addition to these characteristics which favour soft magnetic properties, one has to mention outstanding mechanical properties (elastic limit larger than 2 GPa), and an excellent resistance to corrosion.

However amorphous alloys have serious drawbacks. The metastability of their structure makes them sensitive to aging, and forbids severe thermal treatments or operation at high temperature for fear of recrystallization. The need of including in the chemical composition about 20% of amorphising metalloids lowers the saturation polarization and the Curie temperature. Finally, the preparation by fast quenching leads to large residual stresses which are difficult to eliminate.

A moderate anneal under field gives these materials an induced anisotropy. If the field is applied parallel to the ribbon direction, it gives rise to elongated 180° domains and to a large permeability. With a field perpendicular to the ribbon direction one obtains a fine structure with transverse domains. In the latter case reversible rotations are the preponderant magnetization mechanism, whence weak energy losses in the AC regime.

5.2. MAIN CLASSES OF SOFT AMORPHOUS ALLOYS

5.2.1. High polarization alloys

This concerns iron rich alloys with typical composition $Fe_{82}B_{13}Si_{2.5}C_{2.5}$. Their main advantages are a relatively large saturation polarization, and a reasonable price. On the other hand, their large magnetostriction makes them sensitive to residual stresses, whence relatively reduced permeabilities (see tab. 16.6).

Table 16.6 - Characteristics of some Vitrovac* amorphous ribbons

Alloy	J_s (T)	T_C (°C)	λ_s (10^{-6})	μ_r^{max}	H_c (A.m^{-1})	P (W.kg^{-1})
7505 (Fe based)	1.5	420	30	10^5 (L)	< 4 (L)	10 (T)
6025 (Co based)	0.55	250	< 0.3	6×10^5 (L)	< 0.4 (L)	4 (T)
4040 (Fe-Ni)	0.8	260	8	25×10^4 (L)	< 1 (L)	6 (T)

* *Vitrovac is a trademark of Vacuumschmelze GmbH. The maximum permeabilities are measured at 50 Hz, specific losses at 20 kHz and under 0.2 tesla. Field-annealing procedures: (L) longitudinal, (T) transverse.*

5.2.2. Low magnetostriction alloys

They are characterized by very large permeabilities and very weak losses.

Figures 16.16 and 16.17 show the influence, at room temperature, of the composition on the magnetostriction of iron based amorphous alloys with different cobalt or nickel contents. Magnetostriction vanishes in the vicinity of 70% Co for the first alloys; for the second ones, it is for 65% Ni that the Curie temperature decreases down to room temperature, which also leads to the vanishing of λ_s. These are two typical compositions giving rise to very soft materials (see tab. 16.6).

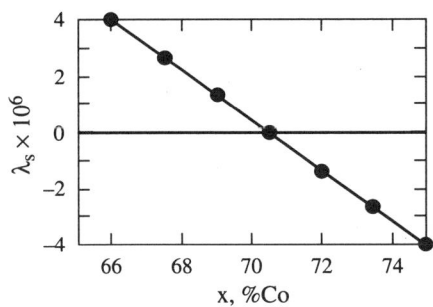

Figure 16.16 - Saturation magnetostriction (λ_s) of the amorphous alloys $Co_xFe_{75-x}Si_{15}B_{10}$ at 20°C as a function of the Co content (x)

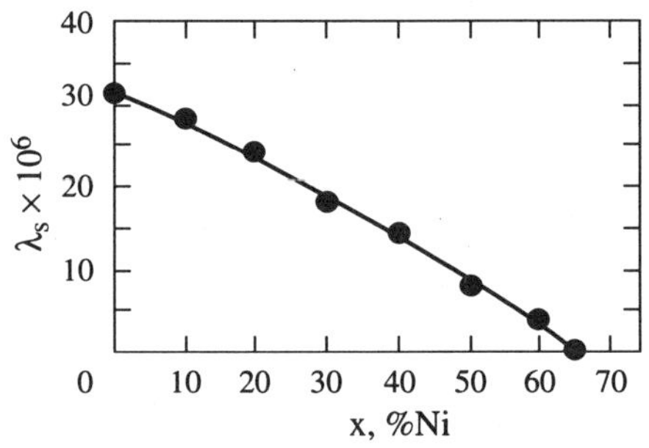

Figure 16.17 - Saturation magnetostriction (λ_s) of the amorphous alloys $Ni_xFe_{80-x}B_{20}$ at 20°C as a function of the Ni content (x)

5.3. APPLICATIONS OF AMORPHOUS ALLOYS

For the moment, these materials do not seem to be able to compete with iron-silicon alloys in large transformers for energy supply at industrial frequencies. The maximum working induction of iron rich amorphous alloys does not exceed 1.4 T, whereas it reaches 1.7 T for the GO Fe-Si alloys. In order to obtain the same voltage, one has to increase the cross-section of the magnetic circuit or the number of turns in the wire windings, which increases weight, volume, and Joule losses. Furthermore, winding large ribbon lengths is a delicate, slow, and expensive operation. However the latter drawback is less serious in countries where the energy supply policy is based on smaller transformers (USA). The Allied Signal Company developed a material called *Powercore,* obtained by gluing together about ten amorphous ribbons. This type of material is not manufactured in Europe.

Amorphous alloys become attractive as soon as the frequency reaches 400 hertz, that used for airborne electronics. The outcome of the competition at medium or high frequency is uncertain, and will probably depend on the improvements simultaneously achieved by competing materials (thin or enriched Fe-Si sheets…).

In general, amorphous alloys are interesting for all applications in which induction varies rapidly, for instance switching power supplies or inductive devices (saturable inductance, choke coils…).

The amorphous high permeability metals compete with the other families of materials with vanishing anisotropy: iron-nickel alloys, ferrites. They are also used in differential circuit breakers, electromagnetic sensors, and shieldings in which their (elastic) flexibility is valuable (*Metshield* type products).

Tailoring magnetostriction through composition offers interesting possibilities. In recording heads, one seeks vanishing magnetostriction, and one simultaneously takes advantage of the high permeability and the weak sensitivity to mechanical stresses and vibrations originating from the head mechanisms. On the other hand force sensors make use of a larger magnetostriction (see chap. 18).

6. *NANOCRYSTALLINE MATERIALS*

Nanocrystalline alloys are the most recent among soft magnetic materials [17]. They are prepared in two stages. One first makes, by the usual melt spinning technique, an amorphous ribbon which is then subjected to a recrystallization annealing around 500-600°C. The latter generally leads to disastrous outcomes for a soft amorphous metal: permeability collapse, possible material destruction. This is not the case for the alloys under consideration, with typical composition $Fe_{73.5}Cu_1Nb_3Si_{13.5}B_9$. Although some details of the recrystallization mechanism are still puzzling, one generally considers that copper atoms, weakly soluble in the crystalline phase, favour the segregation of many germs of this phase, whereas niobium atoms act as growing inhibitors. Other refractory elements such as Cr, V, Mo, Ta, W can play the same role as Nb [18, 19]. One thus obtains a material with a mixed structure in which an amorphous matrix coats crystalline grains. The volume ratio of the crystalline phase ranges from 50 to 80%.

The average grain size decreases with the inhibitor element content, and, on the contrary, increases with the annealing temperature. It usually ranges from 5 to 20 nm.

Analyses by X rays, electron diffraction, and atomic probe show that the crystalline phase is body centred cubic with a DO_3 type superstructure. It is mainly composed of Fe-Si in which the Si fraction can be larger than 20%, all the silicon being concentrated in the grains. The amorphous phase contains the remaining iron, as well as boron and niobium, the contents of the latter elements being close to 30, and 5 to 15%, respectively.

6.1. ELECTROMAGNETIC CHARACTERISTICS

6.1.1. Polarization

Figure 16.18 shows the thermal variation of the saturation polarization J_s of a material with typical composition [20]. Close to 1.3 tesla at room temperature, it decreases when temperature increases in two stages which reflect the crystalline-amorphous mixed structure of the material. Assuming that the bulk polarization is the sum of the contributions of both phases, one finds that they are about 1.3 T (crystal) and 1.2 T (amorphous). The Curie temperatures are close to 600 and 320°C, respectively.

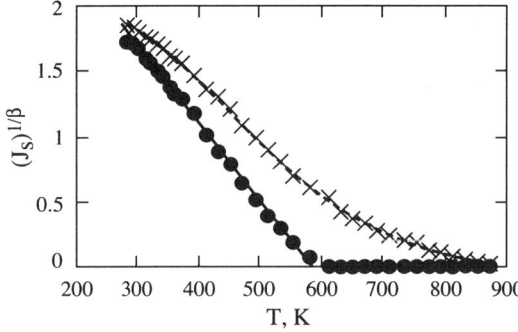

Figure 16.18 - Thermal variation of the saturation polarization (J_S) of the alloy $Fe_{73.5}Cu_1Nb_3Si_{13.5}B_9$ in the amorphous (●), and nano-crystalline (×) states (after [20])

The vertical axis shows $J_S^{1/\beta}$, $\beta \approx 0.36$ being the critical exponent (see chap. 10) of the polarization, given by:
$$J_s(tesla) = J_0\{(T_C - T)/T\}^\beta$$

6.1.2. Anisotropy

The amorphous phase is magnetically isotropic, whereas the anisotropy constant K_1 of the cubic Fe-Si phase reaches 10 kJ.m^{-3} for a Si content close to 20%. A recrystallization anneal performed at too high a temperature (> 600°C) gives rise to a harmful precipitation of very anisotropic Fe_2B.

6.1.3. Magnetostriction

The saturation magnetostriction λ_S of the amorphous alloy FeCuNbSiB is large and positive, of the order of 20×10^{-6}, and very weakly dependent on the silicon content x. It strongly decreases after recrystallization. It also decreases when x increases, and vanishes around x = 15.5% (fig. 16.19). This behavior can be interpreted by assuming that the bulk magnetostriction results from both crystalline and amorphous contributions. One finds, for the Fe-Si phase, a coefficient λ_S that is negative and close to -5×10^{-6}, in good agreement with the determinations of λ_S for a polycrystalline sample of Fe-Si with the same composition.

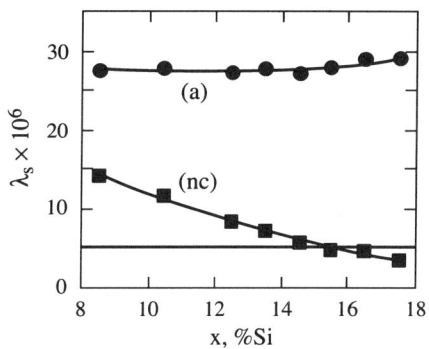

Figure 16.19 - Saturation magnetostriction (λ_s) at 20°C of $Fe_{73.5}Cu_1Nb_3Si_xB_{22.5-x}$ alloys as a function of the silicon content in the amorphous (a) and nanocrystalline (nc) states, after [21]

6.1.4. Resistivity

The resistivity of these materials is of the same order of magnitude as that of the amorphous ones, namely 1 to 1.5×10^{-6} Ω.m.

6.1.5. Uniaxial induced anisotropy

If the recrystallization takes place in a magnetic field, the material acquires an induced uniaxial anisotropy K_u of the order of 10 to 100 J.m^{-3} which mainly depends on composition (in particular on the boron and silicon contents). It is also possible to field-anneal the material after it has recrystallized. The induced anisotropy, weaker than in the previous case, then strongly depends on the annealing temperature. On account of their order of magnitude, one could at first sight think that the induced anisotropies are negligible with respect to the magnetocrystalline anisotropy of the Fe-Si phase ($\approx 10^4$ J.m^{-3}). In fact, K_u has to be compared to the *effective average anisotropy* of the material which, as we will see, is much smaller.

In a ribbon without stresses, Kerr effect observations reveal large domains separated by curved 180° walls, quite similar to the domains in amorphous alloys [21]. This indicates a weak effective anisotropy at the domain and wall scale. On the other hand any stress, even weak, be it macroscopic or localized around an inclusion, shows up through an increased number of small domains with irregular shapes. These observations are at first sight rather surprising for a material with vanishing anisotropy. In fact, the crude model which consists in linearly averaging the magnetostrictions of the crystalline and amorphous phases does not account for the complexity of the local magnetoelastic behavior associated with the coexistence of the two phases [22].

6.2. RANDOM ANISOTROPY MODEL

We are now focusing on the understanding of the reason why a mixed material does not have the anisotropy of the Fe-Si phase. The main concept is that of *random anisotropy*, initially introduced for rare earth based amorphous alloys with strong local anisotropy, and which can easily be adapted to nanocrystalline materials [23, 24]. In the absence of exchange interaction, each magnetic moment would orient independently of each other along the local easy magnetization direction. The exchange energy imposes parallelism of moments at a scale shorter than the characteristic length $L = \sqrt{A/K}$, where A, and K are the exchange and anisotropy constants, respectively. If L is larger than the correlation length of the anisotropy, the magnetic moments cannot follow the fluctuations of the easy magnetization direction.

The question is which value of K one has to use in the calculation of L. The random anisotropy model assumes that it is the average of the local anisotropy over a volume of the order of L^3. Finally one deals with a self consistent model: the effective anisotropy determines the scale over which integration of the local random anisotropy takes place. The application of this model to nanocrystalline materials is presented below. The correlation length of the anisotropy is in this case of the order of the average diameter D of the crystalline grains. Considering the schematic nature of the model, all dimensionless numerical constants are deliberately considered as being equal to unity.

Let K_1 be the anisotropy constant of the crystalline phase. A cubic volume of edge L contains $N = (L/D)^3$ grains.

If the anisotropy directions of the grains are randomly distributed, the central limit theorem predicts for this volume a resulting anisotropy with zero average value, and with a standard deviation $K = K_1/\sqrt{N} = K_1 (D/L)^{3/2}$. Introducing this value, itself depending on L, into the expression for L, leads to $L = A^2 K_1^{-2} D^{-3}$, and then to $K = K_1^4 A^{-3} D^6$. The main result is the very fast variation of the effective anisotropy, *following the sixth power of the grain size*. With the following orders of magnitude: $A = 10^{-11}$ J.m^{-1}, $K_1 = 10$ kJ.m^{-3}, D = 10 nm, $J_s = 1$ T, a very weak effective anisotropy is predicted: $K = 10$ J.m^{-3}. The coercive field, which is the field able to overcome this effective anisotropy, is given by $J_s H_c = K$, i.e. $H_c = K_1^4 A^{-3} D^6 J_s^{-1}$.

A key assumption in this model is the continuity of exchange, which makes it possible to take a mean approach to the mixed material. Its disappearance entails that of soft magnetic properties. This is in particular the case on heating. The amorphous phase with the lowest Curie temperature tends to become paramagnetic, meaning a vanishing of the exchange coupling. The coercive field thus passes from a fraction of an ampere per meter to several tens of ampere per meter between room temperature, and 400°C.

Beyond its schematic character, the random anisotropy model presents a limitation already mentioned in [24]: it does not account for dipolar effects, which are of the utmost importance in soft materials. A more complete assessment of the involved energies must include the magnetostatic term associated with the non uniformity of the magnetization vector. This term breaks the random distribution of moment directions by favouring an increase in the length of the correlation volumes along the local magnetization direction. The filtering effect of the random anisotropy is then modified, probably in a favourable direction on account of the integration volume increase.

Figure 16.20 illustrates the effect of the correlation length of the random medium, and shows how two opposed strategies lead to very soft materials. One consists in averaging the fluctuations at very short scale, 10 nm in nanocrystalline materials, less than 1 nm in amorphous metals. On the contrary, the largest grains give lower coercive fields in classical crystalline materials.

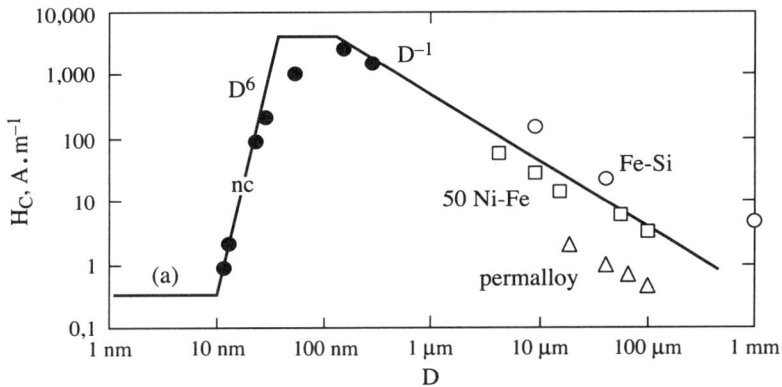

**Figure 16.20 - Coercive field (H_c) as a function of the grain size (D)
for different families of soft materials (after [21])**

Amorphous (a), nanocrystallized (nc) and crystallized materials (Fe-Si, 50FeNi and Permalloy)

6.3. APPLICATIONS OF NANOCRYSTALLINE MATERIALS

Their high permeability and their weak energy loss in ac regime (see tab. 16.7) make them valuable for safety devices (differential circuit breakers), sensors, high frequency transformers (at least up to 100 kHz), filtering inductances… The possibility to play with the value and the direction of the uniaxial anisotropy gives nanocrystalline alloys the same versatility as Fe-Ni-Mo crystallized materials or cobalt rich amorphous metals. With respect to the latter, they exhibit the advantages of a larger polarization, and of good thermal stability of their characteristics thanks to a relatively high Curie temperature of the Fe-Si phase, as well as a probably better time stability. On the other hand they are very brittle.

**Table 16.7 - Main characteristics of a nanocrystalline material with typical
composition $Fe_{73.5}Cu_1Nb_3Si_{13.5}B_9$**

B_s (T)	H_c (A.m^{-1})	μ_r^{max} (DC)	μ_r^{max} (50 Hz)	μ_r^{max} (1 kHz)	L (W.kg^{-1})
1.25	0.5	$> 8 \times 10^5$	$> 5 \times 10^5$	10^5	40

The losses L are measured at 100 kHz and in 0.2 tesla.

Nanocrystalline alloys with other compositions are still laboratory materials but they could become attrative for applications thanks to their large saturation polarization.

That of the $Fe_{60}Co_{30}Zr_{10}$ alloy reaches 1.6 tesla, and that of $Fe_{91}Zr_7B_2$ is larger than 1.7 tesla. It is possible to obtain, by sputtering, nanocrystalline films of Fe-Hf-C-N with a polarization of 1.7 tesla.

> *We now present the **applications** of soft magnetic materials. We will treat energy transformation and actuators. Magnetism allows to convert one form of energy into another one, so that mechanical, electrical, and magnetic energies can be present in the same device.*

7. ENERGY CONVERSION AT INDUSTRIAL FREQUENCIES (50-400 Hz)

We consider here the sinusoidal AC regime, and we will use capital letters for the RMS values of the electric quantities, and capital letters with the subscript "$_m$" for the maximum values of magnetic quantities.

We will show all the interest of soft materials by focusing on the conversion of magnetic energy into electric energy. This conversion takes place through windings, across which the user gets electromotive forces linked to magnetic quantities by relationships of the type:

$$e = -d\Phi_t / dt \qquad (16.4)$$

Here, e is the electromotive force induced across a winding enclosing a total induction flux Φ_t.

The coil producing the flux Φ_t is called *primary winding*, or simply *primary*, and will be marked with the subscript 1, the subscript 2 being used for the *secondary winding*, or simply *secondary*. In what follows we will assume that the coils are wound around magnetic cores, the characteristics of which we will discuss. The secondary, closed on a load of finite impedance, carries a current I_2 whereas the voltage across it takes the value U_2. The power supplied to the user being proportional to the product $U_2 I_2$, we will use materials liable to maximise the ratio $(U_2 I_2) / V$, where V is the machine volume. Since I_2 is itself determined by the amplitude of the voltage applied to the load, the quantity to be controlled is actually U_2.

Without load, i.e. in open secondary circuit so that I_2 is zero, and assuming that the induction field enclosed by the winding is uniform, the voltage u_2 is given by (16.4), and its RMS value can be written:

$$U_2 = n_2 \, S \, \omega \, B_m / \sqrt{2} \qquad (16.5)$$

where S is the coil cross-section, n_2 the turn number of the secondary, ω the angular frequency, and B_m the amplitude of the induction in the secondary. As n_2 and S specify the volume, the weight, and the cost of the secondary, one will try to minimize these quantities by using materials liable to produce large inductions.

With a load, the current I_2 works against the electromotive force being set up. By operating at the primary level, one tries to maintain the voltage U_2 close to its unloaded level. The choice of high permeability materials allows to optimize the coupling between the primary and the secondary, and to limit the current I_1 required to maintain U_2.

It thus appears that *power applications require materials with a high saturation induction, together with the highest possible permeability*. Cost is also an important element. Therefore, power applications (some tens of watts in hi-fi amplifiers, about one gigawatt for distribution) at industrial frequencies typical of electrical engineering (50-400 Hz) mainly use iron based materials.

The dissipation arising from the soft material can also be an important point. One commonly calls *iron losses* this dissipation source. These losses are harmful for several reasons:

◆ the heat produced must be evacuated. The machine size increases, and additional devices (fans...) introduce nuisances: fan noise, pollution risks associated with the cooling fluids... Also, the machine becomes more complicated. This becomes a concern especially in the case of large alternators, the rotors of which are cooled by hydrogen circulation inside a leakproof enclosure.

◆ iron losses, as well as losses in ancillary cooling devices, strongly increase operating costs: this factor is determinant in the case of distribution transformers.

7.1. DISTRIBUTION TRANSFORMERS

In transformers, the primary and secondary are fixed, so that a single magnetic circuit is used. The current flowing in the secondary produces a feedback flux which is transmitted to the primary. The primary current amplitude then increases, in order to maintain the induction flux amplitude constant, as required by the primary voltage source, assumed to operate at constant amplitude. The secondary voltage amplitude is finally maintained to within a few percents, with the primary automatically adapting the current the more efficiently the better the magnetic coupling. One thus seeks to minimize flux leakage between the two coils. This is the reason why they are generally coaxial. The magnetic circuit is closed, which optimizes the magnetic coupling between the coils, and especially decreases the primary current in the case of unloaded operation.

In transformers, in which there are no mechanical losses, iron losses contribute up to 50% to the total losses, the remainder arising from Joule dissipation in the coils.

Table 16.8 - Influence of the power factor cos φ on the transformer efficiency η
for different apparent powers S [2]

S (kVA)	100	1,000	10,000	100,000
η (%) (cos φ = 1)	97	98	98.5	99
η (%) (cos φ = 0.8)	96.3	97.5	98.1	98.8

In fact, and in spite of the high efficiencies reached in current transformers (see tab. 16.8), iron losses take a special importance in the case of distribution transformers which, coupled to the network 24 hours a day, dissipate them all the time. The cost estimate of these losses is rather difficult, because the kilowatt production cost varies with the distribution network load, but even more because these losses must be accumulated over the lifetime of a transformer, i.e. about 30 years. However one can roughly estimate the total iron losses to amount to 50% of the manufacturing cost [26]. This justifies the large research effort invested in optimizing the magnetization properties of transfomer sheets, as well as the manufacturing processes of distribution transformers [27].

These losses are related to eddy currents in the magnetic core. These currents work against the flux variations, so that an increase in operating frequency gives rise both to a power loss increase and to a decrease in the permeability of the material (see § 2.4 of this chapter). At the metallurgical level, eddy currents can be limited by enriching iron with silicon (3% in mass). Among other effects, this increases the resistivity by a factor 4.5 (48×10^{-8} Ω.m for the 3.2% Si Fe-Si alloy compared to 11×10^{-8} Ω.m for pure iron). As eddy currents also depend on the thickness of the material, the magnetic circuits are made by stacking sheets electrically insulated from each other. This makes the use of Fe-Si alloys possible up to frequencies of the order of some hundreds of Hz, characteristic of air borne electrical technology.

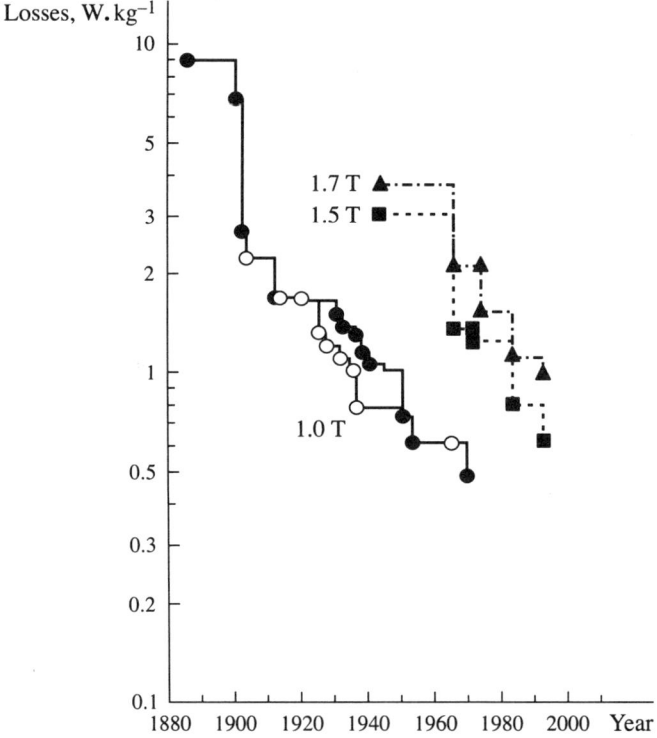

Figure 16.21 - Evolution of transformer sheet performance [28]

Figure 16.21 shows, in semi logarithmic coordinates, the performance evolution of transformer sheets, estimated in terms of losses at 60 Hz for a sinusoidal induction B, of given amplitude. Although the transformer was invented in 1883 by Lucien Gaulard, it is only in 1903 that the first substantial improvements, through silicon enriching, appeared in sheet manufacturing. In 1945 a second technological breakthrough took place with the introduction of the Goss texture. These improvements result in an increase of the maximum working induction of transformer sheets, from 1 T to 1.7 T.

Figure 16.22 reflects this increase of the specific power of a 100 MVA transformer. The improvement seems less dramatic than on the previous figure, because it results from many factors among which the intrinsic sheet properties are only one component.

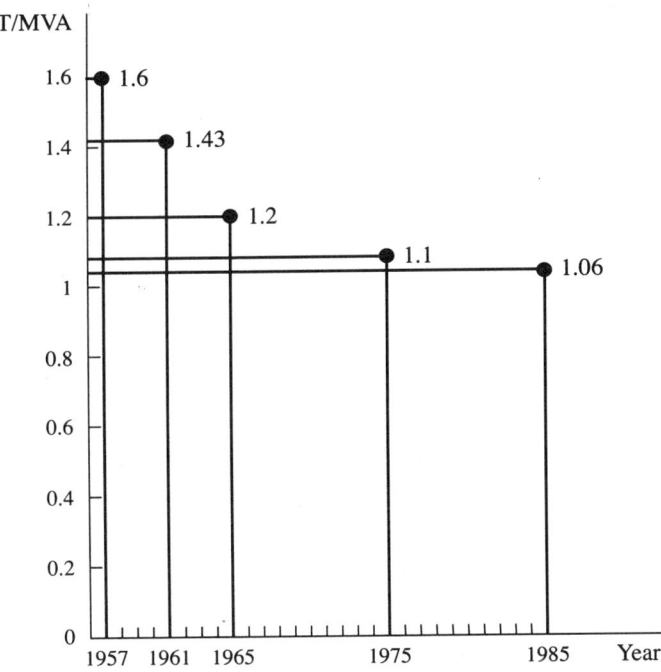

Figure 16.22 - Weight evolution (in tons per MVA)
of a 100 MVA - 225 / 63kV three phase transformer [29]

In particular, it should be noticed that, referred to the same amount of material, the iron losses in a transformer are, as shown in figure 16.23, always larger than those measured on an isolated sheet [26].

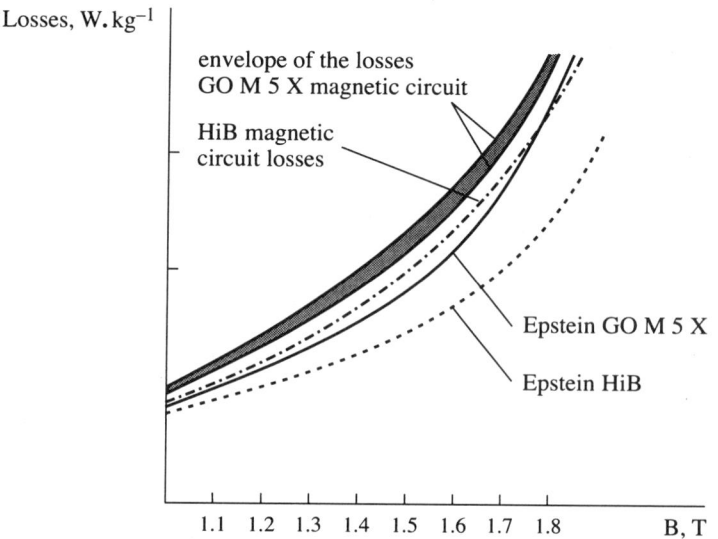

Figure 16.23 - Comparison between the losses measured on a Epstein permeameter and on a transformer core [26]

This is related, among others, to the mechanical stresses which appear when sheets are stacked, as well as to the presence of overlap boundaries between sheets, which gives rise to more or less sophisticated stacking strategies such as those presented in figure 16.24 (ref. [30], p. 236).

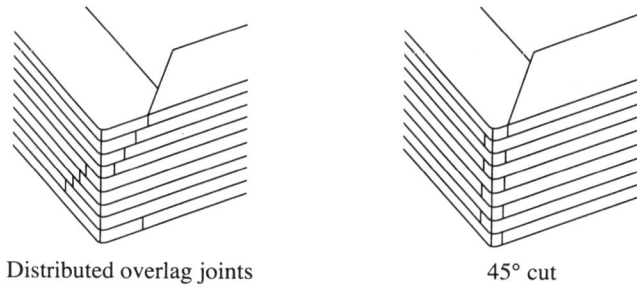

Distributed overlag joints   45° cut

Figure 16.24 - Two stacking modes for magnetic circuits

These considerations lead, in a prospective approach, to reconsidering the manufacturing strategies adopted so far. The stacking processes could be replaced by winding techniques. Such methods were initially considered for single phase transformers based on iron based amorphous alloys (see § 5.3 of this chapter, as well as references [30] p. 263 and [31]), but they cannot be applied for three phase transformers with conventional geometry. Once integrated in a process based on triangular geometry transfomers (figure 16.25), they could give rise to new generations of transformers in which the constantly improving intrinsic sheet performance [32] would be better used [27].

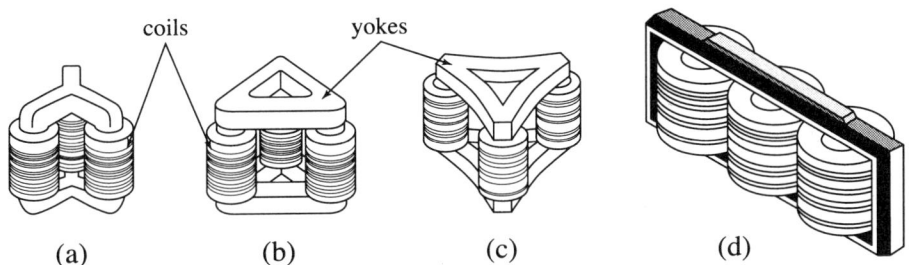

coils **yokes**

(a)   (b)   (c)   (d)

Figure 16.25 - Various geometries of three-phase transformers

Triangular wound (a) and (b), triangular stacked (c), traditional stacked (d) structures

7.2. ROTATING MACHINES

Electromechanical conversion is based on the interaction between primary and secondary fields, the rotation of the moving part, the *rotor*, being an element of this interaction. The fixed part of the machine is the *stator*. We will here illustrate the ways of using soft materials by studying different structures, and we will take the motor torque as the comparison criterion between the various solutions (fields will be produced either by magnets or by currents; there will be, or not, soft materials...)

The variety of studied structures will lead us to use a quite general formalism based on calculating the energy of each system.

7.2.1. A refresher on magnetic system energy

The energy of a field system is the energy spent to bring together the different field sources, assumed to be initially infinitely far from each other, through a thermodynamically reversible process, i.e. without energy dissipation. One can include the energy required to build up each source. One has therefore to consider the work provided by the operator against the work of magnetic forces, and to include the energy which may be required to maintain the sources.

Let us for instance consider the insertion of an element C into a field source S. If C is a circuit carrying a current I, then the work of magnetic forces in the course of a small displacement of the source in the field can be easily calculated, and one finds:

$$dW_m = I \, d\Phi_{S \to C} \qquad (16.6)$$

On the other hand, the electromotive force induced in C by the displacement is: $e = -d\Phi_{S \to C}/dt$. It is counterbalanced by the generator which provides the current flowing in C, and will supply, in order to maintain the current constant, an additional energy:

$$dW_{genC} = (d\Phi_{S \to C}/dt) \, I \, dt = dW_m \qquad (16.7)$$

If the source itself consists of a circuit carrying a current I_s, the energy variation of the system includes the energy supplied by the generator to maintain it constant:

$$dW_{genS} = (d\Phi_{C \to S}/dt) \, I_s \, dt = (d\Phi_{S \to C}/dt) \, I \, dt = dW_m \qquad (16.8)$$

If C is a magnetic moment $\mathbf{m}$, the work of the magnetic force when inserting the source into the field $\mathbf{B}$ will be: $dW_m = \mathbf{m} \cdot d\mathbf{B}$. One easily generalises this expression to a continuous distribution of magnetic moments in a volume V by using the magnetic polarization $\mathbf{J} = \mu_0 (d\mathbf{m}/dV)$; the work of the magnetic force when this volume is moved in a field $\mathbf{H}$ will then be written as:

$$dW_m = \int_V \mathbf{J} \, d\mathbf{H} \, dV \qquad (16.9)$$

7.2.2. Application to the study of some structures

In order to limit calculations, we will use the following approximations: cylindrical symmetry (rotors and stators assumed to be infinite along the z axis), perfect magnets (uniform and environment-independent magnetization), and soft materials with infinite permeability (H equal to zero inside).

Magnet rotor associated with coil stator

The rotor with radius R has a uniform polarization $\mathbf{J_R}$, perpendicular to the rotation axis. The coil is positioned on the internal face of the stator with radius R + g, assumed to be non magnetic, and it carries a current equivalent to a surface density of the form $\mathbf{i_s}(\alpha) = i_s (\cos \alpha) \mathbf{u_z}$. Such a current distribution is the ideal one that one tries to realize in actual machines.

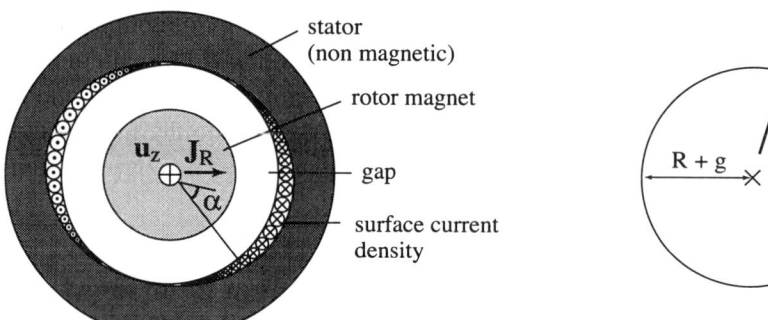

Figure 16.26 - Principle of a rotating
machine with magnetised rotor

Figure 16.27
Equivalent stator

The energy calculation of the system leads us to study the fields produced by the two sources. Both calculations can be treated with a single formalism, because it can be shown that the studied stator can be replaced by a cylinder with radius R + g, which we will call equivalent stator, with a polarization $\mathbf{J_C} = \mu_0 i_s \mathbf{u}$. We will study it, as well as the rotor, in the frame of the Coulomb representation.

One then shows that the field generated by an infinite cylinder with radius R carrying the polarization $\mathbf{J}$ perpendicular to its axis can be written, in cylindrical coordinates, as:

$$r < R \qquad \mathbf{H} = -\mathbf{J}/2\mu_0 \qquad (16.10)$$

$$r > R \qquad \mathbf{H} = \frac{\mathbf{J}}{2\mu_0} \frac{R^2}{r^2} (\cos \theta \mathbf{u_r} + \cos \theta \mathbf{u_\theta}) \qquad (16.11)$$

$$\sigma_1(M_1) = \frac{J_R}{\mu_0}\cos\theta_R$$

$$\theta_C = \phi + \theta_R$$

$$\sigma_2(M_2) = \frac{J_C}{\mu_0}\cos\theta_C$$

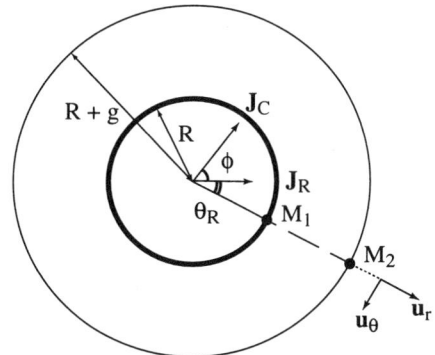

The rotor position is given by the angle ϕ between the polarizations J_C and J_R. σ_1, and σ_2: surface densities of magnetic charges at the periphery of the rotor and of the equivalent stator of figure 16.27, respectively.

Figure 16.28 - Coulomb representation of the problem

One then shows that the field generated by an infinite cylinder with radius R carrying the polarization **J** perpendicular to its axis can be written, in cylindrical coordinates, as:

$$r < R \qquad \mathbf{H} = -\mathbf{J}/2\mu_0 \tag{16.10}$$

$$r > R \qquad \mathbf{H} = \frac{\mathbf{J}}{2\mu_0}\frac{R^2}{r^2}\left(\cos\theta\mathbf{u}_r + \cos\theta\mathbf{u}_\theta\right) \tag{16.11}$$

One can therefore calculate the fields due to both sources. The indices R and C meaning the field generated by the rotor and the stator current, respectiveley, one obtains:

$$r < R \qquad \mathbf{H_R} = -\frac{\mathbf{J_R}}{2\mu_0} \;;\; \mathbf{B_R} = \mu_0\mathbf{H_R} + \mathbf{J_R} \tag{16.12}$$

$$r > R \qquad \mathbf{H} = \frac{\mathbf{J_R}}{2\mu_0}\frac{R^2}{r^2}\left(\cos\theta_R\mathbf{u}_r + \sin\theta_R\mathbf{u}_\theta\right)\;;\; \mathbf{B_R} = \mu_0\mathbf{H_R} \tag{16.13}$$

$$r < R+g \quad \mathbf{B_C} = \mu_0\left(-\frac{\mathbf{J_C}}{2\mu_0}\right) + \mathbf{J_C} = \frac{\mathbf{J_C}}{2}\;;\; \mathbf{H_C} = \frac{\mathbf{B_C}}{\mu_0} \tag{16.14}$$

$$r > R+g \quad \mathbf{H_C} = \frac{\mathbf{J_C}}{2\mu_0}\frac{(R+g)^2}{r^2}\left(\cos\theta_C\mathbf{u}_r + \sin\theta_C\mathbf{u}_\theta\right)\;;\; \mathbf{B_C} = \mu_0\mathbf{H_C} \tag{16.15}$$

Comment - *In what follows, we only consider the energy terms liable to be ϕ dependent. We will thus neglect the energies involved in setting up the sources.*

The total energy E of the system is equal to the mechanical energy W_{ope} supplied by the operator to introduce the stator coil into the rotor field, increased by the energy W_{gen} required to maintain the stator currents. The first energy term is equal and opposite to the work W_m of magnetic forces, whereas the second term is equal to W_m (see eq. 16.7). The system energy is therefore equal to zero.

The generator work is written, considering the stator coil, as: $W_{gen} = \int \Phi(\alpha)dI(\alpha)$ where Φ is the rotor induction flux enclosed by a stator turn element carrying a current dI, and defined by the angle α (fig. 16.26).

Some elementary calculations lead to the equivalent expression:

$$W_{gen} = \int_{eq\,stator} (\mathbf{J_C}/\mu_0)\mathbf{B_R}dV$$

where $\mathbf{B_R}$ is the induction due to the rotor at the place of the volume dV of the equivalent stator. Taking into account the field expression for an infinite cylinder, one finally obtains: $W_{gen} = (\mathbf{J_C J_R}/2\mu_0) V_R$ where V_R is the rotor volume. For a small rotation, one has $dE = 0 = dW_{gen} + dW_{ope}$. One then deduces the value of the electromagnetic torque: $\Gamma_m = -(J_C J_R/2\mu_0)V_R \sin\phi$.

Contribution of the stator flux closure circuit

The stator is now made of a material with infinite permeability. We thus have a third field source (induced source) which becomes magnetised so that its internal field is zero. It is equivalent, with respect to the field generated in the stator, to a cylinder of radius $R+g$ carrying the polarization:

$$\mathbf{J_S} = -\mathbf{J_C} - \left(\frac{R}{R+g}\right)^2 \mathbf{J_R}.$$

The index S being used for the field originating from the stator magnetization, one obtains:

$$r < R+g \quad \mathbf{H_S} = -\frac{\mathbf{J_S}}{2\mu_0} \; ; \quad \mathbf{B_S} = \mu_0\mathbf{H_S} \tag{16.16}$$

$$r > R+g \quad \mathbf{H_S} = \frac{\mathbf{J_S}}{2\mu_0}\frac{R^2}{r^2}(\cos\theta_S\mathbf{u_r} + \sin\theta_S\mathbf{u_\theta}) \; ; \quad \mathbf{B_S} = \mu_0\mathbf{H_S} + \mathbf{J_S}_{actual} \tag{16.17}$$

The magnetization state of the stator is a function of the angle ϕ. One is thus led to take into account the stator formation energy in the total system energy. The latter is easily calculated, the energy spent in inserting a *current* source into a system only made of *magnetised material* sources being zero. One can thus choose first to bring the magnetised stator near the rotor, and then to bring in the stator coil.

One obtains: $\qquad E = -\frac{1}{2}\int_{eq\,stat} \mathbf{J_S H_S}dV - \int_{eq\,stat} \mathbf{J_S H_R}dV = 0.$

The generator work can be written:

$$W_{gen} = \int_{eq\,stat} \frac{\mathbf{J_C}}{\mu_0}(\mathbf{B_R} + \mathbf{B_S})dV = \frac{J_C J_R}{\mu_0} V_{rotor}$$

and the resulting electromagnetic torque is: $\Gamma_m = -(J_C J_R/\mu_0) V_{rotor} \sin\phi$.

The system efficiency is twice larger than that of the previous system, because the coupling between the rotor and the stator coil is twice as large as previously, since the stator acts as a mirror which amplifies the flux in the region enclosed by the coil. This effect fully occurs only insofar as the soft material can be considered as having a very high permeability. This requires the stator to have a covering thick enough that the material is not saturated.

Rotor field created by a current

The above structures are limited to the range of small machines for many reasons, the most simple of which is the difficulty to make large volume magnets. The rotor field is thus usually generated by a coil at the periphery of a rotor which will be assumed once more to have infinite permeability. The rotor polarization $\mathbf{J_R}$ and that of the stator equivalent to the real stator $\mathbf{J_S}$ are always determined by the need to cancel the field in the materials. By introducing $u = [R/(R+g)]^2$ one obtains:

$$\mathbf{J_R} = \frac{2\mathbf{J_{CS}} + \mathbf{J_{CR}}(1+u)}{1-u} \tag{16.18}$$

$$\mathbf{J_S} = -\frac{2\mathbf{J_{CR}}u + \mathbf{J_{CS}}(1+u)}{1-u} \tag{16.19}$$

$\mathbf{J_{CS}}$ and $\mathbf{J_{CR}}$ are the polarizations of the cylinders equivalent to the stator and rotor currents, respectively. The system energy is now written:

$$E = -\frac{1}{2}\underbrace{\int \mathbf{J_R}\mathbf{H_R}dV}_{\text{rotor}} - \frac{1}{2}\underbrace{\int \mathbf{J_S}\mathbf{H_S}dV}_{\text{eq stat}} - \underbrace{\int \mathbf{J_S}\mathbf{H_R}dV}_{\text{eq stat}} + \underbrace{\int \frac{\mathbf{J_{CR}}\mathbf{B_{CS}}}{\mu_0}dV}_{\text{rotor}} \tag{16.20}$$

One obtains:
$$E = \frac{2}{\mu_0}\mathbf{J_{CS}}\mathbf{J_{CR}}\frac{V_{\text{rotor}}}{1-u}.$$

The work supplied by the generator is:

$$W_{\text{gen}} = \underbrace{\int \left(\frac{\mathbf{J_{CS}}}{\mu_0}\right)(\mathbf{B_{CR}} + \mathbf{B_S} + \mathbf{B_R})dV}_{\text{eq stat}} + \underbrace{\int \left(\frac{\mathbf{J_{CR}}}{\mu_0}\right)(\mathbf{B_{CS}} + \mathbf{B_S} + \mathbf{B_R})dV}_{\text{rotor}}$$

i.e.: $W_{\text{gen}} = (4/\mu_0)\mathbf{J_{CS}}\mathbf{J_{CR}}V_{\text{rotor}}/(1-u)$. During a small rotation, the energy variation can be written: $dE = dW_{\text{gen}}/2 = dW_{\text{gen}} - \Gamma_m d\Phi$, hence:

$$\Gamma_m = \frac{1}{2}\frac{dW_{\text{gen}}}{d\Phi} = -\frac{2\mathbf{J_C}\mathbf{J_R}}{\mu_0(1-u)}V_{\text{rotor}}\sin\phi \tag{16.21}$$

The torque is again multiplied by a factor 2 with respect to the previous case. This factor 2 is related to the mirror amplification effect associated with the presence of the soft rotor core. *The factor $1/(1-u)$ suggests that the machine efficiency is the larger the closer to 1 is u,* i.e. the smaller is the airgap g. This is simply related to the above mirror effect, which is more efficient the larger the rotor-stator coupling.

In fact, this result has to be qualified because the mirror effect fully takes place only when the soft material has a large permeability, and this condition is especially sensitive at the interfaces, the regions where charges appear. Now, near the internal surface, the stator polarization can be written $\mathbf{J_{Sactual}} = -\mathbf{J_S}$. The equivalent stator polarization therefore reflects the surface polarization of the real stator, and must be controlled, as well as the rotor polarization. One can choose a maximum polarization value J.

An expansion to first order in g/R of equations (16.18), and (16.19) thus gives:

$$\mathbf{J_R} = \mathbf{J_S} = (R/g)(\mathbf{J_{CS}} + \mathbf{J_{CR}}) < \mathbf{J} \qquad (16.22)$$

One can illustrate the optimization of the various parameters by the semi-quantitative calculation of the maximum torque produced by the machine, obtained for $\phi = \pi/2$; equation (16.22) gives the limiting condition:

$$(R/g)^2 \left(\mathbf{J_{CS}}^2 + \mathbf{J_{CR}}^2\right) = \mathbf{J}^2 \qquad (16.23)$$

This limiting condition can be reached only on large enough machines: on small machines ($P \lesssim 1$ kW), there is not enough space for the number of ampere-turns required to reach very high induction levels: 0.5 T to 0.6 T is a normal induction level, which is obtained with an air gap as small as possible in agreement with the above condition. On the largest machines one has to respect criterion (16.23), so that, setting $\mathbf{J_{CS}} = \lambda \mathbf{J_{CR}}$, equations (16.21), and (16.23) give:

$$\Gamma_{max} = \left(\frac{V_{rotor}\, g\, J^2}{\mu_0 R}\right)\left(\frac{\lambda}{1 + \lambda^2}\right).$$

For a machine of given size, the torque increase can be achieved only by increasing the air gap, and the maximum torque per unit volume is obtained when $\mathbf{J_{CR}} = \mathbf{J_{CS}}$. The example of two three phase synchronous machines rotating at 3,000 rpm [33], the characteristics of which are shown in table 16.9, illustrates the validity of these principles.

Table 16.9 - Comparison between two alternators

P (MW)	P/(S$_{rot}$ L$_a$) (MW . m^{-3})	g/R$_{rot}$	I$_n$ (A)	J$_n$ (A)	λ
293	72	0.15	9,951	2,187	1.03
900	120	0.20	24,056	6,172	1.04

Both alternators are bipolar synchronous machines with smooth poles.
P = alternator power; S_{rot} = rotor cross-section; L_a = stator active length; g = air gap; R_{rot} = rotor radius; I_n = nominal phase current; J_n = nominal rotor current.

7.2.3. The main classes of machines

We have seen how magnetic interactions are used to produce a motor torque. In order to obtain a permanent rotation of the rotor, one has to maintain the angular shift of the fields.

A first solution consists in producing a rotating stator field. This is generally achieved by means of a system of three properly positioned three-phase coils. The rotor can be supplied with a DC current. In this case, the machine adopts a steady-state angular velocity equal to that of the stator field, itself proportional to the frequency of the stator current. One has a so called *synchronous machine*, the main characteristic of which is that its angular velocity is load independent. This property is used for electricity production, and this class of machines includes the most powerful machines: the turbo-alternators used in nuclear generating stations (see fig. 16.29).

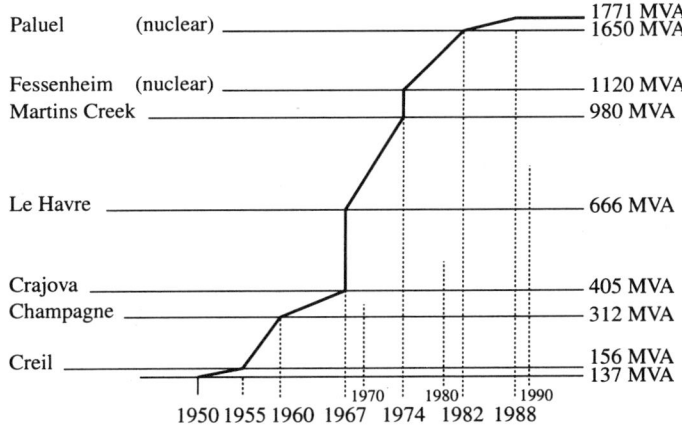

Figure 16.29 - Unit power evolution of turbo-alternators [34]

It is possible, while retaining the rotating stator field principle, to short the rotor circuit. It is the stator field rotation that induces in the rotor coil a reaction current. In this case, the rotor does not rotate at the angular velocity of the stator field. The resistive torque determines, in the steady state, the speed gap, which never exceeds a few per cents. This constitutes an *asynchronous machine*, much used for all the usual motor applications, for small power (from 10 W) as well as for large installations (up to 25 MW), on account of its robustness, its unbeatable manufacturing cost, and the ease of implementation compared to a synchronous machine.

The field production strategy is quite different in DC machines. Here the stator field does not rotate: it is generated by coils supplied with DC current, or by permanent magnets. The angle shift between the rotor and stator fields is kept up by altering, depending on the rotor position, the rotor current supply. This is achieved by means of a collector, a device fixed on the rotor and consisting of many conductor blades, delicate in operation, and with fast wear. A DC current machine is therefore expensive to manufacture (2 to 2.5 times the price of an asynchronous machine with the same power), and it requires careful maintenance. These drawbacks are counterbalanced by a variable speed working range, and they are used in the domain of electric traction, for rolling mill motors in the steel industry, and in many common use devices (leisure electronics, automobile...). Note that their use in the large power range tends to decrease, since variable frequency power supplies allow, thanks to more or less sophisticated feedback systems, the use of synchronous and asynchronous machines in variable speed applications such as railway traction.

7.2.4. Soft magnetic materials used in rotating machines

The rotor core and stator yoke are usually made of magnetic sheet stackings (ref. [30], p. 223). This arrangement, as for transformers, restricts eddy currents due to flux variations (for more details, see § 2.3 of this chapter). This choice also results from manufacturing considerations, as the stacking of a variable number of sheets makes it possible to manufacture machines with different lengths, and therefore different powers, without notably altering the assembly line.

The small machine range (P < 1 kW) escapes these simple rules, because the lowest cost criterion for instance sometimes leads to magnetically degraded solutions. In this category one finds magnetic circuits obtained by compressing iron powder, with relative permeability smaller than that of magnetic sheets (typically 100 instead of 1,000).

Rotors of large power turbo-alternators also escape these rules. They are made in bulk form for several reasons:

♦ on the one hand, the rotor of a synchronous machine sees a static field, since it rotates at the same speed as the stator field. Actually, it is impossible to avoid the presence of spurious harmonic rotating fields. On large enough machines, they are shielded at the rotor surface by a damping device made with short-circuited conductor bars. Eddy currents thus cannot develop in the rotor.

♦ on the other hand, mechanical resistance considerations become critical for these machines: at 3,600 rpm, the angular velocity of a bipolar synchronous machine fed at 60 Hz, the rotor diameter is limited to 1.2 m on account of the centrifugal force which tends to break it; the peripheral parts are submitted under these conditions to an acceleration larger than 8,000 g. The machine power will in fact be determined by its length: a 600 MW turboalternator rotating at 3,000 rpm will be 6 m long in its active part, i.e. 11 m between bearings. Taking into account the high values of the angular velocity and of the length / diameter ratio, one obtains rather low critical (resonance) speeds (700 rpm and 2,400 rpm for the two first critical velocities in the above quoted example) which impose rigorous balancing of the machine [35].

These different requirements lead to manufacturing monolithic cast rotors. Modern casting techniques provide an elasticity limit of the order of 650 to 750 N/mm², from nickel-chromium-molybdenum alloyed steels. The largest rotors are found in the 4 pole turbo-alternators of nuclear generating plants (230 tons for a 1,500 MW machine).

8. ACTUATORS

The world of actuators includes all the applications in which the induced circuit response is accompanied by a motion, the characteristics of which are at least as important as the energetic aspect. Actuators can be classified into two main classes, according to the dominant function:

♦ accuracy of the final position of the moving parts, for applications associated with robotics (stepping machines), with clocks and watches (watch mechanisms), etc.

♦ dynamic performance, for applications associated with electric control (relays, circuit breakers), hi-fi (loudspeaker actuators), etc.

In the first case, the performance of the machine is mainly determined by its structure, more than by the material itself. One will thus mainly use Fe-Si magnetic circuits. In the second case, one seeks to minimize the response time. For applications which are not too critical (relays...), one keeps a Fe-Si magnetic circuit.

A search for more definite aspects of performance can lead the designer to decrease the movable part inertia, possibly by suppressing the soft magnetic core. This is the case of some small rotating machines [36]. One can also work on the material itself, and choose ferrite circuits.

The actuator can be of different types:

♦ electromagnetic actuator: the moving part consists of a magnet;

♦ electrodynamic actuator: the moving part is wound, and the field due to the static part originates from a magnet;

♦ reluctant actuator: the movable part is made of a soft magnetic material; the magnetic circuit reluctance varies with its position;

♦ hybrid actuator: one associates one or several magnets with the reluctant actuator.

Finally, we mention the magnetostrictive actuators which will be treated in chapter 18.

8.1. AN EXAMPLE OF A ROTATING ACTUATOR: STEPPING MACHINE WITH VARIABLE RELUCTANCE

Stepping machines are characterized by an incremental motion of the moving part, the equilibrium position being determined by the field configuration generated by the stator, which is discretized. The shift between two successive equilibrium positions is the step. The moving part is made out of a soft magnetic material shaped in such a way that the circuit reluctance changes with its position. Torque is generated by this effect. The stator, because it is submitted to a flux that changes in direction and amplitude, is obviously laminated. The rotor, as in a synchronous machine, rotates at the same speed as the stator field, and can consist of a single bulk part.

However, the pulse character of the control and therefore of the stator field gives rise to rotor flux variations. Here too laminated structures will often be used in order to reduce harmonic losses. When each phase is supplied sequentially, torque calculation, if it is performed in the same way as we did for machines, would lead to studying the energy of a system consisting of a "current" type source, and two "magnetised soft material" type sources, the rotor and stator. Unfortunately, it is impossible to go very far in this calculation on account of the impossibility to simply determine the stator and rotor magnetizations, which furthermore are not uniform, because of the geometrical complexity. It is thus better to calculate the system energy by figuring out a different process which here can be:

♦ bringing together the rotor, the stator and the stator coil, the latter carrying initially no current, and the stator and rotor being assumed demagnetised,

♦ progressively setting up the coil current.

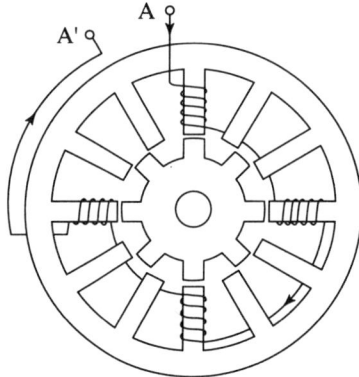

Figure 16.30 - Cross-section of a stepping reluctant motor with 24 steps per turn, three stator phases (only one phase is shown) [37]

The first stage requires no energy. Setting up the current requires an energy supplied by the generator, which is written as a function of the total flux Φ_t and of the coil current i:

$$W_{gen} = \int i d\Phi_t \tag{16.24}$$

The problem is treated approximately by assuming that the total flux in the coil is a univocal function of the current and of the rotor-stator relative position, described for instance by the angle θ. One thus eliminates the hysteretic aspect associated with the magnetic state of the system, and this allows an analytical approach. This assumption means that all the energy spent by the generator is transformed into magnetic energy, without loss. One will write:

$$\Phi_t = L(\theta, i)i \; ; \; E = W_{gen} = \int i d[L(\theta, i)i] \tag{16.25}$$

Moreover if saturation phenomena are neglected, flux becomes a linear function of the current, and one obtains:

$$E = L(\theta) i^2 / 2 \tag{16.26}$$

The variation of the system energy with the angle θ is written:

$$dE = (1/2)(dL/d\theta) d\theta \, i^2.$$

Besides: $dE = dW_{ope} + dW_{gen} = -\Gamma_m d\theta + (dL/d\theta) d\theta \, i^2.$

The torque of the machine is thus: $\Gamma_m = (1/2)(dL/d\theta) i^2.$

Let us notice that, while these approximations provide a compact expression for the torque, the difficulty mentioned above remains, and the problem is shifted to the calculation of the inductance L. Modern solution methods broadly use computers and numerical techniques. It is however possible to proceed in a semi-analytic way. Using the reluctance formalism, the flux closure circuit is represented by a juxtaposition of independent elements, and this problem, while it is too complex to be attacked globally, is thus split.

Note on the reluctance formalism - *The formalism of magnetic reluctance $\mathcal{R}_m$ formally establishes the relationship between the magnetomotive force $\mathcal{E}$ produced by a source (current circuit or permanent magnet), and the resulting flux Φ, associated with the configuration of the magnetic circuit in which flux lines spread. This formalism (see § 1.5 in chap. 2) implies that the circuit can be described as the juxtaposition of regions characterized by a linear relationship between H and B, the permeability coefficient being variable from place to place. Each region is then called a circuit element, and one can establish the formal analogies between an electric and a magnetic circuit, as presented in table 16.10.*

Table 16.10 - Analogies between electric, and magnetic circuits [37]

Relationship, quantity	Electric circuit	Magnetic circuit
Field	E	H
Response	j	B
Specific relationship of materials	$j = \rho^{-1}E = \sigma E$	$B = \mu H$
Flux conservation	div $j = 0$	div $\mathbf{B} = 0$
Characteristic flux	$i = \int_S j dS$	$\Phi = \int_S \mathbf{B} dS$
Potential difference	$u = \oint E dl$	$\mathcal{E} = \oint \mathbf{H} dl = \int_S j dS = Ni$
Ohm's law	$u = Ri$	$\mathcal{E} = \mathcal{R}_m \Phi$
Resistance, reluctance	$R = \int_C \dfrac{dl}{\sigma S}$	$\mathcal{R}_m = \int_C \dfrac{dl}{\mu S}$
Connections in series	$R_{eq} = \sum_k R_k$	$\mathcal{R}_{m_{eq}} = \sum_k \mathcal{R}_{m_k}$
Connections in parallel	$R_{eq}^{-1} = \sum_k R_k^{-1}$	$\mathcal{R}_{m_{eq}}^{-1} = \sum_k \mathcal{R}_{m_k}^{-1}$

As an example, one can describe by means of this formalism an electromagnet with n turns carrying a current i, wound around a soft magnetic circuit characterized by a permeability μ, a length l, a cross-section S, and an airgap of length g. Within the crude approximation which consists in neglecting magnetic flux broadening in the gap, one easily obtains, by applying Ampere's law :

$$\mathcal{E} = ni = (\mathcal{R}_f + \mathcal{R}_g)\Phi \qquad (16.27)$$

where $\mathcal{E}$ is the magnetomotive force, Φ the magnetic flux induction, $\mathcal{R}_i$ the reluctance of the "iron" element: $\mathcal{R}_i = l/(\mu S)$, and $\mathcal{R}_g$ the reluctance of the "airgap" element $\mathcal{R}_g = g/(\mu_0 S)$.

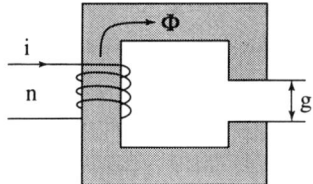

Figure 16.31
Illustration of reluctance

Comments

♦ Depending on the arrangement of the different elements (series or parallel) one can be led to describe them in term of reluctance or *permeance*, its reciprocal ($\Lambda = 1/\mathcal{R}$).

♦ One can, exactly as for electric circuits, associate with the element k the potential difference $\mathcal{R}_k \Phi$. This comes from the fact that, in magnetic conductors at equilibrium, the field **H** is such that **curl**(**H**) = 0, and thus **H** can be obtained from a scalar potential which can be defined to within a constant, and which can be written U. The magnetomotive force is thus none else than the magnetic potential difference applied to the system.

♦ Because equipotential surfaces are perpendicular to flux lines, a circuit element will be better described as a region of constant permeability, bounded by a flux tube, and two equipotentials. The above analytical expression for the reluctance of an element has similarities with the electrical resistance of a cylindrical conductor $R = \rho l / S$, and is based on the flux lines in the element being parallel. This is a special situation, and one can, exactly as for the resistance calculation, consider other situations. The basic approach consists in using, for the element under consideration, the quite general relationships:

$$\mathcal{E}_1 - \mathcal{E}_2 = \int_1^2 \mathbf{H} dl \; ; \; \mathbf{B} = \mu \mathbf{H} \; ; \; \Phi = \int \mathbf{B} d\mathbf{S} = \text{Cte} \; ; \; \mathcal{E}_1 - \mathcal{E}_2 = \mathcal{R}\Phi$$

♦ Finally, and in agreement with the previous comment, it must be pointed out that one can describe the elements only if the flux line geometry is known. This is one of the main difficulties in this approach, because the existence of leakage flux leads to real situations sometimes very far from the over-simplified situations treated above, and difficult to handle in an analytic way. One is thus led to correct the initial representation by adding empirical elements (see [37], p. 66). Let us note that a rigorous approach consists in treating the problem in the most general way by means of numerical simulation software such as *Flux 2D* and *Flux 3D,* developed by the "Laboratoire d'Electrotechnique de Grenoble (LEG)", and marketed by the Cedrat company. Of course, the software price decrease enhances the interest for the latter approach.

On the other hand, the permeability of a magnetic material is itself induction dependent. Flux calculations thus impose an iterative process, a permeability being initially ascribed to each element, and then corrected at each iteration depending on the obtained fluxes.

This approach can be illustrated [37] in the case of the stepping motor of figures 16.32 and 16.33: the permeances that are difficult to calculate correspond to stray fluxes associated with magnetic potential sources (branches shown by broken lines in the right hand part of figure 16.32), and to the air gaps (branches shown as dotted lines in the right hand part of figure 16.32). In this case, the field lines can be represented as linear segments in the gap, and by arcs ending at the rotor and stator teeth, which makes an analytical approach possible.

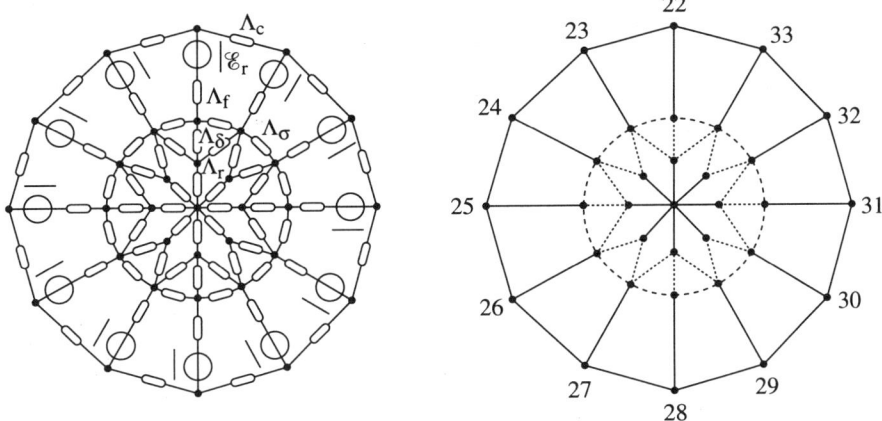

*Figure 16.32 - Formal equivalent scheme of the stepping motor
of figure 16.30 [37]*

Λ_c: *stator yoke permeance;* Λ_f: *stator tooth permeance;* Λ_δ: *gap permeance;*
Λ_r: *rotor tooth permeance;* Λ_σ = *stray permeance between two stator teeth.*

*Figure 16.33 - Approximation
for the field lines in the gap [37]*

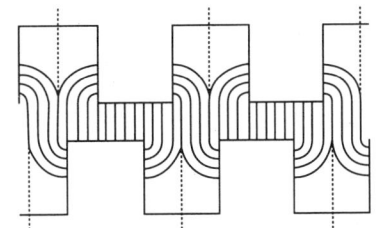

Once the permeances have been calculated, there is no problem in solving the circuit by using specialized software. It is absolutely necessary to take into account saturation effects in the modelling (ref. [37], p. 247). These effects mainly occur in the tooth zones, and one can try to account for them by ascribing to the "tooth" elements a reduced permeability coefficient.

Figure 16.34 reveals the incomplete character of this approach, the teeth being themselves inhomogeneously magnetised, which gives rises to local saturation effects.

A more accurate evaluation of these local effects leads to increasing the number of elements describing the problem (fig. 16.35). Unfortunately one is rapidly faced with a very large number of elements, which makes solving the problem more difficult.

The study of the stepping machine thus reveals some general characteristics associated with reluctant systems:

♦   the torque, or the force, being proportional to the square of the current, there is no linearity of the electromechanical conversion, and the system is ill suited to analog information transmission (as in loudspeakers for instance).

♦   in order to obtain a large torque, the circuits must have large inductance variations. This leads to large flux variations and local saturation phenomena, which drastically affect the amplitude of the torque and force, and which are difficult to take into account. These systems are more difficult to treat than classical machines. Reference [37] contains many other examples of actuators.

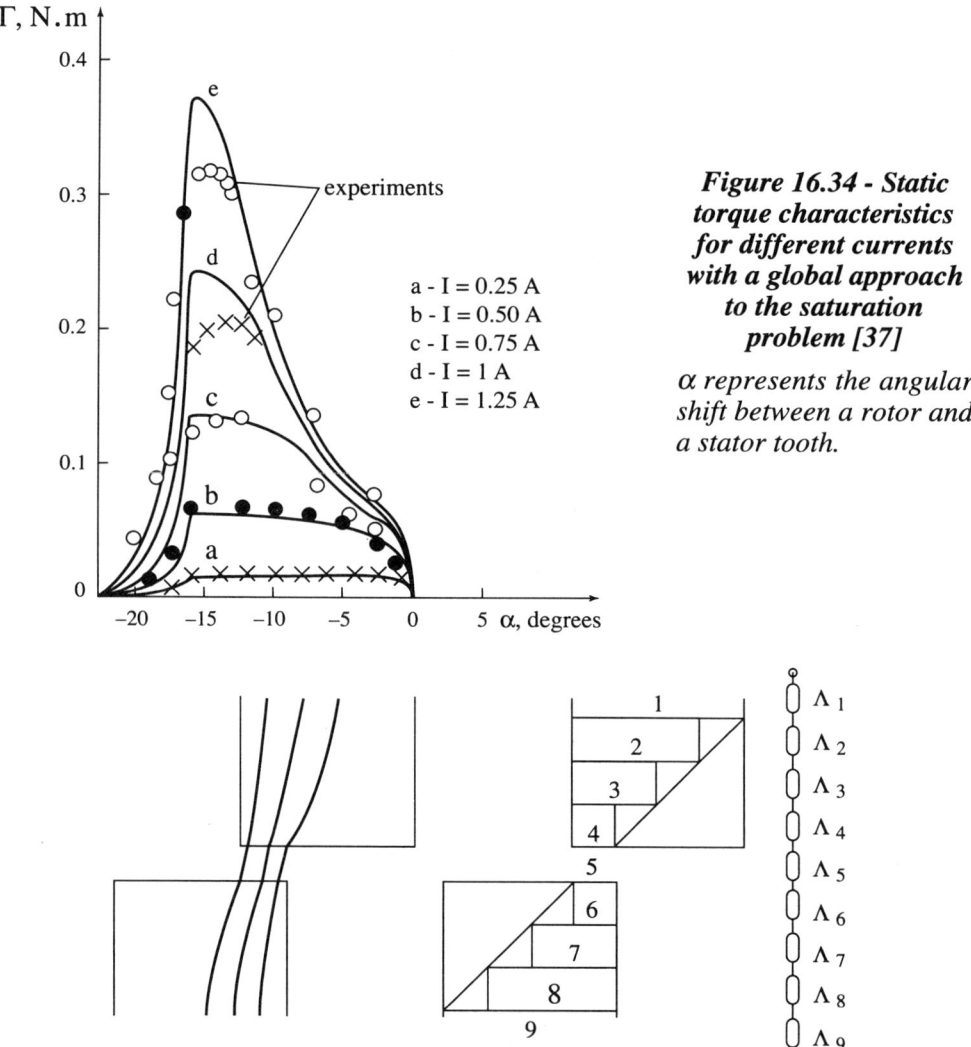

experiments

a - I = 0.25 A
b - I = 0.50 A
c - I = 0.75 A
d - I = 1 A
e - I = 1.25 A

Figure 16.34 - Static torque characteristics for different currents with a global approach to the saturation problem [37]

α represents the angular shift between a rotor and a stator tooth.

Figure 16.35 - Field line concentration phenomenon associated with a tooth shift [37]: modelling through partition into regions with the same permeability

8.2. *CHOICE CRITERIA, ORDERS OF MAGNITUDE*

As mentioned above, the function as well as the cost of the actuator can guide some choices, but size is often a key factor in the choice of the technology: because the torque or the force result from the position dependence of the system energy, the sources allowing optimization of the stored magnetic energy density, i.e. systems with permanent magnets, will be chosen for small actuators.

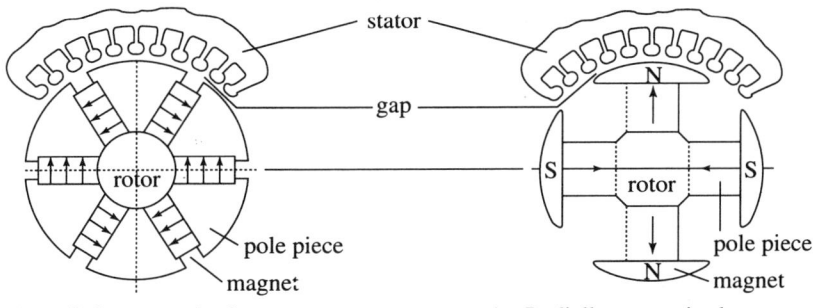

a - Orthoradially magnetised magnets   b - Radially magnetised magnets

Figure 16.36 - Structure of a synchronous motor with permanent magnets [36]

In order to determine an order of magnitude of the characteristic size below which the permanent magnet solution is preferable, consider a cylinder with infinite length, with radius R, made of an infinitely soft magnetic material. This cylinder, mobile around its axis, is equipped with a bipolar field system which can be a crown of peripherical magnets, supposed to model an arrangement such as sketched in figure 16.36-b,. It can also be a coil carrying a surface current density $\mathbf{j_s}$ as in § 7.2.2, which gives rise to an equivalent magnetization $\mathbf{J_C} = \mu_0 \mathbf{j_s}$ (fig. 16.37).

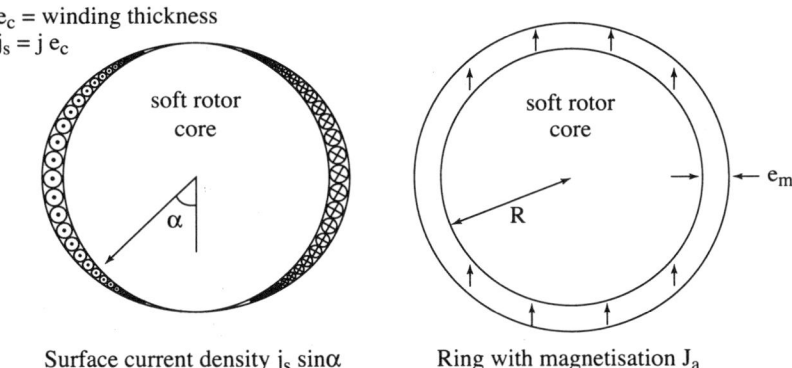

Surface current density $j_s \sin\alpha$ Ring with magnetisation J_a

Figure 16.37 - Comparison between the magnet solution, and the wound solution

We will consider optimistically that the stator geometry is such that at the equilibrium position, given by the rotor angle $\theta = 0$, the energy of the field system is zero, and that in the maximum energy position ($\theta = \pi/4$), the rotor flux closes completely in air.

Setting E for the maximum energy per unit length of the system, the average torque per unit length will be: $\Gamma = 4E/\pi$. From the discussion in § 7.2.2 we obtain for a wound system, marked with subscript "$_c$":

$$\Gamma_c = \frac{J_C^2}{\mu_0} R^2 = \mu_0 \, j_s^2 \, R^2 = \mu_0 \, j^2 \, e_C^2 \, R^2 \qquad (16.28)$$

where j represents the volume current density in the coil of thickness e_C, and is equivalent to the surface current density j_s.

The analytical study of the magnetised system is more complex than that of the wound system. We will assume that the magnets give rise to a uniform polarization $\mathbf{J_m}$ in a crown of thickness e_m. The average torque per unit length is then written:

$$\Gamma_m = (J_m^2/\mu_0) \, 2 \, R \, e_m \qquad (16.29)$$

In order to compare the two solutions, we can consider values typical of low power electrical technology (P < 1 kW): $j = 20$ A.mm^{-2} (ref. [30], p. 224), $e_C = 8$ mm, $J_m = 0.4$ T (cheap ferrite magnets), $e_m = 5$ mm.

The magnet based solution become advantageous for rotors with radius smaller than 4 cm.

Comments - *The current density that we have considered for the calculation of the torque produced by the reluctant machine can seem large. It is generally assumed that a copper conductor in a non ventilated coil withstands a current density ranging between 3 and 5 A.mm^{-2}. The choice of large current densities is justified by the scaling laws, which provide the evolution of key parameters (efficiency, heating, specific power...) as a function of the machine size under scale transformations.*

As an example, we can focus on the problem of heating. The acceptable limit is set by the thermal behavior of the insulators, hence it is the same for two machines built from the same materials. For small machines, heating mainly originates from Joule losses, and heat exchange mainly occurs by convection. One then obtains, with a convection coefficient α characteristic of the external surface area of the system:

$$\Delta T = \rho \, j^2 \, V / \alpha \, A \qquad (16.30)$$

ΔT represents the temperature difference between the machine and the external medium, whereas V, j, ρ, and A are the coil volume, the volume current density, the resistivity, and the exchange surface area, respectively.

One can use relative quantities, determined with respect to a reference machine. They will be written with an asterisk in order to limit the risks of confusion with the true quantities. Therefore only parameters liable to vary with the machine size contribute. In reduced units, equation (16.30) becomes:

$$(\Delta T)^* = (j^*)^2 \, l^* \qquad (16.31)$$

where l* is the reduced size.

In order to maintain heating constant, one will thus impose: $j* = (1*)^{-1/2}$, which justifies the current density increase in small machines. It can then be shown (ref. [37], p. 94) that the mechanical power P_m, and the induction B vary as:

$$(P_m)* = (1*)^4 \quad \text{and} \quad (B)* = \sqrt{1*} \qquad (16.32)$$

Size reduction thus leads to an induction decrease, so that the magnetic circuit is under-used. In small reluctant machines, one will thus choose to increase the size of the coils at the expense of the magnetic circuit volume (ref. [30] p. 224).

Thus a detailed study of the scaling laws is required to optimize the different types of machines, and choose the best suited technology. A detailed study of the use of scaling laws for reluctant or polarised systems will be found in reference [37].

9. ENERGY TRANSFORMATION IN POWER ELECTRONICS

Magnetic components used in power electronics perform smoothing, transformation, energy storage, control-power interface (pulse transformers, magnetic amplifiers), etc. functions.

9.1. SMOOTHING INDUCTANCES, AND COMPONENTS FOR INDUCTIVE ENERGY STORAGE

Smoothing inductances protect the electric networks from the harmonics generated by faulty power installations. They are also used downstream of an electric supplying installation (chopper, undulator, rectifier…), in order to warrant to the user a clean energy source.

If the waves to be filtered are low frequency (harmonics of the 50 Hz frequency), the winding is generally made around a non oriented Fe-Si sheet core. Characterizing the material by its average permeability $\mu = \Delta B / \Delta H$, determined for a current variation amplitude ΔI, one obtains with this arrangement $L = \mu n^2 S / l$, where n is the coil turn number whereas S and l are the cross-section and the length of the magnetic circuit, respectively.

It can be noticed that the inductance magnitude is directly linked to the permeability, which depends on ΔH, and hence on ΔI. L is therefore poorly defined, and this is the reason why the magnetic circuit includes an airgap (fig. 16.38). By introducing demagnetising fields, this limits core saturation risks, and allows the designer to reduce the cross-section.

Figure 16.38 - An air gap allows the cross-section of the yoke for a smoothing inductance to be reduced [38]

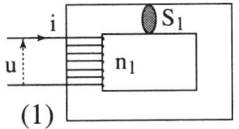
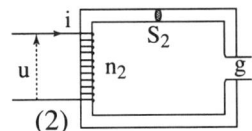

One also increases the coil turn number, which is optimized from dissipation considerations (ref. [38] p. 525). The apparent permeability of the circuit is then, to a first approximation: $\mu = \mu_0 l / g$, where g is the airgap thickness, so that the inductance can finally be written $L = \mu_0 n^2 S / g$. The inductance value is thus only determined by the size, the turn number, and the circuit geometry.

In circuits designed to work with non zero average current, the key parameter is the differential permeability, and the requirement of an airgap is still more critical. It allows moving the average working point of the magnetic circuit to a non saturated zone (fig. 16.39).

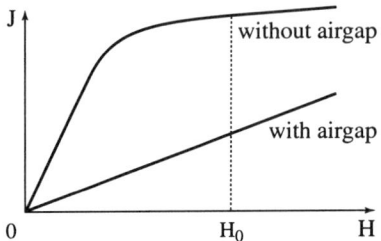

Figure 16.39 - Effect of the air gap on the shift of the working point of the material

The same technology is found for inductive storage devices used in *fly-back* type switching power supplies. The key device in these supplies is the transformer, which works in a rather special way, since the windings do not simultaneously carry current.

In fact, the primary first stores magnetic energy, which is then released by the secondary. This storage function is better achieved the lower the value of the inductance associated with the magnetic circuit. One can consider a coil to be supplied by a step of voltage U for a duration T limited by the chopping frequency. Assuming that the initial current in the coil is zero, the stored energy is: $E = U^2 T^2 / 2L$. It is proportional to the reciprocal inductance.

Here too, as for smoothing inductances, an airgap in the magnetic circuit will be made in order to limit the device inductance.

9.2. REQUIREMENTS ASSOCIATED WITH HIGH FREQUENCY

Working frequency increase allows, for a given voltage level, to reduce the magnetic circuit cross-section, and therefore to increase the power density in the installations. It also allows to decrease the volume of filtering devices [39], and the most extreme conditions are found in the power supplies based on chopping (choppers, switching power supplies), with working frequencies ranging from 20 kHz to some MHz.

In this case the Mn-Zn type soft ferrite based materials are the most appropriate (see § 4 of this chapter, and chap. 17). Their weak saturation magnetization (typically 0.4 T) is compensated by the large resistivity (typically some Ω.m). These ferrites thus cover the operating range from 1 kHz to several hundreds of kHz.

For switching power supplies working at the higher frequencies (typically 1 MHz), materials with distributed airgap are preferentially used [40, 41]. The resonance inductances here have relatively small values (typically 1 to 10 µH), and require, in a magnetic circuit made from a high permeability ferrite type material, a large air gap. The latter introduces radiation harmful for the neighbouring devices, and even more for the inductance coil the turns of which are located near the airgap. These windings see an induced voltage large enough to destroy, through heating, the insulator of the wound wire, and accordingly the inductance (fig. 16.40).

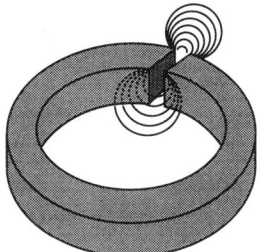

Figure 16.40 - Radiation near the airgap [40]

More generally, compressed powders should be considered as soon as small inductance values are considered. One proceeds by compressing magnetically soft particles in a non magnetic matrix. The magnetic poles that appear at the particle / matrix interface warrant a weak relative permeability of the compressed material, so that the airgap is no more necessary. On the other hand, the particles must be electrically insulated from each other, in order to minimize eddy currents: the particles can thus consist of high saturation magnetization material such as pure iron or its alloys. The electrical insulation is achieved by the binder (compaction of iron powders), and very often complemented by the insulation of the particles themselves (carbonyl iron). Let us specify that the temperature range for use of these powders is mainly determined by the thermal behavior of the organic binder, and is typically limited to the range [–55°C, 125°C].

There exists a large variety of powder grades; table 16.11 shows the characteristics of some of them.

Table 16.11 - Characteristics of some compressed materials

Nature of the powder	J_{sat} (T)	μ_r	T_C (°C)	ρ (Ω.m)
Iron[*]	1 to 1.4	40 to 100	760	10^{-2}
Carbonyl iron[*]	1.6	11 to 25	750	$> 5 \times 10^5$
$Fe_{50}Ni_{50}$[**]	1.5	14 to 160	450	
$Fe_{17}Ni_{81}Mo_2$[**]	0.7	14 to 550	450	
Fe based amorphous alloy[**]	1.5		392	

*: Saphir documentation. **: FEE documentation.

9.3. CONTROL-POWER INTERFACING

In static converters, one endeavours to electrically isolate the control part from the power part by means of pulse transformers. The latter are characterized by two important parameters [42]: the maximum pulse length, and the response time.

9.3.1. Maximum pulse length

The magnetic circuit, the primary winding of which is submitted to a voltage step E, is threaded by an induction flux Φ such that $n_1[\Phi(t) - \Phi(0)] = Et$.

The secondary voltage is $U_2 = n_2(d\Phi/dt) = (n_2/n_1)E$ as long as the magnetic circuit is not saturated. In case of saturation U_2 becomes zero.

Moreover, the pulse is generally unipolar, so that the magnetization state corresponding to the starting time is the remanent magnetization (fig. 16.41).

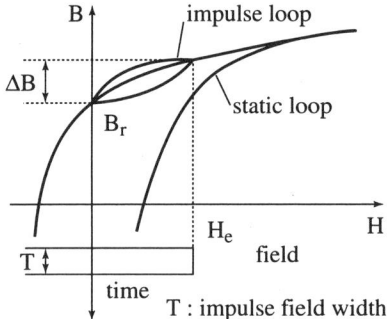

Figure 16.41 - Pulse loop [43]

The time during which the pulse is transmitted is thus larger the higher the quantity $B_s - B_r$ ($B_s \approx J_s$, $\mu_0 H$ being negligible). One therefore uses magnetic circuits producing skewed loops (see § 3 of this chapter, and tab. 16.12).

Table 16.12 - Magnetic characteristics of some alloys
with skewed hysteresis loops [43]

Nominal composition mass%	J_s (T)	H_c (A.m^{-1})	$\Delta B = B_m - B_r$ (T)	μ_Δ*
$Fe_{35}Ni_{65}$	1.25	3	0.9	4,000
$Fe_{45}Ni_{55}$	1.5	10	1.2	5,000
$Fe_{20}Ni_{80}$	0.3	1.5	0.5	10,000

* *The pulse permeability $\mu_\Delta = (\mu_0)^{-1}(\Delta B/\Delta H)$ is measured with rectangular field pulses 10 μs wide. These alloys are 0.05 mm thick ribbons.*

9.3.2. Response time

The second important parameter is the response time, defined as the time necessary for the secondary voltage to increase from 10% to 90% of the final value, following a step type input voltage. Obviously this response time must be as small as possible. It mainly depends on the primary/secondary leakage inductances, which thus imposes a careful winding of the transformer. The rise times are typically of the order 0.1 to 0.3 μs [42].

9.3.3. *Comment on the static switch*

Static converters contain power components (thyristors, transistors…) which, integrated in more or less complex devices, are used as static switches. An alternative method consists in using magnetic amplifiers. The latter consist of an inductance connected in series with the load of the power circuit supplied with an AC current (see ref. [38] p. 544, and ref. [44]).

The inductance impedance being much larger than that of the load Z, one cancels the voltage applied to the load through a dividing effect. An additional winding, DC supplied, allows to shift the working point of the magnetic circuit to a saturated zone, and to cancel the dividing effect. For these devices one chooses materials with rectangular loops which allow a quasi-instantaneous transition between the two open/closed states.

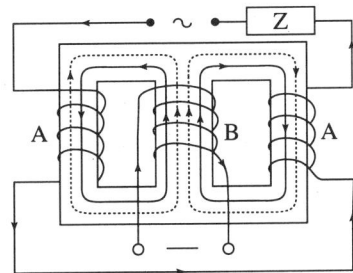

Figure 16.42 - Principle of the magnetic amplifier [38]

— flux due to B
······ flux due to A

The characteristics of some alloys with rectangular loops are reported in table 16.13.

Table 16.13 - Magnetic characteristics of some alloys with rectangular loop [43]

Composition mass%	Trade name	B_r/B_m	J_s (T)	H_c (Am^{-1})	μ_{max}
$Fe_{50}Ni_{50}$	Rectimphy	0.97	1.5	9	100,000
$Fe_{15}Ni_{80}Mo_5$	Pulsimphy	0.95	0.8	1.5	100,000
$Fe_{49}Co_{49}V_2$	Phymax	0.92	2.35	20	50,000
$Fe3\%Si/Goss$	…	0.90	2.0	9	70,000

EXERCISES: *STUDY OF A VOLTAGE STABILIZING DEVICE USING A SATURABLE INDUCTANCE*

Lower-case characters designate the instantaneous values of electric quantities. Capitals are used for rms values. Complex quantities have a horizontal dash above the letter.

We want to eliminate the fluctuation of the rms amplitude U_r of the supply voltage across a load of impedance Z. For this purpose one intends to make the circuit shown in figure 16.43, in which the load is fed via a parallel LC circuit connected to a transformer with a transformation ratio $n_1 / n_2 = 1$. The supplied voltage u_r is applied to the parallel LC circuit.

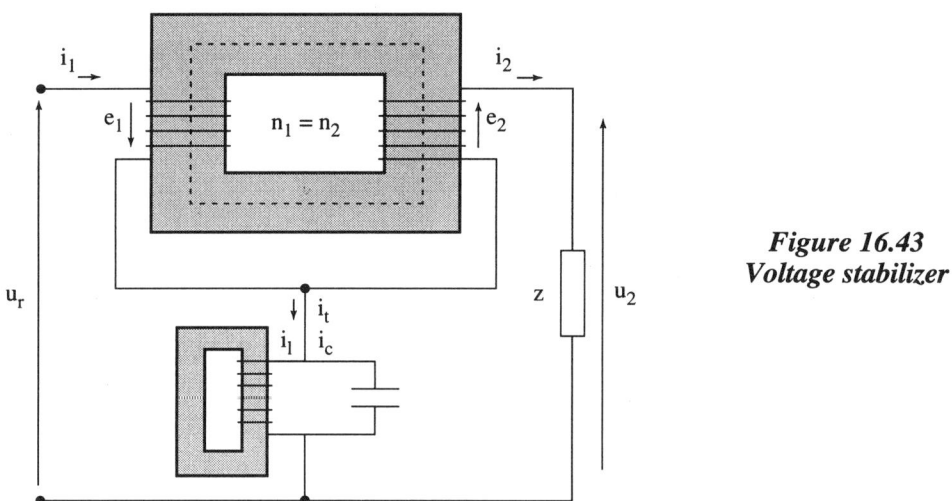

Figure 16.43
Voltage stabilizer

E.1. STUDY OF THIS APPLICATION, AFTER [45]

We will limit ourselves here to the study of the stabilizing effect of our circuit in open-circuit operation, i.e. with the load disconnected ($i_2 = 0$, $i_t = i_1$).

E.1.1. Nominal conditions

Determine the relationship satisfied by L and C so that i_1 is zero for an input voltage with nominal effective amplitude $U_{rn} = 220$ V. The parallel LC circuit is then said to work at anti-resonance.

Note - *Recall the expressions of the impedance of an inductance L and of a capacitance C supplied by a wave of angular frequency ω: $Z_L = jL\omega$, $Z_C = 1/(jC\omega)$, $j^2 = -1$.*

The magnetic circuit of inductance L is liable to saturate, so that the L value is a function of the current i_L. The function $I_L(U_r)$ is approximately characterized by two straight line segments, representing operation for a non saturated magnetic circuit and for a saturated magnetic circuit:

$$U_r = 0 \qquad U_r = 190 \text{ V} \qquad U_r = 250 \text{ V}$$
$$I_L = 0 \qquad I_L = 0.075 \text{ A} \qquad I_L = 0.615 \text{ A}$$

Determine the value of the nominal inductance L_n for the voltage U_{rn}, and deduce the value of C so that the above condition is satisfied ($f = \omega/(2\pi) = 50$ Hz).

How large is then the rms value U_2 of the voltage u_2?

E.1.2. Study of the effect of a voltage fluctuation ΔU_r

Determine the variation ΔI_L associated with a variation ΔU_r with amplitude smaller than 30 V.

Determine also ΔI_C, and show, possibly by using a Fresnel diagram, that finally

$$\overline{I_1} = j(C\omega - \alpha)\Delta \overline{U_r} \quad \text{with } \alpha = 9 \times 10^{-3} \ \Omega^{-1}$$

The current i_1 magnetises the transformer circuit. If L_{tr} is the transformer inductance seen. by the primary in its open-circuit operation, express the value of $\overline{E_1}$ and show, by taking into account the winding directions, that finally $\overline{E_2} = -jL_{tr}\omega\overline{I_1}$ (we here neglect the leakage fluxes between the primary and secondary windings). Deduce the value of L_{tr} required to maintain $U_2 = U_{2n}$.

E.2. REALIZATION OF THE SATURABLE INDUCTANCE

To make the magnetic circuit of the inductance we use a non oriented FeSi sheet the $B_{max}(H_{max})$ curve of which is plotted in figure 16.44 for frequency 50 Hz. The inductance curve $U_r(I_L)$, and the $B_{max}(H_{max})$ variation of the material are also plotted in reduced units in figure 16.45.

Determine the working point (H_{maxn}, B_{maxn}) of the material under nominal operation ($U_{rn} = 220$ V).

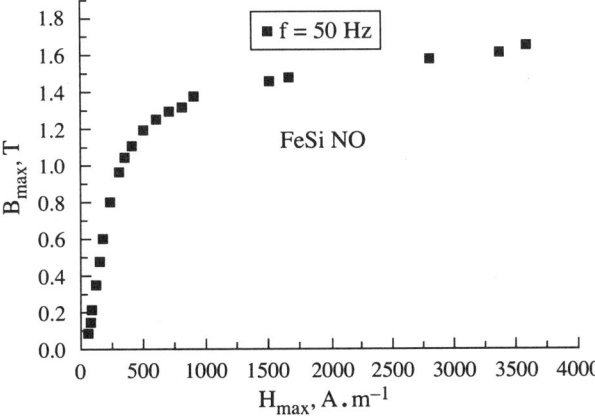

Figure 16.44 - $B_{max}(H_{max})$ curve of FeSi NO sheets

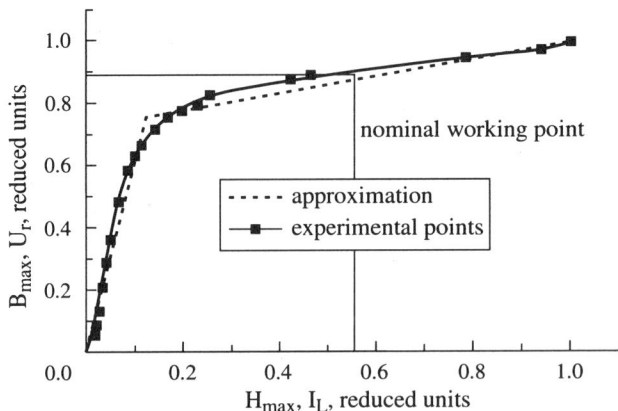

Figure 16.45 - The previous curve in reduced units: $B_{max}(H_{max})$, and $U_r(I_L)$

The magnetic circuit of the saturable inductance is made according to the sketch of figure 16.46.

We focus on the operation under nominal conditions ($U_{rn} = 220$ V, $I_{Ln} = 0.345$ A). The cross-section of the differents branches is such that the magnetic material works everywhere with the same maximum induction B_{maxn}.

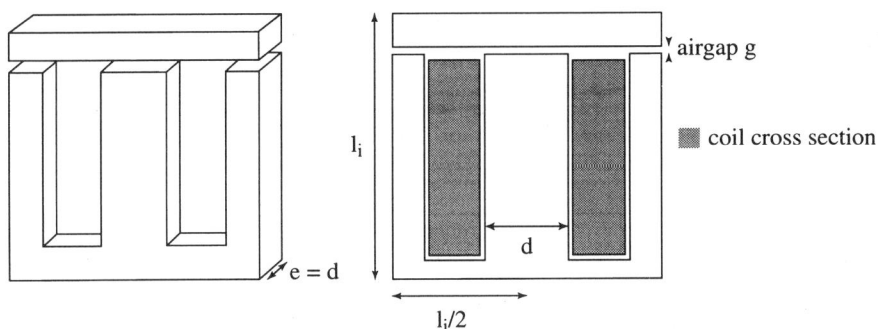

Figure 16.46 - Magnetic circuit of the saturable inductance

By using Faraday's and Ampere's laws, set up the equation system:

$$U_{rn}\sqrt{2} = n S_i \omega B_{max\,n} \qquad\qquad S_f = d\,e = d^2 \qquad\qquad (1)$$

$$B_{max\,n} \frac{3\,l_i}{\mu_v}(1+\lambda) = n I_{Ln}\sqrt{2} \qquad \lambda = \frac{2\,g\,\mu_n}{3\,l_i\,\mu_0} \qquad \mu_n = B_{max\,n}/H_{max\,n} \qquad (2)$$

Experimental analysis shows that at 50 Hz the iron losses can be expressed by:

$$B_{max} < 1.2\ \text{T} \qquad\qquad P_1 = 3.40\ B_{max}^{1.76}\ \text{W.kg}^{-1}$$
$$1.2\ \text{T} < B_{max} < 1.7\ \text{T} \qquad\qquad P_2 = 2.43\ B_{max}^{3.68}\ \text{W.kg}^{-1}$$

Knowing the specific weight of the FeSi alloy ($m_v = 7,800$ kg.m^{-3}) deduce, in nominal operation, the total power lost by the magnetic core as a function of S_i and l_i.

The coil is made with a cylindrical wire with a filling ratio k = 0.6. Express the resistance R of the coil as a function of k, the coil cross-section S_w, the average length of a turn l_m, the resistivity of copper ρ, and n.

Using (1), and writting $l_m = \pi\, l_i / 2$, show that in nominal operation the total power lost is expressed as:

$$P_t = l_i\, (a\, S_i + (b/S_w)/S_i^2)$$

Give the analytical expression of the coefficients a and b.

Deduce a condition on S_i so that the device works in the most economical conditions, and give the analytical expression for the total losses in these conditions.

Numerical evaluation

The resistivity of copper under the (warm) operating conditions is :
$\rho = 2.2 \times 10^{-8}\ \Omega.m$. Take $l_i = 5$ cm.

Determine, for the three cases $\delta = 0.1$ mm, $\delta = 0.2$ mm, and $\delta = 0.3$ mm, the quantities: S_i, S_w, n, s_w (copper wire cross-section), P_t.

Knowing that a copper wire in a non cooled stacking withstands a current density of 5 A.mm^{-2}, and that the winding must fit into the allocated volume, determine the solution which seems the best.

SOLUTIONS TO THE EXERCICES

S.1.1. $\overline{U_r} = jL\omega\overline{I_L} = \dfrac{\overline{I_C}}{jC\omega}$ and $\overline{I_L} + \overline{I_C} = 0$ gives $LC\omega^2 = 1$

$$L_n\omega = \frac{U_{rn}}{I_{Ln}} = 220\Big/\left[0.075 + \frac{0.615 - 0.075}{250 - 190}(220 - 190)\right] = 638\,\Omega$$

$L_n = 2.03$ H $C = 5$ μF $U_2 = 220$ V

S.1.2. $\Delta I_L = \dfrac{0.615 - 0.075}{250 - 190}\,\Delta U_r = 9 \times 10^{-3}\,\Delta U_r$

$\Delta I_C = C\omega\,\Delta U_r$

$\overline{I_I} = \overline{I_t} = \overline{\Delta I_L} + \overline{\Delta I_C}$ has, as well as φ, a phase shift of $-\pi/4$ with respect to U_r

$\overline{E_1} = -jL_{tr}\omega\overline{I_I} = -jn_1\omega\overline{\phi}$

$\overline{E_2} = -jn_2\omega\overline{\phi} = \overline{E_1}$

$\overline{U_2} = \overline{U_r} + \overline{E_2} = \overline{U_{rn}}$ whence

$L_{tr}\,\omega\,(9 \times 10^{-3} - C\omega) = 1;\ L_{tr} = 0.428$ H.

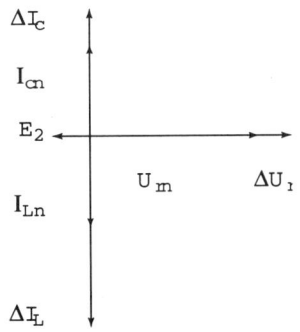

S.2. The nominal working point of the magnetic circuit satisfies:

$$H_{maxn} = 2,000 \text{ A.m}^{-1} \; ; \; B_{maxn} = 1.5 \text{ T}$$

The total power lost by the magnetic circuit is:

$$P_i = 2.43 \, B_{maxn}^{3.68} \, m_v \, 3 \, l_i \, S_i$$

On the other hand $R = \dfrac{n^2 \, \rho \, l_m}{k \, S_w}$. Hence:

$$P_t = l_i \, (a \, S_i + (b \, / \, S_w) \, / \, S_i^2) \qquad a = 7.29 \, B_{maxn}^{3.68} \, m_v = 2.53 \times 10^5 \text{ SI}$$

$$b = \frac{\pi \, \rho \, U_{rn}^2 \, I_{Ln}^2}{k \, \omega^2 \, B_{max \, n}^2} = 2.99 \times 10^{-9} \text{ SI}$$

The minimum loss will be obtained for $S_i = \left(\dfrac{2b}{aS_w}\right)^{1/3} = 2.87 \times 10^{-5} \, S_w^{-1/3}$

Under these conditions we will have:

$$P_t = l_i \left(2^{1/3} + 2^{2/3}\right) \frac{a^{2/3} b^{1/3}}{S_w^{1/3}} = 16.4 \, l_i \, S_w^{-1/3}$$

Numerical values: $l_i = 5$ cm

g (mm)	S_i (cm^2)	S_w (cm^2)	n	s_w (mm^2)	P_t (W)
0.1	5.98	1.1	1,104	0.06	11.4
0.2	4.14	3.33	1,595	0.126	7.86
0.3	3.17	7.42	2,083	0.214	6.01

We see that too small an air gap gives rise to an iron volume too large with respect to the volume of the coil which, because the wire also has too small a cross-section, carries too large a current density.

Too large an air gap leads to too bulky a winding.

With the 0.2 mm air gap we obtain an available coil cross-section equal to $(l_i - d) \, (l_i \, / \, 2 - 3d \, / \, 4) = 2.89$ cm^2 which, after some minor adjustments, is suitable for the calculated coil.

REFERENCES

[1]   H.R. HILZINGER, *J. Magn. Magn. Mater.* **83** (1990) 370.

[2]   G. BERTOTTI, *J. Magn. Magn. Mater.* **54-57** (1986) 1556.

[3]   J.E.L. BISHOP, *J. Magn. Magn. Mater.* **49** (1985) 241.

[4]   R.H. PRY, C.P. BEAN, *J. Appl. Phys.* **29** (1958) 532.

[5]   G. BERTOTTI, V. BASSO, M. PASQUALE, *IEEE Trans. Magn.* **Mag-30** (1994) 1052.

[6]   J.C. BAVAY, J. VERDUN, Alliages Fer-Silicium, *Techniques de l'Ingénieur* **D2110-2111** (1992).

[7]   K. SATO, A. HONDA, K. NAKANO, M. ISHIDA, B. FUKUDA, T. KAN, *J. Appl. Phys.* **73** (1993) 6609.

[8]   M. ISHIDA, K.SENDA, K. SATO, M. KOMATSUBARA, *J. Studies in Applied Electromagnetics* **4** (1996).

[9]   N. TSUYA, K.I. ARAI, K. OHMORI, H. SHIMANAKA, *IEEE Trans. Magn.* **Mag-16** (1980) 728.

[10]   E. FERRARA, F. FIORILLO, M. PASQUALE, A. STANTERO, M. BARICCO, J. DEGAUQUE, M. FAGOT, B. VIALA, E. DU TRéMOLET DE LACHEISSERIE, J.L. PORTESEIL, *J. Magn. Magn. Mater.* **133** (1994) 366.

[11]   S. CROTTIER-COMBE, S. AUDISIO, J. DEGAUQUE, C. BERAUD, F. FIORILLO, M. BARICCO, J.L. PORTESEIL, *J. Phys. IV* **C5** (1995) 1045.

[12]   G. COUDERCHON, J.F. TIERS, *J. Magn. Magn. Mater.* **26** (1982) 196.

[13]   J. SMIT, H.P.J. WIJN, Ferrites (1961) Bibliothèque Technique Philips, Dunod, Paris.

[14]   P. BEUZELIN, Ferrites doux, *Techniques de l'Ingénieur* **E1760** (1987).

[15]   J.F. RIALLAND, *Techniques de l'Ingénieur* **D2400** (1989).

[16]   P. DUWEZ, *Trans. Am. Soc. Metals* **60** (1967) 607.

[17]   Y. YOSHIZAWA, S. OGUMA, K. YAMAUCHI, *J. Appl. Phys.* **64** (1988) 6044.

[18]   Y. YOSHIZAWA, K. YAMAUCHI, *Mater. Sci. Eng.* **A133** (1991) 176.

[19]   U. KöSTER, U. SCHüNEMANN, M. BLANK-BEWERSDORFF, S. BRAUER, M. SUTTON, G.B. STEPHENSON, *Mater. Sci. Eng.* **A133** (1991) 611.

[20]   G. HERZER, *IEEE Trans. Magn.* **Mag-25** (1989) 3327.

[21]   R. SCHäFER, A HUBERT, G. HERZER, *J. Appl. Phys.* **69** (1991) 5325.

[22]   G. BORDIN, G. BUTTINO, A. CECCHETTI, M. POPPI, *J. Magn. Magn. Mater.* **150** (1995) 363.

[23]   R. HARRIS, M. PLISCHKE, M.J. ZUCKERMANN, *Phys. Rev. Lett.* **31** (1973) 160.

[24]   R. ALBEN, J.J. BECKER, M.C. CHI, *J. Appl. Phys.* **49** (1978) 1653.

[25]   G. SéGUIER, F. NOTELET, Electrotechnique industrielle (1985) Technique et documentation, Lavoisier, Paris.

[26]   G. LE ROY, J. DEBERLES, *J. Magn. Magn. Mater.* **26** (1982) 83.

[27]   A.J. MOSES, M. YASIN, M. SOINSKI, *J. Magn. Magn. Mater.* **133** (1994) 637.

[28]   T. BARRADI, Thesis (1990) Ecole Supérieure d'Electricité, Gif sur Yvette.

[29]   J.P. ARTHAUD, P. MATUSZEWSKI, L. LATIL, *Revue Générale d'Electricité* **11** (1992) 52.

[30]   P. BRISSONNEAU, Magnétisme et matériaux magnétiques pour l'électrotechnique (1997)
        Hermès, Paris.

[31]   A. NAFALSKI, D.C. FROST, *J. Magn. Magn. Mater.* **133** (1994) 617.

[32]   M. KOMATSUBARA, Y. HAYAKAWA, T. TAKAMIYA, M. MURAKI, C. MAEDA,
        M. ISHIDA, M. MORITO, *J. Phys. IV* **8** (1998) 467.

[33]   J.F. HEUILLARD, *Techniques de l'Ingénieur* **D3 II** (1982) 495.

[34]   J.L. SEICHEPINE, *Revue Générale d'Electricité* **11** (1992) 19.

[35]   G. RUELLE, *Techniques de l'Ingénieur* **D3 II** (1993) 530.

[36]   S. ALLANO, *Techniques de l'Ingénieur* **D3 III** (1995) 720.

[37]   M. JUFER, Electromécanique (1995) Presses polytechniques et universitaires romandes,
        Lausanne.

[38]   Encyclopédie des sciences industrielles; Electricité, généralités et applications (1973) Quillet,
        Paris.

[39]   J.P. FERRIEUX, F. FOREST, Alimentations à découpage, convertisseurs à résonance (1994)
        Masson, Paris.

[40]   D. CHIARANDINI, J.M. FESTE, *Electronique* **15** (1992) 55.

[41]   J.P. FESTE, *Electronique* **12** (1991) 144.

[42]   C. GLAIZE, *Techniques de l'Ingénieur* **D3 I** (1989) 120.

[43]   G. COUDERCHON, *Techniques de l'Ingénieur* D2 I (1994)130.

[44]   G.M. ETTINGER, Magnetic Amplifiers (1957) Methuen & Co, London.

[45]   J.P. SIX, P. VANDEPLANQUE, Exercices et problèmes d'électrotechnique industrielle (1995)
        Technique et documentation, Lavoisier, Paris.

CHAPTER 17

SOFT MATERIALS
FOR HIGH FREQUENCY ELECTRONICS

Electronics is different from power electrical engineering in two major respects. The range of frequencies involved is typically restricted to 1 kHz in power electrical engineering, whereas it extends to the microwave regime, beyond 1 GHz, in electronics. The excitation level of magnetic materials is near saturation in power electrical engineering, leading to very large non-linear effects, while it remains usually much lower in electronics, where the linear regime prevails since the signals have low amplitude. While power electrical engineering mainly uses ferromagnetic alloy sheets, good conductors of electricity, the favorite materials of electronics are ferrimagnetic oxides, with insulator or semiconductor electrical character.

We first introduce complex susceptibility and permeability, then describe the details of the physical mechanisms that effectively govern the frequency variation of the magnetization processes. We then consider the case of materials saturated by a static magnetic field, which are therefore in a single domain state where only rotation mechanisms come into play: this brings in the phenomena of gyrotropy and resonance. We finish with a presentation of the materials and of some applications:

♦ *In the radiofrequency range, the usual materials are spinel ferrites, in particular manganese zinc (Mn-Zn) and nickel zinc (Ni-Zn) ferrites. Their main applications are in inductive linear components. We will see two examples.*

♦ *In the microwave range, ferrimagnetic garnets, in particular yttrium iron garnet (YIG), are used, along with the spinels. We will describe non-reciprocal devices, resonators and microwave absorbers.*

1. COMPLEX SUSCEPTIBILITY AND PERMEABILITY

We describe the response of a magnetic material to a weak magnetic field which varies rapidly in time. In chapters 2 and 3, we already examined the response to a weak DC or slowly variable field: we were then led to the simple notions of static susceptibility and permeability.

If the field **h**, while still weak, is rapidly variable, we accept that the response of the material, viz its induced magnetization **m**, cannot follow the excitation instantly. This is the dynamic regime, and it is clear that now a new description is necessary.

1.1. ISOTROPIC COMPLEX SUSCEPTIBLITY AND PERMEABILITY

Let us first consider the simple case where magnetization is colinear to the field that creates it. A convenient description of the dynamic response of such a material to a weak excitation field can be based on a general property of linear systems. If, in such a system, the excitation is harmonic, i.e. sinusoidal, with angular frequency ω, then the response is also sinusoidal, with the same angular frequency. In the present case, if $h = a \cos \omega t$, then $m = b \cos(\omega t + \varphi)$, with the amplitude ratio b/a and the phase shift φ depending only on ω for a given material under given conditions (in particular temperature).

We note that h(t) and m(t) can each be considered as the real part of a complex number, respectively $a \exp j\omega t$ and $b \exp[j(\omega t + \varphi)]$ which can also be expressed as $b \exp(j\varphi) \exp j\omega t$. The terms multiplying $\exp j\omega t$ in these expressions are called the *complex amplitudes*, and they are denoted respectively as $\overline{h}$ and $\overline{m}$ (however this notation will be dropped further on, whenever the context clearly indicates that we deal with a complex amplitude). Here we thus have $\overline{h} = a$ and $\overline{m} = b \exp(j\varphi)$.

The complex susceptibility χ is now defined as the ratio:

$$\chi = \overline{m}/\overline{h} = (b/a) \exp(j\varphi) \tag{17.1}$$

χ is a function of ω through its modulus b/a and its argument φ, which characterise the material alone, and we can also write:

$$\chi = \chi' - j\chi'' \tag{17.2}$$

where χ' and χ'' are of course functions of ω. This notation for χ is in fact the most frequently used one. We will understand later on why the imaginary part of χ is denoted as $-\chi''$.

The definition of the complex relative permeability follows immediately from that of the susceptibility (see eq. 2.52 in chap. 2):

$$\mu = 1 + \chi = \mu' - j\mu'' \quad ; \quad \mu' = 1 + \chi' \quad ; \quad \mu'' = \chi'' \tag{17.3}$$

Note that, throughout this chapter, μ is used insteadof μ_r.

1.1.1. Physical meaning of χ', χ'', μ' and μ''

Through the definition of complex susceptiblity, the instantaneous magnetization m(t) induced by a field $a \cos \omega t$ is the real part of the product $(\chi' - j\chi'') a [\cos \omega t + j\sin \omega t]$, hence:

$$m(t) = a [\chi' \cos \omega t + \chi'' \sin \omega t] \tag{17.4}$$

Thus χ' gives the component of m that is in phase with the excitation field, while χ'' gives the component lagging by 90° in phase.

The instantaneous power provided by the field to unit volume of the material is then, from chapter 2:

$$P(t) = \mu_0 h(t) \, dm/dt = \mu_0 \omega a^2 \cos \omega t \, (-\chi' \sin \omega t + \chi'' \cos \omega t) \quad (17.5)$$

and its average value
$$<P> = (1/2) \mu_0 \omega \chi'' a^2 \quad (17.6)$$

This average value is thus non zero : it is positive, since it necessarily corresponds to dissipation inside the material. The parameter χ'' is thus itself positive, which was the purpose of the notation in (17.2), and it characterises the *magnetic losses*. All these considerations extend of course to induction and to permeability due to relations (17.3).

We will see that, from a practical point of view, the ratios below play a specially important role :

$$\tan \delta_m = \mu''/\mu' \qquad 1/\tan \delta_m = Q_m$$
$$M_e = \mu' \, Q_m = \mu'/\tan \delta_m = \mu'^2/\mu'' \quad (17.7)$$

δ_m is the *magnetic loss angle*, Q_m the magnetic quality factor of the material and M_e its *figure of merit*.

1.1.2. General dynamic regime

Knowing the functions $\chi'(\omega)$ and $\chi''(\omega)$ or $\mu'(\omega)$ and $\mu''(\omega)$, we can predict the response to any time-dependent excitation field, provided we remain in the linear range. Consider a field h(t) for which we only assume that $h(t) = 0$ for $t < 0$. Under some practically very unrestrictive conditions, h(t) can be written as a Fourier integral [1]:

$$h(t) = \frac{2}{\pi} \int_0^\infty h'(\omega) \cos \omega t \, d\omega \quad (17.8)$$

where $h'(\omega)$ is given by the inverse transform:

$$h'(\omega) = \frac{2}{\pi} \int_0^\infty h(t) \cos \omega t \, dt \quad (17.9)$$

Thus the exciting field appears as an (infinite) sum of elementary harmonic excitations. Each of these elementary excitations produces, from what we saw above, an elementary response :

$$dm(\omega) = h'(\omega) \, d\omega \, [\chi'(\omega) \cos \omega t + \chi''(\omega) \sin \omega t] \quad (17.10)$$

and, from the superposition principle which by definition all linear systems obey, the material's response can finally be described as :

$$m(t) = \int_0^\infty h'(\omega) \, d\omega \left[\chi'(\omega) \cos \omega t + \chi''(\omega) \sin \omega t\right] \quad (17.11)$$

This complex susceptibility formalism is thus indeed a complete description of the dynamic properties of a magnetic material.

1.1.3. Kramers-Kronig relations

The mere assumption of linearity already led us to the huge simplification provided by the description of all the dynamic properties through a single function, albeit complex, of ω, $\chi(\omega)$ or $\mu(\omega)$. Another assumption, which is still more general because it is the causality principle (the cause comes before the effect) shows that, in $\chi(\omega)$ and in $\mu(\omega)$ the real and imaginary parts are not independent functions. If either of them is known between $\omega = 0$ and infinite ω, then the other one is perfectly determined. After Kramers and Kronig [2], we have:

$$\chi'(\omega) = \frac{2}{\pi} \int_0^\infty \frac{\chi''(\overline{\omega})}{\overline{\omega}^2 - \omega^2} \, \overline{\omega} \, d\overline{\omega} \qquad (17.12)$$

$$\chi''(\omega) = \frac{-2}{\pi} \int_0^\infty \frac{\chi'(\overline{\omega})}{\overline{\omega}^2 - \omega^2} \, \omega \, d\overline{\omega} \qquad (17.13)$$

The immediate consequence of these relations is that it is really sufficient to know a *single* real function, for example $\chi'(\omega)$ or $\chi''(\omega)$, to have a complete description of the dynamic properties of a magnetic material. However, this function must be known over the *whole* spectrum, and this somewhat reduces the practical effect of the Kramers-Kronig relations.

1.2. ANISOTROPIC MATERIALS. COMPLEX SUSCEPTIBILITY AND PERMEABILITY TENSORS

In the general case, the induced magnetization **m** is no more colinear to the field **h** which produces it. It can be shown that, in a rectangular coordinate system (1, 2, 3), the complex amplitude components of magnetization $\overline{m}_1$, $\overline{m}_2$, $\overline{m}_3$ are related to the complex amplitude components of the field $\overline{h}_1$, $\overline{h}_2$, $\overline{h}_3$, through the matrix relation:

$$\begin{pmatrix} \overline{m}_1 \\ \overline{m}_2 \\ \overline{m}_3 \end{pmatrix} = \begin{pmatrix} \chi_{11} & \chi_{12} & \chi_{13} \\ \chi_{21} & \chi_{22} & \chi_{23} \\ \chi_{31} & \chi_{32} & \chi_{33} \end{pmatrix} \begin{pmatrix} \overline{h}_1 \\ \overline{h}_2 \\ \overline{h}_3 \end{pmatrix} \qquad (17.14)$$

Matrix χ_{ij}, all terms of which are normally complex functions of ω, expresses the relation between **m** and **h** in an arbitrarily chosen coordinate system.

This is the representation, in a particular system, of a much more general linear relation, which is independent of the coordinate system in which it is described. This is referred to as a tensor relation [3]. It is written in the symbolic form

$$\mathbf{m} = \chi \mathbf{h} \qquad (17.15)$$

where χ is the complex susceptibility tensor.

There is no difficulty in defining the complex permeability tensor μ by writing:

$$\mu = 1 + \chi \qquad (17.16)$$

This implies the convention that **1** is the tensor that transforms any vector into itself. We recall there is a one-to-one correspondence between the operations defined on tensors and on their representative matrices. For example, the matrix representing the product of two tensors is the product of the representative matrices.

The complex susceptibility tensor has some specific properties. An obvious property is that each component of the representative matrix, whatever the axis system chosen, satisfies the Kramers-Kronig relations. It is clear also that the representative matrix can take on a simple form if the material features remarkable symmetries, and if the set of axes is adequately chosen.

Finally, as was pointed out in chapter 2, it can be shown that, in the static case ($\omega = 0$), the matrix is necessarily symmetric with repect to its first diagonal. Therefore there is a special set of axes, called the principal axis system, in which the representative matrix is diagonal. The existence of principal axes, at finite frequency too, can be derived for a material without remanent magnetization and not submitted to a static magnetic field. We will have the opportunity of checking these properties on the basis of concrete physical models later on.

2. MEASURING THE COMPLEX SUSCEPTIBILITY AND PERMEABILITY

The preferred method for measuring χ or μ depends on the isotropic or anisotropic character of the material, on the geometry of available or possible samples and of course on frequency. Before embarking on the actual description of the methods, it is worthwhile introducing the simple, and useful notion of external susceptibility and permeability, and the so-called reciprocity theorem.

2.1. INTERNAL AND EXTERNAL SUSCEPTIBILITIES

The susceptibility we discussed till now can be qualified as internal, because it relates the magnetization **m** to the field **h** which actually prevails within the material. However, many measuring techniques and many practical devices involve a given distribution of the external field applied to the sample. Let us consider an ellipsoid-shaped specimen, with axes Ox, Oy, Oz, submitted to the uniform field $\mathbf{h}_0$ produced by external sources. From elementary magnetostatics as discussed in chapter 2, it can be shown that the induced magnetization **m** and the internal field **h** are uniform too, and verify the relation:

$$\mathbf{h} = \mathbf{h}_0 - \mathbf{N}\mathbf{m} \qquad (17.17)$$

where **N** is the demagnetising field tensor.

This tensor is represented in the system Oxyz by a diagonal matrix. Left-multiplying both sides of relation (17.17) by the susceptibility tensor χ, we obtain:

$$\mathbf{m} = \chi\,\mathbf{h}_0 - \chi\,\mathbf{N}\,\mathbf{m} \tag{17.18}$$

hence:

$$\mathbf{m} = (1+ \chi\,\mathbf{N})^{-1}\chi\,\mathbf{h}_0 \tag{17.19}$$

This relation defines the *external susceptibility* χ_e:

$$\chi_e = (1+ \chi\,\mathbf{N})^{-1}\chi \tag{17.20}$$

χ_e is not an intrinsic parameter. It characterises a sample of the material with a well-defined shape.

2.1.1. External susceptibility in a simple case

Consider the frequent special case where the principal axes of internal susceptibility and of the demagnetising field tensor coincide :

$$[\chi] = \begin{pmatrix} \chi_1 & 0 & 0 \\ 0 & \chi_2 & 0 \\ 0 & 0 & \chi_3 \end{pmatrix} \text{ and } [N] = \begin{pmatrix} N_1 & 0 & 0 \\ 0 & N_2 & 0 \\ 0 & 0 & N_3 \end{pmatrix}.$$

Relation (17.20) takes the same form if it is expressed in terms of the representative matrices. We then find:

$$([1] + [\chi][N])^{-1}[\chi] = \begin{pmatrix} \chi_1/(1+\chi_1 N_1) & 0 & 0 \\ 0 & \chi_2/(1+\chi_2 N_2) & 0 \\ 0 & 0 & \chi_3/(1+\chi_3 N_3) \end{pmatrix}$$

Note that, if the material is isotropic ($\chi_1 = \chi_2 = \chi_3 = \chi$), this is not true for external susceptibility, except for a sphere where $N_1 = N_2 = N_3 = 1/3$.

We note that external susceptibility, measured along a principal direction both of the material and of the ellipsoidal sample, is given by the scalar relation:

$$\chi_e = \chi/(1 + \chi\,N) \tag{17.21}$$

where χ is the (internal!) susceptibility of the material along the same principal direction.

2.1.2. Effect of the demagnetising field on the drifts and losses

In formula (17.21) the internal and external susceptibilities can be simply replaced by the corresponding permeabilities, provided the moduli of χ and χ_e are much larger than unity, which is generally the case. We then have:

$$\frac{1}{\mu_e} = \frac{1}{\mu} + N \tag{17.22}$$

Now consider a small variation $\Delta\mu$ of μ around a nominal value (a variation that could, e.g. be due to a change in temperature).

We then have:

$$\frac{\Delta\mu_e}{\mu_e^2} = \frac{\Delta\mu}{\mu^2} \qquad (17.23)$$

This relation shows that the relative variations $\Delta\mu_e/\mu_e$ for μ_e are smaller than those of μ by a factor μ_e/μ. In other words, the demagnetising field acts to increase the *stability* of permeability, at the expense of its absolute value. This is a classical feedback effect, as electronics engineers know it. It explains why ferrite manufacturers always indicate the drift coefficients in the form $(1/\mu^2)\,d\mu/dP$ and not $(1/\mu)\,d\mu/dP$, where P is the parameter on which μ depends, e.g. temperature.

If $\mu = \mu' - j\mu''$ with $\mu'' \ll \mu'$, the above relation yields directly

$$\frac{\mu''_e}{\mu'^2_e} = \frac{\mu''}{\mu'^2} \qquad (17.24)$$

which shows that the product $\mu'_e Q_e$ remains equal to $\mu' Q = M_e$ (eq. 17.7). The figure of merit is thus conserved when a demagnetising effect is introduced, and indeed this is what makes this parameter M_e interesting.

2.1.3. Skin effect and dimensional resonance

All the above magnetostatic considerations on the effect of the demagnetising field rest on the assumption that electromagnetic propagation effects are negligible, at least on the scale of the sample. While this condition is easily fulfilled for insulators with low dielectric constant, the situation is different for metals or insulators with very high dielectric constant.

For metals, we saw in chapter 13 that electromagnetic effects appear at very low frequencies for samples in the usual size range. These effects in particular restrict the penetration of magnetic field into the bulk of the sample (skin effect). Here again, the measured permeability is the external one (some authors call it the effective permeability), but the magnetostatic formula (17.20) is of course no more valid. We reproduce below the formulas giving the skin depth δ in a magnetic material with (internal) permeability μ and electrical resistivity ρ as well as the even more important expression of the external permeability μ_e for a thin plate, with thickness e, of such a material. Table17.1 gives some typical orders of magnitude of δ for conductors. The skin effect phenomenon only becomes important in practice if the skin depth is near, or less than, the thickness e of the plate.

$$\delta = \sqrt{2\rho/\mu\mu_0\,\omega} \qquad \alpha = (1-j)(e/\delta) \qquad \mu_e = \mu\left[\tan\left(\frac{\alpha}{2}\right)\Big/\frac{\alpha}{2}\right] \qquad (17.25)$$

More generally, the magnetostatic approach fails when the sample dimensions are comparable to, or larger than, some *characteristic length*. For conductors, as we just saw, the characteristic length is the skin depth δ for the frequency f used.

***Table 17.1 - Typical values of the skin depth or of the half-wavelength
in various materials***

Material	Frequency		1 Hz	50 Hz	1 MHz	1 GHz
Copper	$\mu = 1$	$\rho = 2 \times 10^{-8}\ \Omega.m$	6.6 cm	0.93 cm	66 µm	2.1 µm
Soft iron	$\mu = 1,000$	$\rho = 10^{-7}\ \Omega.m$	0.5 cm	1.5 mm	5 µm	–
Mn-Zn Ferrite	$\mu = 5,000$	$\rho = 10^2\ \Omega.m$	50 m	–	5 cm	–
Ni-Zn Ferrite	$\mu = 1,000$	$\varepsilon = 15$	37,000 km	–	37 m	–

For an insulator, the characteristic length is the *wavelength* λ at this frequency f:

$$\lambda = c / \left(f \sqrt{\varepsilon \mu} \right) \tag{17.26}$$

where c is the velocity of light in vacuo, ε the dielectric constant of the material, and μ its relative magnetic permeability. Typically, as $\lambda/2$ approaches a sub-multiple of one of the dimensions d of the sample, a *resonance* in the external permeability, called a dimensional resonance, is observed [4]. Table 17.1 gives typical values of $\lambda/2$ for an insulating ferrite (Ni-Zn). Here again, we note that, to avoid these perturbations, and remain in the range of applicability of magnetostatic formulas, the condition $d \ll \lambda/2$ will have to be valid.

2.2. RECIPROCITY THEOREM

Consider a coil carrying current I. It creates at any point P in space a field $\mathbf{H_0}$ which can be expressed as:

$$\mathbf{H_0} = \mathbf{C_H} I \tag{17.27}$$

where $\mathbf{C_H}$ is the vector field coefficient at P (see chap. 2). If a point dipole, with magnetic moment $\mathfrak{m}$ (not to be mixed up with magnetization $\mathbf{m}$), is placed at point P, the reciprocity theorem indicates that the flux Φ produced by the dipole across the coil is:

$$\Phi = \mu_0 \mathbf{C_H} \mathfrak{m} \tag{17.28}$$

This theorem was derived in chapter 2, and is most familiar in the community dealing with magnetic recording [5]. We now show that it is also very useful in measurement problems.

2.3. MEASURING BY PERTURBATION OF A COIL

This technique can be applied to a sample of magnetic material that can be approximated by a small ellipsoid. We assume it to be an insulator. Let L be the inductance of the coil in the absence of the sample. Let us show that introducing the sample leads to a variation ΔL in the inductance. Let I be the complex amplitude of the current in the coil. The field at the sample position is $\mathbf{C_H} I$, and it therefore induces in the sample the magnetization $\mathbf{m} = \chi\ \mathbf{C_H} I$.

From the reciprocity theorem, if we denote by V the sample volume, the flux $\Delta\Phi$ across the coil from the sample is then:

$$\Delta\Phi = \mu_0\, \mathbf{C}_H\, \mathfrak{m} = \mu_0\, \mathbf{C}_H\, \chi_e\, \mathbf{C}_H\, I\, V \qquad (17.29)$$

The variation ΔL we are interested in is thus:

$$\Delta L = \Delta\Phi/I = \mu_0\, \mathbf{C}_H\, \chi_e\, \mathbf{C}_H V \qquad (17.30)$$

If we define a reference system Oxyz such that Ox is aligned with $\mathbf{C}_H$, the above relation gives:

$$\Delta L = \mu_0\, \mathbf{C}_H^{\,2}\, \chi_{11}^{\,2}\, V \qquad (17.31)$$

The measurement is sensitive to only one component of the external susceptibility matrix.

In practice, the most prevalent situation remains that of an isotropic material with (internal) susceptibility χ. Using equation (17.21), we then have

$$\Delta L = \mu_0\, \mathbf{C}_H^{\,2}\, V\, \chi/(1 + N\,\chi) \qquad (17.32)$$

A valuable regime is $N\chi \ll 1$. Then:

$$\chi = \Delta L/\mu_0\, \mathbf{C}_H^{\,2}\, V \qquad (17.33)$$

To fulfill the condition $N\chi \ll 1$, the sample should be very elongated along the measuring direction, hence it should be a rod, or a thin plate.

The analysis above was based on the assumption of an almost point-like specimen. However, it can obviously be extended to the case of samples with non-negligible size, insofar as the field coefficient $\mathbf{C}_H$ of the measuring coil can be made homogeneous over large enough a volume. This can be obtained with long solenoids, with n' windings per metre. From chapter 2, $\mathbf{C}_H$ is along the solenoid axis, and its modulus $\mathbf{C}_H$ is n'.

Formula (17.33) shows that the measurement of susceptibility and of permeability is turned into the measurement of a variation in self inductance. Several methods are applicable. The most usual one nowadays involves a digital impedance meter, and actually is based on the measurement of a change in impedance ΔZ. Thus formula (17.33) is recast into:

$$\chi = \Delta Z/j\omega\,\mu_0\, \mathbf{C}_H^{\,2}\, V \qquad (17.34)$$

2.4. TWO COIL MEASUREMENT

This technique can be considered as a variant of the above method, insofar as it consists in perturbing through the sample to be measured not the inductance of a single coil, but the mutual induction coefficient between two coils.

The reciprocity theorem shows that:

$$\Delta M = \mu_0\, \mathbf{C}_{H1}\, \chi_e\, \mathbf{C}_{H2}\, V \qquad (17.35)$$

ΔM is the variation in the mutual induction coefficient observed when the sample, with external tensor susceptibility χ_e and volume V, is placed at point P where the vector field coefficients of the two coils are respectively C_{H1} and C_{H2}.

A setup that is very much used in practice includes two coils. The first coil, B_1, called the field coil, carries an exciting current I_1. The second coil, B_2, actually consists of two oppositely wound windings in series, so that the mutual induction coefficient with no sample, M, between B_1 and B_2, cancels, as shown on figure 17.1.

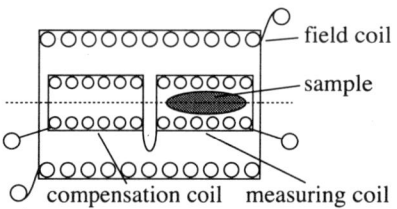

Figure 17.1 - Two coil measurement, with the measuring winding and the compensation winding in the secondary

The voltage measured across B_2 in the absence of sample is of course zero. The sample is placed in *one* of the windings that make up the measuring coil B_2. The voltage V_2, proportional to ΔM, hence to one component of susceptibility (or a linear combination of components), is then measured, as indicated by relation (17.35). It is worth noting in particular that this method can provide the non-diagonal components of the susceptibility matrix.

2.5. *FREQUENCY LIMITATIONS IN COIL METHODS*

The two methods described above are restricted in frequency by the phenomenon of coil resonance. Any coil has a distributed parasitic capacitance, which turns it into a true parallel oscillator circuit, characterised by some resonance frequency f_B. The "magnetostatic" formulas above are only valid for frequencies f much smaller than f_B. Crudely put, the resonance frequency grows with the reciprocal of the length of wound wire, hence as the reciprocal of the number of windings for a given coil size.

In both above methods, a relevant measurement of *sensitivity* is a ratio like $\omega \, \Delta L / \Delta \chi$ or $\omega \, \Delta M / \Delta \chi$ which are increasing functions of frequency and of the number of windings. In the search for maximum sensitivity while satisfying the constraint $f \ll f_B$, one is thus led to adapting the windings to the measurement frequency: the rule is to use many windings for low frequencies, few windings for high frequencies, down to the limiting case of the *single winding* coil.

2.6. *MEASURING TOROIDAL SAMPLES*

This is actually the most classical technique, but it only applies to isotropic materials (at least in the plane perpendicular to the torus axis). It can be considered as an extreme variant of the coil perturbation method where, on the one hand the sample occupies the whole volume that is subject to the excitation field, and, on the other hand, the demagnetising field coefficient is rigorously zero.

The sample is thus a torus, preferably with rectangular cross-section, with internal diameter d_1, external diameter d_2 and height h along its axis. This geometry can be used over a very large range of frequencies. For radio-frequency measurements (below 100 MHz), n turns of isolated copper are distributed very regularly over the perimeter of the torus. It is desirable to fulfill the conditions:

- $d_2 - d_1 \ll d_2 + d_1$, so that the internal and external perimeters are practically equal to the mean perimeter $l = \pi(d_1 + d_2)/2$,

- and $l/n \ll Min\{h, (d_2-d_1)/2\}$, expressing the fact that the distance between two neighbouring turns remains small with respect to the smaller dimension of the section of the toroid. This condition is most important in the case of samples with small permeability.

In this situation, it can be shown that the complex impedance of the coil is given by:

$$Z = R + j X = R_b + R_m + j \omega L$$
$$R_m = \omega \mu'' \mu_0 n^2 S/l \qquad (17.36)$$
$$L = \mu' \mu_0 n^2 S/l$$

where $S = (d_2 - d_1)h/2$ is the torus section and R_b the ohmic resistance of the coil. We see that μ' can be simply calculated from the measured inductance L, while determining μ'' from the measured resistance R implies knowing R_b. In practice, the problem arises only in the case of very weak losses, i.e. when $\tan \delta \ll 1$. One solution is to make the winding with several strands of wire, which leads to nearly frequency-independent resistance. R_b is then determined independently through a simple DC measurement.

Using a toroidal sample, one can also use a two-winding method. In this case, the complex permeability is obtained by measuring a transfer impedance rather than a conventional impedance, or a mutual inductance rather than an inductance. Here again, the method can be considered as an extreme variant of two coil perturbation, in which the sample occupies the whole volume that is submitted to the excitation field.

The transfer impedance Z_{21} is defined by $V_2 = Z_{21} I_1$, where V_2 is the voltage across the secondary (with n_2 turns) when the primary (including n_1 turns) carries current I_1.

We see that this method in fact consists in imposing a field (via the current in the primary winding) and in measuring the induced flux (via the voltage across the secondary).

This remark is important, for it indicates that the secondary does not have to be regularly spread over the perimeter of the torus.

It can be shown that:

$$Z_{21} = -j \omega \mu \mu_0 n_1 n_2 S/l = -j \omega M \qquad (17.37)$$

where the mutual induction coefficient between the two windings, $M = \mu \mu_0 n_1 n_2 S/l$ is complex, as is μ. Hence both its modulus and its phase must be measured.

This can be done using a lock-in amplifier or a numerical impedance meter in the *three-connector* configuration (for low frequencies), or a vector network analyser. An important advantage of this two-coil method is that it does not involve the coils' resistance: it is therefore well suited to the measurement of low losses.

At high frequencies, all these methods based on multi-turn windings on tori stumble on the limitations imposed by the coils' resonances. It is then mandatory to switch to line techniques.

Here, the torus with height h fills the available section of a coaxial air line (fig. 17.2).

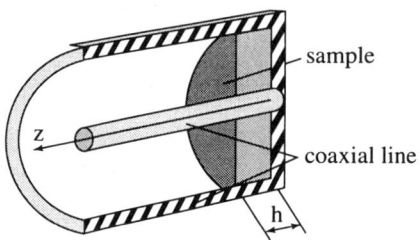

Figure 17.2 - Measurement of permeability using a coaxial line

It is in contact with a shorting plane that terminates the line, and which we take as the origin of z. The impedance Z of the line segment filled with the sample is measured on plane z = h. The only condition that must be fulfilled for evaluation to remain simple is then: h << λ_m, where λ_m is the wavelength in the material at the used frequency.

Z then takes the form:

$$Z = j\mu (2\pi h / \lambda) Z_c \qquad (17.38)$$

where Z_c is the characteristic impedance of the line (usually 50 Ω) and λ the wavelength in vacuo. Both the modulus and the phase of Z, or both its real and imaginary parts, are measured using a vector network analyser. The calculation of μ' and μ" is straightforward through complex relation (17.38).

3. ELEMENTARY SUSCEPTIBILITY MECHANISMS

We now go into the detail of the special mechanisms governing complex susceptibility and permeability in ferri- and ferromagnetic materials.

We already saw in chapter 3 that the evolution of the technical or effective magnetization of a ferro- or ferrimagnet under the influence of an applied field involves two mechanisms: the growth of some domains at the expense of others (wall mechanism), and the rotation of spontaneous magnetization within each domain (rotation mechanism).

We recall that *technical or effective magnetization* is an *average* of local magnetization, or possibly of its variation only, over a *volume much larger than that of the domains or grains* comprising the material. We will return to this definition when we deal with the problem of homogenization, in § 4. In the following, we consider spontaneous magnetization as a vector with rigid (constant) modulus, hence as an object with two degrees of freedom.

We also note that the two mechanisms discussed here are not really distinct. The motion of a wall implies a rotation of magnetization, as discussed in chapter 5. As we will see later, this distinction is valuable nevertheless.

3.1. ROTATION MECHANISM

In the part of chapter 7 devoted to atomic magnetism, we saw that an elementary angular momentum is associated with an elementary magnetic moment, and that they are colinear. Corresponding to the magnetization $\mathbf{M}$ of the material there is thus a density of angular momentum $\mathbf{\pounds}$ which can be written:

$$\mathbf{\pounds} = -\mathbf{M}/\gamma \qquad (17.39)$$

where γ is a constant factor characterising the moment carrier, called the gyromagnetic factor. The existence of the angular momentum gives the magnetization of each elementary volume of magnetic material the properties of an *ideal gyroscope*.

What do we mean by "ideal gyroscope"? A real mechanical gyroscope consists of a solid, with rotation symmetry around an axis, which is given an angular velocity $\boldsymbol{\omega_s}$ around this axis. The angular momentum associated with vector $\boldsymbol{\omega_s}$ will be called spontaneous, and is given by:

$$\mathbf{\pounds_s} = I_s\,\boldsymbol{\omega_s} \qquad (17.40)$$

where I_s is the moment of inertia of the solid with respect to the considered axis.

Assume the center of mass P of the gyroscope is fixed. An ideal gyroscope is the limiting case of a real gyroscope for which the moment of inertia with respect to any axis through P goes to zero, but where simultaneously ω_s tends toward infinity so that the angular momentum $I_s\omega_s$ remains finite and equal to $\mathbf{\pounds_s}$. When the axis of revolution itself rotates around P with finite angular velocity, the angular momentum remains rigorously equal to the spontaneous momentum $\mathbf{\pounds_s}$. We further know that the rotation of a solid around a fixed point P, whether it is an ideal gyroscope or not, obeys the equation:

$$d\mathbf{\pounds_P}/dt = \boldsymbol{\Gamma_P} \qquad (17.41)$$

$\mathbf{\pounds_P}$ is the angular momentum with respect to P, and $\boldsymbol{\Gamma_P}$ the moment with respect to this point of the forces applied to the solid.

In an ideal gyroscope, momentum $\mathbf{\pounds_P}$ reduces to the momentum $\mathbf{\pounds_s}$ with fixed modulus, so that the equation of motion is:

$$d\mathbf{\pounds_s}/dt = \boldsymbol{\Gamma_P} \qquad (17.42)$$

This equation can immediately be adapted to magnetization $\mathbf{M}$ by using (17.39):

$$d\mathbf{M}/dt = -\gamma\,\boldsymbol{\Gamma} \qquad (17.43)$$

where $\mathbf{M}$ is a vector with constant modulus, and $\boldsymbol{\Gamma}$ is now the density of torque acting on the magnetization.

3.1.1. Isotropic rotation (or uniaxial symmetry)

We consider a region, within one domain, where the spontaneous magnetization **M** at rest has a position defined by $\mathbf{M} = \mathbf{M_s}$.

This rest position can be imposed by the anisotropy energy alone, and in this case $\mathbf{M_s}$ is along one of the easy magnetization axes (see chap. 3). Anisotropy energy may also be complemented with the interaction energy between magnetization and a static field **H**. **H** is the field that effectively prevails within the material. It can originate either from an external source, or from a demagnetising effect, or from a combination of both. In the presence of **H**, the Oz axis is no more necessarily an equilibrium position for **M**, except in special cases. We consider precisely such a case: the material's anisotropy is assumed to be uniaxial, with axis Oz, and the possible static field **H** is parallel to Oz.

Then a first contribution to $\mathbf{\Gamma}$ is a restoring torque $\mathbf{\Gamma_r}$ which exists as soon as **M** differs from $\mathbf{M_s}$. Its direction is perpendicular to the $(\mathbf{M_s}, \mathbf{M})$ plane, and its modulus Γ_r is independent on the azimut of this plane. This is the reason for the term "isotropic rotation".

In chapter 3, we saw that the uniaxial anisotropy energy can be expanded to lowest order under the form: $U_A = (1/2)\mu_0 \mathbf{H_A}\mathbf{M_s}\sin^2\theta$, where $\mathbf{H_A}$ is the anisotropy field. This leads to:

$$\mathbf{\Gamma_r} = -dU/d\theta = (H_A\cos\theta + H)\mu_0 M_s\sin\theta \qquad (17.44)$$

If $\mathbf{H_s}$ is the vector parallel to $\mathbf{M_s}$ (fig. 17.3) with modulus $H_s = H_A\cos\theta + H$, the restoring torque can be written as:

$$\mathbf{\Gamma_r} = \mu_0\,\mathbf{M}\times\mathbf{H_s} \qquad (17.45)$$

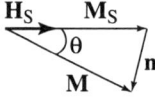

Figure 17.3

In the general case, $\mathbf{H_s}$ is a function of the rotation θ via the contribution of uniaxial anisotropy. However, $\mathbf{H_s}$ becomes independent of θ to first order when θ remains small.

Free motion

In the free motion of magnetization, $\mathbf{\Gamma}$ is by definition reduced to the restoring torque Γ_r, and the equation of motion can be written:

$$d\mathbf{M}/dt = -\gamma\mu_0\mathbf{M}\times\mathbf{H_s} \qquad (17.46)$$

It can be checked that, if the magnetization has been rotated from its rest position, at time zero, by an angle θ, the free ensuing motion is a precession of **M** on a cone with opening angle 2θ at angular velocity:

$$\omega = \gamma\mu_0 H_s = \gamma\mu_0(H_A\cos\theta + H) \qquad (17.47)$$

ω becomes independent of θ only when the latter is small, or when $H_A = 0$. In the latter case, Larmor's formula is retrieved (see chap. 4).

Damping

The above precession continues indefinitely as soon as the magnetization is set free after having been rotated from its rest position. Actually, experiment shows that the magnetization returns to its rest position $\mathbf{M_s}$ after a number of revolutions, and therefore the initial energy is not conserved. There is thus a dissipation effect, and this must be accounted for directly in the equation of motion (17.46) through an extra torque. Several formulations have been proposed. We will only discuss the most intuitive one, due to Gilbert [6]. It is natural, in analogy with viscous damping as used in mechanics, to consider the magnetization vector as a *rigid rod*, with length M_s, whose *end* is braked by a *force* proportional to its speed. This implies writing the friction torque in the form:

$$\mathbf{\Gamma_f} = -(\alpha/\gamma M_s)\,\mathbf{M} \times d\mathbf{M}/dt \tag{17.48}$$

where α is a dimensionless coefficient, called the damping coefficient. This yields the classical form of Gilbert's equation for free motion:

$$d\mathbf{M}/dt = -\gamma\mu_0\,\mathbf{M} \times \mathbf{H_s} + (\alpha/M_s)\,\mathbf{M} \times d\mathbf{M}/dt \tag{17.49}$$

It can be checked that the solution to this equation is indeed a damped precession, in which the end of vector $\mathbf{M}$ returns to the rest position through a spiral inscribed on the sphere with radius M_s.

Linearization - Polder's tensor

Equation (17.49) describes free motion. We note it is non linear in $\mathbf{M}$ because $\mathbf{H_s}$ depends on $\mathbf{M}$ (see above for the definition of $\mathbf{H_s}$), and also because of the damping term.

Now assume there is an excitation field $\mathbf{h}$ within the material. $\mathbf{h}$ differs from the static field $\mathbf{H}$ whose effect was included in the restoring torque. In particular, $\mathbf{h}$ can vary rapidly in time, and it is often qualified as dynamic.

We insist on the fact that it is the field actually prevailing in the material at the point considered. Just as for the case of $\mathbf{H}$, it may originate from an external source, or from a demagnetising effect, or from a combination of both.

Consider for the moment the case of zero damping (eq. 17.46). $\mathbf{h}$ simply adds to $\mathbf{H_s}$, and the equation of motion becomes:

$$d\mathbf{M}/dt = -\gamma\mu_0\,\mathbf{M} \times (\mathbf{H_s} + \mathbf{h}) \tag{17.50}$$

This equation remains non linear because of the term $\mathbf{M} \times \mathbf{h}$, and again because $\mathbf{H_s}$ depends on θ, hence on $\mathbf{M}$. However, it can be linearised by taking the so-called small signal assumption.

To this aim, we set $\mathbf{m} = \mathbf{M} - \mathbf{M_s}$, and we assume that we have both $\mathbf{h} \ll \mathbf{H_s}$ and $\mathbf{m} \ll \mathbf{M_s}$. Since by definition $\mathbf{M_s} \times \mathbf{H_s} = 0$, and since from our assumption $\mathbf{m} \times \mathbf{h}$ is an infinitely small quantity of second order, we get :

$$d\mathbf{m}/dt = -\gamma\mu_0\,(\mathbf{M_s} \times \mathbf{h} + \mathbf{m} \times \mathbf{H_s}) \tag{17.51}$$

Our small signal assumption also makes $\mathbf{H_s}$ a vector with constant modulus, equal to $H_A + H_0$.

Let us build a system of orthogonal axes Oxyz by adding to the Oz we already defined two axes Ox and Oy. Because the system has rotation symmetry around Oz, axes Ox and Oy can be chosen arbitrarily in the plane perpendicular to Oz. Let m_1, m_2, m_3 be the components of $\mathbf{m}$, and h_1, h_2, h_3, those of $\mathbf{h}$, with respect to Oxyz. Equation (17.51) projects into:

$$dm_1 / dt + \gamma \mu_0 H_s m_2 = \gamma \mu_0 M_s h_2$$
$$dm_2 / dt - \gamma \mu_0 H_s m_1 = -\gamma \mu_0 M_s h_1 \tag{17.52}$$
$$dm_3 / dt = 0$$

Assume the excitation is sinusoidal, with angular frequency ω, and set:

$$\omega_H = \gamma \mu_0 H_s \qquad \omega_M = \gamma \mu_0 M_s \tag{17.53}$$

Equation system (17.52) provides in particular the response $\mathbf{m}$ in the permanent regime. Using the complex notation, we note that $d/dt = j\omega$, hence:

$$\begin{vmatrix} -\omega_H & j\omega & 0 \\ j\omega & \omega_H & 0 \\ 0 & 0 & 0 \end{vmatrix} \begin{pmatrix} m_1 \\ m_2 \\ m_3 \end{pmatrix} = \omega_M \begin{pmatrix} -h_1 \\ h_2 \\ h_3 \end{pmatrix} \tag{17.54}$$

which is easy to invert, yielding:

$$\begin{pmatrix} m_1 \\ m_2 \\ m_3 \end{pmatrix} = \begin{vmatrix} K & jV & 0 \\ -jV & K & 0 \\ 0 & 0 & 0 \end{vmatrix} \begin{pmatrix} h_1 \\ h_2 \\ h_3 \end{pmatrix} \tag{17.55}$$

with: $K = \omega_H \omega_M / (\omega_H^2 - \omega^2) \; ; \; V = \omega \omega_M / (\omega_H^2 - \omega^2)$ (17.56)

The matrix defined by equation (17.55) is the representation in Oxyz of the rotation susceptibility tensor, also called *Polder's tensor* [7]. It describes an elementary mechanism (rotation), in the absence of damping, under the assumptions of linearity (weak signals) and of uniaxial symmetry. Regarding the last point, we can check that this matrix is invariant for any rotation around Oz. Due to its special form, Polder's matrix can be reduced without loss in generality to its components in the plane perpendicular to $\mathbf{M_s}$.

$$|\chi| = \begin{vmatrix} K & jV \\ -jV & K \end{vmatrix} \tag{17.57}$$

Gyromagnetic resonance - Snoek's limit

For $\omega = 0$, we see that $V = 0$, and the susceptibility matrix reduces to the scalar $\chi_s = \omega_M / \omega_H = M_s / H_s$. For $\omega \to \omega_H$, K and V diverge: this is gyromagnetic resonance. The product of the static susceptiblity by the angular frequency for resonance gives:

$$\chi_s \omega_H = \omega_M = \gamma \mu_0 M_s \tag{17.58}$$

This relation shows that, for given value of M_s, the higher ω_H, the lower will be χ_s. It thus expresses a limitation in the frequency performance of magnetic materials, known as Snoek's limit [8].

Introducing damping into Polder's matrix

It is easy to show that the linearised form of Gilbert's equation [6] under an excitation **h** is:

$$dm / dt = -\gamma\mu_0 (\mathbf{M_s} \times \mathbf{h} + \mathbf{m} \times \mathbf{H_s}) + (\alpha / M_s)(\mathbf{M_s} \times d\mathbf{m} / dt) \quad (17.59)$$

Using the same process as above, we are led to a modified form of Polder's matrix. Just by changing ω_H into $\omega_H + j\alpha\omega$ in formulas (17.56), we get the new components of the matrix:

$$K = \omega_M (\omega_H + j\,\alpha\,\omega) / (\omega_H{}^2 - \alpha^2\omega^2 - \omega^2 + 2j\,\alpha\,\omega\,\omega_H)$$

$$V = \omega\,\omega_M / (\omega_H{}^2 - \alpha^2\omega^2 - \omega^2 + 2j\,\alpha\,\omega\,\omega_H) \quad (17.60)$$

The components K and V thus become complex, and take respectively the forms:

$$K = K' - jK'' \;\; ; \;\; V = V' - jV''$$

The corresponding spectra are schematically represented on figure 17.4.

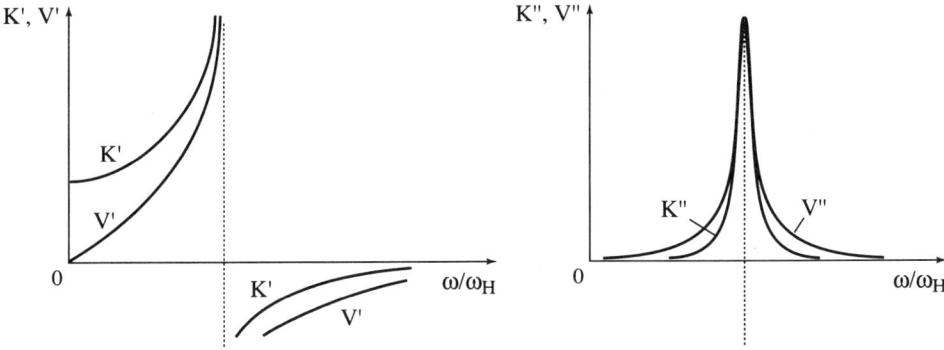

Figure 17.4 - Susceptibilities K and V vs ω / ω_H:
K' and V' are the real parts, K'' and V'' the imaginary parts

3.1.2. Anisotropic rotation

Returning to the beginning of §3.1.1, let us drop the assumption of uniaxial symmetry. $\mathbf{M_s}$ remains, by definition, the magnetization at rest. Let us again build an orthogonal system of axes, which we denote Ox'y'z, with Oz again parallel to $\mathbf{M_s}$. A small rotation of **M** starting from the rest position $\mathbf{M_s}$ can be described just by the components $m_{x'}$ and $m_{y'}$ on Ox' and Oy' of $\mathbf{m} = \mathbf{M} - \mathbf{M_s}$. Since position $\mathbf{m} = 0$ is an equilibrium position, the energy density U associated with the deviation ($m_{x'}$, $m_{y'}$) can necessarily be expanded to lowest order into the form:

$$U / \mu_0 = (1/2) A' m_{x'}{}^2 + (1/2) B' m_{y'}{}^2 + C m_{x'} m_{y'} \quad (17.61)$$

If $m_{x'}$, $m_{y'} \ll M_s$, the expansion can be restricted to these terms. It can be shown that, by choosing another system of axes, obtained through an appropriate rotation about Oz, expression (17.61) can be diagonalised into:

$$U/\mu_0 = (1/2)\ Am_1^2 + (1/2)\ Bm_2^2 \qquad (17.62)$$

where m_1 and m_2 are the components of **m** on the new axes. We insist on the fact that, within our assumptions, this is the most general situation that can be encountered. Equation (17.62) shows that an equivalent restoring field is associated with any magnetization **m**. Its components are:

$$-dU/\mu_0\ dm_1 = -Am_1 \ ; \ \ -dU/\mu_0\ dm_2 = -Bm_2$$

hence it is generally non colinear to **m**. This is why this situation is termed anisotropic rotation.

Under an actually applied field h_1, h_2, we can take the restoring field into account in the calculation of the dynamic response, simply by substituting $h_1 \to h_1 - Am_1$ and $h_2 \to h_2 - Bm_2$ in the right hand side of system (17.54) in §3.1.1.

However, because this substitution accounts for the whole torque acting on magnetization **M**, we must take out from (17.54) the contribution due to H_s which is tantamount to setting $\omega_H = 0$. Finally, we get:

$$\begin{vmatrix} -A\omega_M & j\omega & 0 \\ j\omega & B\omega_M & 0 \\ 0 & 0 & 0 \end{vmatrix} \begin{pmatrix} m_1 \\ m_2 \\ m_3 \end{pmatrix} = \omega_M \begin{pmatrix} -h_1 \\ h_2 \\ h_3 \end{pmatrix} \qquad (17.63)$$

Inverting this system, we find a new susceptibility matrix:

$$|\chi| = \begin{vmatrix} K_1 & jV & 0 \\ -jV & K_2 & 0 \\ 0 & 0 & 0 \end{vmatrix} \qquad (17.64)$$

with $K_1 = \omega_2\omega_M/(\omega_1\omega_2 - \omega^2)$; $K_2 = \omega_1\omega_M/(\omega_1\omega_2 - \omega^2)$; $V = \omega\,\omega_M/(\omega_1\omega_2 - \omega^2)$; $\omega_1 = A\,\omega_M$; $\omega_2 = B\,\omega_M$.

Violation of Snoek's limit

Relations (17.64) reveal a resonance at the angular frequency $\omega_r = (\omega_1\omega_2)^{1/2}$. By setting $\omega = 0$, we check that the susceptibility matrix is diagonal but anisotropic, with $K_1(0) = \omega_M/\omega_1 = 1/A$ and $K_2(0) = \omega_M/\omega_2 = 1/B$.

If $\omega_1 < \omega_2$, $K_1(0) > K_2(0)$, and if we write the product $K_1(0)\ \omega_r$, we obtain:

$$K_1(0)\ \omega_r = \omega_M(\omega_2/\omega_1)^{1/2}.$$

We see that this product is larger than Snoek's limit which characterises isotropic rotation. We will come back to this point later.

Anisotropic rotation with friction

We can add to system (17.63) the linearised term corresponding to Gilbert's torque $(\alpha / M_s) \, d\mathbf{M} / dt \times \mathbf{M}$. This leads to the following components of the susceptibility matrix, in the case of anisotropic rotation with friction:

$$K_1 = \frac{\omega_M (\omega_2 + j \alpha \omega)}{\omega_1 \omega_2 - (1 + \alpha^2) \omega^2 + j \alpha \omega (\omega_1 + \omega_2)}$$

$$K_2 = \frac{\omega_M (\omega_1 + j \alpha \omega)}{\omega_1 \omega_2 - (1 + \alpha^2) \omega^2 + j \alpha \omega (\omega_1 + \omega_2)} \qquad (17.65)$$

$$V = \frac{\omega_M \, \omega}{\omega_1 \omega_2 - (1 + \alpha^2) \omega^2 + j \alpha \omega (\omega_1 + \omega_2)}$$

3.2. WALL MECHANISM

We here consider the special case of a 180° Bloch wall in a material with uniaxial anisotropy. As in the static investigation of chapter 5, we first examine the oversimplified situation of a plane infinite isolated wall.

We take the origin in the middle of the wall, and call Ox the axis perpendicular to the wall, Oz being parallel to the magnetization of one of the domains. We assume for example that the domain to the left of the wall has magnetization $+M_s$, the domain on the right having magnetization $-M_s$ (fig. 17.5-a).

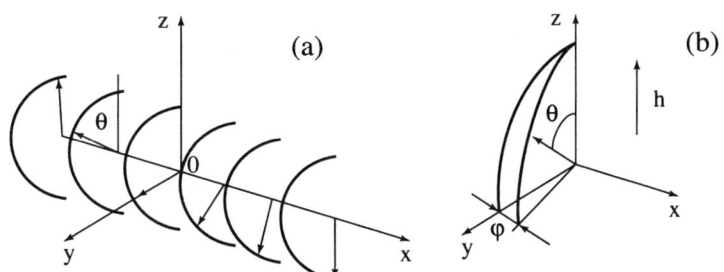

Figure 17.5 - (a) Structure of the Bloch wall at rest
(b) Effect of applying a field h // Oz over a time $t_0 \ll 2\pi / (\gamma \mu_0 h)$

In the wall at rest, $\mathbf{M}$ is parallel to the yOz plane and defined by angle θ.

In chapter 5, using different notations for the axes, we showed that the function $\theta(x)$ satisfies equation:

$$2A \, d^2\theta / dx^2 - 2K_u \sin \theta \cos \theta = 0 \qquad (17.66)$$

which expresses the balance, within the wall, between the torques due to exchange and to anisotropy. A more convenient form of this equation is:

$$\delta^2 \, d^2\theta / dx^2 - \sin \theta \cos \theta = 0 \qquad (17.67)$$

where $\delta = (A / Ku)^{1/2}$ is a parameter with the dimension of a length.

As was shown in chapter 5, δ actually measures, to within a factor π, the *thickness* of the wall at rest. We also note that any function $\theta(x)$ satisfying:

$$d\theta / dx = \pm \sin\theta / \delta \qquad (17.68)$$

is a solution of equation (17.67).

3.2.1. Ballistic behaviour - Döring mass [9]

Let us apply at time $t = 0$ a field $\mathbf{h}$ along $+\mathbf{M_s}$. The magnetization is immediately submitted to a torque $\boldsymbol{\Gamma} = \mu_0 \mathbf{M} \times \mathbf{h}$. This torque is zero in each of the domains, but it is non-zero within the wall. As long as there is no deviation from static equilibrium, this is the only torque acting on $\mathbf{M}$.

Therefore, as we have seen in the preceding section, the result is a precession, in the positive direction, of $\mathbf{M}$ around Oz at angular frequency $\omega_z = \mu_0 \gamma h$.

At time $t \ll 2\pi / \omega_z$, the magnetization thus takes on, at each point, a new orientation defined by θ, which is unchanged with respect to its static value, and by the new angle $\varphi = \omega_z t \ll 2\pi$ (fig. 17.5-b). We note that φ is the same for all of space (including outside the wall, where it loses any significance). We also note that there appears a longitudinal magnetization component M_x which is x-dependent:

$$M_x = M_s \sin\theta \sin\varphi = M_s \varphi \sin\theta \qquad (17.69)$$

Since this situation is out of equilibrium, we of course expect a reaction torque to appear. However, in this new wall configuration (with $\varphi \neq 0$ but uniform), both the exchange and anisotropy energies have remained exactly the same as in the rest configuration.

The reaction torque can thus be related neither to anisotropy nor to exchange. But we can note that the longitudinal magnetization given by (17.69) necessarily produces a demagnetising field: $H_x = -M_x$.

The expected torque is thus due to H_x, and its effect is to introduce a precession of $\mathbf{M}$ around Ox at angular frequency:

$$\omega_x \approx d\theta / dt = \mu_0 \gamma H_x = -\mu_0 \gamma M_s \varphi \sin\theta = -\omega_M \varphi \sin\theta \qquad (17.70)$$

Now assume that, at time $t_0 \ll 2\pi / \omega_z$, we suppress the field $\mathbf{h}$. The precession around Oz immediately stops and angle φ remains fixed at the value $\varphi \sim \mu_0 \gamma H t_0 \ll 2\pi$.

The only remaining motion is thus the precession of $\mathbf{M}$ about Ox in the demagnetising field of the wall. We note:

- that, insofar as φ remains very small, the precession occurs practically in the wall plane zOy,
- that the direction of this precession is such that θ decreases at all points in the wall, i.e. that $\mathbf{M}$ get nearer to $+\mathbf{M_s}$,
- that, from relation (17.70), the angular frequency for precession $\omega_x \sim d\theta / dt$ is distributed in the wall thickness according to figure 17.6-a.

The distribution $\theta(x)$ thus changes as indicated on figure 17.6-b, which suggests a continuous translation of the wall at some velocity v (here from left to right). If this is true, then θ becomes a function of the sole variable $u = x - vt$.

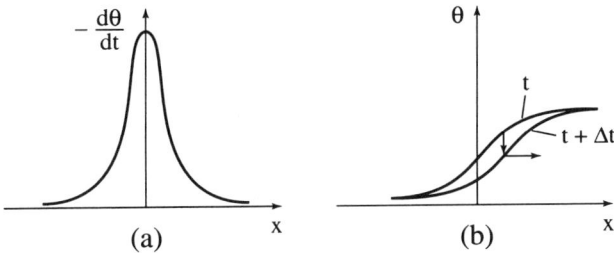

Figure 17.6 - (a) Distribution of the velocity of precession around Ox, from (17.72) - (b) Effect on the distribution $\theta(x, t)$

The precession velocity $d\theta / dt$ thus satisfies relation:

$$d\theta / dt = -v\, d\theta / du \qquad (17.71)$$

and from (17.71),

$$d\theta / du = -(\omega_M\, \varphi \sin \theta) / v \qquad (17.72)$$

This equation describes the structure of the moving wall relative to the set of axes that are linked to it. Equation (17.68) shows that this structure remains identical, to within the small azimut φ, to that of the wall at rest. Identifying, we find the velocity:

$$v = \omega_M \delta \varphi = \omega_M \delta \mu_0 \gamma h t_0 = (\mu_0 \gamma)^2 h M_s \delta t_0 \qquad (17.73)$$

This translation movement at velocity v was produced by a field pulse, with amplitude h and duration t_0. It occurs for infinitely long after this impulse! This behaviour is reminiscent of that of a material point with mass m, moving without friction. It can be called the *ballistic regime*. Of course, it is due to the fact that we did not take damping into account (we implicitly used Larmor's equation).

Before returning to this point, let us show that the analogy between the wall and a material point with mass can be taken quite far. The field h applied during time t_0 produced on the wall a pressure $2\mu_0 h M_s$ hence an impulse per unit area $2\mu_0 h M_s t_0$. Thus the ratio $2\mu_0 h M_s t_0 / v$ defines the equivalent of a mass per unit area, called *Döring's mass* [9]:

$$m_W = 2 / (\mu_0 \gamma^2 \delta) \qquad (17.74)$$

The existence of this inertial mass implies that the moving wall has a surface energy :

$$\gamma = \sigma_W + 1/2\, m_W\, v^2 \qquad (17.75)$$

where σ_W is the surface energy at rest. We have seen that angle φ, hence the longitudinal magnetization M_x, are proportional to velocity v, and we understand that the kinetic energy is actually the magnetostatic energy associated with the demagnetising field $H_x = -M_x$.

3.2.2. Introducing damping - Wall mobility

We now take damping into account by replacing Larmor's equation by Gilbert's equation. The analysis is simplified if we treat the extra torque as a weak perturbation.

Let us assume that the structure of the moving wall is that determined in the absence of damping, and let us express the extra torque: the vector $\mathbf{M}$ with constant module rotates practically in the plane of the wall. The friction torque $(\alpha / M_s) \mathbf{M} \times d\mathbf{M} / dt$ is therefore practically perpendicular to the wall (we still assume that φ is very small) with modulus:

$$(\alpha / \gamma M_s) M_s M_s \partial\theta / \partial t = (\alpha / \gamma) M_s \partial\theta / \partial t \qquad (17.76)$$

This torque can be balanced at all points in the wall by a DC field, with adequate modulus, applied along Oz. The torque produced by such a field is also practically perpendicular to the wall, and its modulus is $\mu_0 h M_s \sin\theta$. Since, due to (17.71), (17.72) and (17.73), we have $\partial\theta / \partial t = (v / \delta) \sin\theta$, both torques will be compensated at all points if the velocity satisfies:

$$v = (\mu_0 \gamma \delta / \alpha) h \qquad (17.77)$$

This solution thus corresponds to a translation motion that is identical to the ballistic motion described above. It therefore conserves the static structure of the wall, to within the small azimut φ. However, field $\mathbf{H}$ must now exist all the time, and there is a linear relation between v and h. The proportionality constant $\mu_0 \gamma \delta / \alpha$ is called the mobility of the wall.

This behaviour is quite analogous to that of a material point with mass, submitted to viscous friction. The magnetic pressure P can be described by relation:

$$P = 2\mu_0 h M_s = \eta_W v \qquad (17.78)$$

η_W is a friction coefficient, defined per unit area.

From (17.77): $\eta_W = 2 M_s \alpha / (\gamma \delta) \qquad (17.79)$

The ratio $m_W / \eta_W = (\alpha \omega_M)^{-1}$ is more meaningful than either of the parameters m_W or η_W, because it is dimensionally a time, and roughly represents the duration over which the ballistic motion of the wall persists after the field is turned off. This duration is the longer the smaller is the damping coefficient. Since $2\pi / \omega_M$ is of the order of one period of the gyromagnetic resonance, and since α is a dimensionless number of the order of 10^{-1} to 10^{-3} or even 10^{-4}, we can remember that the ballistic motion lasts for anything between a few and a few thousands periods. According to this theory, wall friction results from the damping of the rotation as it appears in Gilbert's formulation. We will return to this point later.

3.2.3. Equation of motion for an isolated plane wall

We just showed that an ideal infinite isolated plane wall behaves like a material point with mass m_W subject to viscous friction. We also saw in chapter 6 that, after it is moved from its rest position within a real material, a wall is generally submitted to a restoring pressure which has many origins.

Here we describe this restoring pressure P_r in a purely phenomenological way:

$$P_r = -R_W x \qquad (17.80)$$

where x is the displacement with respect to the equilibrium position and R_W the wall stiffness. From the above discussion, the equation of motion now is:

$$m_W \, d^2x/dt^2 + \eta_W \, dx/dt + R_W \, x = 2\mu_0 M_s h \qquad (17.81)$$

where we made the obvious generalization to the case where the field $\mathbf{h}$ is not colinear to the easy axis. This is a very familiar equation, viz that of a damped harmonic oscillator. In the permanent sinusoidal regime, the solution to (17.81) is expressed in complex notation, with $\omega_W^2 = R_W/m_W$, as:

$$\bar{x} = \frac{2\mu_0 M_s \bar{h}}{R_w - m_w\omega^2 + j\omega\eta_w} = \frac{2\mu_0 M_s \bar{h}}{R_w} \frac{\omega_w^2}{\omega_w^2 - \omega^2 + j\dfrac{\eta_w}{m_w}\omega} \qquad (17.82)$$

We note that this expression describes a resonance at frequency $\omega_W/2\pi$. Such wall resonances have indeed been observed, and they can be considered as the experimental confirmation of a wall mass [10, 11].

3.2.4. Generalization to curved walls

If the wall has a curvature that is small with respect to the reciprocal of its thickness δ, we can assume that its internal structure remains close to that of a plane wall. We therefore continue characterising this curved wall locally through a surface mass m_W and a friction coefficient η_W.

The equation of motion locally becomes:

$$m_W \, d^2x/dt^2 + \eta_W \, dx/dt + R_W \, x = 2\mu_0 M_s (h + h_d) + \Delta P \qquad (17.83)$$

with R_W now representing the strictly local contribution to the restoring pressure.

If the wall is curved, poles generally appear, inducing a demagnetising field. Let h_d be its component along Oz. Finally, in analogy with the surface tension of a liquid, the curvature directly produces a pressure related to surface energy σ_W:

$$\Delta P = \sigma_W (1/R_1 + 1/R_2) \qquad (17.84)$$

where R_1 and R_2 are the principal radii of curvature of the wall at the investigated point. We will later use these results in the case of spherical curvature ($R_1 = R_2 = R$).

3.2.5. Diffusion relaxation

This is the special manifestation of a general phenomenon called the magnetic after-effect. It is described in chapter 6 and in references [12, 13, 14]. These effects are associated with the presence, in the magnetic material, of mobile point defects that have in common the property that they interact with magnetization, hence contribute to the anisotropy energy and therefore to the wall energy.

Because the defects are mobile, it becomes possible –and this is an essential point– to alter the energy of the system both by moving the defects and by changing the orientation of magnetization, or by moving a wall.

This phenomenon can be accounted for in a purely phenomenological way by introducing, in equation (17.81) for the harmonic response of an isolated wall, a frequency-dependent stiffness. At low frequency, the wall moves slowly, and the defects have time to migrate so they constantly minimise the energy of the system: then the stiffness should be rather low. At high frequency, the defects do not move any more, the interaction energy between the defects and the magnetization is far from its minimum, and the stiffness is larger.

Because the defect motion is thermally activated, this phenomenon is strongly temperature dependent.

3.2.6. Disaccommodation

After a thermal demagnetization cycle, or under an AC field, almost all materials feature a slow decrease of their permeability as a function of time, according to a logarithmic law : $\Delta\mu = -D'\log_{10}t$, where D' is a constant (see § 5.3.2 in chap. 6). This phenomenon is ascribed to an increase in the wall stiffness, which again originates in the presence of mobile defects that interact with magnetization.

The only difference with diffusion relaxation as discussed above lies in the relevant characteristic time constants. Here, they are comparatively very large, and, even more important, they are distributed over a very large range of values. It can be shown that the logarithmic law observed originates in this distribution. To characterise this phenomenon, according to relation (17.23), ferrite manufacturers give the disaccommodation coefficient in the form:

$$D = d\mu / (\mu^2 d\log_{10}t) = -\mu^2 D'.$$

4. SUSCEPTIBILITY IN THE DEMAGNETISED STATE: THE HOMOGENIZATION PROBLEM

In the preceding sections, we discussed the harmonic response of an isolated domain or wall to the field effectively acting at this domain or wall, assuming it to be known. We now tackle the more concrete case of a material which we assume to be in the demagnetised state, i.e. in a state where the mean static magnetization is zero.

In this situation, the material is heterogeneous, because it is necessarily divided into domains. This purely magnetic heterogeneity can be complemented by structural heterogeneity. This is the case for ceramics or polycrystalline alloys, which consist both of domains and of grains (crystallites).

Two extreme situations are possible, and are indeed observed in practice :
♦ Each grain contains many domains. This is generally the case of ferrites.
♦ The domains are larger, or even much larger, than the grains. This is often the case in crystallized metal alloys, and even more for nano-crystallised alloys, or in amorphous materials insofar as these can be considered as the ultimate limit of nanocrystalline materials.

It is clear that a volume ΔV arbitrarily selected in the sample around a point M is sufficiently representative of the material at this point (hence independent of ΔV) only if it contains a large number of grains in the first case, of domains in the second case. More precisely, the spatial averages taken over ΔV, <**m**> for the induced magnetization, and <**h**> for the field, then have an accurate meaning, and we can write:

$$<\mathbf{m}> \; = \; \chi_{eff} <\mathbf{h}> \tag{17.85}$$

which defines the average or effective susceptibility of the material. In the general case, this is a tensor. The notations <**m**> and <**h**> are introduced only in relation (17.85) which defines the effective susceptibility. In the following, we return to **m** and **h**. Writing relation (17.85) is the same as defining a homogeneous medium equivalent to the heterogeneous material we are interested in. This is why the theoretical determination of the effective susceptibility is referred to as a homogenization problem.

4.1. THE APPROXIMATION OF ADDITIVITY OR OF THE MEAN

In this type of approach, we assume that each domain, each grain and each wall is submitted to the same field, viz the applied field in the material.

We will criticise this assumption in the next section. It makes it in particular possible to deal with each contribution (rotation or wall displacement), and even with each element (domain, wall or grain) independently.

4.1.1. Snoek's model [8]

Snoek considers only the rotation process. Domain i, with volume V_i, and tensor susceptibility χ_i, submitted only to the field **h**, carries a magnetic moment $\mathbf{m} = V_i \chi_i \mathbf{h}$.

The average magnetization of a macroscopic volume V is then:

$$\mathbf{m} = \sum_i (V_i / V) \chi_i \mathbf{h} = \left(\sum_i c_i \chi_i \right) \mathbf{h} = \chi_{eff} \, \mathbf{h} \tag{17.86}$$

where c_i is the volume fraction of domains having the orientation labelled by subscript i. Thus, in the case of a material with uniaxial anisotropy and no damping, the susceptibility χ_i is that from Polder's approach, and the effective susceptibility χ_{eff} is just the average of χ_i over all orientations of the easy axis. For a uniform distribution of these orientations, the susceptibility then reduces to the scalar:

$$\chi = \frac{2}{3} K = \frac{2 \omega_M \, \omega_H}{3 \left(\omega_H^2 - \omega^2 \right)} \tag{17.87}$$

According to this formula, which neglects rotation damping, the product of the static susceptibility and of the resonance angular frequency is given by:

$$\chi(0)\,\omega_H \;=\; 2/3\,\omega_M \;=\; 2/3\,\gamma\,\mu_0\,M_s \qquad\qquad (17.88)$$

We recover Snoek's limit for the single domain (formula 17.58) to within the factor 2/3. The best feature of this model is that it emphasises, in agreement with observation, that it is not possible, within a family of cubic compounds with constant magnetization, to increase the static permeability and the cutoff frequency simultaneously.

4.1.2. The model of Globus et al. [15]

Globus's model takes into account only the wall mechanism, and it is in fact a wall stiffness model rather than a homogenization model. Nevertheless, since microstructure plays an essential role, it is not inconsistent to deal with it in this section.

Globus assumes that each grain, modelled by a sphere with diameter D, is divided into two domains by a 180° wall which, in the demagnetised state and in zero field, is in a diametrical position. It is also assumed that, in the case of weak excitation, the edge of the wall remains pinned to its rest position. The simplest excitation is then deformation into a spherical cap, characterised by a radius of curvature R and an elongation u, these two variables being related by the purely geometrical relation $R = D^2/8\,u$.

This deformation entails, from equation (17.84), a restoring pressure due to the surface tension, $2\sigma_W/R = 16\,\sigma_W\,u/D^2$, whence a contribution to the wall stiffness per unit area:

$$R_W \;=\; 16\,\sigma_W/D^2 \qquad\qquad (17.89)$$

In the absence of interaction between the grains, the static susceptibility ($\omega = 0$) of the ceramic is then expressed, using (17.89) and (17.82), as:

$$\chi \;=\; (1/4)\,\mu_0\,S'\,D^2\,M_s^2/\sigma_W \qquad\qquad (17.90)$$

where S' is a weighted average wall area per unit volume, taking into account the wall orientation distribution with respect to the applied field.

We check that S' is of the order of $1/D$, so that:

$$\chi \;=\; (1/4)\,\mu_0\,D\,M_s^2/\sigma_W \qquad\qquad (17.91)$$

The model thus predicts, in agreement with experiment (see section 6.1), a linear variation of the static susceptibility and permeability with the grain diameter. However, we note that another interpretation of this behaviour has been more recently suggested [16].

In the general case ($\omega \neq 0$), setting $\omega_W^2 = \sigma_W/4Dm_W$, the model gives for the complex susceptibillity:

$$\chi = \frac{\mu_0 \, D \, M_s^2}{4 \sigma_w} \frac{\omega_w^2}{\omega_w^2 - \omega^2 + j \frac{\eta_w}{m_w} \omega} \tag{17.92}$$

According to Guyot *et al.* [17], it is very often possible to neglect the effect of wall inertia and to take only friction into account. This is equivalent to letting m_w tend to zero, hence ω_w tend to infinity. Setting $\tau = \eta_w / R_w$, the above expression then becomes:

$$\chi = \frac{\mu_0 \, D \, M_s^2}{4 \sigma_w} \frac{1}{1 + j \omega \tau} \tag{17.93}$$

which characterises a special form of dispersion called wall relaxation. This was investigated in detail, especially by Guyot *et al.* [17] in the garnet $Y_3Fe_5O_{12}$ (YIG), a favorite *test material* which is not used in radio frequency techniques, but which we will discuss again in the context of microwave applications. This investigation also confirms that wall friction is not the consequence of rotation damping alone. There obviously are other, not yet identified, mechanisms.

In closing this section, let us mention a difficulty in Globus's model that was noted by several authors [18] : it does not take into account the demagnetising field associated with the curved wall.

4.2. EFFECTIVE MEDIUM MODEL

The effective medium model originally applies to a random composite consisting of n phases which are assumed to be isotropic. From symmetry arguments, it is then clear that the effective susceptibility remains isotropic. For example, in a magnetic composite made up of two phases, with susceptibility χ_1 and χ_2 respectively, theory provides the expression of χ_{eff} as a function of χ_1, χ_2 and of the volume fractions c_1 and $c_2 = 1 - c_1$ of the two constituents.

Rigorously and contrary to the approximation made in section 4.1, the effective susceptibility of a composite cannot be given by an additivity law:

$$\chi_{eff} \neq c_1 \chi_1 + c_2 \chi_2 \tag{17.94}$$

Such a law assumes that each grain in the composite is submitted to the same field. This assumption inevitably leads to an inconsistency since, when they are submitted to the same field, neighbouring grains with different characteristics will feature different magnetizations. The magnetization therefore necessarily fluctuates spatially, on the scale of the grains. Consequently there appear magnetic poles, which in turn imply a fluctuation in the field itself.

This is the real difficulty in the calculation of χ_{eff}, i.e. of what specialists call the homogenization problem.

The so-called effective medium approach [19, 20] provides in many cases an analytic solution to the problem. It was applied to ferrites recently. The appendix at the end of this chapter contains a detailed description of the principle of this method and of its application to the case we are interested in here, viz determining the effective susceptibility of a magnetic ceramic. Here we just state the starting assumptions of this model by Bouchaud and Zerah [21].

It is assumed that each grain is subdivided into many 180° domains, according to figure 17.7, which shows a structure that is invariant on average for any rotation of the grain around the axis Oz, parallel to the magnetization in each domain. In other words, Bouchaud and Zerah [21] assume that the grain is on average isotropic in the plane perpendicular to Oz.

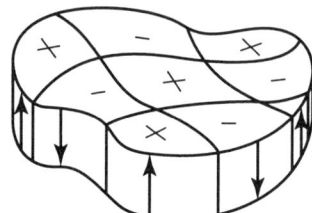

***Figure 17.7 - Domain structure in the grain,
in Bouchaud and Zerah's model***

The effective medium method is used at two levels. First, the effective susceptibility is calculated for the grain, i.e. for a 180° domain population. In a second step, knowing the susceptibility of the grains, the susceptibility is calculated for the ceramic, i.e. for a population of grains with random orientations. If only the rotation mechanism is considered, this provides (see appendix) a very simple result, which is rigorously valid for $|\mu_{eff}| \gg 1$:

$$\mu_{eff} = \frac{1}{2} \sqrt{(1+K)^2 - V^2} \qquad (17.95)$$

where K and V are the components of Polder's matrix, as defined in section 3.1.

If rotation damping is neglected, this yields:

$$\mu_{eff} = \frac{1}{2} \sqrt{\frac{(\omega_M + \omega_H)^2 - \omega^2}{\omega_H^2 - \omega^2}} \qquad (17.96)$$

a particularly simple result, which already explains, at least qualitatively, the most salient experimental facts.

Since the term under the square root is negative between ω_H and $\omega_H + \omega_M$, the complex permeability $\mu_{eff} = \mu_{eff}' - j\mu_{eff}''$ is purely imaginary within this interval, and purely real outside. The corresponding spectrum is sketched on figure 17.8-a. We note that it reproduces, albeit in a very schematic way, a remarkable characteristic of the experimental spectra (see fig. 17.15), viz the existence of a loss peak (μ_{eff}'') with a very asymmetrical shape, involving a maximum at ω_H and a very slow decrease (roughly a $1/\omega$ variation) for $\omega \gg \omega_H$.

Figure 17.8-b gives the spectrum, now calculated with non-zero damping ($\alpha \sim 1$) taken into account. As could be expected, the effect of damping is mainly to smooth down the divergences, without shifting the characteristic frequencies too much. The likeness with the experimental spectra now becomes striking.

Of course the model predicts a Snoek-type limit, albeit, from (17.96), with a factor 1/2 instead of the factor 2/3 given by the additivity model of section 4.1. Detailed comparison [22] between experimental spectra and those predicted by equation (17.95) show, more generally, good agreement provided the wall mechanism may be neglected.

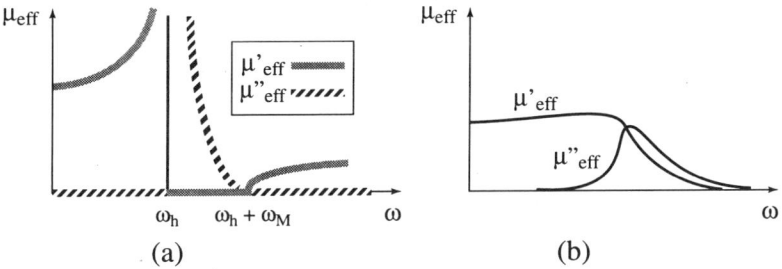

Figure 17.8 - Theoretical spectra, from equation (17.96)
(a) without damping - (b) with damping (schematic)

4.3. WALL CONTRIBUTION AND ROTATION CONTRIBUTION

Actually, each of the models described above puts the emphasis on one special magnetization process: wall motion, or magnetization rotation. But it is certain that, in usual ferrites, both mechanisms contribute, in different proportions, to the permeability. A specially interesting example, which however does not concern a really usual material (at least in radio-frequency applications), is shown in figure 17.9 from Guyot *et al.* [14].

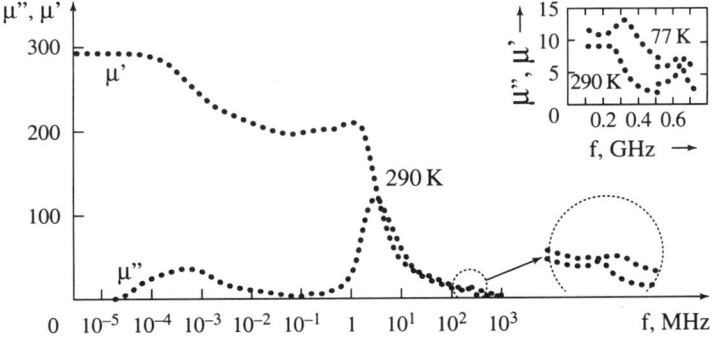

Figure 17.9 - Complex permeability spectrum of a YIG ceramic
The inset is a magnification of the part of the main curve contained in the circle.

This spectrum clearly shows a diffusion relaxation around 800 Hz, wall relaxation (in fact a very damped resonance) around 6 MHz, and finally a weak rotation dispersion beyond 100 MHz.

In Ni-Zn and especially in Mn-Zn ferrites, it is not easy to separate the wall contribution and the rotation contribution, and even less so to quantitatively assess their relative values. This is why this problem is still controversial [18].

An essential, but often not clearly formulated, difficulty is that these two contributions are normally not additive. This is shown in the effective medium model, as introduced in the appendix.

5. SUSCEPTIBILITY IN THE SATURATED STATE MAGNETOSTATIC MODES

We now focus on the situation in which the magnetic material is saturated through the application of a static internal field **H**, often called the polarising field. For simplicity's sake, we assume that the intrinsic anisotropy of the material is negligible, so that only the field **H** contributes to the restoring torque acting on the magnetization. This approximation is often valid in practice, and it makes it possible to deal simultaneously with single crystal and polycrystalline materials.

This situation is much simpler than the one we investigated in the previous section. Not only because magnetization rotation is the only elementary susceptibility mechanism that remains, but also because the composite effects associated with domains disappear.

5.1. CIRCULAR SUSCEPTIBILITIES AND PERMEABILITIES GYROTROPY

We start out by considering an unbounded medium submitted to an internal static field **H** parallel to Oz in an orthogonal set of axes Oxyz.

This situation is reminiscent of that studied in section 3.1.1, but we must realise that the whole medium is now a single domain, and that, less importantly, H_s reduces to H.

If, in the linear system (17.55), we perform the change in variable:

$$m_+ = m_1 + j m_2 \; ; \; m_- = m_1 - j m_2 \qquad (17.97)$$

we obtain: $$m_+ = (K + V) h_+ = \chi_+ h_+ \; ; \; m_- = (K - V) h_- = \chi_- h_- \qquad (17.98)$$

with $$h_+ = h_1 + j h_2 \; ; \; h_- = h_1 - j h_2 \qquad (17.99)$$

which is equivalent to diagonalising the susceptibility matrix. The components m_+, m_-, h_+ and h_- are called circular. The significance becomes clear if we investigate, for example, the instantaneous values $m_1(t)$ and $m_2(t)$, so denoted to distinguish them from m_1 and m_2 which, as we recall, are complex components.

Let us write $m_1(t) = m \cos \omega t = \text{Re}(m \exp j \omega t)$, and look at two situations:

♦ $m_+ = m_1 + j\, m_2 = 0$, hence $m_2 = j\, m_1 = j\, m$ and $m_- = m_1 - j\, m_2 = 2m$. We then see that $m_2(t) = -m \sin \omega t$, so that the instantaneous magnetization vector has constant modulus, and rotates in the Oxy plane in the negative direction at angular velocity ω: as expected, this situation corresponds to $m_+ = 0$ and $m_- \neq 0$.

♦ $m_- = 0$. We check that this situation corresponds to $m_+ \neq 0$ and to magnetization rotating in the positive direction.

It is straightforward to go over to the circular components of permeability, which will be useful further on:

$$\mu_+ = 1 + \chi_+ \; ; \quad m_- = 1 + \chi_- \qquad (17.100)$$

Relations (17.98) show that χ_+ and χ_- are normally different, hence that the response of the polarised material to a rotating field depends on the direction of rotation of the field. This phenomenon is called *gyrotropy*. The resonance in internal susceptibility, as discussed in section 3.1.1, corresponds, after (17.56), to $\omega = \omega_H$ and to $K = V \to \infty$.

Therefore, the resonance at ω_H also corresponds to $\chi_+ \to \infty$ while $\chi_- = 0$. The gyrotropy in the system is then a maximum, and we understand the term "gyromagnetic resonance" used in section 3.1.1.

A sinusoidal field applied along one of the axes Ox or Oy can also be decomposed into a positive circular component and a negative circular component. We see that, at resonance, only the positive component will yield a response, since $\chi_- = 0$. The response of the material, when it is excited by a rectilinear field (and not a rotating field) will again be a rotating magnetization.

We will see below that this property is used in so-called YIG filters.

5.2. *UNIFORM MAGNETOSTATIC RESONANCE*

Consider a sample in the shape of an ellipsoid with axes Ox, Oy, Oz, made of a ferro- or ferrimagnetic material with uniaxial anisotropy, with easy magnetization direction Oz, submitted to a static field H (often called the polarising field) parallel to Oz in order to retain the uniaxial symmetry. We specify that H is the internal field in the sample, and we recall that it is related to the external field H_0 (assumed to be also parallel to Oz) by relation $H = H_0 - N_z M$ where N_z is the demagnetising field coefficient of the ellipsoid along direction Oz.

The internal dynamic susceptibility of the sample is thus uniform and described by Polder's matrix (17.57). Here we look into the *external dynamic susceptibility* χ_e.

Going from Polder's matrix to the external susceptibility matrix, we can use transformation (17.20), noting that in $Oxyz$ the matrix of demagnetising field coefficients of the ellipsoid is diagonal, with principal components N_x, N_y, N_z. We can also recast system (17.54) by expressing the components of the internal dynamic field as a function of the external components and of the dynamic magnetization components, and then perform the inversion.

The result is an external susceptibility matrix which takes on exactly the form (17.64) with ω_1 and ω_2 replaced by:

$$\omega_1 = \omega_H + N_x \omega_M \qquad \omega_2 = \omega_H + N_y \omega_M \qquad (17.101)$$

with $\omega_H = \gamma \mu_0 H$ and $\omega_M = \gamma \mu_0 M_s$. The shape effect described by the demagnetising field coefficients thus turns out to be identical to the effect of an anisotropy. This is why one often refers to shape anisotropy. The basic reason for this analogy is immediately understood if we consider the restoring torque to which the magnetization is submitted when it deviates from a principal axis of the ellipsoid.

Consider for simplicity's sake the case of an ellipsoid, with rotation symmetry around Oz ($N_x = N_y = N$), and let us rotate the magnetization away from axis Oz by a small angle $\Delta\theta$, for example in plane zOx.

Apart from the known contributions due to H and to the intrinsic anisotropy of the material, we immediately check that there does indeed appear an extra restoring torque $N(M_s.\Delta\theta) M_s = H_F M_s \Delta\theta$. This shows that $H_F = NM_s$ plays exactly the same part as a uniaxial anisotropy field.

Let us return to the general ellipsoid. The components K_1, K_2 and V of matrix (17.64) diverge, in the absence of damping, at angular frequency $(\omega_1.\omega_2)^{1/2}$. Since we know that $\omega_H = \gamma \mu_0 H = \gamma \mu_0 (H_0 - N_z M_s)$, we obtain the famous formula by Kittel [23] which yields the angular frequency for external susceptibility resonance for an ellipsoid:

$$\omega_R = \gamma\mu_0 \sqrt{\left[H_0 + (N_x - N_z)M_s\right]\left[H_0 + (N_y - N_z)M_s\right]} \quad (17.102)$$

This resonance is called magnetostatic because the demagnetising effects play the major role. We will see later how this phenomenon can be evidenced and used for applications. In practice, the geometries used are only the sphere and the thin film, the latter being described as a very oblate ellipsoid with rotation symmetry.

For the sphere we have $N_x = N_y = N_z = 1/3$, hence the simple result:

$$\omega_R = \gamma\mu_0 H_0 \qquad (17.103)$$

For the thin film, there are two extreme situations: perpendicular polarization and parallel polarization.

In perpendicular polarization: $N_z = 1$, $N_x = N_y = 0$, whence:

$$\omega_R = \gamma\mu_0 (H_0 - M_s) \qquad (17.104)$$

In parallel polarization, Oy is chosen as the perpendicular to the plane of the film: $N_x = N_z = 0$; $N_y = 1$.

$$\omega_R = \gamma\mu_0 [H_0 (H_0 + M_s)]^{1/2} \qquad (17.105)$$

5.3. *NON UNIFORM RESONANCE: MAGNETOSTATIC MODES*

The resonance angular frequency ω_R of the external susceptibility tensor χ_e of an ellipsoid corresponds to a non-zero solution of equation:

$$\chi_e^{-1} \mathbf{m} = 0 \qquad (17.106)$$

where $\mathbf{m}$ is a uniform vector whose components oscillate at frequency $\omega_R / 2\pi$.

Taking a concrete example, let us consider a sphere, and assume that, in the absence of external excitation, the magnetization is rotated from its equilibrium direction Oz, e.g. through inclination in the plane zOx by the same small angle throughout the volume of the sphere. If the magnetization is set free at time $t = 0$, the free motion that follows is by definition the solution of (17.106) that satisfies $m_x = M_s \, \Delta\theta$ at $t = 0$. This will be called the uniform magnetostatic mode.

This mode can be considered as the solution to a self-consistency problem, schematically shown on figure 17.10-a. This only describes the fact that, in free motion, the dynamic magnetization $\mathbf{m}$ is produced *via* the *internal susceptibility* χ by the demagnetising field $-\mathbf{Nm}$ alone. This approach suggests an interesting question. If $\mathbf{m}$ is not uniform, the resulting demagnetising field is a solution of the equations of magnetostatics for a given distribution of magnetization, and it is not uniform either. Are there special distributions of $\mathbf{m}$ which satisfy the generalised self-consistency diagram shown in 17.10-b?

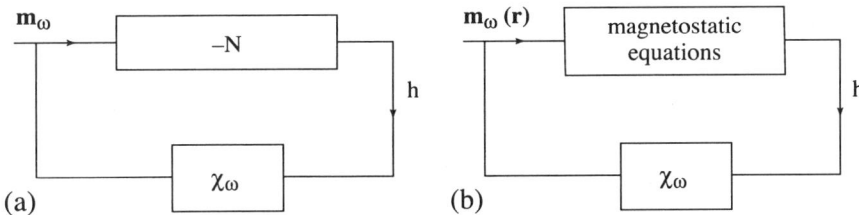

Figure 17.10 - Self-consistency diagram
(a) for the uniform mode in an ellipsoid - (b) for non-uniform modes

The answer was given by Walker as early as 1957 [24]. Indeed there is an infinite number of such distributions, according to the very general rule stating that, in any linear dynamic system, the number of modes is equal to the number of degrees of freedom (which is infinite for a continuum).

A magnetostatic mode is thus characterised by a profile, i.e. some spatial distribution of the magnetization oscillation (defined to within a multiplying factor), and by an angular eigen-frequency.

The uniform mode for an ellipsoid thus appears just as a special magnetostatic mode, whose profile is a uniform distribution, and whose angular eigenfrequency is given by (17.102). Figure 17.11 suggests, after Walker [24], the type of profile corresponding to a non-uniform mode for a sphere.

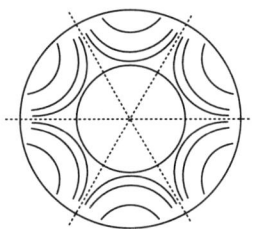

***Figure 17.11 - An example of the profile for
a non-uniform magnetostatic mode in a sphere,***
after Walker [24]

*This is the instantaneous distribution (for $\omega t = m.2\pi$,
m an integer) of the dynamic magnetization vector in
the equatorial plane.*

The magnetostatic mode concept was introduced here for samples bounded in all three dimensions. It generalises to geometries that are unbounded along at least one dimension, e.g. along an axis Oz. The mode profile then has the shape of a wave propagating along Oz: $\mathbf{m} = \mathbf{m}(x, y) \exp j(\omega t - kz)$.

This is called a propagation mode. It is characterised by a specific *transverse profile* $\mathbf{m}(x, y)$ and by a dispersion relation $\omega(k)$.

The magnetostatic propagation modes in the plane of a thin film were investigated for various configurations [25]. If Oy is the perpendicular to the film, there are two dimensions along which the sample is assumed to be infinite (at least on the scale of the wavelengths $2\pi/k$ involved), viz Ox and Oz. The investigated modes propagate along one of these directions, e.g. Oz, and $\mathbf{m}$ is independent of the other one (x). The characteristic profile then reduces to a distribution along the thickness only, $\mathbf{m}(y)$, and the dispersion relation to a function of one single variable $\omega = \omega(k_z)$.

Finally, the notion of magnetostatic mode generalises also to samples assumed to be unbounded in all three directions of space. The profiles are plane waves which are expressed as a function of the coordinate vector $\mathbf{r}$ and of the wave-vector $\mathbf{k}$ by:

$$\mathbf{m} = \overline{\mathbf{m}} \exp j \left(\omega t - \mathbf{k} \cdot \mathbf{r}\right) \qquad (17.107)$$

The dispersion relation then takes a particularly simple form. Let Oz be the axis carrying both the polarising field and the spontaneous magnetization. Since the system has rotation symmetry around Oz, vector $\mathbf{k}$ is defined through its modulus k and its angle θ with respect to Oz. It can then be shown that:

$$\omega = [\omega_H (\omega_H + \omega_M \sin^2\theta)]^{1/2} \qquad (17.108)$$

where we recall that $\omega_H = \gamma \mu_0 H$ and $\omega_M = \gamma \mu_0 M_s$.

The angular eigenfrequency of the modes propagating along direction θ is thus independent of k. For $\theta = 0$ and $\theta = \pi/2$, we recover the angular eigenfrequency of the uniform modes for a thin film polarised respectively along its perpendicular and in its plane. We note that, in relations (17.104) and (17.105), H_0 is the external field.

The set of magnetostatic modes of the form (17.107) thus occupies in the (ω, k) plane a band of angular frequency $\{\omega_H, [\omega_H (\omega_H + \omega_M)]^{1/2}\}$ which, as we will see, plays the basic role in the damping of uniform resonance. Magnetostatic modes are used in pratice, either as a way of investigating materials, or in devices which we will briefly describe below.

5.4. THE ROLE OF EXCHANGE INTERACTION: SPIN WAVES

We saw, especially in the investigation of domain walls, that a spatial variation in the local direction of magnetization entails a torque density related to exchange energy. This torque adds to that due to the dynamic demagnetising field, and it alters the dispersion relations. It can in particular be shown that (17.108) becomes:

$$\omega_k = \{[\omega_H + k^2\omega_M A/\mu_0 M_s^2][\omega_H + \omega_M(\sin^2\theta + k^2 A/\mu_0 M_s^2)]\}^{1/2} \qquad (17.109)$$

The dispersion curves corresponding to $0 < \theta < \pi/2$ occupy in the (ω, k) plane a band bounded by two parabolas (fig. 17.12). In this band, we can define, albeit in a rather vague way, the existence areas for magnetostatic modes ($k \to 0$) and for spin waves ($k \to \infty$).

As we already mentioned above, this band, which is here modified by the exchange interaction, plays an essential part in the physical mechanisms of rotation damping. This problem is dealt with in many papers and books [26, 27]. The main mechanism in the damping of uniform rotation is the transfer of energy from this mode to higher order modes, through a cascade of coupling events.

Figure 17.12 - Dispersion relations of plane magnetostatic waves and of spin waves

Condition for coupling of uniform resonance to higher order modes. The line with length K_m shows the upper boundary of the spectrum of inhomogeneities.

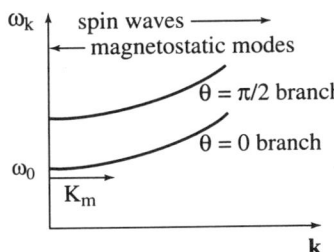

The final result is the excitation of spin waves with very short wavelengths, and crystal lattice vibrations (heating). In particular, reference [27] contains a detailed description of the various coupling effects between modes as defined in an ideal situation. An essential factor is the existence, even within a single crystal material, of static inhomogeneities such as local variations in anisotropy, various defects, pores, inclusions... These inhomogeneities act, in the presence of uniform magnetization oscillations, as *secondary sources* whose spatial distribution can be described by a spectrum of wave-vectors **K**.

The condition for a mode with wave-vector **k** to be excited by a uniform oscillation at angular frequency ω_0 (which can be described by a wave-vector with modulus $k = 0$) can then be written $\omega_k = \omega_0$ and $\mathbf{k} = \mathbf{K}$. This condition is very schematically pictured through the construction in figure 17.12, where the horizontal line with ordinate ω_0 has for its length the upper boundary of the spectrum describing the inhomogeneities. We see that coupling exists only in a band of frequency, roughly between ω_H and $[\omega_H(\omega_H + \omega_M)]^{1/2}$. Experiment shows that damping is indeed larger in this band than outside (fig. 17.13).

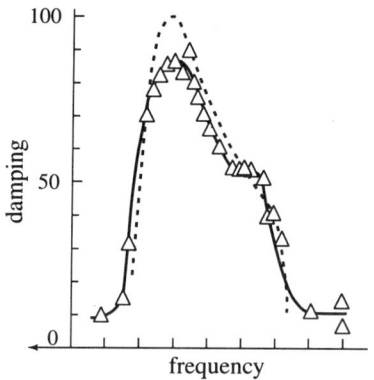

Figure 17.13 - Rotation damping vs frequency in a substituted YIG ceramic (YIG: Ca, V), after Patton [27]

The triangles are experimental points, while the dashed curve is a theoretical curve. The units are arbitrary on both axes.

This is called the *two-magnon* mechanism. It is not consistent with Gilbert's phenomenological description, at least in its general form (17.49). On the other hand, expressions (17.60) remain valid, provided coefficient α is allowed to be frequency-dependent.

6. AN OVERVIEW OF RADIO-FREQUENCY MATERIALS AND APPLICATIONS

The soft materials that are most used in the radiofrequency range (typically below 100 MHz) are ceramics (polycrystals) of mixed oxides, with very diverse formulas, designated under the generic name of ferrites. We stressed, at the beginning of this chapter, that the major advantage of ferrites over metallic alloys, in particular those of power electrical engineering, is that they are insulators or semiconductors. This eliminates or at least alleviates the induced current or skin effect problem (see for example tab. 17.1 and the related section).

Many books were already dedicated to the various aspects, basic or applied, of ferrites [28, 29, 30, 31]. The reader should refer to them for a description more thorough than this short overview, as well, of course, as to the part of chapter 4 in the present book that deals with ferrimagnetism.

In radiofrequency applications, the relevant characteristic is almost always the complex permeability $\mu' - j\mu''$ in the demagnetised state, without static polarising field. We discussed above the mechanisms that govern this permeability. Here, we will just describe the observed behaviours, insofar as they are strictly useful to the device designer.

Two families of materials are mostly used in radiofrequency components: spinel ferrites, as encountered in section 5 of chapter 16, and planar hexaferrites.

6.1. SPINELS

Originally, the term spinel designates a cubic crystal structure type, that of the natural mineral $MgAl_2O_4$, which is not magnetic. The general formula of spinels is $X^{2+}(Y^{3+})_2(O^{2-})_4$, and their structure is represented on figure 17.14. This structure consists in a face centred cubic lattice of oxygen ions, the smaller metallic ions being intercalated in interstices with octahedral (16) or tetrahedral (8) symmetry, respectively called d and a sites [28].

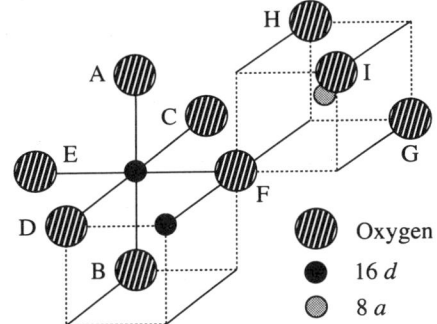

Figure 17.14 - The spinel structure

The unit cell contains 8 a sites, with tetrahedral symmetry (surrounded by atoms F, G, H and I), and 16 d sites with octahedral symmetry (surrounded by atoms A, B, C, D, E and F).

Lodestone, or *magnetite* which we mentioned in chapter 1, is a *natural* magnetic spinel (see chap. 24) with formula $Fe_3O_4 = Fe^{2+}(Fe^{3+})_2O_4$. The antiferromagnetic coupling between neighbouring Fe^{2+} and Fe^{3+} ions gives rise to the ferrimagnetism of this compound. A structural, electronic and magnetic transition, the Verwey transition, occurs at 120 K. This oxide is a ferrimagnet, with a Curie temperature of 858 K. The molecular formula of this ionic compound can be written $(Fe^{++}O^{--})(2Fe^{+++}3O^{--})$. Thus, the magnetism of magnetite arises from the Fe^{++} ions (whose free ground state is $^5D_4$) and Fe^{+++} ions (free ground state $^6S_0$), which are ferrimagnetically coupled. Half of the Fe^{3+} ions and the Fe^{2+} ions are located on the 16 d sites (with octahedral symmetry) while the remaining Fe^{3+} ions are on the 8 a sites with tetrahedral symmetry. This is said to correspond to an *inverse* spinel, to distinguish it from the *direct* spinel for which *all* the Fe^{2+} ions would be on a sites, and the Fe^{3+} would be on d sites. In a free ion model, we thus expect magnetite to have structure $\uparrow(6\mu_B)\uparrow(5\mu_B)\downarrow(5\mu_B)$ leading to a resulting moment of $g_J J = 6\mu_B$ per molecule. This is in utter disagreement with the spontaneous magnetization measured at low tempereature, only 4.1 μ_B per formula, very close to the value expected for complete quenching of the orbital angular momentum (g = 2, S = 2, hence a magnetic moment 4 μ_B). Furthermore, the g factor is near 2, which confirms that only spin contributes to the magnetic moment.

In practice, the soft spinels used in inductive components are all *synthetic* materials, deriving from magnetite through the substitution of iron atoms by other elements (Ni, Mn, Zn, Mg, Li, etc.).

We will just describe briefly the properties of the two most important families, the mixed nickel-zinc ferrites (for short Ni-Zn) with formula $Ni_xZn_{1-x}Fe_2O_4$, and the manganese-zinc (Mn-Zn) family with formula $Mn_xZn_{1-x}Fe_2O_4$.

The magnetic properties, in particular the static permeability, as well as the electrical properties, are very sensitive to the exact composition, in particular the oxygen stoechiometry. They are also sensitive to annealing and quenching processes, due to the inversion phenomenon we already mentioned in connection with magnetite, because it can be total or partial [29]. We must also not forget the influence of the ceramic microstructure (porosity and grain size).

However, in general, the Mn-Zn ferrites are characterised by a rather large static permeability –referred to as giant permeability– and a rather low electrical resistivity (of the order of 10^4 Ω.m). On the other hand, the Ni-Zn ferrites have more moderate permeability, but they are almost perfect insulators.

The resistivity of Mn-Zn compounds can be very notably increased through calcium substitution, while retaining the property of giant permeability [32]. Unfortunately, this is only a composite effect: calcium concentrates at the grain boundaries, and it creates there an isolating film while the resistivity in the bulk of the grain remains practically unchanged. This special microstructure leads to an *anomalously high apparent dielectric constant*, hence to exceptionally low dimensional resonance frequencies.

The Mn-Zn ferrites are thus normally restricted to low frequency applications while, at least as far as electrical limitations are concerned, the Ni-Zn ferrites can cover the whole radiofrequency range.

Figure 17.15, after Smit and Wijn [33], gives the real (μ') and imaginary (μ") permeabilities as a function of frequency f between 0.1 MHz and 4 Ghz, for Ni-Zn ferrites with variousNiO/ZnO molar ratios.

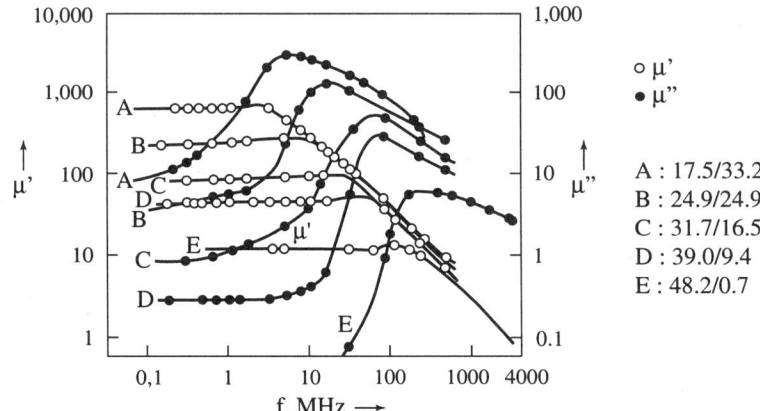

Figure 17.15 - Permeability spectra of Ni-Zn ferrites
for various molar ratios NiO/ZnO, after [33]

μ' is seen to remain constant and equal to its static value μ_s up to a *cut-off frequency* f_c beyond which μ' decreases; while a large imaginary component μ" appears. Generally, f_c is thus the upper limit to the operating frequency for the material. The remarkable point in this set of curves is that the product $\mu_s f_c$ is practically constant, which means that it is impossible to have both a strong static permeability and a high maximum operating frequency. This law is known as *Snoek's limit,* and its origin was explained above.

Figure 17.16, after Slick [34], shows the variation of the ratio μ"$/\mu$'^{2}, the reciprocal of the figure of merit defined in section 1.1.1, as a function of frequency, for various commercial Mn-Zn and Ni-Zn ferrites.

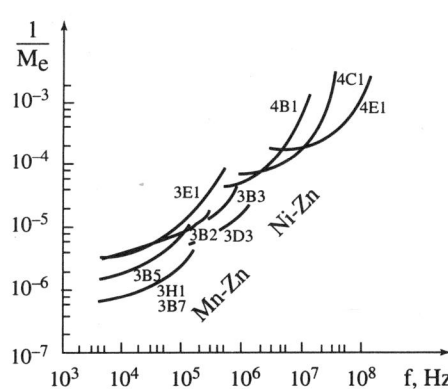

Figure 17.16
Reciprocal of the figure of merit
1 / M_e = tan δ/μ' vs frequency
for two families of spinel ferrites,
after [34]

This graph shows in particular that there definitely are preferred operating range for each of the two families of compounds: roughly above 1 MHz for Ni-Zn, and below for Mn-Zn.

We mentioned above the influence of the grain size on the static permeability. The experimental results of Globus *et al.* [15] indicate a linear dependence of this parameter as a function of the average grain diameter in the ceramics. Other effects are related to the microstructure, but these are the most significant for theoretical interpretation, as we saw in section 4.1.

6.2. PLANAR HEXAFERRITES

The hexaferrites [28] are so called because they crystallise in the hexagonal system. This entails quite remarkable magnetocrystalline anisotropy properties, which distinguish them sharply from the cubic ferrites and in particular the spinels.

Solid state chemists classify hexaferrites in four phases, M, Z, Y and W which are distinguished both by their compositions and by their structures (tab. 17.2). For the user, it is sufficient to consider two families, the easy-axis hexaferrites and the easy-plane hexaferrites (see chap. 3, § 2.3).

Table 17.2 - Hexaferrites

Phase	Formula	Anisotropy
W	$BaMe_2Fe_{16}O_{22}$	Composition dependent
Y	$Ba_2Me_2Fe_{12}O_{22}$	Easy plane - Me = Zn, Co, Mn, Mg, Ni
Z	$Ba_3Me_2Fe_{24}O_{41}$	Composition dependent
M	$BaFe_{12}O_{19}$	Easy axis

Till now, only the easy-axis hexaferrites, the prototype of which is baryum hexaferrite $BaFe_{12}O_{19}$, have been used in industrial applications –but they occupy the major market of consumer magnets (chap. 15).

Here we focus on the easy-plane materials, because these are magnetically soft. These compounds actually feature six equivalent easy axes in the basal plane of the hexagonal cell, and these are all rest orientations for the magnetization. With respect to such a rest position, the orientation of magnetization can be described by two angles, θ measuring the deviation out of the basal plane, and φ, the azimutal angle, measuring the deviation from the easy axis within the basal plane.

If θ and φ are small, the anisotropy energy can be expanded (see eq. 17.62) in the form $U = (1/2)\mu_0 M_s^2 (A\varphi^2 + B\theta^2) = (1/2)\mu_0 M_s (H_\varphi \varphi^2 + H_\theta \theta^2)$ where H_θ and H_φ are anisotropy pseudo-fields. We thus deal with a situation of anisotropic rotation, and we saw that this leads to overcoming Snoek's limit (see § 3), all the more the larger is the ratio $B/A = H_\theta / H_\varphi$. For example, in compound $Ba_2Zn_2Fe_{12}O_{22}$, this ratio is of the order of 3,000, and we expect an increase factor $(3,000)^{1/2} \sim 55$.

Actually, although they have long been given a trade name (ferroxplana), these materials have not yet really come out of the laboratory. We nevertheless think they should be discussed, because it can be hoped that progress in ceramics preparation methods, in particular for oriented ceramics, will make it possible at last to reach the theoretical results.

6.3. APPLICATIONS IN LINEAR INDUCTIVE COMPONENTS

The main linear applications of ferrites are summarised in table 17.3, after Slick [34]. We describe two of them in detail: the ferrite antenna and the inductance.

Table 17.3 - Main linear radiofrequency applications of ferrites

Ferrite	Component	Function	Frequency	Required property
Mn-Zn	inductance	filtering	< 1MHz: Mn-Zn	large μ Q product
Ni-Zn	"	"	< 100 MHz: Ni-Zn	"
Ni-Zn	transformer	impedance adaptation	< 500 MHz	large μ
Ni-Zn	antenna	radio receivers	< 15 MHz	large μ, small ε
Mn-Zn	deflection circuit	cathode tubes	< 100 kHz	large μ, large M_s

Before the advent of ferrites, the loop antenna was the prevalent device in radio receivers for long, medium and even short waves (up to 15 Mhz). Its principle is simple, as shown on figure 17.17-a. It is a flat coil with area S, made up of n turns. The open circuit output voltage of this antenna, when placed in an induction, with amplitude b_0, due to an electromagnetic wave with angular frequency ω, has amplitude $n \omega S b_0$. The sensitivity of the antenna is directly proportional to its area S, which leads to a fairly cumbersome device. These air loops have been replaced for several decades by ferrite antennas, the principle of which is shown on figure 17.17-b.

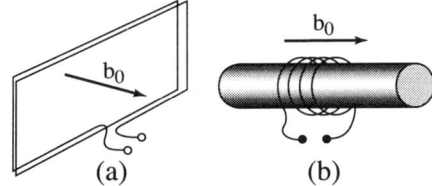

Figure 17.17 - Principle of the loop (a)
and of the ferrite (b) antennas

(a)   (b)

If the bar (with permeability $\mu \gg 1$) is described as a prolate ellipsoid with demagnetising field coefficient N along its axis, its external permeability μ_e is, from (17.20):

$$\mu_e = 1 + \chi_e \sim \chi_e = (\mu - 1)/\{1 + N(\mu - 1)\} \sim 1/N \qquad (17.110)$$

We assume here that μN remains much larger than 1 although N is assumed to be small. The induction b created in the bar by the external induction b_0 parallel to its axis is thus:

$$b = (\mu_0/N)(b_0/\mu_0) = b_0/N \qquad (17.111)$$

Since $N \ll 1$, we see that the ferrite bar operates as a flux multiplier, or more exactly as a flux concentrator. Thus, for given sensitivity, the air loop with area S can be replaced by a ferrite bar with cross-section S.N. As indicated in table 17.3, the preferred materials for this application are nickel zinc ferrites, mainly because of their high resistivity. We note that this flux concentration principle can also be applied to other devices, notably to magnetometers.

Inductances (L) are among the basicvcomponents of electronics. An essential application, in association with capacitors with capacitance C, is band-pass or selective filtering.

The principle of the selective filter takes advantage of the resonance of an LC circuit (fig. 17.18-a). The transmission curve (the ratio of the output to the input voltage vs frequency) of such a filter is schematically represented on figure 17.18-b. It can be shown that the relative width $\Delta f/f$ of the transmission peak, defined as indicated on figure 17.18-b, is given by:

$$\Delta f/f = 1/Q \qquad (17.112)$$

where Q is the quality factor of the resonant circuit. Q is limited by the losses in the condenser and in the inductance, but the main limitation usually stems from the latter, so that we can practically take $Q = L\omega/R$ with L the inductance of the coil and R its total resistance.

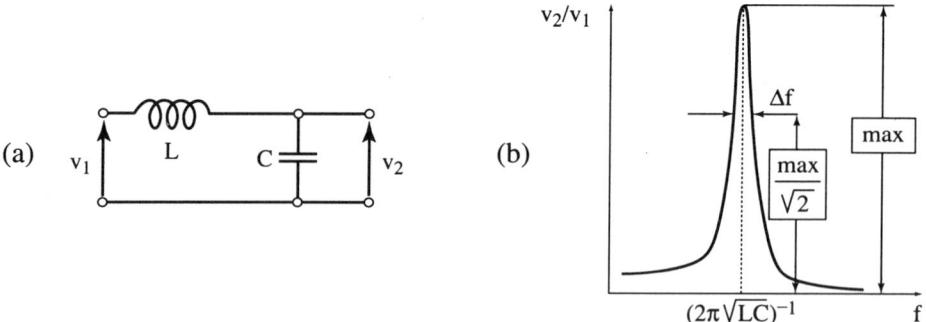

Figure 17.18 - Selective filter *(a) diagram - (b) transmission curve*

Making highly selective filters thus implies especially optimising the quality factor of the inductance. We will see that this leads to using a ferrite circuit or core.

We take the value L of the inductance and the frequency f as fixed, and we assume the geometry to be determined too. The winding is supported by a toroid with cross-section S, average perimeter l (we take for simplicity's sake $l/2\pi \gg (S)^{1/2}$), in which an airgap with thickness e can be introduced. Let p be the average perimeter of a turn, and s the total cross-section of copper available for the current in the coil. Apparently the only free parameters are the number of windings in the coil n, the width e of the airgap, and the permeability μ of the torus. Actually, the two last parameters are not independent, because magnetic circuit theory shows that a torus with permeability μ, with average perimeter l, involving a gap with thickness g, is equivalent to a homogeneous torus (without air gap) with permeability μ_e (called the effective permeability) given by:

$$(\mu_e)^{-1} = \mu^{-1} + g/l \qquad (17.113)$$

By comparing this relation to (17.22), we see that the air gap produces the equivalent of a demagnetising effect, with coefficient $N = g/l$. This is therefore a way of increasing the effective quality factor $Q_e = \mu'_e/\mu''_e$, at the expense of μ'_e since in this case the figure of merit $\mu'_e Q_e$ is conserved (see § 2.1.2).

To demonstrate the value of the magnetic circuit, we compare two coils with equal inductance, one involving n_1 turns on a torus with permeability one, the other with n_2 turns on a torus (possibly with an air gap) with effective permeability μ_e.

According to formula (17.36) giving L, we see that $n_2 = n_1/(\mu'_e)^{1/2}$. For constant total copper cross-section s, the cross-section of the wire is therefore s/n_1 in the first coil, and s/n_2 in the second. If ρ is the resistivity of the wire, the corresponding resistances are $R_1 = \rho\, n_1^2\, p/s$ and $R_2 = \rho\, n_2^2\, p/s = R_1/\mu'_e$. Thus the use of a magnetic torus makes it possible, at constant total volume, to decrease the ohmic resistance of the coil by the factor μ'_e.

However, from equation (17.36), the total resistance of the inductance also includes a contribution due to the magnetic losses:

$$R_m = \omega \mu''_e \mu_0 n^2 S/l = L\omega/Q_e,$$

where Q_e is the effective quality factor defined above. Finally, the quality factor of the inductance can be expressed as:

$$Q = L\omega/[R_1/\mu'_e + L\omega/Q_e] = 1/[(\mu'_e Q_1)^{-1} + \mu'_e/M_e] \quad (17.114)$$

where M_e is the figure of merit, and $Q_1 = L\omega/R_1$.

We see that as μ'_e is altered at constant M_e (by acting on the air gap ratio g/l), the quality factor Q goes through a maximum when:

$$\mu'_e = \mu_{opt} = (M_e/Q_1)^{1/2} \quad (17.115)$$

We note that this corresponds to $R_m = R_1/\mu'_e = R_2$, hence to equal contributions to the losses from the copper and the magnetic material.

The maximum value of the quality factor is $Q = (1/2)(Q_1 M_e)^{1/2} = Q_e/2 = Q_2/2$, where $Q_2 = L\omega/R_2$. A numerical example is given in Exercise 4 at the end of this chapter.

7. OVERVIEW OF MICROWAVE MATERIALS AND APPLICATIONS

The materials used for microwave technology (tab. 17.4) are described in detail by Von Aulock [29], and more recently by Nicolas [35]. The spinels and hexaferrites, which we very partly described in section 6, are complemented by the ferrimagnetic garnets, discovered in Grenoble at the end of the 1950's by Bertaut and Forrat [36].

Table 17.4 - Main materials used in microwave technology

Spinels	(Mg-Zn) Fe_2O_4, (Mn-Zn)Fe_2O_4, (Ni -Zn) Fe_2O_4, $Li_{0,5}Fe_{2,5}O_4$
Hexaferrites	type M: Ba $Fe_{12} O_{19}$ and substituted (easy axis)
	type Y, Z (easy plane)
Garnets	YIG: $Y_3Fe_5O_{12}$, (Ga, Al, Cr, In, Sc) substituted YIG and rare-earth
	(La, Pr, Nd, Sm, Eu, Gd, Tb, Dy, Ho, Er, Yb, Lu) substituted YIG

7.1. THE FERRIMAGNETIC GARNETS

The basic formula of the ferrimagnetic garnets is $R_3Fe_5O_{12}$ where R is a rare-earth element (with a restriction, however, on the ionic radius –see ref. [29]) or yttrium. These compounds crystallise in the cubic system, but their structure is much more complicated than for spinels. They are described in reference [29], which also provides a rather complete description of the chemical, structural and magnetic properties of this family, which turns out to be extremely rich when the many possible substitutions to the basic formula are taken into account.

Among all these compounds, yttrium iron garnet $Y_3Fe_5O_{12}$, usually designated by the acronym YIG, plays a dominant role because its rotation damping is particularly small, which leads to very narrow gyromagnetic resonance lines.

The resonance is normally investigated on spheres obtained from a ceramic or from a bulk crystal, by operating at fixed excitation frequency and varying the static polarising field. This simplifies the problem both of the generation of the excitation signal and of the microwave circuitry (actually, this is a bit less true now, with the advent of digital network analysers).

The sphere can be placed in a waveguide closed by a shorting piston at distance d (close to half the wavelength) from it, and one measures the reflection coefficient (fig. 17.19). Its variation vs polarising field H in the neighbourhood of resonance takes the shape of an absorption peak, with a width at half maximum ΔH that characterises the damping. In YIG at 9 GHz, values of $\mu_0\Delta H$ on the order of 0.3 mT in ceramics and 0.03 mT in single crystals are frequently encountered.

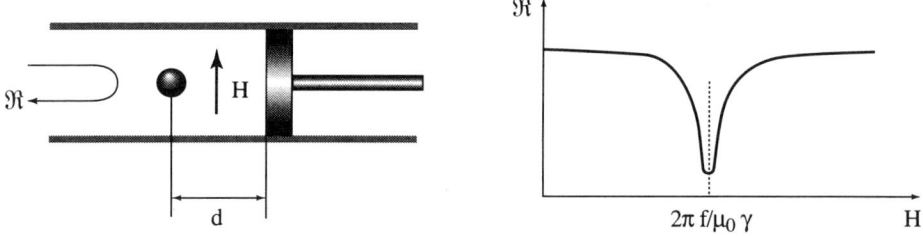

***Figure 17.19 - Measuring the width of the uniform resonance line
on ferrite spheres***

It is often more significant to characterise the resonance through its *quality factor* $H/\Delta H$, where H is the resonance field at the frequency used. Knowing that, for YIG, $\gamma = 28$ GHz/T, we obtain, for f = 9 GHz, $\mu_0H = 0.35$ T whence Q ~ 1,000 for ceramics and Q ~ 10,000 for single crystals. These high values of the quality factor are absolutely essential in the applications to filters and oscillators which we describe below.

7.2. *MICROWAVE APPLICATIONS*

The so-called non-reciprocal devices are the most specific, and also the most classical application of ferrites in the microwave range. They are described by Waldron [4]. We discuss them briefly here, and also describe a more recent application, the YIG resonator.

7.2.1. *Non-reciprocal devices*

All these devices use the gyrotropy of magnetised ferrites, i.e. the difference between the circular permeabilities μ_+ and μ_-. Here we just illustrate the principles by first introducing the microwave Faraday effect. We know that, in an isotropic medium with relative dielectric constant ε and permeability μ, a plane electromagnetic wave with rectilinear polarization propagates (without change) at phase velocity $c\,(\varepsilon\mu)^{-1/2}$ where c is the speed of light in vacuo. In this respect, plane waves with rectilinear polarization are the electromagnetic (degenerate) eigenmodes of the isotropic medium.

In a ferrite magnetised along Oz, the two eigenmodes have *opposite circular polarizations*. In other words, the fields that propagate without change along Oz are rotating, one in the positive direction, the other in the negative direction, and their phase velocities are respectively $c\,(\varepsilon\mu_+)^{-1/2}$ and $c\,(\varepsilon\mu_-)^{-1/2}$. This is referred to in optics as circular birefringence. It is important to note that the directions of circular polarization are here defined with respect to the magnetization direction in the ferrite, and not with respect to the propagation direction of the wave.

Consider a plate of ferrite with thickness d, magnetised perpendicular to its faces. A plane rectilinear wave that strikes the plate at normal incidence can always be considered as the sum of two rotating fields, one rotating in the positive direction, the other in the negative direction (fig. 17.20-a).

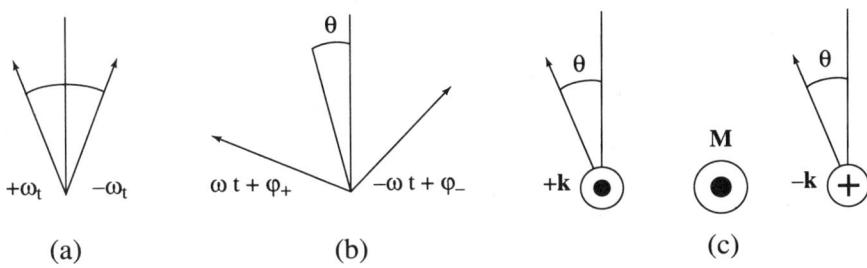

Figure 17.20 - Microwave Faraday effect: *(a) field on the entrance face (b) field on the exit face - (c) non-reciprocal rotation*

These two fields propagate independently in the ferrite, and thus undergo different phase shifts, viz:

$$(\varepsilon\mu_+)^{1/2}\omega d/c \;=\; k_+ d \;=\; \varphi_+ \quad \text{and} \quad (\varepsilon\mu_-)^{1/2}\omega d/c \;=\; k_- d \;=\; \varphi_-.$$

On the exit face of the plate, their combination reconstructs a rectilinear wave whose polarization direction has turned by an angle:

$$\theta \;=\; (\varphi_+ - \varphi_-)/2 = (1/2)(\omega d\varepsilon^{1/2}/c)[(\mu_+)^{1/2} - (\mu_-)^{1/2}]$$

in the positive direction since $\mu_+ > \mu_-$, hence $\varphi_+ > \varphi_-$ (see fig. 17.20-b).

As a result, the magnetised ferrite plate with thickness d rotates the polarization of a rectilinear wave by a well defined angle θ, proportional to d and to $(\mu_+)^{1/2} - (\mu_-)^{1/2}$.

The direction of rotation with respect to the oriented axis Oz is independent of the propagation direction. Thus a rectilinear wave propagating in direction $-Oz$ again has its polarization rotated in the positive direction, as in figure 17.20-c.

On the other hand, for an observer linked to the *wave-vector* **k** of the rectilinear incident field, the direction of rotation changes when **k** is reversed. In this respect, the device is a non-reciprocal rectilinear polarization rotator.

Although the orders of magnitude are not at all the same (the rotation is of the order of 10^{-4} radian per wavelength in visible light optics, and of the order of a radian per wavelength in the microwave range), the phenomenon is quite similar to the Faraday effect in optics (see chap. 13), whence the name microwave Faraday effect.

Application to the circulator

Consider the device represented on figure 17.21. It includes, apart from a ferrite disk magnetised perpendicular to its plane, two polarization separators.

Such separators are commonplace components in optics (Wollaston prism). They split a single incident beam into two beams with perpendicular rectilinear polarizations. Conversely, they make it possible to recombine two distinct incident beams with perpendicular polarizations into a single beam. The complete device is a multipole with four inputs / ouputs, numbered from 1 to 4 as indicated on figure 17.21-b. The device is tuned so that the Faraday rotation in the ferrite cylinder is exactly 45°.

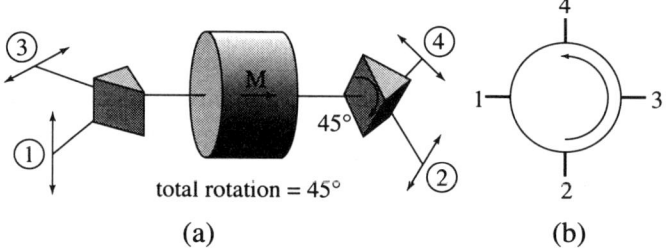

$$(a) \qquad\qquad\qquad (b)$$

Figure 17.21 - Principle of a Faraday effect circulator (a)
and corresponding diagram (b)

The truth table is then as indicated on table 17.5. It indicates in particular that a signal entering at 1 exits exclusively at 2. If it enters at 2, it exits at 3, etc., whence the diagram in figure 17.21-b and the name "circulator" for this kind of device.

Table 17.5 - Truth table for the four input / ouput circulator

Output / Input	1	2	3	4
1	0	0	0	1
2	1	0	0	0
3	0	1	0	0
4	0	0	1	0

The qualities required of the material for such an application are obviously as large as possible a difference between μ_+, and μ_- (to keep the device small), and as small as possible a propagation attenuation. The difference $\mu_+ - \mu_-$ increases as resonance is approached, but losses then appear too (μ_+ and μ_- become complex). Thus the ferrite should be polarised fairly far away from gyromagnetic resonance.

Actually, microwave technology usually does not deal with plane waves propagating in free space, except for some millimeter wave applications. It rather deals with waves confined in guides. Nevertheless, the basic principle of a circulator in a waveguide remains the use of gyrotropy. figure 17.22-a after Waldron [4] shows for example how Faraday rotation can be used in a comparatively direct way in a circular guide section containing a ferrite bar, and coupled to inputs/outputs consisting of rectangular guides.

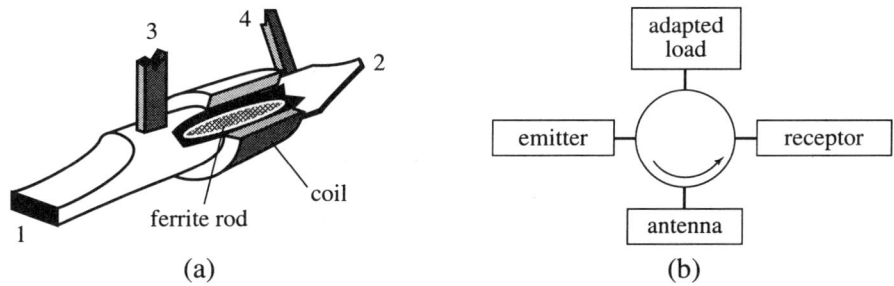

(a)   (b)

Figure 17.22 - Guide version of the Faraday effect circulator (a)
Use of a circulator in a microwave emitter/receiver (b)

Here, the rectangular guides themselves play the role of polarization separators mentioned in the above example. Their size is determined so that they propagate only the mode (denoted H10) for which the electric field is perpendicular to the major faces. By setting the Faraday rotation in the circular section containing the ferrite to 45°, we recover the truth table of table 17.5. Figure 17.22-b also illustrates a classical use of a circulator in a microwave emitting/receiving equipment, using a single antenna for emission and reception. The circulator makes it possible both to ensure the emission and reception signals remain separated, and to let the oscillator operate on a constant load, a condition for good stability [4].

Isolator

This is a multipole with two inputs/outputs, characterised by the very simple truth table of table 17.6: the systems passes the incident power only one way, here from 1 to 2.

Table 17.6 - Truth table of the isolator

Output/ Input	1	2
1	0	0
2	1	0

The circulator is based on gyrotropy or, in the language of optics, circular birefringence. In the isolator, we can speak of circular dichroism. Here, use is made of the difference between the imaginary parts of the circular permeabilities. In order to maximise this difference, the device is set at gyromagnetic resonance. Figure 17.23 illustrates one application mode of this general principle to an isolator in a rectangular guide, operating in the fundamental mode (H10). In this mode, the magnetic field is parallel to the major faces of the guide, i.e. parallel to the plane of figure 17.23. It is generally elliptically polarised, with the rotation direction changing in the middle of the major faces. Furthermore, these rotation directions change when the propagation direction of the mode changes. It thus suffices to fill half (or even a fraction of half) the guide with ferrite magnetised perpendicular to the major faces of the guide to make the device's attenuation sensitive to propagation direction. At resonance, the difference can be huge, with typically an attenuation by a factor of 1,000 in the isolating direction, and a transmission of 75% of the power in the passing direction.

Isolator are commonly used in microwave circuitry to connect a generator and its load, in particular in measuring instruments.

Figure 17.23 - Isolator in a rectangular guide
Polarization of the magnetic field in the H10 mode (left) unsymmetric filling with magnetised ferrite (right).

7.2.2. Tunable resonator

This device very directly uses uniform resonance in a ferrite sphere, typically 1 mm in diameter, submitted to a static polarising field H_0 (see section 5.2). A straightforward application is selective filtering (fig. 17.24).

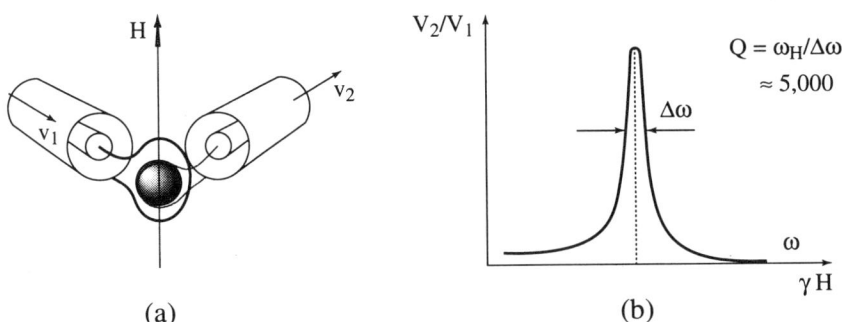

Figure 17.24 - Tunable selective filter (a) and its transmission curve (b)

Since the ferrite used in this component is almost exclusively YIG, it is commonly referred to as a YIG filter [37].

The coaxial line filter of figure 17.24-a also uses the fact that this resonance mode is circularly polarised. The coupling loops are in perpendicular planes, and their mutual induction appears only in the immediate neighbourhood of resonance, whence the tranmission curve of figure 17.24-b. Such filters commonly feature relative selectivities $\Delta f / f$ of the order of 2.10^{-4} ($Q = 5,000$), and their main interest is that they are electronically tunable, in practice over a range than can extend from 1 H to over 10 GHz (tuning just involves changing the current in the polarising coil!).

We finally mention that easy axis hexaferrites such as $(Ba, Sr) Fe_{12} O_{19}$ are used in this type of application for operating in the millimeter range ($f > 30$ GHz). Their specific interest in this area is their very high anisotropy field (for $BaFe_{12}O_{19}$, $\mu_0 H_A = 1.7$ T), which decreases the polarising field required for a given resonance frequency.

7.2.3. Microwave absorbers

Research on microwave absorbers is at present active. Its motivation is primarily military (stealth), but it could have significant by-products in civilian areas (environment and public health).

Consider a metallic surface, which we consider in a first approach as an infinite perfectly conducting plane. We know that such a plane reflects an incident electromagnetic wave totally, at least in the radio- and microwave frequency ranges.

Is it possible to coat the metal with a special film (this will be called the screen), as light and thin as possible, and with the result that the reflection coefficient vanishes? This is a classical stealth problem, and the most elegant solution involves a magnetic material.

Here we only examine the simple case of normal incidence of a harmonic electromagnetic wave with angular frequency ω, using the formalism of wave impedance [38].

We thus assume that the metal plane is coated with a magnetic film with thickness h and complex permeability $\mu(\omega) = \mu' - j\,\mu''$. This film also inevitably has a dielectric constant $\varepsilon > 1$. The impedance at the free surface of the magnetic coating (as seen from the incidence side) is given by the classical formula [38]:

$$Z = j\,Z_M \tan(\beta h),$$

where $\beta = \omega\,(\mu_0\,\varepsilon_v\,\mu(\omega)\,\varepsilon)^{1/2}$ is the propagation constant in the magnetic material, while $Z_M = (\mu_0\,\mu(\omega)/\varepsilon_0\,\varepsilon)^{1/2}$ is the wave impedance in the same material.

The reflection coefficient R on the coated plane is then $R = (Z - Z_0)/(Z + Z_0)$, with $Z_0 = (\mu_0/\varepsilon_0)^{1/2}$ the wave impedance in vacuo, with numerical value 377 Ω. We see that cancelling R implies making $Z = Z_0$. Since we are looking for a solution involving a very thin coating, we will assume that $\beta h << \pi/2$. Expanding the tangent to first order, we thus get:

$$Z = j\,\omega\mu_0\mu(\omega)h = j\,\omega\mu_0 h\,\mu' + \omega\mu_0 h\,\mu''.$$

We notice that if $\mu'' \gg \mu'$ the impedance practically reduces to a real part $Z \approx \omega \mu_0 h \mu''$. We can thus cancel the reflection coefficient by making $\omega \mu_0 h \mu'' = Z_0$, which means $\mu'' \beta_0 h = 1$, with $\beta_0 = \omega (\mu_0 \varepsilon_0)^{1/2}$ the propagation constant in vacuo. This leads to the practical relation $h = \lambda / (2 \pi \mu'')$, with λ the wavelength of the electromagnetic wave in vacuo at the frequency considered.

We note that if $\mu''(\omega)$ varies as $1/\omega$, hence as does λ, the condition can be satisfied at all frequencies, and the reflection coefficient is cancelled for all ω.

As we mentioned above, it so happens that the spectrum of ferrites (especially the nickel-zinc ferrites) approximately satisfies these conditions: above a cutoff frequency which corresponds to its maximum, $\mu''(\omega)$ actually varies as $1/\omega$, while the real part μ' decreases much faster, and soon becomes negligible with respect to μ''. Recall that this behaviour can be explained –as discussed above– by Bouchaud and Zerah's model.

Ferrites thus make it possible to make thin wide-band screens. For example a screen with an absorption band centered on 400 MHz typically has a thickness of 3 mm, since in the best materials $\mu'' = 40$ at 400 MHz ($\lambda = 0.75$ m).

It is clear from the above discussion that the energy of the incident wave is dissipated in the magnetic material (hence the importance of parameter μ''). This is of course no surprise, but we can wonder whether the same result can be obtained with an absorbing dielectric. The answer is contained in the above calculation, but a simple qualitative argument explains better why this approach cannot lead to a thin screen. Near a conducting (uncoated) plane, in normal incidence, all fields are tangential. The electric field is thus necessarily zero while the magnetic field is maximum. A thin dielectric film has no influence –to first order– on the reflection coefficient.

APPENDIX: THE EFFECTIVE MEDIUM METHOD

A1 - PRINCIPLE OF THE METHOD

In a two-phase composite, with volume fractions respectively c_1 and $c_2 = 1 - c_1$, we consider a particular grain. We assume it to be spherical and to belong for example to phase 1, and we want to calculate the field $\mathbf{h_1}$ acting within this grain if we know the average field $\mathbf{h}$ to which the composite is submitted. In the effective medium approach, the medium surrounding the chosen grain is described as a homogeneous medium with precisely the effective susceptibility we want to find. We are thus brought back to a classical problem in magnetostatics. The internal field $\mathbf{h_1}$ is colinear to the average field $\mathbf{h}$, and the moduli are related by:

$$h_1 = \frac{3 \mu_{\mathrm{eff}}}{\mu_1 + 2 \mu_{\mathrm{eff}}} h \qquad (17.116)$$

where we use the permeabilities $\mu_{\mathrm{eff}} = 1 + \chi_{\mathrm{eff}}$ and $\mu_1 = 1 + \chi_1$.

The magnetization in the grain of interest is thus:

$$m_1 = \frac{3\mu_{eff}\chi_1}{\mu_1 + 2\mu_{eff}}h \qquad (17.117)$$

The same calculation can be performed for a grain of phase 2, and the average calculation then expressed:

$$m = c_1m_1 + c_2m_2 = 3\mu_{eff}\left(\frac{c_1\chi_1}{\mu_1 + 2\mu_{eff}} + \frac{c_2\chi_2}{\mu_2 + 2\mu_{eff}}\right)h \qquad (17.118)$$

But, from the definition of χ_{eff}, we must also have $m = \chi_{eff}h$, whence a self-consistency relation between the permeabilities:

$$\frac{\mu_{eff} - 1}{3\mu_{eff}} = \frac{c_1(\mu_1 - 1)}{\mu_1 + 2\mu_{eff}} + \frac{c_2(\mu_2 - 1)}{\mu_2 + 2\mu_{eff}} \qquad (17.119)$$

from which we can derive μ_{eff} as a function of μ_1, μ_2, c_1 and $c_2 = 1 - c_1$. We check in particular that, if $\mu_1 = \mu_2 = \mu$, we regain the expected result $\mu_{eff} = \mu$.

A2 - APPLICATION TO A POLYCRYSTALLINE FERRITE : SUSCEPTIBILITY AND PERMEABILITY OF A GRAIN

We start from the assumptions formulated in section 4.2 of this chapter. In order to calculate the effective susceptibility of a multi-domain grain, we must modify the approach while retaining its principle, because there are two new facts.

First, the dimensionality of the problem is reduced to 2. The problem of a spherical inclusion is replaced by that of a cylindrical inclusion because, from our assumption of average invariance in rotation, each domain can be considered as a cylinder with axis along Oz. The second difference is that the response of the inclusion (a particular domain) is not described by a simple scalar susceptibility, but by a matrix (Polder's matrix) which is described, in a set of rectangular axes Oxyz, by the well-known equation (17.55).

Consider first the problem of a two-dimensional, two-phase composite, the two-dimensional analogue of the problem at the beginning of this appendix. It can be shown that relation (17.116) is replaced by:

$$h_1 = \frac{2\mu_{eff}}{\mu_1 + \mu_{eff}}h \qquad (17.120)$$

h_1 is the internal field in a cylindrical inclusion of phase 1, under the assumption that the average field $\mathbf{h}$ is applied in the plane Oxy. We note that the media are only required to be isotropic in this plane. However, this formulation still remains unsuited to the tensor character of the susceptibility of the inclusion which, in the case we are interested in, is a domain. To solve this difficulty, we rewrite (17.120) as:

$$\mathbf{h_1} = a\,\mathbf{h} + b\,\mathbf{m_1} \qquad (17.121)$$

which shows $\mathbf{h_1}$ to be the sum of two contributions: a $\mathbf{h_1}$ is the empty cavity field, i.e. the field that would be observed if $\mathbf{h}$ was applied in the medium while forcing the magnetization in the inclusion to be zero. We note that this field is indeed colinear to $\mathbf{h}$ since the medium is isotropic and the cavity cylindrical.

Contribution b $\mathbf{m_1}$ is the one that would be observed if $\mathbf{m_1}$ was imposed without a field being applied to the medium. Here again, it is obvious that this reaction field remains colinear to $\mathbf{m_1}$.

We finally note that, if the medium outside the inclusion is vacuum, then the field b $\mathbf{m_1}$ is nothing else than the classical demagnetising field.

Formulation (17.121) applies to the case of an inclusion with tensor susceptibility within a matrix which remains isotropic. As it naturally applies also to the situation where all media are isotropic, relation (17.120) can be used to calculate a and b. a is immediately obtained by setting $\mu_1 = 1$ in (17.120):

$$a = 2\,\mu_{eff}/(1 + \mu_{eff}) \tag{17.122}$$

And, since relation (17.120) describe a situation where the applied field $\mathbf{h}$ and magnetization $\mathbf{m}$ simultaneously exist, b is obtained from the difference:

$$b\,\mathbf{m} = \left(\frac{2\mu_{eff}}{\mu_{eff} + \mu_1} - \frac{2\mu_{eff}}{\mu_{eff} + 1} \right) \mathbf{h} \tag{17.123}$$

with
$$\mathbf{h} = \frac{\mu_{eff} + \mu_1}{2\mu_{eff}\,(\mu_1 - 1)}\,\mathbf{m_1} \tag{17.124}$$

This last equation is directly deduced from equation (17.120) using $\mathbf{m_1} = (\mu_1 - 1)\,\mathbf{h_1}$. It implies that $b = -1/(\mu_{eff} + 1)$ hence finally:

$$\mathbf{h_1} = \frac{2\mu_{eff}}{\mu_{eff} + 1}\,\mathbf{h} - \frac{1}{\mu_{eff} + 1}\,\mathbf{m_1} \tag{17.125}$$

We note that coefficients a and b depend only on the matrix and not on the inclusion.

We are now in a position to calculate the effective susceptibility of the magnetic grain for directions in the plane Oxy. We will assume that this effective susceptibility is isotropic (which results from symmetry), and we project relation (17.125) onto the axes. To retain notation μ_{eff} for the effective permeability of the ceramic, we here denote as μ_R the effective permeability of the *grain*:

$$\begin{aligned}
(1 + \mu_R)\,h_{1x} + m_{1x} &= 2\mu_R\,h_x \\
(1 + \mu_R)\,h_{1y} + m_{1y} &= 2\mu_R\,h_y
\end{aligned} \tag{17.126}$$

Using Polder's matrix and setting $P = 1 + K + \mu_R$, we get:

$$\begin{aligned}
P\,h_{1x} + j\,Vh_{1y} &= 2\mu_R\,h_x \\
-j\,Vh_{1x} + P\,h_{1y} &= 2\mu_R\,h_y
\end{aligned} \tag{17.127}$$

Introducing the direction cosines α and β of **h** in the plane Oxy, we obtain, by solving (17.127):

$$h_{1x} = 2\mu_R h \frac{\alpha P - j\beta V}{P^2 - V^2}$$

$$h_{1y} = 2\mu_R h \frac{\beta P + j\alpha V}{P^2 - V^2}$$

(17.128)

Then m_{1x} and m_{1y} can be obtained by using Polder's matrix again. Actually, since the grain is known to be isotropic globally, we only need to calculate the projection of **m** on **h**, i.e. $\alpha m_{1x} + \beta m_{1y}$. We find without difficulty:

$$\alpha m_{1x} + \beta m_{1y} = \frac{2\mu_R \left(KP - V^2\right)\left(\alpha^2 + \beta^2\right)}{P^2 - V^2} h$$

$$= \frac{2\mu_R \left(KP - V^2\right)}{P^2 - V^2} h$$

(17.129)

We see that the response of a given domain in the direction of the applied field is really independent of the direction of this field, provided it is in the Oxy plane. The susceptibility of the grain is immediately deduced, and it is obviously isotropic in the plane, in agreement with our initial assumption:

$$\chi_R = \mu_R - 1 = 2\mu_R \frac{\left(KP - V^2\right)}{P^2 - V^2}$$

(17.130)

The above relation is a third degree equation in μ_R (see the expression of P in terms of μ_R). However, we note it admits the obvious solution: $\mu_R + 1 = 0$, which leads to P = K. This solution is not physically satisfactory, but it allows us to reduce the degree of equation (17.130).

By factoring out the quantity P – K, equation (17.130) reduces to:

$$P^2 - 2(K+1)P + V^2 = 0$$

(17.131)

Hence the required solution, chosen by taking into account the condition that the static permeability must be positive:

$$P = K + 1 + \sqrt{\left(K + 1\right)^2 - V^2}$$

(17.132)

and the effective permeability of the grain:

$$\mu_R = \sqrt{\left(K + 1\right)^2 - V^2}$$

(17.133)

This result can also be established rigorously through a statistical treatment of Laplace's equation. It is remarkable that the effective medium theory, although it is just an approximation, provides the exact solution in this special case!

The complete permeability matrix for the grain in system Oxyz can now be written in the form:

$$|\mu_G| = \begin{pmatrix} \mu_R & 0 & 0 \\ 0 & \mu_R & 0 \\ 0 & 0 & \mu_P \end{pmatrix} \tag{17.134}$$

which involves, without expressing it explicitly, the wall permeability μ_P. We note that, in the chosen axis system, the matrix does take this diagonal form.

We can now move over to the second step, viz the calculation of the effective susceptibility of the ceramic considered as a compact assembly of spherical grains, each characterised by a permeability matrix such as (17.134) but with principal axes that change randomly from one grain to the next.

We again treat this problem using the effective medium method, and therefore return to the problem of a spherical inclusion which was dealt with at the beginning of the section, but this time for an anisotropic inclusion. We must therefore rewrite relation (17.116) under the form (17.121). We use the same method as for the cylindrical inclusion. The result is:

$$h_i = \frac{3\mu_{eff}}{1 + 2\mu_{eff}} h - \frac{1}{1 + 2\mu_{eff}} m_i \tag{17.135}$$

This relation allows the determination of the components m_{ix}, m_{iy}, m_{iz} induced along the principal axes of the anisotropic grain i by the applied field, with components along these axes $\alpha_i h$, $\beta_i h$, $\gamma_i h$. As the ceramic must at the end be found isotropic, it suffices to know the projection of the induced magnetization $\alpha_i m_{ix} + \beta_i m_{iy} + \gamma_i m_{iz}$ on the applied field direction.

The magnetization of the ceramic is now obtained by taking the average $<m_i>$ of m_i over all grains, knowing that: $<\alpha_i^2> = <\beta_i^2> = <\gamma_i^2> = 1/3$, and that $<\alpha_i\beta_i> = <\alpha_i\gamma_i> = <\beta_i\gamma_i> = 0$.

We thus find, calling χ_{eff} the required effective susceptibility:

$$\frac{\chi_{eff}}{1 + \chi_{eff}} = \frac{2\chi_R}{3 + \chi_R + 2\chi_{eff}} + \frac{\chi_P}{3 + \chi_P + 2\chi_{eff}} \tag{17.136}$$

This equation is normally of third degree in χ_{eff}. But here again there is an obvious (non physical) solution, which makes it possible to reduce the search for the physical solution to the root of a second degree equation:

$$3 + 2\chi_{eff} = 0 \ ; \quad \chi_{eff} = -3/2 \tag{17.137}$$

After some manipulation, we obtain:

$$2\chi_{eff}^2 + (3 - \chi_R)\chi_{eff} - 2\chi_R - (1 + \chi_R)\chi_P = 0 \tag{17.138}$$

Hence the physical solution (selected using the criterion $\chi_{\text{eff}} > 0$ for χ_R, and χ_P real and positive):

$$\chi_{\text{eff}} = (1/4)\left[\chi_R - 3 + \sqrt{(\chi_R - 3)^2 + 16\chi_R + 8\chi_P(1 + \chi_R)}\right] \quad (17.139)$$

We note however that, in the general case, the rotation and wall contributions are not additive. However, if χ_P and χ_R are small, a first order expansion of (17.139) yields $\chi_{\text{eff}} = (2/3)\chi_R + (1/3)\chi_P$, which corresponds exactly to an average of the susceptiblity tensor, taken over all grains without taking interactions into account.

If we assume the rotation contribution to be negligible, we obtain a strange but not unexpected result:

$$\chi_{\text{eff}} = (3/4)\left(\sqrt{1 + (8/9)\chi_P} - 1\right) \quad (17.140)$$

We see in particular that χ_{eff} does not grow proportionally to χ_P: for $\chi_P \gg 1$, χ_{eff} is proportional to $(\chi_P)^{1/2}$. Here again, the effect originates from interactions between grains which can obviously not be taken into account in additivity models.

If, conversely, we neglect the wall contribution, we find:

$$\chi_{\text{eff}} = \frac{\chi_R - 3}{4} + \frac{\chi_R + 3}{4}\sqrt{1 + \frac{4\chi_R}{(\chi_R + 3)^2}} \quad (17.141)$$

In this formula, the square root remains close to one for all values of χ_R: it goes through a maximum near 1.15 for $\chi_R = 3$, and tends toward 1 for χ_R tending to zero or infinity. An approximation to (17.141) which is in particular good for $\chi_R \gg 3$ is therefore $\chi_{\text{eff}} = \chi_R/2$ whence, using (17.133):

$$\chi_{\text{eff}} = \frac{1}{2}\sqrt{(K + 1)^2 - V^2} \quad (17.142)$$

If we take for K and V expressions (17.10), valid in the absence of damping, this formula becomes:

$$\mu_{\text{eff}} = \frac{1}{2}\sqrt{\frac{(\omega_H + \omega_M)^2 - \omega^2}{\omega_H^2 - \omega^2}} \quad (17.143)$$

EXERCISES

E.1. Using the reciprocity theorem, derive formulas (17.36).

E.2. Using the reciprocity theorem, derive formula (17.38). We recall that the characteristic impedance of a coaxial air line with radii R_1 and R_2 is given by $Z_c = (1/2\pi)(\mu_0/\varepsilon_0)^{1/2}\ln(R_2/R_1)$. Generalise this formula to the case of a torus with radii R'_1 and R'_2 which does not completely fill the line $(R_1 < R'_1 < R'_2 < R_2)$.

E.3. Calculate the demagnetising field energy associated with longitudinal magnetization in the wall, and recover Döring's mass (eq. 17.74).

E.4. Assume you want to make a 1 mH inductance, with copper windings, using a toroidal geometry with the following geometrical constraints: $S = 5 \times 5$ mm^2; $p \sim 4 \times 5$ mm; $l = 100$ mm; $s \sim 100 \times 0.1$ mm$^2 = 10$ mm^2. The operating frequency is 100 kHz. Choose the material, and calculate the optimal air gap thickness. Take $\rho = 2 \times 10^{-8}$ Ωm.

SOLUTIONS TO THE EXERCISES

S.4. The operating frequency immediately determines the choice of a Mn-Zn ferrite with figure of merit $M_e = 0.4 \times 10^6$ (from fig. 17.16).

Calculation yields $n_1 = [L / (\mu_0 S / l)]^{1/2} = 1{,}730$, the resistance $R_1 = \rho \, n_1^2 \, p / s = 120$ Ω, and finally the quality factor in air $Q_1 = L\omega / R_1 = 5.3$. This leads to the optimal effective permeability $\mu_{opt} = (M_e / Q_1)^{1/2} = 300$, the quality factor $Q_m = M_e / \mu_{opt} = 1{,}300$, and the quality factor of the inductance $Q = Q_m / 2 = 650$. To determine the thickness g of the air gap, we first note that the permeability of the chosen material at the relevant fequency is much larger than $\mu_{opt} = 300$, hence $\mu_{opt} = l / g$, and $g = 330$ mm.

REFERENCES

[1]   J. IRVING, N. MULLINEUX, Mathematics in Physics and Engineering, (1959) Academic Press, New York-London, 592.

[2]   H.A. KRAMERS, *Phys. Z.* **30** (1929) 522;
        R. DE L. KRöNIG, *J. Opt. Soc. Amer.* **12** (1926) 547.

[3]   J.F. NYE, Physical Properties of Crystals: their representation by tensors and matrices (1987) Oxford Univ. Press, New York.

[4]   R.A. WALDRON, Ferrites: principes et applications aux hyperfréquences (1964) Dunod, Paris.

[5]   J.C. MALLINSON, The foundation of magnetic recording (1987) Academic Press, New York-London.

[6]   T.A. GILBERT, Equation of motion of magnetization, Armor Research Foundation Rep. No 11 (1955) Chicago, USA.

[7]   D. POLDER, *Phil. Mag.* **40** (1949) 100.

[8]   J.L. SNOEK, *Physica* **14** (1948) 207.

[9]   W. DöRING, *Z. für Naturforschung* **3a** (1948) 374.

[10]   C.M. SRIVASTAVA, S.N. SHRINGI, R.G. SRIVASTAVA, G. NANADIKAR, *Phys. Rev.* **B14** (1976) 2052.

[11]   C.M. SRIVASTAVA, O. PRAKASH, R. AYIAR, *Phys. Stat. Sol.* **A64** (1981) 787.

[12]   L. NéEL, *J. Phys. Rad.* **13** (1952) 249.

[13]   L. NéEL, *J. Phys. Rad.* **15** (1954) 225.

[14]   M. GUYOT, T. MERCERON, V. CAGAN, A. MERSEKHER, *Phys. Stat. Sol.* **A106** (1988) 595.

[15]   A. GLOBUS, *J. Phys.* 38, CI.-1(1977).

[16]   M.T. JOHNSON, E.G. WISSER, *IEEE Trans. Magn.* **26** (1990) 1987.

[17]   M. GUYOT, V. CAGAN, *J. Magn. Magn. Mater.* **17** (1982) 202.

[18]   J. SMIT, Proc. 4th Intern. Conf. on ferrites, San Francisco, part. I (1984), F.F.Y. Wang Ed. American Ceramic Society, Columbus, USA.

[19]   D.J. BERGMAN, *Phys. Report* **43** (1978) 377.

[20]   R. LANDAUER, *in* Electrical transport and optical properties of inhomogeneous media, *AIP Conf. Proc.* No 40 (1978), J.C. Garland & D.B. Tanner Eds, AIP, Woodbury, New York.

[21]   J.P. BOUCHAUD, P.G. ZERAH, *Phys. Rev. Lett.* **63** (1989) 1000.

[22]   J.P. BOUCHAUD, P.G. ZERAH, *J. Appl. Phys.* **67** (1990) 5512.

[23]   C. KITTEL, *Phys. Rev.* **73** (1948) 155.

[24]   L.R. WALKER, *J. Appl. Phys.* **29** (1958) 318.

[25]   D. STANCIL, Theory of magnetostatic waves (1993) Springer Verlag, Berlin-New York.

[26]   M. SPARKS, Ferromagnetic relaxation theory (1964) McGraw-Hill, New York.

[27]   C.E. PATTON, Microwave resonance and relaxation *in* [30], 575.

[28]   J. SMIT, H.P. WIJN, Ferrites (1959) John Wiley & Sons, New York.

[29]   W. VON AULOCK, Handbook of microwave ferrite materials (1965) Academic Press, New York-London.

[30]   D.J. CRAIK, Magnetic oxides (1975) John Wiley & Sons, New York.

[31]   R. VALENZUELA, Magnetic ceramics (1994) Cambridge Univ. Press.

[32]   C. GUILLAUD, *Proc. IEEE* **104B** (1957) 165.

[33]   J. SMIT, H.P.J. WIJN, *Adv. Electr.and Electr. Physics* **6** (1954) 69.

[34]   P.I. SLICK *in* Ferromagnetic materials, Vol. 2 (1980) 189, E.P. Wohlfarth Ed., North Holland, Amsterdam.

[35]   J. NICOLAS, *ibid.*, 243.

[36]   E.F. BERTAUT, F. FORRAT, *C.R. Acad. Sci. Paris* **242** (1956) 382.

[37]   J. HELSZAJN, YIG resonators and filters (1985) John Wiley & Sons, New York.

[38]   C. VASSALO, Electromagnétisme classique dans la matière (1980) Dunod, Paris.

Chapter 18

Magnetostrictive materials

Most of the magnetoelastic effects described in chapter 12 have industrial applications: controlled thermal expansion alloys, actuators and sensors. In this chapter we present the most widely used magnetostrictive materials as well as their principal industrial applications.

1. The Invar and Elinvar family

1.1. Controlled thermal expansion alloys

In fabricating complex mechanical assemblies for use under extreme temperature conditions, one has to consider the compatibility of the thermal expansion of the various components. Typical examples include cryogenic liquid containers, the inner and outer walls of which are at very different temperatures, and metallic pieces soldered to materials with low thermal expansion coefficients (e.g. metal-glass solder). Exploiting the strong positive volume magnetostriction of certain iron alloys, metallurgists have developed a whole family of alloys with tuneable thermal expansion, i.e. the thermal expansion coefficients of which can have values much lower than those of classic alloys.

The archetype of these alloys is *Invar®*, of composition $Fe_{65}Ni_{35}$, discovered in 1896 by C.E. Guillaume, which shows practically zero thermal expansion at room temperature. Some characteristics of three Invar-type alloys, commercially developed by Imphy SA, are shown in table 18.1.

Invar is used when a material with very low thermal expansion at ambient or low temperatures is needed. The applications of Invar are very diverse, ranging from enormous methane tankers to tiny parts for metrology, and including such applications as high definition colour TV masks.

The remarkable physical properties of this alloy are related to the extreme sensitivity of its electronic structure to interatomic distances. The iron atoms have the remarkable property, in the face centred cubic alloy, of existing in two different spin states: one spin state conforms to Hund's rules, and gives rise to a high value of magnetic moment (2.2 to 2.5 μ_B) and to a larger crystalline cell parameter while the other spin state gives rise to a smaller magnetic moment (0.8 to 1.5 μ_B) and a smaller cell parameter.

Table 18.1 - Principal physical characteristics of crystallised alloys with tuned thermal expansion (Imphy Ugine Précision)

Properties	Invar	Dilver 0	Dilver P_0
α_T, 10^{-6} K^{-1}	1^a, 0.6^b, 1.9^c, 4.7^d	8.9^a, 9.8^c, 10.6^e, 11.4^f	6.5^a, 5.5^c, 4.9^e, 7.5^f
Specific mass, $kg.m^{-3}$	8130	7500	8250
Tensile stress, MPa	600	800	730
(" after anneal)	(500)	(570)	(550)
Vickers hardness	200	230	220
(" after anneal)	(130)	(180)	(150)
Melting temperature, K	1723	1723	1723
Thermal conductivity, $W.m^{-1}.K^{-1}$	10.5	12.1	17.5
Specific heat, $J.K^{-1}.kg^{-1}$	510	500	500
Electrical resistivity, $\mu\Omega.m$	0.75	0.65	0.49
Curie temperature, K	503	843	698
Magnetic polarization, T	1.6	–	1.6

α_T is measured between –100 and 0°C (a), 0-100°C (b), 100-200°C (c), 200-300°C (d), 300-400°C (e), 500-600°C (f).
The other properties are measured at room temperature.

When heating these alloys from low temperature, we progressively populate the second state, and the resulting cell contraction roughly compensates the normal thermal expansion: in figure 18.1 we see that the volume of the alloy is almost invariable between 0 and about 400 K, an observation which gave rise to the name Invar, a registered trademark of Imphy Ugine Précision.

The Invar effect is an exchange magnetostriction effect but it is the value of the atomic magnetic moment, more so than the exchange integral, which changes with distance.

The Invar alloy is still the reference alloy in this domain, but many other families of alloys showing similar properties have been since discovered. These new alloys are special steels with *tuneable thermal expansion*, and have an ever growing number of applications: metrology, high resolution television, metal-glass solder, methane tankers, etc. [1].

These alloys possess a giant volume magnetostriction, and consequently a forced magnetostriction which is about 50 times greater than that of other ferromagnetic substances ($\partial V / V \partial H = 1.2 \times 10^{-10} / A.m^{-1}$), which can cause problems: to secure perfect dimensional stability, the invar must be shielded from all magnetic fields or else replaced by an antiferromagnetic alloy which, to a first approximation, is insensitive to magnetic fields. However, antiferromagnetic substances with a linear thermal expansion coefficient of less than 4×10^{-6} K^{-1} have not yet been discovered.

The temperature dependence of the linear thermal expansion coefficient α_T of the Fe$_{65}$Ni$_{35}$ Invar alloy is compared with that of Ni in figure 18.1: while anomalies occur for both Ni (representative of "normal" magnetic metals and alloys) and Invar at around T$_C$, an additional anomaly occurs at lower temperature in the Invar. Note the dramatic reduction in α_T below the T$_C$ of Invar: however the thermal expansion coefficient is close to zero just near 100 K and 290 K.

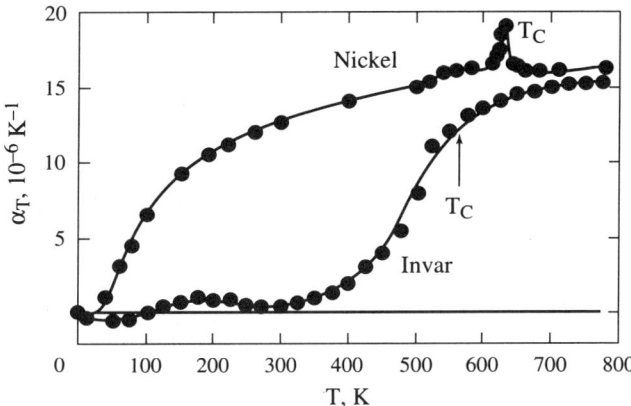

Figure 18.1 - Temperature dependence of the linear thermal expansion coefficient of Invar, *after [2],* **and nickel,** *after [3]*

The Dilvers are used for fracture-free glass-metal soldering in electronic assemblies submitted to significant thermal variations above room temperature.

Certain Fe based amorphous alloys have thermal expansion coefficients which also remain low due to their strong positive volume magnetostriction: e.g. ribbons of Metglas® 2605SC have a linear thermal expansion coefficient $\alpha_T = 5.9 \times 10^{-6}$ K^{-1}, which is half that of Co based Metglas® 2705M. This weak expansibility occurs together with remarkable mechanical and magnetic properties, as can be seen in table 18.2.

1.2. ALLOYS WITH CONSTANT ELASTIC MODULI

Giant second order magnetoelastic effects were also observed in the family of Invar alloys, again by C.E. Guillaume, the discoverer of Invar and a Nobel Laureate for physics. In 1920 he discovered **Elinvar**, an alloy of Fe containing 36% nickel and 12% chromium, which has a face centred cubic structure leading to a Young's modulus which does not vary with temperature: in this case it is the magnetic contribution to the elastic constant which compensates the normal thermal variation of this constant. A number of alloys exhibit this "Elinvar effect".

Such alloys are industrially interesting and can be used in precision engineering, e.g. in watchmaking technology, to make springs the stiffness of which remains unaffected by thermal variations [1].

For applications, the objective is not always to have thermal invariance of Young's modulus, $\partial Y / \partial T = 0$. The aim can be the invariance of the specific property exploited in a given application.

For example, if we use the mechanical resonance frequency of a bar of length ℓ in a longitudinal vibration mode $f_0 = (1/2\ell)(Y/\rho)^{1/2}$, it would be convenient to make f_0 independent of temperature. This leads to: $\partial \ln(Y)/\partial T = -\partial \ln(\ell)/\partial T$, since the specific mass ρ varies as ℓ^{-3}. In general, Young's modulus Y decreases as temperature increases while ℓ increases: the two effects can thus compensate, leading to the thermal stability of f_0.

2. MAGNETOSTRICTIVE MATERIALS FOR ACTUATORS

2.1. MATERIALS WITH HIGH ANISOTROPIC MAGNETOSTRICTION

Materials with a high anisotropic magnetostriction (Joule magnetostriction) are sought to make all sorts of actuators. They must produce relatively large deformations in the smallest excitation fields possible. As magnetostriction is an even effect with respect to magnetization, it is necessary to polarise the material with a static magnetic field so as to operate near the inflection point of the $\lambda(H)$ characteristic, called the *average working point*.

We recall that the coefficients s_{ij}^H, μ_{kl}^σ and d_{mn}, defined (see eq. 12.41) around this working point, may be functions of the applied field and depend on the previous history of the material. Values found in literature are then average values.

Characteristic data of diverse magnetostrictive materials are compared in table 18.2. k_{33} and d_{33} are defined in § 9.1 of chapter 12.

We see that certain rare earth based alloys discovered around 75 (in grey in tab. 18.2) show magnetostrictive deformations 50 to 100 greater than those of conventional materials discovered in the 1950's and 1960's (nickel and alfer); we will now see why and how.

2.2. TERFENOL-D

This alloy deserves particular attention for two reasons: firstly, it is the best candidate for a room temperature magnetostrictive actuator (in 1990 it set the world record for sonar acoustic power density); secondly, it is the product of model research in the domain of magnetic materials.

**Table 18.2 - Curie temperature, and room temperature physical properties
of some magnetostrictive materials**

Substance	T_C (K)	J_s (T)	λ_s (10^{-6})	ρ_e ($\mu\Omega m$)	k_{33}^{max}	d_{33}^{max} (mA^{-1})
Nickel	631	0.63	−36	0.07	0.31	-3.1×10^{-9}
Alfer (Fe 13%Al)	773	1.3	+40	0.9	0.32	7.1×10^{-9}
45 Permalloy (Fe 65%Ni)	713	1.6	+27	0.6	0.17	–
2V-Permendur (2V, 49Fe, 49Co)	1,253	2.4	+70	0.3	0.26	–
Ni 4%Co	683	0.68	−31	0.1	0.50	–
Magnetite (Fe$_3$O$_4$)	853	0.61	+40	10^2	0.36	–
Nickel Ferrite (NiFe$_2$O$_4$)	863	0.33	−33	$>10^{10}$	0.20	–
Ni$_{0.98}$Co$_{0.02}$Fe$_2$O$_4$	863	0.33	−32	$>10^{10}$	0.38	–
DyZn [001] at 4.2 K	140	anisotr.	(+5,300)	0.5	–	–
Terfenol (TbFe$_2$)	698	1.1	+1,750	–	0.35	–
Terfenol-D* (Tb$_{0.3}$Dy$_{0.7}$Fe$_2$)	653	1.0	+1,100	0.6	0.75	57×10^{-9}

* *Note that d_{33} is very sensitive to stress for Terfenol-D (57×10^{-9} m . A^{-1} under
0 MPa, but only 10×10^{-9} m . A^{-1} under 40 MPa).*

Originally, engineers charged with designing magnetostrictive actuators were
interested in two classes of materials:

♦ 3d metals and alloys which are capable of working at room temperature, and under
   relatively low magnetic fields, but have too weak a magnetostriction to compete
   with piezo-electric ceramics,

♦ 4f metals which show giant magnetostrictions, but very low Curie temperatures,
   complex magnetic structures, and such high magnetocrystalline anisotropies that
   large magnetic fields would be required to produce noticeable deformation.

It appeared that it might be possible to obtain materials with high magnetostriction
even at room temperature, and under reasonable excitation, by alloying rare earths
with 3d metals such as iron and/or cobalt.

A.E. Clark discovered the alloy TbFe$_2$, also known as *Terfenol* (**Ter**bium, **Fe**, **N**aval
Ordnance **L**aboratory), which combines a relatively high Curie temperature due to the
iron and a giant magnetostriction due to the terbium. However, this alloy's high
magnetocrystalline anisotropy limits its use. Soon after, in 1975, Clark prepared a new
alloy Tb$_{0.3}$Dy$_{0.7}$Fe$_2$, known as *Terfenol-D* (D for **D**ysprosium), which maintains a
giant magnetostriction and a relatively high Curie temperature, but which can be
saturated in moderate magnetic fields. Saturation is possible for this composition
because, to first order, the contributions of terbium and dysprosium to the magnetocrystalline anisotropy cancel at room temperature. We note the strong anisotropy of its
magnetostriction: $\lambda^{\epsilon,2}$ is 2.4×10^{-3}, while $\lambda^{\gamma,2}$ is at least one hundred times weaker.

From equation (12.21), that is: $\lambda_s = (4/15)\lambda^{\gamma,2} + (2/5)\lambda^{\varepsilon,2}$, we see that an *isotropic* polycrystalline Terfenol-D sample will attain just 40% of $\lambda^{\varepsilon,2}$. It thus seems preferable to prepare textured samples, favouring the [111] direction if possible. This has proved difficult, and samples used on an industrial scale have, up to recently, been textured bars with orientation [11$\overline{2}$]. Two techniques have been developed to prepare samples: the modified Bridgman (MB) technique, and the "Free Standing Zone" or FSZM technique. Both techniques produce twinned bicrystals with the [11$\overline{2}$] direction in the twin plane, which produces significant internal stress at the twin boundary, and somewhat reduces the d_{33} coefficient of magnetostrictivity. The maximum deformation $\Delta\lambda = (1/6)\lambda^{\gamma,2} + (5/6)\lambda^{\varepsilon,2}$ expected for such a texture is of the order of 2×10^{-3}. Much effort is being made to improve the performance of this alloy and to reduce its fabrication cost: [111] oriented bars have been prepared in China by the Czochralski method in a cold crucible and in levitation [4], and more recently the Japanese company TDK brought to the market bars produced from a powder route which show significant magnetostrictivity in weak fields [5].

Due to their texture, there is some confusion about the definition of *magnetostriction coefficients* for materials with "giant magnetostriction". λ_s is by definition the coefficient measured on a non-textured polycrystalline sample, given by equation (12.21) cited above.

In fact, manufacturers often give as the "magnetostriction coefficient" the value $\lambda_{//}$ of the longitudinal magnetostriction measured in the [11$\overline{2}$] direction in a pre-stressed grain oriented sample, which is markedly higher than the value of λ_s. Finally, dynamic deformation under resonance conditions can greatly overcome the theoretical static limit, as we saw in § 9.3 of chapter 12: a dynamic deformation of 3.5×10^{-3} was observed under resonance for Terfenol-D, although λ_s is only 1×10^{-3}.

It is the high energy density of Terfenol-D which allows it to compete with ceramic piezo-electrics. It is instructive to compare the relative displacement of the ends of bar-shaped samples of laminated ceramic PZT and Terfenol-D as a function of the pressure applied by the actuator (fig. 18.2).

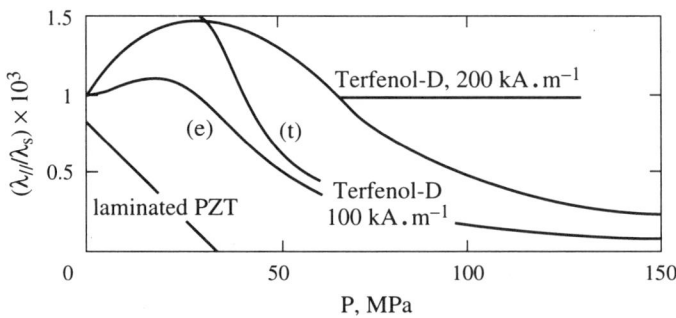

Figure 18.2
Comparison of the reduced deformation of two identically shaped bars of Terfenol-D, and laminated ceramic PZT, as a function of the applied pressure P

The available energy is given by the area under the curve: thus the Terfenol-D compares favourably as shown for the experimental curves at H = 100 and 200 kA.m^{-1}. The theoretical curve (t), given for H = 100 kA.m^{-1}, is calculated in exercise 3 *neglecting magnetocrystalline anisotropy*, which explains why it does not describe correctly the behavior at low stress (theory predicts that all compressive stresses less than 35 MPa increase $\lambda_{//}$ from λ_s to $(3/2)\lambda_s$ but experiment shows that a field of at least 200 kA.m^{-1} must be applied if $\lambda_{//}$ is to approach such a value).

Until recently, at least eight companies had commercialised Terfenol-D: two in Europe, Johnson & Matthey - Rare Earth Products in Great Britain, and Feredyn Europe in Sweden which have since ceased production, two in the USA, Etrema Products (subsidiary of Edge Technologies), and EDO, the most important producer of piezoelectric ceramis, and four in Japan, TDK, NKK, Toshiba and Sumitomo. The standard products are cylindrical bars, tubes and disks. The properties vary slightly according to the preparation technique used, the sample's history and the operating conditions. Some average physical properties of products supplied by three producers are given in table 18.3.

Table 18.3 - Physical properties of Terfenol-D alloys at 20°C

Properties	Units	Etrema a	Reacton b	Magmek 86 c	Magmek 91 c
Specific mass (ρ)	kg.m^{-3}	9,250	9,250	9,100	7,300
Linear thermal expansion coefficient (α_T)	10^{-6} K^{-1}	12	12	12	
Compressibility modulus (κ)	GPa	90			
Young's modulus (Y^H)	GPa	25-35		26.5^d	22^d
Young's modulus (Y^B)	GPa		50	55	
Velocity of sound (v^H)	m.s^{-1}	1,720		1,720^d	1,740
Velocity of sound (v^B)	m.s^{-1}		2,450	2,450	
Compressive strength	MPa	700		300	250
Tensile strength	MPa	28		28	120
Specific heat (C_p)	J.K^{-1}.kg^{-1}	320-370			
Thermal conductivity	W.m^{-1}.K^{-1}	10.5			
Electrical resistivity (ρ_e)	$\mu\Omega$.m	0.6	0.6	0.6	10^4
Curie temperature (T_C)	K	653		660	
Magnetic polarization ($\mu_0 M_s$)	T	1			
Relative permeability (μ_{33}^τ/μ_0)	–	5 - 10	10	9.3	2 - 12
Relative permeability ($\mu_{33}^\varepsilon/\mu_0$) ^{e}	–			4.5	
Joule magnetostriction ($\lambda_{//}$)	10^{-3}	1.5 - 2	1.5	1.4 - 1.8	1.1
Magnetomechanical coupling (k_{33}^{max})	–	0.7 - 0.8	0.5 - 0.8	0.72	
Static magnetostrictivity (d_{33}^{max})	nm.A^{-1}	24^f 57^g	12	17	4
Dynamic magnetostrictivity (d_{33}^{max})	nm.A^{-1}		6^h		
Elastic energy density	kJ.m^{-3}	14 - 25		14 - 25	11

^{a} *Etrema Products-Edge (USA)* ^{b} *Johnson-Matthey (UK)* ^{c} *Feredyn AB (Sweden)* ^{d} *Demagnetised state* ^{e} $\mu_{33}^\varepsilon/\mu_0 = \mu_{33}^\tau(1 - k_{33}^2)/\mu_0$ ^{f} *MB alloy* ^{g} *FSZM* ^{h} *alloy under 45 GPa.*

We note that the electrical resistivity of Terfenol-D is in general weak, except for the Magmek-91 composite which is specially prepared for high frequency applications. This composite also possesses a much higher tensile strength (120 MPa compared to 28 MPa), but these improved characteristics are achieved at the cost of a loss of magnetostrictivity ($d_{33} = 4$ nm.A^{-1} instead of 10 to 60 nm.A^{-1}). Finally we note that the prototype [111]-oriented single crystal bars produced in China [4] have an even higher magnetostriction reaching 75 nm.A^{-1}.

Research into terfenol alloys remains active, and it is probable that a greater range of transducer bars will be available in the future. The coefficient of magnetomechanical coupling k_{33}^{max}, which is an important parameter, now reaches values of 0.75, which is much higher than values achieved in the past.

It must be remembered that the parameters quoted in table 18.3 correspond to the optimum values recorded at 20°C; the magnetostrictivity decreases rapidly as we move away from this temperature, and also if we pre-stress the transducers, a treatment which is generally desired due to their fragility.

Figure 18.3 shows the thermal variation of the magnetostriction of a bar of Terfenol-D submitted to a compressive stress of 19 MPa: in weak fields, from 20 to 40 kA.m^{-1}, the magnetostriction remains almost constant between 0°C and 90°C, but this is not the case in higher fields [6].

The compensation between the positive magnetocrystalline anisotropy of dysprosium and the negative magnetocrystalline anisotropy of terbium is perfect at only one given temperature, which explains why the performance of this alloy is strongly reduced away from this optimum temperature. Nevertheless, the exceptional qualities of these materials justify their use despite the need for crude temperature regulation.

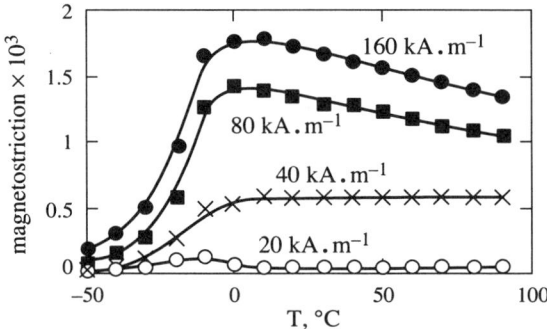

Figure 18.3 - Thermal variation of the magnetostriction of Terfenol-D
These curves, which are very sensitive to stress, were measured on bars submitted to a pre-stress of 19 MPa.

Too high an excitation frequency can also degrade the above quasi-static performance. Due to the skin effect related to eddy current losses, the dynamic magnetostrictive characteristics, such as the apparent magnetomechanical coupling coefficient, may depend on the shape of the active element if the frequency becomes higher than a characteristic frequency f_c, which, for a cylinder of diameter d and electrical conductivity γ, is given by:

$$f_c = 2 / (\gamma \pi \mu_{33}^\varepsilon d^2) \tag{18.1}$$

where μ_{33}^ε is the magnetic permeability at constant deformation. The use of a composite of low electrical resistivity is then recommended.

In conclusion, at room temperature, Terfenol-D is comparable with ceramic PZT because of its strong saturation deformation and its high power density. Moreover, for low frequency applications, the lower velocity of sound (2,450 m.s^{-1} for Terfenol-D compared to 3,100 m.s^{-1} for PZT) allows resonance to be achieved with shorter bars. It is also better than electrostrictive materials such as PMN-PT, the good performance of which disappears at 40°C (their Curie temperature).

The principal disadvantages of Terfenol-D seem to be its hysteresis, a relative sensitivity to temperature, the necessity of magnetically polarising them and their high cost; finally we mention the fragility of the stoichiometric alloy under traction. The remedy is to hold the samples under compression and to remain slightly under-stoichiometric in iron: this is why the composition of alloys available on the market is $Tb_{0.3}Dy_{0.7}Fe_x$ with x varying from 1.90 to 1.95 without a great variation in magnetostrictive performance.

2.3. USE OF TERFENOL-D IN ACTUATORS

Owing to their strong magnetostriction, Terfenol-D alloys are able to produce significant force, and generate rapid, precise movement with considerable power. The main industrial applications of Terfenol-D are linear actuators, which in essence are magnetostrictive bars, polarised by a static magnetic field, and usually submitted to a compressive stress, which elongate under the influence of a quasi-static or dynamic excitation field.

Some concrete examples will illustrate the precautions to be taken in selecting the most appropriate material for a given application: electro-valves (fuel injection, cryogenic applications…), micro-pumps (heads of ink jet printers), automatic tool positioning with wear compensation (machine tools), active vibration damping, fast relays, gears, auto-locking actuators (robotics), rapid shutters, automatic focusing (optics), hooping under field (when a bar of Terfenol-D elongates, its diameter decreases). More detailed information can be obtained in a recently published reference book [7].

2.3.1. Linear actuator

An example of a linear actuator is shown in figure 18.4. We see here how much the magnetic and mechanical aspects intimately overlap: it is thus best to design the entire system so as to optimise performance. Software has been developed for this purpose.

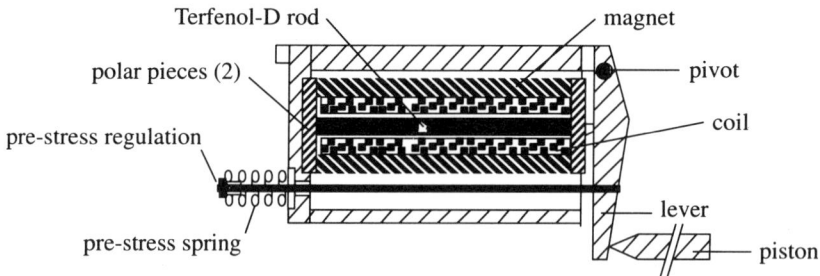

Figure 18.4 - An example of a linear actuator
(Documentation Etrema Products, Ames, IA, USA)

2.3.2. Differential actuator

Another prototype, the differential actuator, deserves to be mentioned because starting from two rectilinear movements, it can produce rotational motion of the mobile axis (fig. 18.5).

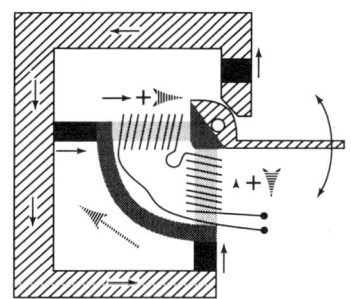

Figure 18.5 - Differential actuator

Static flux is shown by black arrows, dynamic flux by the hashed arrows. Soft magnetic material: hashed volumes; Terfenol-D bar: in light grey (under coils); magnets: in black; insulating soft magnetic materials: in dark grey [8].

Nevertheless, the motion is very restricted. The static (polarization) and dynamic (excitation) magnetic fluxes follow different paths, so it is possible to observe, in two bars placed at right angles, a dynamic magnetic field which reinforces the static field for one bar and opposes the magnetic field for the other. Thus, the first bar elongates while the second shortens, which generates the desired rotational motion.

To allow magnetostrictive bars to remain permanently under stress, it is also possible to use two aligned bars which work symmetrically: when the first elongates, the second shortens. This "push-pull" assembly is particularly well adapted to active position control since, after expansion, one of the two elements is always returned to its contracted position by the other which then expands.

2.3.3. Wiedemann effect actuators

Another very particular type of actuator exploits the direct Wiedemann effect: it consists of a spiral spring made from a magnetostrictive wire which is coated by a winding which allows it to be longitudinally magnetised.

When a current circulates in the spring, the latter is submitted to a helical magnetic field (see § 5 of chapter 12), and experiences a twisting due to the Wiedemann effect, which generates a linear displacement of one end of the spring if the other end remains fixed (fig. 18.6). Such an actuator was designed and actually used as a micropositioner of heavy optical pieces in a telescope.

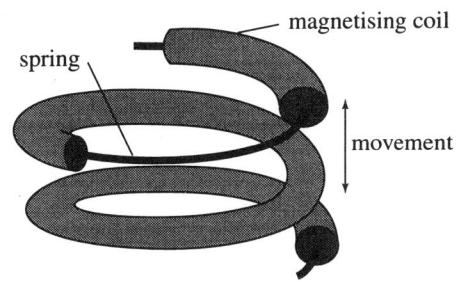

Figure 18.6
Wiedemann effect actuator [9]

2.3.4. *Magnetostrictive linear motor*

This motor consists of just a bar of Terfenol-D and a tube of the same diameter, which acts as the stator: as Joule magnetostriction develops at constant volume, it is possible to move the bar up and down the tube by applying a magnetic field, and then to block the bar in a given position by suppressing the magnetic field. What is happening is that as the bar elongates its diameter decreases under the magnetic field, and then increases again when the field is suppressed. This explains the total absence of backlash in this type of motor designed by L. Kiesewetter in Berlin: the mobile part is blocked in the absence of exciting currents. The movement is thus generated by a magnetic field applied along a small length starting at one end of the bar: the diameter decreases and the length of bar which experiences the field is free to elongate. As the field is moved to the other extremity of the bar, the deformed zone also moves, and finally, when the magnetic field is suppressed, the bar regains its initial length but has been displaced by a distance h in the direction opposite to that in which the field was moved, as shown in figure 18.7.

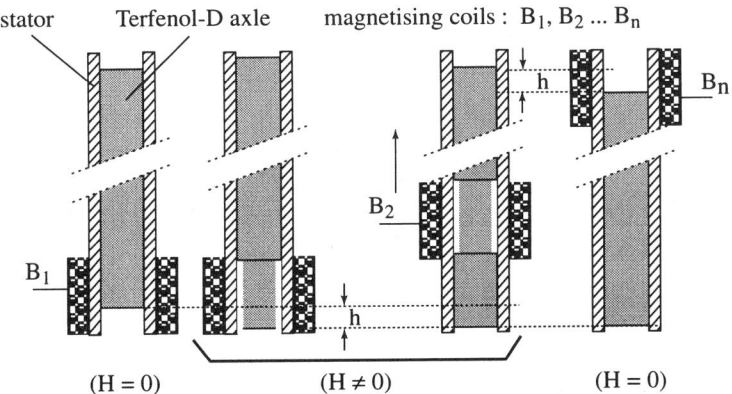

Figure 18.7 - *Kiesewetter's linear magnetostrictive motor*

The axle, made of Terfenol-D, advances like a caterpillar, by crawling.
It can support very heavy loads, even when stationary [10].

Repeating the cycle several times, we have a linear stepper motor. Reverse motion is produced by applying the field in the opposite direction. In practice, the field is created successively by n coils B_1, B_2, B_n, placed side by side. For a 10 cm long bar of magnetostriction 10^{-3}, a displacement of 10 μm can be produced by exciting ten 1 cm segments successively. With 10 cycles per second, the bar is moved with a velocity of 0.1 mm per second. The length of the run is limited by the geometry of the device, more specifically by the length of the stator. The maximum admissible load depends on the mechanical performance of the stator: e.g. such a motor could be used to position loads of more than a ton with a 1 micron precision (see exercise 4 at the end of the chapter).

2.3.5. Magnetostrictive rotary motors

It is possible to create rotary stepper motors by combining two out of phase linear movements according to the principles of friction motors or four stroke electric motors. Most of these motors have two essential characteristics in common, a very *strong torque* (even at zero velocity), and a *total absence of backlash*.

Friction motors comprise two perpendicular linear actuators. One can put the stator in contact with a rotor, and when the contact is made, the second actuator displaces the rotor. The first actuator then retracts, and the second can return to its rest position without moving the rotor. Actually, the two movements are simultaneous, and the point of contact of the stator describes an ellipse. A motor torque of 2 Nm has already been achieved with such a motor. Significant progress is still expected, but the main drawback of this type of motor is the friction wear of the surfaces in contact, as in the case of piezoelectric motors.

The four stroke rotary electric motor works in a different way:
1. actuators grip the rotor,
2. then they are pushed by a second assembly of actuators,
3. the grip of the first actuator is released, freeing the rotor,
4. and the second assembly of actuators return to their rest position.

The motor is then ready for a new cycle. A 1991 prototype was able to produce a torque of 12.2 Nm on the rotor [11].

Mixed motors, which use magnetostrictive and piezo-electric materials at the same time, have also been constructed. The mechanical performances of these two classes of materials are very different and may be complementary, in particular in the domain of shear resistance.

In conclusion, we expect strong torques, even at low speeds, for magnetostrictive motors. At zero velocity the motor maintains its position without the least bit of backlash. Comparison with piezo-electric motors is not easy because of the extensive research being carried out in both domains. However, we can predict that magnetostrictive motors, because of their higher costs, will be reserved for high energy density applications, e.g. aerospace devices.

Specific power ratings for the major motor types of the 1990's are of the order of 20 W.kg^{-1} for a standard electric motor, 50 W.kg^{-1} for a servo-motor, 80 W.kg^{-1} for a piezo-electric based ultrasonic motor, and 100 W.kg^{-1} for a magnetostrictive motor. The comparison of such different devices is not easy, as it should also take account of the weight and bulkiness of the energy sources used.

2.3.6. Sonar

The transmission of electromagnetic waves in salty ocean water encounters such attenuation that it is impossible to imagine guiding or locating an obstacle with the aid of radar. Thus, in 1935 submarines started using the sonar technique which involves emitting acoustic waves and measuring the time it takes the echoes to return to deduce the distance from a reflecting object. At first nickel was used, as it could act as the emitter, due to Joule magnetostriction, and at the same time as the receiver, due to the inverse effect. When nickel became scarce during the war, it was replaced by an aluminium iron alloy named "Alfenol" or "Alfer". After the second world war, magnetostrictive materials were replaced by ceramic piezo-electrics. Today, there is renewed interest in the use of magnetostrictive devices, and prototypes made with Terfenol-D have broken the world record for emitted power density.

Emitters usually consist of pre-stressed bars, with one extremity connected to an impedance adapter which couples the bar to the sea water. The polarization is induced by permanent magnets [12].

3. SENSOR MATERIALS

Research has been active in the domain of magnetostrictive sensors ever since the discovery of the Wiedemann effect torque-meter by Kobayosi in 1929, and it has experienced an unprecedented expansion with the discovery of Metglas 2605SC. The best materials for magnetostrictive sensors are not those with giant magnetostriction, but those which can convert elastic energy into magnetic energy, or *vice versa*, with an efficiency that approaches 100% so as to minimise signal loss.

This condition thus favours materials with the *highest possible value of the magnetomechanical coupling coefficient and the lowest possible losses*. The parameter to be optimised is not λ_s, but $\partial\lambda/\partial H$, or the magnetostrictivity d_{33}, which also characterises the sensitivity of the material's magnetization to stress. Metallic nickel –although very vulnerable to shock– was used as a sonar detector during the second world war, but was then replaced by piezo-electric ceramics. In the 1980's, a metallic glass with excellent magnetoelastic properties, Metglas® 2605SC, the characteristics of which are given below, appeared as a serious competitor for piezo-electric ceramics.

The magnetostrictivity $\partial\lambda/\partial H$ can be increased by annealing the material. This releases internal stresses and increases the initial magnetic permeability, without necessarily modifying the value of the saturation magnetostriction, λ_s (see fig. 12.15). In the case of amorphous materials, an anneal at temperatures *less than* the recrystallization temperature T_X, can sometimes induce a structural relaxation and thus slightly modify λ_s, while an anneal at a temperature *above* T_X always strongly modifies λ_s, which is sensitive to local atomic order.

Note - *There are experimental circumstances when the above demands do not apply. Realising it can lead to a more economical design. Later we will see how to construct a torque sensor based on the inverse Wiedemann effect. In this device, the sensor contains a magnetic head which is situated some tenths of mm from the shaft to allow torque measurement even when the shaft is rotating: thus, most of the magnetic reluctance is concentrated in the air-gap. Whether the material has a permeability of 10,000 or 300, and whether the magnetomechanical coupling coefficient k is 0.99 or 0.75, the sensor's response curve will be practically the same!*

The solution is to use a shaft made of ordinary steel, and to measure the modulation of its permeability under stress: generally the sensitivity will be sufficient.

3.1. METGLAS 2605SC

Today's best isotropic magnetostrictive material is an iron rich metallic glass, Metglas® 2605SC, produced by *Allied Signal Inc.*. This is the material most suited for making high sensitivity magnetoelastic sensors. It is an $Fe_{81}B_{13.5}Si_{3.5}C_2$ alloy prepared by melt spinning: a stream of molten alloy is projected onto a cold wheel where it is cooled by a thousand degrees in one thousandth of a second. This freezes the alloy in the liquid state (it is amorphous, i.e. non-crystallised). The ribbons have widths of 10 to 100 mm, and a thickness of about 20 μm.

Table 18.4 contains some properties of this alloy which is frequently used in all sorts of sensors. Its electrical resistivity is 20 times greater than that of nickel, for a comparable magnetostriction which is however positive. Note its very interesting elastic properties, its weak thermal expansion coefficient, and its good corrosion resistance.

Annealing the ribbon in a magnetic field applied in its plane, but perpendicular to the ribbon length, for 10 minutes at 642 K, induces a very weak magnetic anisotropy ($K = 35$ J $\cdot$ m^{-3}) with the easy direction of magnetization along the direction of the applied field. A longitudinal magnetic field of the order of 50 A $\cdot$ m^{-1} is sufficient to compensate for this weak anisotropy: thus a magnetomechanical coupling coefficient close to 1 is observed ($k_{33} = 0.97$).

Table 18.4 - Physical properties of a Metglas 2605SC ribbon annealed under magnetic field

Properties (symbols)	Units	Annealed Metglas 2605SC
Density (ρ)	kg $.$ m^{-3}	7,320
Crystallization temperature (T_x)	K	753
Linear thermal expansion (α_T)	10^{-6} K^{-1}	5.9
Young's modulus (Y^H)	GPa	25
Young's modulus (Y^B)	GPa	200
Vickers hardness under 50g (H_v)	–	880
Tensile elastic limit	MPa	700
Thermal conductivity	W $.$ m$^{-1}.$K^{-1}	9
Electrical resistivity (ρ_e)	$\mu\Omega.$m	1.35
Curie temperature (T_C)	K	643
Magnetic polarization ($J_s = \mu_0 M_s$)	T	1.35 (H = 80 A$.$m^{-1}) 1.61 (satn.)
Initial relative permeability (μ_{33}^τ / μ_0)	–	20,000 (80,000 at σ_{zz} = 1 MPa)
Maximum relative permeability (μ^{max})	–	300,000
Magnetostriction coefficient (λ_s)	10^{-6}	30
Magnetomechanical coupling coeff (k_{33}^{max})	–	0.97 (H = 50 A$.$m^{-1})
Static magnetostrictivity (d_{33}^{max})	nm$.$A^{-1}	1,000

3.2. SENSORS BASED ON INVERSE MAGNETOELASTIC EFFECTS

The inverse Joule magnetoelastic effect (see § 6.2 of chap. 12) is exploited in simple *force*, *percussion* and *pressure* sensors. The geometry of these sensors can be infinitely varied, but the working principle remains the same: pick-up coils measure the flux variations generated by changes in stress applied to a weakly magnetised Metglas ribbon.

3.2.1. Force sensor

A ribbon of metallic glass, of composition $Co_{75}Si_{15}B_{10}$, which has a negative magnetostriction λ_s, is suspended and maintained vertical by the application of a weak tensile force (fig. 18.8-a). An exciting coil is used to create an AC magnetic field at the centre of this ribbon. Two pick-up coils detect the induced flux variations. The hysteresis loop corresponding to this "free ribbon" situation is shown in figure 18.8-b, where it is compared with the loop measured when a supplementary stress is applied in addition to the initial tensile stress ("loaded ribbon"). The flux variation in the pick-up coils, which is related to the decrease in the permeability of the ribbon, is significant around zero field. Thus such sensors have a very high sensitivity, but are also very sensitive to all variations in excitation current: this is the reason why it is preferable to operate near the "pseudo-saturation" value of induction, B_{max}. This markedly reduces the sensor's sensitivity to current variations; figure 18.8-c shows the characteristic response of such a sensor.

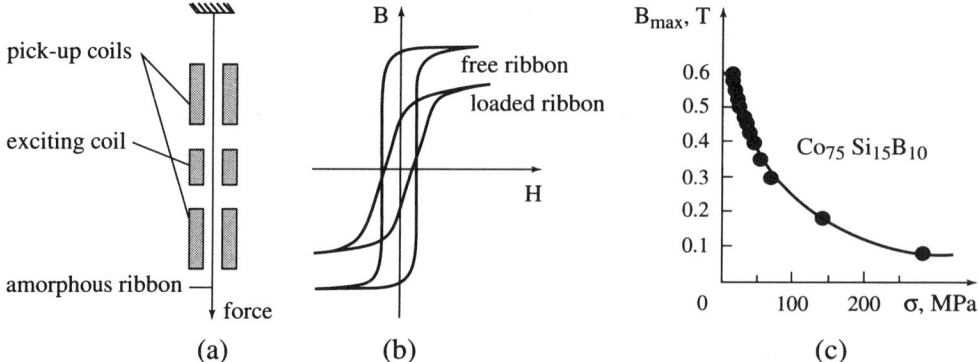

Figure 18.8 - Force sensor in which pick-up coils measure the flux variations generated by changes in applied stress [13]

3.2.2. Ultrasensitive magnetometry

Based on a very different principle, a sensitive magnetometer can be made from a ribbon of Metglas deposited on a piezo-electric ceramic substrate oscillating at resonance; under the effect of this oscillatory stress, the magnetization of the ribbon is modulated with an amplitude proportional to the external magnetic field. When the magnetic field is modulated at a low frequency F, it is possible to detect the AC component by lock-in detection: the ultimate sensitivity of the machine, which varies with frequency $(1/F)^{1/2}$, reaches that of ultrasensitive "flux gate" type magnetometers, i.e. of the order of 10 pT at 1 Hz.

3.2.3. Inverse Wiedemann effect torquemeter

The inverse Wiedemann effect can be exploited to measure, without contact, and from a distance, the torque exerted on a ferromagnetic, and magnetostrictive shaft. Under the influence of the torsion generated by the torque, such a shaft will have an anisotropic distribution of mechanical stresses, zero along and at ±90° to the torque direction, maximum and of opposite sign at ±45°. If a torsion to the right increases the magnetic permeability at +45° it will reduce it at –45°, while a torsion to the left has the inverse effect (see § 6.3 in chap. 12).

A magnetic bridge circuit consists of an excitation arm (P_1, P_2) and a detection arm (S_1, S_2). Flux closure occurs at the shaft, and the detection coils deliver an alternating signal which is proportional to the torsion. The signal cancels out in the absence of a torque, i.e. when the bridge is balanced. Many types of such *torquemeters* exist. The magnetic excitation-detection circuit is separated from the shaft by an airgap, thus it is possible to make measurements on mobile shafts: this type of sensor is interesting for, among others, the automobile industry (fig. 18.9).

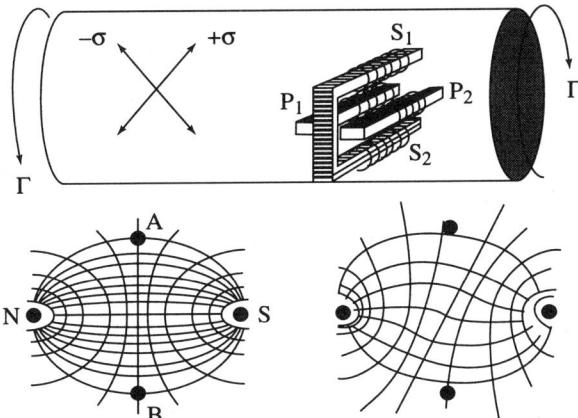

Figure 18.9 - Principle of the "Torductor" torquemeter and schematic diagram of the equipotential lines before and after deformation [14]

If the shaft is non-magnetic, or if the distribution of stresses at the surface of the bar is too inhomogeneous, it is possible to stick Metglas ribbons to the surface which act as the active element of the sensor.

We now mention the multisensor of Tyren and Lord, designed to simultaneously measure the torsion torque and angular velocity of a rotating shaft [15]. This sensor comprises an amorphous ribbon, stuck to the surface of a shaft, which is sensitive to torque related stresses; when its permeability varies, the inductance of a coil, wound around the ribbon and connected in series with a capacitance, is modified. The LC circuit thus established is entirely passive, and turns with the axle to which it is attached. A stationary emitter-detector measures the resonance frequency of the LC circuit: measuring the frequency variation we can calculate the intensity of the torsion torque while the intensity of the received signal, modulated at each turn of the shaft, gives the angular velocity. Moreover the same emitter, working at many different frequencies, can control many sensors located at different points of the rotating part.

3.3. SENSORS BASED ON DIRECT MAGNETOELASTIC EFFECTS

3.3.1. Magnetostrictive magnetometers

The direct Joule magnetostriction effect can be used to detect and measure extremely small magnetic field variations: the US Naval research labs are very active in this area of high resolution magnetometry. Various techniques can be used: e.g. modulation of the optical path in an interferometer in which a bar of Terfenol-D, which is sensitive to the field variations, is inserted; or variation, as a function of the intensity of the magnetic field, of the index of refraction of an optic fibre which is covered with a layer of Metglas [16, 17].

The sensitivity of such devices can reach 10^{-3} to 10^{-4} A.m^{-1}.

3.3.2. Position detector

The direct Wiedemann effect has been used to generate a local *shear* which then propagates along a wire or rod. For example, a ring shaped permanent magnet can slide along the outside of a ferromagnetic tube. When a current impulse goes through the tube, a shear occurs at the level of the ring (where the resulting magnetic field becomes helical for a short instant). This deformation pulse then propagates along the tube, and is detected at the tube end. A measure of the transit time allows this device to be used as a *position sensor*, e.g. the Captosonic device commercialised by Equipiel (fig. 18.10).

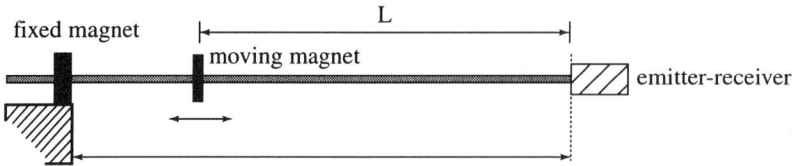

Figure 18.10 - The "Captosonic" position sensor [18]

4. INTEGRATED ACTUATORS AND SENSORS

The development of integrated microsystems has given rise to active research into magnetoelastic effects in thin films.

The aim is to achieve the greatest possible deflection in the smallest magnetic field possible: thus the aim is to find a material which has, at the same time, a strong magnetoelastic coupling coefficient $b^{\gamma,2}$, and as weak as possible an anisotropy field, that is a low magnetic anisotropy and a high magnetization, so as to have a high derivative $\partial b(H)/\partial H$. The most important parameter is thus $b(H)$ (or the coefficient $b^{\gamma,2}$), and not the function $\lambda(H)$ (or the coefficient $\lambda^{\gamma,2}$) as with bulk materials (see § 4.6 of chap. 12).

Studies are still at the exploratory stage, but it already seems that amorphous rare earth-cobalt thin films have much better magnetostrictive characteristics than amorphous thin films of Terfenol, which are based on rare earths and iron. The latter show a notable sperimagnetism (see fig. 4.21), and consequently, a very low magnetization.

Amorphous Tb-Co films have quite a high magnetostriction, but a relatively low magnetization, so that it was proposed to create an artificial structure consisting of alternating layers of Tb-Co and Fe-Co: the Fe-Co alloy is crystallised and has a high magnetization. Provided the alternating layers are not too thick, a strong magnetic coupling may occur between the different Fe-Co layers thus reinforcing the exchange field in the amorphous magnetostrictive layers, and both magnetizations thus rotate together: this multilayered material thus possesses both a notable magnetostriction and a low saturation field.

$Fe_{0.75}Co_{0.25}/Tb_{0.27}Co_{0.73}$ (5.5 nm / 4.5 nm) multilayers can have significant values of magnetostriction in weak fields (3.5×10^{-4} deformation in a field of under 80 kA.m^{-1}), as can be seen in figure 18.11, where their high performances are compared with those of amorphous $Tb_{0.27}Co_{0.73}$.

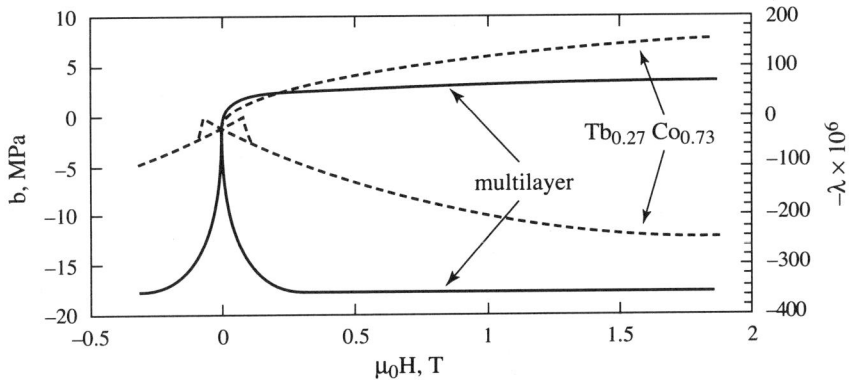

**Figure 18.11 - Field variation of the magnetoelastic function b(H)
defined by equation (12.23) for two thin films**

Comparison between a crystallised thin film of $Tb_{0.27}Co_{0.73}$ and a multilayer of composition $Fe_{0.75}Co_{0.25}/Tb_{0.27}Co_{0.73}$ (4.5 nm/5.5 nm). The right hand scale is the magnetostriction that would be observed for a bulk material with the same magnetoelastic coupling assuming a shear modulus of 50 GPa [19].

Even more spectacular performances were obtained for multilayers of composition $Fe_{0.75}Co_{0.25}/Tb_{0.18}Co_{0.82}$ (6.5 nm / 4.5 nm) which have a magnetoelastic coupling coefficient $b^{\gamma,2} = -44.5$ MPa ($\lambda^{\gamma,2} = 890 \times 10^{-6}$), and a derivative, under a field of 2 mT, of $\partial |b^{\gamma,2}| / \mu_0 \partial H = 4,800 \text{ MPa.T}^{-1}$ [20].

The deformations achieved with such multilayers are now sufficient to plan making micro-actuators, which can be controlled from a distance by simply applying a magnetic field. These devices will find applications in surgery or in medicine (micro-pumps for the controlled delivery of medicine internally, micro-scalpels, etc.), or in electronics, e.g. to make rapid response micro-switches.

A thin film magnetostrictive actuator could work in torsion mode (something which is not possible with piezoelectric materials): this mode lends itself well to a compensation of the thermal derivative [21]; nevertheless, the mechanical realization of such a device remains difficult.

5. CONCLUSIONS AND PERSPECTIVES

Magnetostrictive materials have a bright future in sensor applications, principally because of the possibility to sense the control signal from a distance and without wires, which is not the case for piezo-electric ceramics. For the same reason, magnetostrictive micro-actuators and micro-sensors are beginning to be used in medical applications, though research in this area is just beginning.

More classical magnetostrictive actuators (motors, sonar devices) must compete with piezoelectric ceramics, which are less expensive, and which are continuously being improved. However, there exists a potential market in space applications where mass reduction is paramount and in very low frequency underwater applications.

EXERCISES

E.1. The forced magnetostriction of an Invar alloy is giant (see § 1.1 of this chapter: $\partial V / V \partial H = 1.2 \times 10^{-10} / A.m^{-1}$). Given the linear thermal expansion coefficient of this alloy between 0°C and 100°C (tab. 18.1), calculate the strength of the magnetic field which is sufficient to produce in a bar of Invar a magnetostrictive expansion equal to the expansion that would result if the bar were heated by 10 degrees from the temperature of melting ice. Deduce the precautions to be taken so that Invar is used under optimum conditions.

E.2. We would like to build a gaussmeter (device for measuring magnetic field strengths) using an optical interferometer to measure the elongation of a bar of Terfenol-D which is 20 cm long and correctly polarised. What will be the maximum displacement in the bar for a magnetic field variation of $1\ A.m^{-1}$? Discuss the range of applications of such a gaussmeter.

E.3. Calculate the theoretical variations (fig. 18.2) in the relative deformation $\lambda_{//} / \lambda_s$ of a bar of Terfenol-D as a function of applied pressure for $H = 100\ kA.m^{-1}$. Suppose that $\lambda_s = 10^{-3}$.

E.4. In order to obtain sheets of paper with a constant thickness, a papermaker would like to position, to within a micron, a one ton roller to flatten out the paper paste. Knowing that the movement of the actuators should be ±10 µm about the equilibrium position, to take account of the mechanical wear of the moving pieces, propose a magnetostrictive solution to this problem. Data in figure 18.2 and table 18.3 may be useful.

E.5. Is it possible to design a balance based on the principle of figure 18.8 and using Metglas 2605SC as the active element?

SOLUTIONS TO THE EXERCISES

S.1. A magnetic field of $150\ kA.m^{-1}$ would give a *linear deformation* of 6×10^{-6} which is equivalent to a heating by 10°. It is thus advisable to avoid bringing a powerful magnet near Invar parts if an excellent dimensional stability is to be guaranteed.

S.2. Table 18.3 shows that the best Terfenol-D material (Etrema) can have a magnetostrictivity of $57 \times 10^{-9} / A \cdot m^{-1}$. A 20 cm bar is thus elongated by 11.4 nm for a field $1 A \cdot m^{-1}$, i.e. 1/50th of a fringe. It is possible to detect 1/1,000th of a fringe with a good quality interferometer, thus this type of gaussmeter could detect a static magnetic field of $50 mA \cdot m^{-1}$ ($\mu_0 H = 60$ nT).

S.3. The theoretical curve of figure 18.2 could be calculated by minimising the total energy, $E = E_{el} - \mu_0 M_s H \cos \varphi - (3/2) \lambda_s \sigma (\cos^2 \varphi - 1/3)$ of a bar subjected to a magnetic field H, and a compressive stress σ, both applied along the length of the bar: $\partial E / \partial \varphi = 0$.
Two solutions exist, $\cos \varphi = -\mu_0 M_s H / 3 \lambda_s \sigma$, and $\varphi = 0$.
Applying equation (12.20) with $\varphi = \theta$, under zero stress, the demagnetised state being isotropic, $\lambda_0 = 0$, and the application of a field saturates the material ($\varphi = 0$), we thus observe $\lambda_{//} = \lambda_s$. Under weak compressive stress, $P = -\sigma$, the demagnetised state is characterised by $\varphi = \pi / 2$, and thus $\lambda_0 = -\lambda_s / 2$. Thus there is a discontinuity in the variation of $\lambda - \lambda_0$ as a function of P at $P = 0$. In the saturated state, $\varphi = 0$, and we thus observe $\lambda_{//} = \lambda_s - \lambda_0 = (3/2) \lambda_s$. Finally, when the stress increases beyond the value for which $\mu_0 M_s H = 3 \lambda_s P$, the cosine becomes less than one, and it is the second solution which holds, then:
$$\lambda_{//} = \lambda - \lambda_0 = (3/2) \lambda_s \cos^2 \varphi = (3/2) \lambda_s (\mu_0 M_s H / 3 \lambda_s P)^2 \qquad (18.2)$$
which varies as $1 / P^2$. The difference with the experimental curve is explained by the fact that this simple model ignores all anisotropy effects.

S.4. The bearings carrying the roller must rest on 2 actuators which each carry 500 kg, i.e. 5 kN. Figure 18.2 shows that –under a maximum magnetic field of $100 kA \cdot m^{-1}$– the maximum magnetostriction is 1.12×10^{-3}, and thus the minimum length of the actuators, h, is $(20 / 1.12 \times 10^{-3})$ μm, i.e. 18 mm. A pressure of 20 MPa must be exerted to obtain the desired movement, imposing a cross-section of $S = 2.5$ cm^2 for each bar. The necessary volume of terfenol-D is thus 2×4.5 cm^3. It is possible to change the dimensions, increasing the length and reducing the cross-section, provided that the pressure remains less than the compression limit (700 MPa): for $l = 25$ mm, $P = 35$ MPa which gives a cross-section of 1.43 cm^2 and thus a volume of 2×3.58 cm^3. Finally for $l = 44.5$ mm, $P = 50$ MPa the cross-section is 1 cm^2, and the volume of terfenol-D is 2×4.45 cm^3. Thus the best would be to use two bars of about 25 mm in length and 13.5 mm in diameter, so as to minimise the mass of terfenol-D.

S.5. No, because the magnetostriction of 2605SC is positive, and the sensor in figure 18.8 works only for negative magnetostriction.

REFERENCES

[1]   G. BERANGER, F. DUFFAUT, J. MORLET, J.F. TIERS, Les Alliages de Fer et de Nickel
       (1996) Technique et documentation, Lavoisier, Paris.

[2]   Technical document (1994) Imphy SA.

[3]   T.G. KOLLIE, *Phys. Rev.* **B16** (1977) 4872.

[4]   G.G. WU, X.G. ZHAO, J.H. WANG, J.Y. LI, K.C JIA, W.S. ZHAN, *Appl. Phys. Lett.* **67**
       (1995) 2005;
       X.G. ZHAO, G.G. WU, J.H. WANG, K.C JIA, W.S. ZHAN, *J. Appl. Phys.* **79** (1996)
       6225.

[5]   T. MORI, T. NAKAMURA, H. ISHIKAWA, *Actuator 98*, VDI-VDE-Technologiezentrum
       Informationstechnik GmbH (1998) Berlin.

[6]   A.E. CLARK, Proc. 3rd Intern. Conf. on New Actuators, VDI-VDE-Technologiezentrum
       Informationstechnik GmbH (1992), 127, Berlin.

[7]   E. DU TRéMOLET DE LACHEISSERIE, Magnetostriction: Theory, and Applications of
       Magnetoelasticity (1993) C.R.C. Press, Boca Raton, USA.

[8]   T. CEDELL, L. SANDLUNG, M. FAHLANDER, Proc. 2nd Intern. Technology Transfer
       Congress, VDI-VDE-Technologiezentrum Informationstechnik GmbH (1990), 156, Berlin.

[9]   V.I. AKSININ, V.V. APOLLONOV, S.A. CHETKIN, V.V. KIJKO, A.S. SAVRANSKI, Proc.
       3rd Intern. Conf. on New Actuators, VDI-VDE-Technologiezentrum Informations-technik
       GmbH (1992), 147, Berlin.

[10]   L. KIESEWETTER, Proc. 2nd Intern. Conf. on Giant Magnetostrictive Alloys (1988),
       C. Tyren Ed., Marbella.

[11]   J.M. VRANISH, D.P. NAIK, J.B. RESTORFF, J.P. TETER, *IEEE Trans. Magn.* **27** (1991)
       5355.

[12]   F. CLAEYSSEN, N. LHERMET, R. LE LETTY, J.C. DEBUS, J.N. DECARPIGNY,
       B. HAMONIC, G. GROSSO, *Proc. Undersea Defence Technology* **93** (1993) 246, Microwave
       Exh. and Pub. Ltd. Ed., London.

[13]   J. SEEKIRCHER, B. HOFFMANN, *Sensors and Actuators* **A21-A23** (1990) 401.

[14]   C.H. TYREN, D.G. LORD, Sensor, European patent PCT/SE90/00444.

[15]   H. WINTERHOFF, E.A. HEIDLER, *Technisches Messen* **50** (1983) 461.

[16]   J.H. WANDASS, J.S. MURDAY, R.J. COLTON, *Sensors and Actuators* **19** (1989) 211.

[17]   M.D. MERMELSTEIN, A. DANDRIDGE, *Appl. Phys. Lett.* **51** (1987) 545.

[18]   J.F. PEYRUCAT, *Mesures* **51** (juin 1986) 43.

[19]   J. BETZ, Thesis (1997) Université Joseph Fourier, Grenoble, France.

[20]   E. QUANDT, A. LUDWIG, J. BETZ, K. MACKAY, D. GIVORD *J. Appl. Phys.* **81** (1997)
       5420.

[21]   J. BETZ, K. MACKAY, J.C. PEUZIN, B. HALSTRUP, N. LHERMET, *Actuator 96*
       Conference Proc. **5** (1996) 283, Axon Technology Consult Gmbh, Bremen, Germany.

CHAPTER 19

SUPERCONDUCTIVITY

Superconductiviy is a quantum phenomenon. It shows up through strange physical properties which arouse curiosity but also lead to interesting applications. The presence of superconductivity in a book on magnetism is justified: first, because superconductor substances are generally strongly diamagnetic, but also because superconductors are used both to generate relatively intense magnetic fields (superconducting magnets), and to detect very weak magnetic inductions (SQUIDs).

1. INTRODUCTION

H.K. Onnes observed for the first time in 1911, in Leiden, that the electrical resistance of mercury suddenly drops to zero below 4.2 K. During the following decades, a large number of metals having the same property at low temperature were discovered with the noticeable exception of the best conductors: noble metals, and alkali metals [1, 2, 3]. A superconductor is thus a material with zero electrical resistivity, but this phenomenon appears only below a critical temperature T_c, which was always very low $(T_c < -250°C)$ before 1986. Then came superconductors with higher critical temperatures, but still well below room temperature [4]. In 1933, Meissner revealed that superconducting materials expel magnetic induction **B**, i.e. **B** = 0 inside a superconductor. It can be considered as a perfect diamagnetic material with a susceptibility equal to –1 (fig. 19.1-a).

$$\mathbf{B} = \mu_0 (\mathbf{H} + \mathbf{M}) = 0 \qquad \mathbf{M} = -\mathbf{H} \qquad \chi = -1 \qquad (19.1)$$

In fact, this effect is only observed for weak magnetic fields. These two properties $(\rho = 0$ and $\chi = -1)$ were long thought to characterize the superconducting state. However, they cannot be deduced from each other. Infinite conductivity does not imply perfect diamagnetism: it only imposes zero electrical field inside a superconductor but, according to Maxwell's equations, the magnetic induction then cannot be time dependent.

If only infinite conductivity characterized such a material, it should, when cooled under field, trap the magnetic induction. But it actually expels magnetic induction. This is therefore a distinct fundamental property. An important consequence of this second property is that the state of the material does not depend on the transformations it underwent but only on the final state. The superconducting state is thus a thermodynamic state.

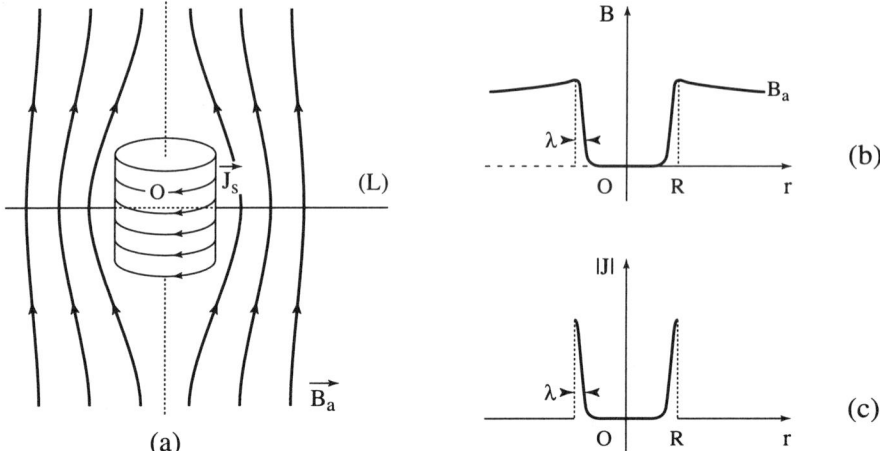

Figure 19.1 - Meissner effect

(a) Modification of the induction lines when a superconductor is introduced into a uniform induction B_a. At each point B is the sum of B_a and of the induction due to the surface distribution of supercurrents, which generate an induction $-B_a$ in the material. Note that, when the material is in its normal state, it is very weakly magnetic ($\chi = 0$) so that $B = B_a$ outside as well as inside the material.
(b) Amplitude of B along the straight line (L). In the material B decreases exponentially over the penetration depth λ.
(c) Amplitude of the supercurrent density along line (L).

2. DEFINITION OF SUPERCONDUCTIVITY

In fact, superconductivity can be defined neither as a state of zero resistance nor through flux expulsion (the Meissner effect) from inside the material. In some cases, which we will see in the following, resistance can be non zero and magnetic induction can enter a superconductor.

The superconducting state is a macroscopic quantum state defined by the existence of a unique wave function $\psi(\mathbf{r})$ for all the electrons. This function is called the order parameter. The situation has to be contrasted with that of electrons in a normal metal, in which each electron has its own wave function because two electrons, according to the Pauli principle, cannot be in the same state. In the superconductor, electrons can experience an effective attractive interaction via lattice vibrations. They then form electrons pairs, called Cooper pairs. Such pairs, being made up of two fermions, can, for our qualitative explanation, be considered as bosons. They can therefore all be in the same state, whence a unique wave function for all pairs. This situation is analogous to that of a photon flux which is described by a plane wave, for instance $\psi(\mathbf{r}) = e^{i\mathbf{kr}}$.

The fundamental characteristic of superconductivity is the existence of this wave function, which was first introduced phenomenologically by the Russian theorists Ginzburg and Landau. This wave function is a complex quantity which can be written:

$$\psi(r) = |\psi(r)| \, e^{i\varphi(r)} \qquad (19.2)$$

$|\psi(r)|^2$ then represents a quantity proportional to the number of Cooper pairs or to the intensity of superconductivity, whereas $\varphi(r)$ is the phase. The phase gradient is related to the supercurrent, as can be seen by analogy with the plane wave, for which $\mathbf{k} = \boldsymbol{\nabla}\varphi$.

The de Broglie relation connects the velocity v of the particle, with mass m, to k:

$$\mathbf{v} = \frac{\hbar\mathbf{k}}{m} = \frac{\hbar}{m}\,\boldsymbol{\nabla}\,\varphi \qquad (19.3)$$

As electrons are charged particles, the existence of a velocity implies a current, which is thus a supercurrent. The state without current is a state with $\mathbf{k} = \mathbf{0}$. The phase $\varphi(r)$ of $\psi(r)$ is the same throughout the superconductor. The phase here corresponds to a macroscopic quantity, in contrast to the electronic wave function of a normal metal.

In a superconductor, the presence of impurities cannot change the wave function of the whole set of Cooper pairs, and does not lead to resistance.

All the properties of superconductors are based on the fact that electrons have an effective attractive interaction. This interaction, which can counterbalance the Coulomb repulsion between electrons, can be visualized in the following way: an electron locally attracts neighbouring ions but, as it moves rapidly, the polarization it has created remains after it is gone because the ions are heavier, hence relax slowly. This locally positive region can then attract another electron. This attraction process (the basis of the BCS theory, after Bardeen, Cooper and Shrieffer) was accepted as the explanation for the superconductive properties before the discovery of superconducting cuprates by Bednorz and Müller. This discovery brings up again the problem of the origin of the attractive interaction in the so called high critical temperature superconductors.

3. SOME FUNDAMENTAL PROPERTIES OF SUPERCONDUCTORS

3.1. THERMODYNAMIC CRITICAL FIELD AND CRITICAL CURRENT

Since the superconducting state is a thermodynamic state, its energy is lower than that of the normal state. Expelling magnetic induction from a volume V costs energy $V\,B^2/2\mu_0$. If this energy is larger than the energy gained by the material as it becomes a superconductor, it will remain normal in order to minimize its energy. There therefore exists a critical magnetic induction $B_c = \mu_0 H_c$ which destroys the superconducting state, and which is called the thermodynamic critical induction. In simple superconducting metals this induction is very small, on the order of 10^{-2} tesla. This means that only a current of limited magnitude, called the critical current, can go through a superconductor without switching it back, through the magnetic field it produces, to the normal state [5]. Obviously, superconductors would never be used in practice for carrying current if other materials, with higher critical inductions, had not been discovered.

3.2. CHARACTERISTIC LENGTHS

Two lengths characterize a superconductor: the penetration depth λ of the magnetic field, and the coherence length ξ. The former originates from the fact that, in order to cancel the magnetic induction B inside a superconductor, supercurrents are produced at the surface. These currents circulate over some thickness λ, called the penetration length or penetration depth of the magnetic field, which typically ranges from some nanometers to a micrometer (fig. 19.1-b). The second length, called the coherence length ξ, is the distance over which the number of Cooper pairs can vary. It ranges from a fraction of nm to some hundreds of nm. If $\xi > \lambda$, the superconductor is said to be of type I, and this is the case for the majority of simple metals. If $\xi < \lambda$, the superconductor is of type II, and we will see the difference in properties that follow.

4. PHYSICAL EFFECTS RELATED TO THE PHASE

4.1. FLUX QUANTIZATION IN A RING

A surprising property is the quantization of the magnetic flux threading a superconducting ring. The value of the induction flux enclosed by a superconducting ring can only be an integral number of times the flux quantum Φ_0 given by:

$$\Phi_0 = h/2e = 2 \times 10^{-15} \text{ weber} \tag{19.4}$$

where e is the electron charge. This flux quantum occurs in many properties and, because it is extremely small, it will allow to measure extremely small fluxes, fields, voltages, etc. This property comes from the fact that quantum mechanics gives for the expression of the velocity of Cooper pairs (with charge 2e), in the presence of an induction $\mathbf{B} = \mathbf{curl\,A}$:

$$\mathbf{v} = \left(\hbar\, \boldsymbol{\nabla}\, \varphi - 2e\,\mathbf{A} \right) / m \tag{19.5}$$

The density of supercurrent is thus:

$$\mathbf{J} = 2e\left| \Psi(r) \right|^2 \mathbf{v} = \left(2e/m \right) \left| \Psi(r) \right|^2 \left(\hbar\, \boldsymbol{\nabla}\, \varphi - 2e\,\mathbf{A} \right) \tag{19.6}$$

By writing that the current is zero inside the ring if its thickness is larger than 2λ (fig. 19.2) because screening currents circulate only near the surface, over a thickness which does not exceed λ, we obtain:

$$\hbar \, \nabla \, \varphi = 2e\,\mathbf{A} \qquad\qquad (19.7)$$

Consider a closed circuit C inside the ring. At any point, the previous relationship leads to:

$$\int_C \hbar \, \nabla \, \varphi \, dl = \int_C 2e\,\mathbf{A} \, dl \qquad\qquad (19.8)$$

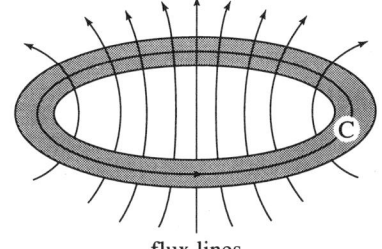

Figure 19.2 - Magnetic flux enclosed by a superconducting ring

flux lines

When the observation point has circulated around the ring, the phase can only vary by a multiple of 2π since one must find the same wave function. On the other hand, the curl of the vector potential around the circuit C is equal to the flux Φ enclosed by C.

One obtains $2\pi\,n\,\hbar = 2e\,\Phi$, i.e.:

$$\Phi = n\,\Phi_0 \qquad\qquad (19.9)$$

The flux enclosed by the ring is a multiple of the flux quantum Φ_0. It is the sum of the flux Φ_{ext} of the applied induction and of the flux Φ_s due to the supercurrents in the ring:

$$\Phi = \Phi_{ext} + \Phi_s \qquad\qquad (19.10)$$

Because the external flux is not quantized, the supercurrents adjust so that flux quantization is always satisfied.

This formula also shows that the vector potential $\mathbf{A}$ is coupled to the phase. The magnetic field thus makes it possible to alter the phase of the order parameter exactly as a medium with a different refractive index affects the phase of a beam in optics. One can thus get supercurrents to interfere by forcing a phase difference through a magnetic field. This is the principle of the dc SQUID, i.e. **D**irect **C**urrent **S**uperconducting **Qu**antum **I**nterference **D**evice.

4.2. JOSEPHSON EFFECT

If we consider two isolated superconductors, each of them is described by a wave function, and these functions are not related to each other. Brian Josephson showed that the wave functions of two superconductors become coherent as soon as there is a coupling, even weak, between them, i.e. there exists a phase coherence between the order parameters [6]. For instance, if two superconductors are separated by an insulating barrier with thickness on the order of the nanometer (this barrier is called a

Josephson junction, see fig. 19.3), the two wave functions are no more independent, and the phases of the two superconductors are related.

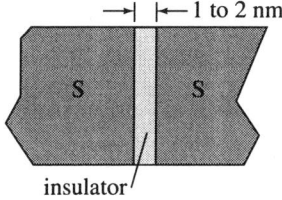

Figure 19.3 - Josephson junction

A supercurrent can then flow, without voltage, through the barrier between the two superconductors. Tunneling occurs for the Cooper pairs in the same way as for electrons, but only insofar as the current does not exceed a critical value, typically much smaller than the critical current of the superconductor. The current is related to the phase gradient, hence here to the phase shift, γ, between the two superconductors. Because the phase is defined to within a multiple of 2π, the current is a periodic function of γ with period 2π. Josephson showed that the current flowing across the junction between the two superconductors is given by:

$$I = I_0 \sin \gamma \qquad (19.11)$$

If the injected current is larger than I_0, the junction become resistive, and a voltage appears. This voltage is a characteristic of the junction, given by:

$$V_t = R I_0 \qquad (19.12)$$

where R is the junction resistance. For the metals used in these junctions, such as lead or niobium, this voltage is of the order of 2.5 meV.

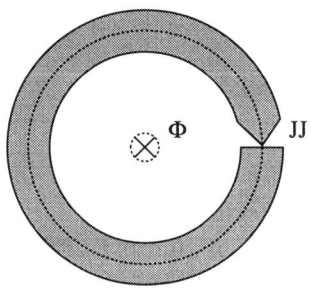

Figure 19.4 - Superconducting ring with a single Josephson junction: the rf SQUID

Consider a ring with a Josephson junction (fig. 19.4). The phase shift across the junction can be calculated as in the previous section, and we obtain:

$$\gamma = \Phi_B - \Phi_A = \frac{2e}{h} \int_C \mathbf{A} . d\mathbf{l} \qquad (19.13)$$

If the junction thickness is neglected, the integral gives the flux enclosed by the ring, whence:

$$\gamma = -2\pi \Phi / \Phi_0 \qquad (19.14)$$

The current across the junction is then given by equation (19.11).

4.3. AC JOSEPHSON EFFECT

Josephson also considered the case when there is a voltage across the junction, for instance that which would be given by a flux variation in the ring with a junction considered above. The induced voltage is given by $-d\Phi/dt$. On account of the relationship between the flux and the phase, we obtain:

$$V = \frac{\Phi_0}{2\pi} \frac{d\gamma}{dt} \qquad (19.15)$$

Although it was derived for a very special case, this relationship is quite general, and shows that in a Josephson junction submitted to a constant voltage V there is an alternating current:

$$I = I_0 \sin \omega_j t \qquad (19.16)$$

$\omega_j = 2\pi V / \Phi_0$ is the Josephson (angular) frequency. For a voltage $V = 1\ \mu V$, this leads to a frequency of 484 MHz, corresponding to a wavelength of 620 µm. This frequency-voltage relationship can be interpreted as follows: the transfer of a Cooper pair from one side of the junction to the other involves an energy of $V \times 2e$, hence it is related to the emission of a photon with energy $\hbar\omega_j = V \times 2e$.

4.4. DYNAMICS OF A SHUNTED JUNCTION

A Josephson junction exhibits a hysteretic current-voltage characteristic. When the current increases from zero and crosses the value I_0, the voltage suddenly jumps to a value different from zero, but returns to zero only if the current decreases to a value much smaller than I_0. This hysteresis must be eliminated in applications such as the SQUID. This is obtained by shunting the junction with a resistor. The junction is thus characterized by its critical current, with its capacitance C, and the shunt resistance R in parallel. The current can be written as (fig. 19.5-a):

$$I = C\frac{dV}{dt} + I_0 \sin \gamma + \frac{V}{R} \qquad (19.17)$$

which can be rewritten, using (19.15), as:

$$2\pi\Phi_0 C\frac{d^2\gamma}{dt^2} + \frac{2\pi\Phi_0}{R}\frac{d\gamma}{dt} = I - I_0 \sin\gamma = -\frac{1}{2\pi\Phi_0}\frac{dU}{d\gamma} \qquad (19.18)$$

with:

$$U = \frac{\Phi_0}{2\pi}(I\gamma + I_0 \cos\gamma) \qquad (19.19)$$

We can easily understand the junction dynamics by realizing that equation (19.18) describes the motion of a ball, the position of which is given by γ, moving on a corrugated roof (fig. 19.5-b and 19.5-c).

The term involving C corresponds to the mas of the ball, that involving R corresponds to friction. The average slope of the roof is given by I. If the slope is small, $I < I_0$, the ball remains in a hollow, and oscillates across it. The average value of the velocity, i.e.

of $d\gamma/dt$, is zero, and thus there is no voltage across the junction. When the current is larger than I_0, the slope of the roof is such that the ball rolls down along the roof. $d\gamma/dt$ becomes different from zero, and a voltage appears across the junction. This voltage increases with the current I.

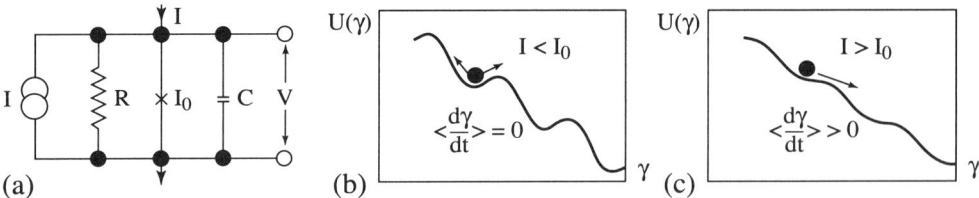

Figure 19.5 - (a) Josephson junction shunted by a resistor
(b) and (c) Model of an inclined corrugated roof for $I < I_0$ and $I > I_0$

5. SQUIDS

The SQUID is the most sensitive magnetic field sensor. It is a flux-voltage converter. It turns a small, hardly detectable, flux variation into a measurable voltage variation. The output voltage is a periodic function of the applied flux, with period Φ_0. One can detect an output signal which corresponds to a flux much smaller than Φ_0. SQUIDs can thus measure any physical quantity which can be transformed into a flux, such as a magnetic field or its gradient, a current, a voltage, a displacement or a magnetic susceptibility. SQUIDs combine two physical phenomena, flux quantization in a superconducting ring, and the Josephson effect. There exist two types of SQUIDs, the dc SQUID (direct current) involving two Josephson junctions, and the rf SQUID (high frequency) which has only one junction. Although rf SQUIDs are more common on the market, dc SQUIDs are more sensitive.

5.1. THE DC SQUID

It consists in a superconducting ring which is interrupted in both branches by a thin film of an insulator (on the order of 1 nm), the Josephson junction (fig. 19.6).

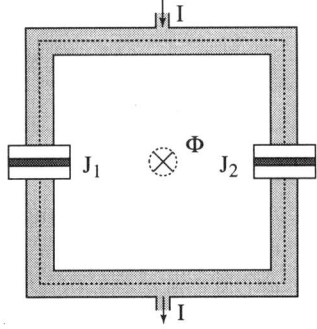

Figure 19.6 - The dc SQUID

The current that can pass through a SQUID without resistance is a function of the flux enclosed by the ring. This variation takes place over a flux scale Φ_0.

The SQUID principle is very simple. Consider a symmetrical device in the absence of any external field. If a current I is injected into the SQUID, this current splits equally into the two branches. We can increase the current without the appearance of a voltage up to an intensity that is two times the critical current of one junction. Now assume that we apply a very small magnetic field. As the flux enclosed by the ring must be an integral number of flux quanta, a screening current that cancels the applied flux will appear, so that the total flux is zero.

In one branch, this current I_s will run in the same direction as that of the current injected into the SQUID, and in the other branch it will run in the opposite direction (fig. 19.7). The SQUID will develop a voltage as soon as the current $I/2 + I_s$ in one of the branches is equal to I_0. The critical current of the SQUID is thus lowered by $2I_s$. Let us now increase the external magnetic field. If the flux exceeds $\Phi_0/2$, the SQUID, instead of screening the external flux, will prefer to increase it up to Φ_0.

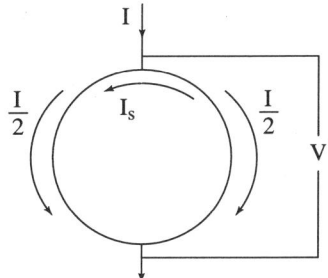

Figure 19.7 - dc SQUID with its bias and screening currents

The screening current will thus run in the opposite direction. It is then easy to see that the screening current will have the variation shown in figure 19.8. It reverses every time the flux is a half integer number of flux quanta. The critical current of the SQUID will oscillate as shown in figure 19.9.

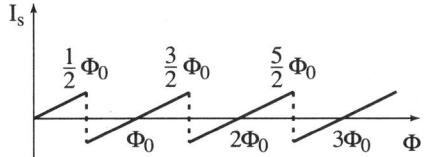

Figure 19.8 - Screening current as a fonction of the applied magnetic flux

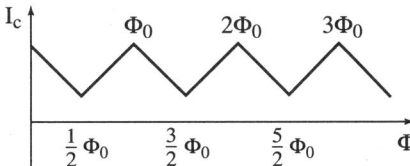

Figure 19.9 - Critical current as a fonction of the applied magnetic flux

It will be a minimum for all half integral multiples of Φ_0. The I(V) characteristics are shown in figure 19.10-b for two values of the applied flux.

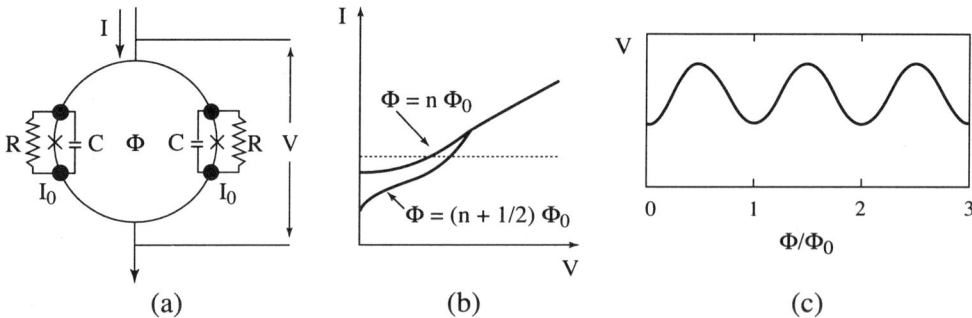

Figure 19.10 - (a) A dc SQUID - (b) I-V characteristics
(c) V vs Φ / Φ₀ for constant bias current I_B

The upper curve corresponds to an applied magnetic flux equal to an integral number of flux quanta, and the lower one to a half integral number. When the magnetic flux is continuously increased the curve oscillates between these two limits. If a current slightly larger than the critical current is injected into the SQUID, the latter is always resistive. The voltage across the SQUID is then a periodic function of the applied magnetic field, with period equal to a flux quantum. The measurement of an applied flux is based on this relationship. We can approximately calculate the relationship between the variations in voltage and in flux. The former is given by Ohm's law $\Delta V = R \, \Delta I$.

The critical current variation is twice that of the screening current. As the screening current is equal to the applied flux divided by the self inductance L of the ring, we have:

$$2I = 2\Delta\Phi / L \tag{19.20}$$

Keeping in mind that the SQUID's resistance is half that of one of the junctions, we find:

$$\Delta V = (R / L) \, \Delta\Phi \tag{19.21}$$

The resistance and the self inductance of the ring being known, this formula allows us to calculate $\Delta\Phi$ as a function of ΔV.

5.2. THE RF SQUID

The rf SQUID consists of a ring involving a Josephson junction. We showed that the phase difference across the junction is given by equation (19.14), and the current in the ring by equation (19.11).

$$I = -I_0 \sin 2\pi \, (\Phi / \Phi_0) \tag{19.22}$$

The flux enclosed by the ring is the sum of the applied flux and of the flux LI created by the current in the self inductance L of the ring:

$$\Phi = \Phi_{ext} + L\,I \tag{19.23}$$

The magnetic behavior of such a ring depends critically on the value of the quantity:

$$\beta = 2\pi\frac{L\,I_0}{\Phi_0} \qquad (19.24)$$

Figure 19.11 shows the irreversible behavior for $\beta > 1$. For $\beta < 1$, the magnetic behavior is reversible, and the supercurrents introduce only a weak modulation around the applied flux. The rf SQUID uses the hysteretic behavior, with β ranging from 3 to 6.

The external flux includes a flux Φ_a to be measured, and a radiofrequency flux obtained by inductively coupling the ring to a radiofrequency coil belonging to a parallel LC circuit (fig. 19.12). The losses in the device are a periodic function of the flux Φ_a, with period Φ_0, and appear as voltage. The voltage variation is directly related to the flux variation. This setup thus measures only variations in flux or in magnetic induction, or field gradients.

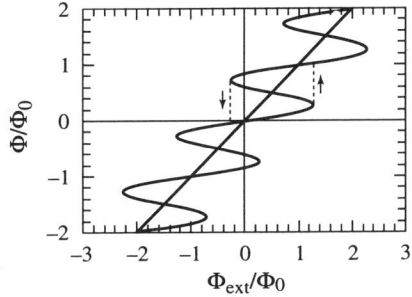

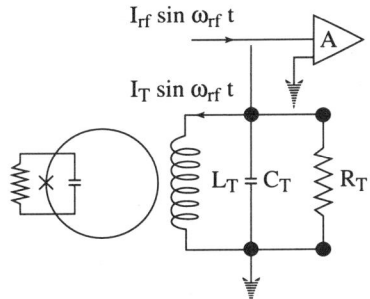

Figure 19.11 - rf SQUID: total flux Φ (normalized to Φ_0) as a function of Φ_{ext} / Φ_0 for $LI_0 = 1.25\ \Phi_0$

Figure 19.12 - rf SQUID inductively coupled to a parallel LC circuit

5.3 THE SQUID IN PRACTICE

dc SQUIDs are made by photolithography or by electron beam lithography. The main problem is the requirement to inductively couple an input coil to the SQUID. SQUIDs are generally made of niobium or lead. The shunt resistance is of the order of a few ohms, the capacitance of the order of the pF, and the self inductance some fractions of nH. In almost all applications, the SQUID is used with a feedback circuit as a zero flux sensor. Figure 19.13 shows three different, frequently encountered, geometries.

The sensitivity of the rf SQUID used in the hysteretic mode is generally limited by the rf amplifier noise, because the transfer factor H is relatively small:

$$H = dV/d\Phi \qquad (19.25)$$

where V is the output voltage. H is of the order of $10\ \mu V/\Phi_0$, and the amplifier noise V_N of the order of $10^{-9}\ V.Hz^{-1/2}$.

The flux sensitivity is thus typically $\Delta\Phi = V_N/H = 10^{-4}\ \Phi_0.Hz^{-1/2}$.

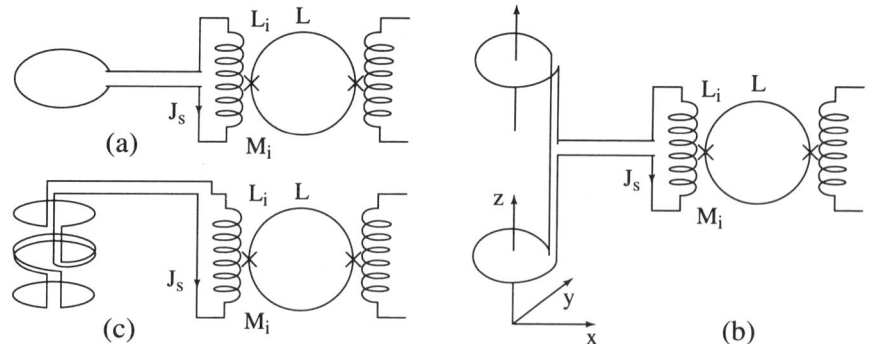

Figure 19.13 - Superconducting flux transformers using flux locked SQUIDs
(a) magnetometer - (b) first order gradiometer - (c) second order gradiometer

For a dc SQUID, the transfer factor is larger:

$$H = V/\Phi_0 \qquad\qquad (19.26)$$

and reaches 1 mV/Φ_0. The amplifier noise is negligible compared to that from the SQUID. Thermal fluctuations are then important. They impose two requirements on the SQUID parameters: the coupling energy between the two superconductors must be large compared to $k_B T$, and the standard deviation of the flux due to thermal noise must be small compared to Φ_0. In practice this imposes that L be smaller than $\Phi_0^2/5\,k_B T$, i.e. L smaller than 15 nH at 4.2 K.

6. TYPE-I AND TYPE-II SUPERCONDUCTORS

6.1. CRITICAL FIELDS AND VORTICES

Another consequence of flux quantization is the existence of two types of superconductors. Expelling the magnetic induction B corresponds to an energy $VB^2/2\mu_0$ where V is the sample volume. For very weak inductions, this energy is smaller than that gained when switching to the superconducting state.

For larger values of the magnetic energy, i.e. when the applied field is large enough, the material returns to the normal state. At this point one has to distinguish two types of superconductors. In *type-I superconductors*, for the previously defined thermodynamical critical field H_c (§ 3.1), the material suddenly returns to the normal state, which is non magnetic (or very weakly magnetic) (fig. 19.14-a). In *type-II superconductors*, for a certain field range, the material is in the so called *mixed state*. The flux progressively penetrates into the sample between a lower critical field H_{c1} and an upper critical field H_{c2}, above which the material is in the normal non magnetic state (fig. 19.14-b).

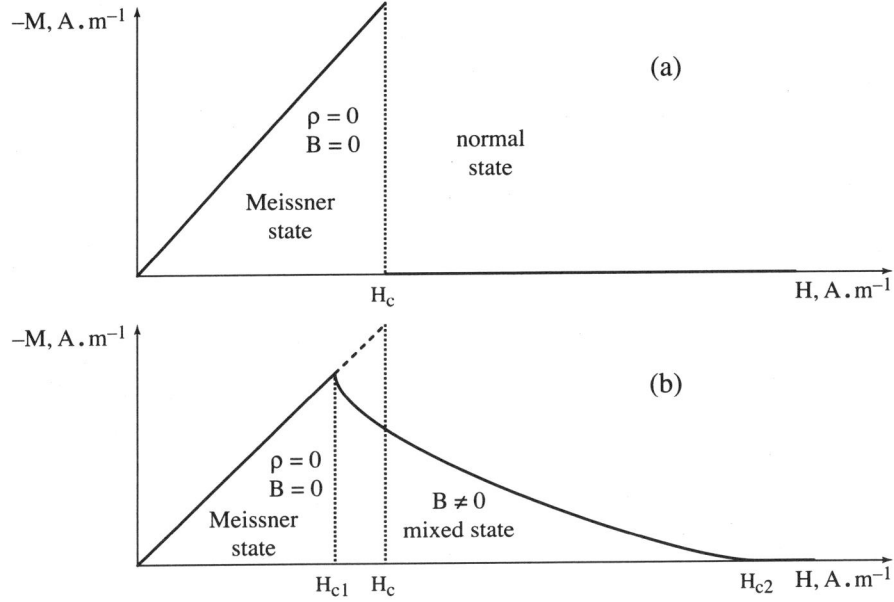

**Figure 19.14 - Magnetization as a function of the magnetic field for
(a) a type-I superconductor and (b) an ideal type-II superconductor**

In the mixed state the magnetic energy lost is much smaller than for total expulsion. It implies the creation of tubes (called flux tubes, vortex lines or vortices) of normal material in which flux can pass. Each tube corresponds to a flux quantum Φ_0, and has a radius of the order of λ (fig. 19.15).

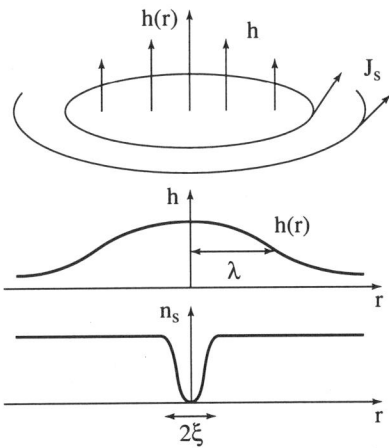

Figure 19.15 - Cross-section of an isolated vortex in a type-II superconductor

The magnetic field is maximum near the center. h decreases away from the vortex axis because of the screening that occurs over a region with radius ~ λ (the penetration depth). The density of super-conducting electrons n_s is reduced only in a small core region of radius ξ. ($n_s = 0$ corresponds to the normal state) (after [3])

The core of the vortex, corresponding to the region where the material is in the normal state, has a radius of the order of ξ (fig. 19.15). A rough energy calculation shows that only superconductors with $\lambda > \xi$ can have such a behavior. These superconductors can withstand much larger induction (about a hundred tesla), and therefore, under certain conditions, much larger currents.

We already defined the thermodynamical critical field. For a type-I superconductor, this is the critical field at which the material turns normal. For type-II superconductors, the flux is expelled only below H_{c1}. In the mixed state, there is an induction $B = n \Phi_0$, where n is the number of vortices per unit cross-sectional area, in the superconductor. Each vortex has a core where superconductivity is destroyed over a radius ξ. When the distance between vortices becomes of this order of magnitude, the material returns to the normal state, which defines the second critical field H_{c2}:

$$\mu_0 H_{c2} = \frac{\Phi_0}{2\pi \xi^2} \qquad (19.27)$$

One can show that, near H_{c1}, the distance between vortices is of the order of λ. This implies:

$$\mu_0 H_{c1} \approx \frac{\Phi_0}{\pi \lambda^2} \qquad (19.28)$$

The following relationship between H_{c1}, H_{c2}, and H_c follows:

$$H_{c1} H_{c2} = H_c^2 \ln \frac{\lambda}{\xi} \qquad (19.29)$$

As the critical thermodynamical induction does not vary much in type-II superconductors, and ranges between 0.1 and 1 T, this means that the larger H_{c2} the smaller H_{c1}. For instance, in high T_c superconductors where B_{c2} is of the order of 100 T, B_{c1} is of the order of 10^{-2} tesla.

6.2. THE ABRIKOSOV LATTICE

Between H_{c1} and H_{c2}, flux thus penetrates into the superconductor through vortices. The latter repel each other, but they are forced to enter into the superconductor by the magnetic pressure. Abrikosov showed that they form a triangular lattice (fig. 19.16).

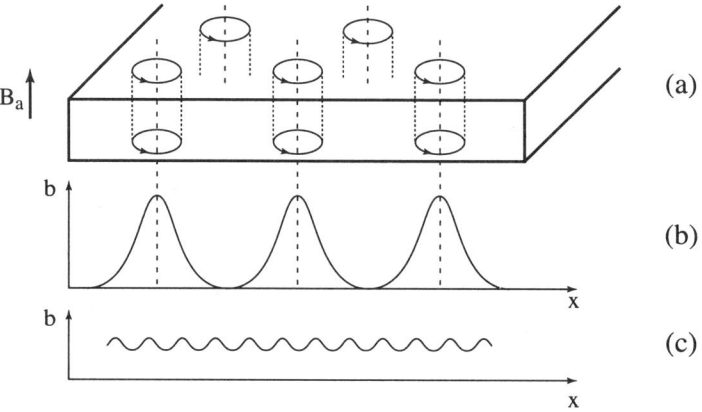

Figure 19.16 - Abrikosov lattice (a) and microscopic magnetic induction
for two different fields: for H slightly above H_{c1} (b) and slightly below H_{c2} (c)
Flux penetration occurs through an increase in the vortex density, i.e. a decrease of
the lattice parameter. Note that near H_{c2} the vortices overlap strongly.

The mixed state is a stable state, and we discussed the magnetization curve for an ideal type-II superconductor. However, experimentally it is difficult to obtain the predicted ideal behavior. Samples have to be carefully prepared and thoroughly annealed (curve A of fig. 19.17). If such a "perfect" sample is submitted to cold working, the behavior becomes completely different, and features an irreversibility in magnetization which increases with the density of defects in the material (curve B of fig. 19.17). The flux has difficulty in penetrating on account of inhomogeneities and defects. We will explain this feature in the next section.

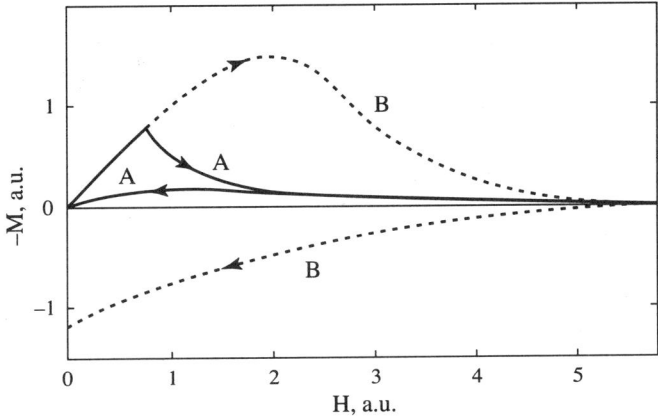

***Figure 19.17 - An example of a hysteresis loop in magnetization measurements
on a type-II superconductor***

*Curve A: annealed material; curve B: after cold-working.
Note that the horizontal and vertical scales are the same.*

6.3. CRITICAL CURRENT IN TYPE-II SUPERCONDUCTORS

Type-II superconductors withstand intense magnetic fields without returning to the normal state. One could thus conclude that they can carry large currents, since a large current is necessary to produce the H_{c2} field. The problem is not so simple. Above H_{c1}, which is small if H_{c2} is large, vortices are submitted to a force if a current passes through the material. This force can be considered as a Laplace force or, in hydrodynamic terms, a Magnus force. Vortices will thus move in the presence of a current. Because of this motion, the flux at a point of the superconductor varies, and, according to Maxwell's equations, an electric field appears in the material. The coexistence of a current and an electric field leads to dissipation, i.e. to resistivity. The superconductor thus has non zero resistivity above H_{c1} (this is the reason why the absence of resistance cannot be used as a definition of a superconductor). Dissipation originates from vortex motion in the presence of current. How can this dissipation be eliminated? By pinning vortices, in order to prevent them from moving.

6.4. VORTEX PINNING

Creating a vortex requires energy since it is necessary to destroy superconductivity in a tube of radius ξ. If there is an inclusion consisting of a material which does not become superconductor, and if the vortex crosses this inclusion, its energy will be decreased since superconductivity does not have to be destroyed in the inclusion.

Shifting the vortex from this position will require a current large enough to generate a force that can move it. Current can thus flow without dissipation as long as the vortex remains pinned. Actually, the whole vortex lattice has to be pinned. The principle remains the same, but calculating the current required to move the lattice, i.e. the critical current, becomes a very complex problem which we will not treat. However, we can now understand the irreversibility observed in the magnetization.

If there are defects in the superconductor, vortices do not enter easily into the superconductor under the effect of magnetic pressure because the pinning centers oppose their displacement. Conversely, once vortices are in the material, they will not easily get out when the magnetic field is decreased, and the equilibrium state will not be reached. There is thus an irreversibility of magnetization. On can link this irreversibility to the current the superconductor can carry without dissipation since both phenomena have the same origin. If ΔM is the difference between the magnetization measured in increasing field and that measured in decreasing field (fig. 19.18), we get the approximate expression:

$$I_c = 2\,\Delta M / d \qquad\qquad (19.30)$$

where d is the size of the sample or, for a granular sample, that of the grains.

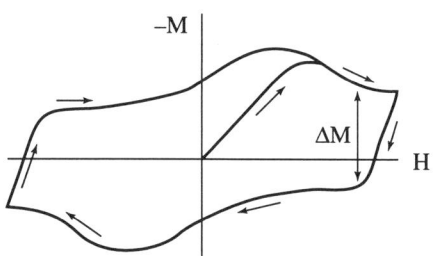

Figure 19.18 - Characteristic magnetization curve of a type-II superconductor

The hysteresis ΔM for a given field is a measure of the critical current density I_c. In equation (19.30), d is the thickness of the sample, or the grain dimension in the case of a polycrystalline ceramic.

The problem of the critical current in type-II superconductors is therefore a technological problem. It is necessary to create defects that pin vortices. For instance, dislocations favour pinning, but the size of the superconducting units also play a role. Commercially available superconducting wires in fact have a very complicated structure. Some superconductors can withstand a current of the order of 10^7 A.cm^{-2}. Comparing this current density to the maximum density a copper wire can withstand, viz of the order of 10^3 A.cm^{-2}, we understand the interest of these materials.

7. SUPERCONDUCTING MATERIALS

Many simple elements in the periodic table are superconductors, with the noticeable exception of alkali, noble and magnetic metals (see tab. 3.2). Among superconducting materials one finds a lot of alloys and intermetallic compounds. For instance NbTi is very much used for wires, and Nb_3Sn and Nb_3Ge have critical temperatures of 18.3 K and 23 K respectively, close to the maximum obtained before the appearance of superconducting cuprates.

In 1986, G. Bednorz and K.A. Müller synthesized a copper oxide $(La_{1-x}Ba_x)_2CuO_4$, which superconducts at 30 K. In 1987 a critical temperature of 92 K was reached in $YBa_2Cu_3O_7$. The appearance of these copper oxides was a revolution as the first one discovered beat all critical temperature records, and liquid nitrogen temperature (77 K) was exceeded a few months later. Further records were obtained with bismuth cuprates (the compound $Bi_2Ca_2Sr_2Cu_3O_{10}$ has a critical temperature of 110 K), and with thallium cuprate (the compound $Tl_2Ca_2Ba_2Cu_3O_{10}$ has a T_c of 125 K). T_c reaches 135 K for $HgBa_2Ca_2Cu_3O_{8+\delta}$ (also called Hg-1223).

These superconducting compounds are oxides containing a number n of successive layers of copper oxide CuO_2 ($n \geq 1$) where superconductivity takes place, and structural blocks, which are essential for charge transfer toward the CuO_2 layers and to give rise to their superconductivity.

8. APPLICATIONS

The large majority of applications [7] use type-II superconductors, the only ones which can withstand large current densities (or large magnetic fields). Technological feats were involved in obtaining superconducting wires long enough to make efficient magnetic field coils. Metallic superconducting wires (NbTi, Nb_3Sn), now in routine use, were developed over the last thirty years. They work at liquid helium temperature (4.2 K), and, in spite of this handicap, their applications progressively extended from the laboratory to industry, advanced research or medical technology. Their main application is the production of very intense magnetic fields, which can also be very stable and homogeneous. The progressive improvement in superconducting wire quality allows to find on the market (Oxford Instruments) superconducting coils reaching 20 T. Beyond this field, there is a risk that the material switches to the normal state, but the use of *high T_c superconductors* (within coils made of classical superconductors) would already allow to reach fields of the order of 23 T. In the near future, as high T_c superconducting wires tolerating current densities beyond $10,000 \; A.cm^{-2}$ begin to appear, one can consider the manufacturing of super-conducting coils working at liquid nitrogen temperature. It is already possible to reach DC fields up to 45-50 T by adding to the field generated by a superconducting coil that created by a copper coil (which dissipates a power of 20 MW). These fields could be reached without loss with non resistive high T_c superconductor coils.

Smaller fields, but homogeneous over large volumes, are used in hospitals for magnetic resonance imaging (MRI) (see chap. 23). Finally stable magnetic fields are required to confine a plasma (research on nuclear fusion) or a particle beam (research on particle physics, CERN): they are produced with superconducting wires. The possibility of reaching intense fields with a smaller volume of material and without losses allowed to build prototypes of rotating machines (motors or generators) operating at liquid helium temperature. Thanks to superconducting materials, many innovations are conceivable, for instance in the field of levitation (magnetic bearings and superconductors for the suspension of rotating machine shafts), or current limiters in power supply lines. For all these applications, the interest of using high T_c superconductors working at 77 K or above is obvious. These applications could extend to all intense magnetic field facilities, and to high power machines. Progress in permanent magnets, on the other hand, concerns all small size applications (motors, information readout and storage, etc.).

In the field of detection, the Josephson effect is widely used in laboratories to measure extremely small (magnetic or electric) signals [8]. The applications of these extremely sensitive sensors extend to the medical field: it becomes possible to detect anomalies in brain activity through magnetic field mapping (magnetoencephalography), which allows to exactly localize the damaged region, in view of a possible operation. Magnetic field mapping associated with heart activity (magnetocardiography) allows to detect certain anomalies in blood circulation or cardiac muscle functioning. This technique is external and hence harmless, and can be applied to detect heart anomalies of a foetus. Finally, detection of anomalies in the earth magnetic field is widely used in the search for natural resources, geothermal springs, sismic faults, etc. Adding a SQUID to a micrometric displacement system (tunneling scanning microscopy) already allowed magnetic object imaging with a resolution of the order of a micrometer. These techniques should spread to all fields where high sensitivity is required, and a high T_c superconductor based SQUID is already marketed.

Applications in electronics and electrical engineering will spread more easily as the working temperature of new superconducting materials increases, and as they can carry higher current densities. In electrical engineering, the use of liquid nitrogen (77 K) is very worthwhile with material that can carry 10,000 A.cm^{-2}. But, just as the manufacturing of high-performance wires from metallic alloys (NbTi, Nb$_3$Sn, etc.) took years, time will be required to produce high-performance wires from materials which naturally exist as ceramics. Finally, the miniaturization of SQUIDs (see the microSQUID in fig. 1.9) leads the way to a new high sensitivity type of magnetometry for samples with sub-micron size.

REFERENCES

[1]   J.P. BURGER, Supraconductivité des métaux, des alliages et des films minces (1974) Masson, Paris.

[2]   M. TINKHAM, Introduction to Superconductivity (1975) McGraw-Hill, New York.

[3]   P.G. DE GENNES, Superconductivity of Metals and Alloys (1966) W.A. Benjamin, New York.

[4]   M. CYROT, D. PAVUNA, Introduction to Superconductivity and High-Tc Materials (1992) World Scientific, Singapore.

[5]   A.M. CAMPBELL, J. EVETTS, Critical Currents in Superconductors (1972) Taylor and Francis, London.

[6]   A. BARONE, G. PATERNO, Physics and Applications of the Josephson Effect (1982) John Wiley & Sons, New York.

[7]   Applications de la Supraconductivité (1990), Observatoire français des techniques avancées Ed., Masson, Paris.

[8]   Superconducting devices (1990), S.T. Ruggiero & D.A. Rudman Eds, Academic Press, New York-London.

CHAPTER 20

MAGNETIC THIN FILMS AND MULTILAYERS

Progress in high vacuum and ultra high vacuum material preparation techniques now allow the preparation of artificial structures consisting of ultra thin films of magnetic materials, as well as multilayer systems containing stacks of different materials, some of which being magnetic. In addition, many techniques are also now available for both the structural and magnetic characterization of these systems.

The thickness of the individual layers may vary from one atomic plane to some tens of nanometres. These thicknesses are comparable to some characteristic length scales in magnetism: the Fermi wavelength, the range of exchange interactions, the domain wall thickness, the mean free path of electrons...

In general, magnetism is very sensitive to the local atomic environment. As it is possible to vary this environment in a very controlled way in thin films and multilayers, unique phenomena, which do not exist in bulk materials, can be observed in these systems.

These phenomena include the possibility to stabilise new crystallographic phases, non-existent or very unstable in bulk form, the appearance of magnetic anisotropy induced by interfacial stress, magnetic coupling effects between adjacent films, giant magnetoresistance, tunnel magnetoresistance.

Apart from their interest from a fundamental point of view, thin films and multilayers find many applications, particularly in the area of magnetic or magneto-optical recording. These include media for both parallel and perpendicular magnetic recording, media for magneto-optical recording, soft magnetic thin films (flux guides, transformers...), magnetostrictive materials (microactuators), and magnetoresistive materials for magnetic field sensors.

1. *FROM CONTROL OF THE LOCAL ENVIRONMENT TO CONTROL OF THE PHYSICAL PROPERTIES*

Interest in the physical properties of magnetic metallic thin films and multilayers is not new. Louis Néel and his collaborators had already embarked on very important pioneering work during the 1950s and 60s in treating Bloch and Néel walls in thin films, magnetostatic coupling between ferromagnetic thin films separated by a non-magnetic layer, surface and interface anisotropies, and coupling between ferromagnetic and antiferromagnetic materials [1].

Nevertheless, the technological progress of the last three decades has given rise to intense research in this field, stimulated by the improved crystallographic and chemical quality of the samples, and also by the improved control over their properties.

The previous chapters, in particular chapter 8, have shown that the magnetism of metals is very sensitive to the local atomic environment. This environment influences both the strength and sign of the exchange interactions, and it determines the local anisotropy of the material. The magnetic properties of itinerant magnetic materials vary dramatically as a function of the environment. For example, the material with composition YCo_2 is non-magnetic when it is a crystalline compound but ferromagnetic when it is an amorphous alloy.

Another spectacular example is that of iron. Natural iron has a body-centred-cubic (bcc) crystallographic structure, and is ferromagnetic under normal conditions. However, a face-centred-cubic (fcc) phase can be stabilised when iron is prepared in thin film form, and it is non-magnetic.

There is some, though relatively little, flexibility for varying the local atomic environment, and thus controlling certain magnetic properties, in bulk materials. For example, it is possible to change the interatomic spacing by applying a uniaxial or hydrostatic stress on a crystal. It is sometimes possible to change the crystallographic structure of a sample by annealing it or irradiating it with heavy ions. Thus, an amorphous alloy may crystallise if it is annealed at a sufficiently high temperature. Certain metals spontaneously change their crystallographic structure as a function of temperature or pressure. It is also possible to modify the local chemical environment by alloying or using particular doping elements. In all these cases, the experimentalist is working on the macroscopic scale, with little control of what happens at the atomic level.

In the preparation of thin films and multilayers, a continuous flux of atoms arrives at, and then condenses on a substrate. To obtain a good crystallographic structure for the deposited thin film, one usually chooses a single crystal substrate with the cell parameter of the surface as close as possible to that of the material being deposited. One may then be able to epitactically grow the film, atomic layer by atomic layer. Under the best growth conditions one can obtain a perfect single crystal sample, also known as a *superlattice*.

For many studies, in particular for those which are more application-oriented, epitactic growth is not necessary. The sample then consists of stacks of different films, which may be polycrystalline or even amorphous. Such samples are called *multilayers*. The growth of multilayers offers many more degrees of freedom for varying the local atomic environment than is possible with bulk sample preparation. For example, in creating interfaces between given materials, it is possible to vary the chemical environment in a controlled fashion.

The cell parameter of a material may be extended or contracted, by a few %, by epitactically depositing it in thin film form onto a substrate with a slightly different cell parameter. The extension or contraction corresponds to the application of a considerable pressure, experimentally unattainable for bulk samples. In forcing a given material to adopt the crystallographic structure of a substrate, one can stabilise, under normal conditions, crystallographic phases which do not exist in bulk form, except in some cases under very extreme conditions. Thus, iron which is normally bcc can be stabilised in an fcc structure by depositing it epitactically on a (100) single crystal of Cu which is naturally fcc. Alternatively, it can also be prepared in a hexagonal-close-packed (hcp) structure by depositing it on a Ru hcp substrate. In a similar way, Co, which is naturally hcp, can be stabilised in an fcc or a bcc phase by depositing it on Cu(100) or GaAs (001), respectively. Many examples of this type can be found in the literature [2].

In addition to being able to alter the local atomic environment, one can also vary the thickness of thin films or individual layers in multilayers to cover the range of the characteristic length scales of magnetism. These length scales are:

♦ *The Fermi wavelength*: $\lambda_F = 2\pi / k_F$ is typically of the order of 0.2 to 2 nm in magnetic metals, where k_F is the Fermi wave vector. This length scale plays an important role for quantum size effects in multilayers. In a metallic thin film, for example, the conduction electrons are confined within the film, which constitutes a quantum well. Reflections against the outer surfaces give rise to standing waves of period $\lambda_F / 2$.

♦ *The exchange interaction length*: depending on the nature of the exchange interaction, the interaction length can vary from one interatomic distance (of the order of 0.2 nm) up to a few tens of interatomic distances (of the order of 10 nm). The exchange interactions which are most frequently encountered are those of the RKKY type, which are transmitted by the conduction electrons of noble metals (e.g. Cu, Ag, Au). These are long range interactions which oscillate (see § 3.1 of chap. 9). There are also collective interactions between the 3d electrons of the magnetic transition elements (Fe, Co, Ni), which are treated within the framework of itinerant magnetism (chap. 8). The interactions between rare earth ions, or between rare earth ions and transition metal ions, are on a shorter scale due to the more localised nature of rare earth magnetism (chap. 8). The so-called super-exchange interactions, which are indirect, can also play a role in multilayers containing oxide, or more generally insulating, layers (chap. 9). All of these interactions play a role in the magnetic ordering phenomena of the multilayers, either by determining the magnetic order within individual layers, or by influencing the magnetic ordering of a layer through interactions with adjacent magnetic layers.

◆ *The domain wall width*: the magnetization of a ferromagnetic material has a spontaneous tendency to subdivide into magnetic domains so as to reduce its magnetostatic energy (chap. 3). The shape and size of the domains results from a balance between the reduction in magnetostatic energy due to the break-up of the magnetization into domains and the increase in the exchange and anisotropy energies due to the creation of domain walls between neighbouring domains (chap. 6). In addition, hysteretic effects are associated with the nucleation of domains and propagation of domain walls, so that the domain configurations usually depend on the magnetic history of the samples.

The domain wall width in thin films is of the same order of magnitude as in bulk samples, ranging from a few nanometres in materials with strong anisotropy to several tens of nanometres in materials with weak anisotropy. Nevertheless, a characteristic aspect of thin films is a change in the nature of the domain wall as a function of the thickness of the magnetic layer. Bloch walls exist in films with a thickness above about 40 nm, while Néel walls exist in thinner films. This is because too much magnetostatic energy would be required to create Bloch walls in very thin films [1];

◆ *The electronic mean free path*: this is the average distance travelled by a conduction electron in a metal between two successive collisions. The mean free path is proportional to the conductivity of the metal. In ferromagnetic transition metals (Fe, Co, Ni, and most of their alloys), it also depends on the direction of the electron spin relative to the local magnetization direction.

The mean free path can vary from a few interatomic distances in metals of high resistivity to some hundreds of nanometres in metals which are good conductors, and contain few growth defects. It plays a key role in all transport phenomena in multilayers (electrical resistivity, thermal conduction, magneto-resistance…).

2. PREPARATION AND NANOSTRUCTURE OF MAGNETIC THIN FILMS AND MULTILAYERS

2.1. INTRODUCTION

Thin film preparation techniques may be divided into two categories: *chemical vapour deposition* (CVD) techniques, and *physical vapour deposition* (PVD) techniques. The choice of technique depends on the nature of the material to be deposited. Chemical techniques, which involve the decomposition of a gas acting as the transport agent at the substrate surface, are mostly used in the preparation of thin-film semi-conductors. More recently, electrolysis has been used to prepare metallic magnetic multilayers by electro-deposition. At present, electrolysis may be used to deposit the following metals: Al, Ag, Au, Cd, Co, Cu, Cr, Fe, Ni, Pb, Pt, Rh, Sn, Zn [3].

Physical deposition techniques include sputtering and molecular beam epitaxy (MBE), which are the most widely used techniques for the preparation of magnetic multilayers. There are very few restrictions on the type of materials that can be deposited (metal or insulator), and on the substrates that can be used. These processes involve three steps:

♦ the emission of atoms or particles from the target,

♦ their transfer to the substrate, and

♦ their condensation onto the substrate.

These steps are described in more detail below. These physical deposition techniques are always carried out in vacuum chambers. The vacuum required depends on the demand for crystallographic and chemical quality of the film to be deposited. Using the kinetic theory of perfect gases, one can estimate the order of magnitude of the time needed to deposit, at room temperature, an atomic plane of a given type of impurities, of partial pressure P. The root-mean-square speed of these impurities in the gas phase is $u = (3k_B T / m)^{1/2}$ where k_B is Boltzmann's constant (see app. 2), T is the temperature of the gas, and m is the mass of the impurity. The density of particles per unit volume is given by $n = P / k_B T$. All particles which arrive at the substrate, of surface area S, during a time dt are considered to be in a volume S.u.dt. About 1/6 of these particles are travelling towards the substrate.

One can deduce that an atomic plane of impurities will be deposited onto the substrate in a time:

$$t = 2 (3m k_B T)^{1/2} / a^2 P \tag{20.1}$$

where a is the cell parameter of the lattice of impurity atoms deposited (about 0.2 nm). For oxygen at a pressure of 10^{-5} Pa, an atomic plane of oxygen will be deposited in about one minute; at 10^{-6} Pa, it would take about 10 minutes; at 10^{-7} Pa, 100 minutes… The base pressure in the chamber, and the deposition rate must thus be chosen as a function of the relative tolerance for impurity atoms in the sample. The conversion factors for units of pressure commonly used for thin film deposition machines are given in appendix 2.

2.2. EXPERIMENTAL TECHNIQUES MOST COMMONLY USED TO PREPARE MAGNETIC THIN FILMS AND MULTILAYERS

2.2.1. Sputtering

Sputtering is the most widely used laboratory technique for preparing thin films. It is a very flexible method, allowing the deposition of both metals and insulators, and many deposition parameters can be easily varied. Moreover, this process is already used industrially for magnetic materials, facilitating the technology transfer required to allow the large scale production of new materials with interesting applications.

The majority of sputtering machines work with a base pressure of 10^{-6} to 10^{-4} Pa. These relatively low pressures are achieved with diffusion or turbo pumps operating with cold traps, or with cryogenic pumps. An inert gas, usually Ar, is introduced into the chamber in a controlled manner, and maintained at a pressure of between 0.05 and 2 Pa. The gas is ionised in a strong electric field, creating a plasma which shows up through luminescence. The positive Ar ions are attracted towards a target of the material to be deposited. The bombardment of the target with these relatively heavy ions results in atoms being torn out of the target, i.e. sputtered away. These atoms travel through the plasma and the neutral gas, and condense on a facing substrate. In the case of metallic targets, the ions may be attracted to the target by applying a constant negative voltage to the target. The target current is constant, and proportional to the number of ions that reach the target per unit time (DC sputtering).

In the case of insulating materials, the DC method cannot be used as the target would very quickly become positively charged, and would therefore repel the positive ions of the plasma. Instead, a radio-frequency voltage (usually 13.56 MHz) is applied to the target in the so-called RF sputtering mode. Due to the large difference in mobility of the ions and electrons of the plasma, a negative potential appears spontaneously on the target. During a fraction of a radio-frequency period the target is bombarded by positive Ar atoms while during the remainder of that period the target is bombarded with the electrons, thus neutralising the charge left by the positive ions. Moreover, the electrons excited by the RF electric field maintain the plasma by ionising further neutral Ar atoms. In the RF mode it is possible to maintain a plasma in an Ar pressure of 0.02 Pa.

There exist a number of variations of the sputtering process, including the triode, diode, diode-magnetron modes or ion-beam sputtering. Magnetron systems are the most widespread, allowing high deposition rates of up to $10 \, nm.s^{-1}$. In these systems the field lines of permanent magnets placed behind the target act to channel the electric charges, and thus concentrate the plasma in the vicinity of the target, resulting in an increased sputtering rate.

Once atoms have been sputtered from the target, they should travel across the plasma and neutral gas to arrive at the substrate. In transit, they may or may not collide with the surrounding Ar atoms, depending on the pressure of Ar atoms, and the target to substrate distance.

The mean free path (mfp) of the atoms may be defined as the average distance travelled before a collision. Suppose there are $n = P/k_B T$ atoms of Ar per unit volume of the chamber, with an effective collision cross-section characterised by an effective diameter d (typically of the order of 0.2 to 0.4 nm). On travelling one unit length, a sputtered atom will, on average, collide with all Ar atoms contained within a cylinder of unit length and diameter 2d. This corresponds to $n \pi d^2$ collisions. Thus the mfp is given by $1/n \pi d^2$. A more rigorous treatment, based on the kinetic theory of perfect gases, gives:

$$\text{mfp} = \frac{1}{\sqrt{2}n \pi d^2} = \frac{k_B T}{\sqrt{2}P \pi d^2} \qquad (20.2)$$

Taking an average atomic diameter of the order of 0.4 nm, at ambient temperature, gives:

$$\text{mfp}\,(cm) = \frac{0.6}{\text{P}\,(Pa)} \qquad (20.3)$$

Target-substrate distances are usually of the order of a few centimetres. Depending on the Ar pressure during deposition, two regimes are possible.

If the pressure is such that the mfp is greater than the target to substrate distance, we are in a direct regime. In this case the sputtered atoms do not collide with the Ar atoms, and thus arrive at the substrate with a rather uniform incidence, and their energy may be significant (several electronvolts).

On the contrary, if the mfp is less than the target to substrate distance, the sputtered atoms experience one or more collisions before eventually arriving at the substrate. In this case, there is a thermalization of the flux of sputtered atoms. The sputtered atoms will be deposited with a more random incidence, and with an energy which is closer to $k_B T$ (i.e. a fraction of an electronvolt).

In general, the best growth conditions are achieved at the cross over between both regimes. If the sputtered atoms arrive at the substrate with too much energy, they can damage the previously deposited layers. On the contrary, if they arrive with too little energy, their mobility on the surface of the substrate will be low, and consequently the surface will be rough. Moreover, if the Ar pressure is too high, the film may be contaminated with various Ar gas impurities.

As with MBE (see below), once the atoms have condensed on the substrate, the quality of the film growth depends very much on the type of substrate used, and on the temperature at which it is maintained. Non-oriented substrates, in particular Si(100) substrates, which are covered with an amorphous oxide layer (SiO_2), are generally used in sputtering. Quite often samples grow in polycrystalline form, with the grain size varying from a few nanometres to some tens of nanometres. These grains usually grow in a coherent fashion, forming a columnar structure across the sample thickness. In general, the growth direction of samples prepared by sputtering corresponds to a stacking of dense planes, i.e. a [111] growth direction for fcc phases, [110] for bcc phases, and [0001] for hcp phases. For more information on thin film preparation methods, the reader is referred to [4].

2.2.2. Molecular beam epitaxy (MBE)

MBE machines are much more sophisticated than sputtering machines, and operate at base pressures which are 2 to 3 orders of magnitude lower (i.e. 10^{-9} to 10^{-7} Pa). The growth chamber is usually equipped with *in situ* structural characterization facilities allowing the growth of very high quality samples. This is an excellent preparation technique for fundamental studies, but may prove difficult to upscale to an industrial level. Nevertheless, MBE is used industrially for the preparation of semi-conductor films, which are easier to grow than epitactic metal films.

In MBE, as in sputtering, the material to be deposited is evaporated, and then condensed onto a substrate. There are many ways to evaporate the material, the simplest being to heat the material in a crucible to near its sublimation or evaporation temperature, by passing a current through an electrical filament surrounding the crucible. Additional heat may be generated by bombarding the crucible with electrons which are thermally emitted by the filament. Another method consists of bombarding the surface of the material with an electron beam created with an electron gun. In this case the heating of the material is much more localised. An advantage of this method is its weak thermal inertia. Another method worth mentioning is Pulsed Laser Deposition (PLD), in which the material is locally heated with a laser beam. Excimer, and NdYaG lasers are the most commonly used lasers.

Once created, the vapour propagates within a certain cone, with part of it travelling towards the substrate. Taking into account the low value of pressure, of the order of 10^{-6} to 10^{-4} Pa during evaporation, the mfp of the atoms in the vapour is very long (more than a few hundred metres). Therefore the atoms suffer no collisions before reaching the substrate, where they very quickly reach thermal equilibrium with the substrate.

In epitaxic deposition, the growth mode is such that the structure of the material deposited has a direct relationship with that of the substrate. In practice, this means the growth of a single crystal film on a single crystal substrate. Thin film growth processes are widely studied both experimentally and theoretically. To achieve epitactic growth, there must be some geometrical compatibility between the crystallographic cell at the substrate surface, and at least one crystallographic plane of the material to be deposited. For example, the crystallographic planes of Fe which are perpendicular to the [001] growth direction match very well those of Au which are perpendicular to the [001] axis, after a rotation through 45° of the Fe cell relative to the Au cell. The reason is that Fe and Au have a cell length ratio very close to $\sqrt{2}$ (see fig. 20.1).

Fe(001)[1 10] // Au(001)[100]

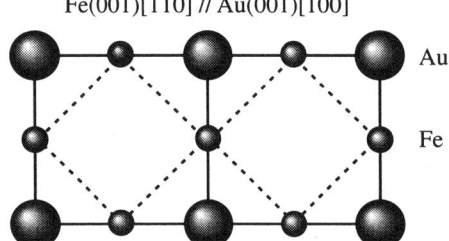

Figure 20.1 - Illustration of the epitactic relationship between a Au substrate and a deposited Fe film

Cell mismatches of a few % are tolerable. In such cases, the two lattices can be matched by creating a dislocation lattice or simply by elastic deformation. The temperature of the substrate is a very important growth control parameter. Usually, an elevated substrate temperature favours epitactic growth by increasing the mobility of the atoms which arrive at the surface (adatoms). However, too high a temperature may promote significant diffusion at the interface.

Three growth modes are identified:

- an atomic layer by atomic layer mode; this is the mode required for epitactic growth (this 2D growth mode is called the Frank-Van der Merwe mode),
- a three dimensional island growth mode (this 3D growth mode is called the Volmer-Weber mode),
- initial formation of a continuous atomic plane followed by island growth on top (this intermediate growth mode is called the Stransky-Krastanov mode).

The balance of the surface, interface, and electronic energies of the various constituents determines which mode prevails [2].

2.2.3. Chemical and structural methods for characterising thin films and multilayers

A distinction is made between *in situ* methods which allow characterization during sample growth, without exposure to air, and *ex situ* methods for which the sample must be taken out of the preparation chamber.

The most common *in situ* methods are reflection high energy electron diffraction (RHEED), low energy electron diffraction (LEED), and Auger spectroscopy. RHEED allows the direct monitoring of the thin film growth during deposition. Layer-by-layer growth can be identified by characteristic diffraction peaks, and by oscillations in the diffraction intensity during deposition. The maxima of the oscillations correspond to the complete filling of an atomic plane. LEED allows an analysis of the crystallographic structure within the plane. Auger spectroscopy allows a chemical analysis of the sample surface to a depth of about 1 to 2 nm.

The principal *ex situ* structural characterization methods are small angle X-ray scattering, X-ray diffraction, transmission electron microscopy and scanning-probe microscopies, in particular atomic force microscopy (AFM), very well suited to characterize the roughness of the surface of a film. The stacking of atomic planes can be visualised with high (atomic) resolution cross-sectional microscopy.

3. MAGNETISM OF SURFACES, INTERFACES AND THIN FILMS

The atomic environment at the surface of a material, or at the interface between two different materials, is strongly modified in comparison to that of bulk material. At a surface, the number of neighbours, known as the coordination number, is reduced and, moreover, the symmetry is not the same as for the bulk. As a result, contraction or dilation of the crystallographic cell occurs. Electron transfer from one material to another or hybridization, whereby the electronic orbitals of adjacent materials overlap, may occur. These phenomena can modify the electronic structures, and therefore the magnetic properties of these structures in comparison with bulk materials [5, 6]. Some examples are given below.

3.1. INCREASE IN THE MAGNETIC MOMENT AT THE SURFACE OF A TRANSITION METAL

Here we are interested in the properties of ultra-clean surfaces, which can only be prepared in ultra-high vacuum systems. Any gas absorption (O, Ar…) or other forms of surface contamination can considerably modify the properties mentioned below. *Ab initio* electronic structure calculations of ferromagnetic transition metals (Fe, Co, and Ni) predict that the magnetic moment is higher at the surface than in the bulk [6]. This is due to a narrowing of the d-band because of the lower surface atom coordination, and results in an increase of the density of states $N(\varepsilon_F)$ at the Fermi level. In the transition metals which are characterised by itinerant magnetism (see chap. 8), the increase of $N(\varepsilon_F)$ leads to an increase of the surface magnetism (Stoner criterion).

For example, the magnetic moment on a (001) surface of a crystal of bcc Fe is 2.96 μ_B / atom while the moment in the bulk is 2.2 μ_B / atom. For a (110) surface, the surface moment is only 2.65 μ_B / atom, which is weaker than that of the (001) surface. This difference results from the difference in coordination between these two surfaces. In a bcc crystal, the (110) planes are the most dense planes, i.e. the planes in which the atoms are closest. In other words, the surface density of atoms of Fe on a (110) surface is higher than that on a (001) surface. As a result, the narrowing of the d-band, and thus the increase in the magnetic moment, is less remarkable for the compact (110) surface than for the (001) surface. The same phenomenon is observed for an fcc crystal, in which the (111) planes are the most dense. For example, a compact (111) surface of Ni has a moment of 0.63 μ_B / atom while the moment in the bulk is 0.56 μ_B / atom. The (001) surface, which is less dense, has a moment greater than 0.68 μ_B / atom.

This tendency to increase the magnetic moment while reducing the co-ordination is more remarkable if we consider a chain of atoms or an isolated atom. On moving from a bulk material to a (001) surface, to a linear chain, and finally to an isolated atom, the magnetic moment of Ni increases from 0.56 to 0.68 to 1.1, and finally to 2.0 μ_B / atom. For Fe, these values are 2.25, 2.96, 3.3 and 4.0 μ_B / atom. The magnetic moment approaches the free atom value as the dimensionality is reduced.

Increases in the moment at the surface have been experimentally observed in some systems by *in situ* magnetic characterization techniques based on spin-polarised low energy electron diffraction (SPLEED) and scanning electron microscopy with polarization analysis (SEMPA).

In addition, a significant structural relaxation effect often occurs at the surface of a crystal. This relaxation directly influences the magnetic properties. For example, to minimise its total energy, the (001) surface of a crystal of Fe approaches, by about 4%, the atomic plane immediately below it, resulting in a reduction, by 6%, in the surface magnetic moment, compared to the non-relaxed state.

3.2. SURFACE MAGNETISM IN MATERIALS
NOT POSSESSING A MOMENT IN THE BULK

Due to the increase in the density of states at a surface, certain materials which are non-magnetic in bulk form may become magnetic at the surface, or when deposited as thin films on a substrate. This is the case for vanadium [6]. Surface magnetism may also appear in certain antiferromagnetic materials due to lack of compensation at the surface. Another particular case is Cr, which displays a modulation of the atomic moment along the [100] axis (spin density wave). A magnetic order exists in each (001) plane, but these planes are coupled antiferromagnetically. Therefore, there is no net bulk moment. Nevertheless, the (001) surfaces are ferromagnetically ordered, giving a magnetic moment which is greater than the moment of the atoms in the bulk. Experimentally, these ferromagnetic planes can be observed with a scanning tunnelling microscope (STM) using a chromium dioxide tip [7]. CrO_2 is a half-metallic material. The densities of states at the Fermi level for spin up $\uparrow$ and spin down $\downarrow$ electrons are very different. As a result, the tunnel current between the Cr surface and the CrO_2 tip depends on the local orientation of the magnetization. It has been observed that, at every monoatomic step on the (001) surface of Cr, the magnetic moment changes from one direction to the opposite direction.

3.3. EFFECTS INDUCED BY THE SUBSTRATE
ON THE MAGNETISM OF ULTRATHIN EPITACTIC FILMS

While in the last section we considered free surfaces of semi-infinite crystals, we will now consider ultrathin films (of one or a few atomic planes) epitactically grown on substrates which are of a different material than the film. The magnetic properties of the thin film depend strongly on the nature of the substrate for many reasons. If the cell matching between the substrate and the deposited film is not perfect, both materials will be deformed, to a greater or lesser extent, depending on their respective rigidities and thicknesses. This results in a variation of the cell parameter of the deposited material (by up to a few %) causing a change in its magnetic properties. In general, a contraction of the cell results in a broadening of the electronic band, and thus a reduction in the moment and in the magnetic stability. This is similar to what happens when pressure is applied to a bulk material. On the contrary, a dilation of the cell tends to increase the magnetic moment. Another structural effect of the substrate that is sometimes observed is the stabilization of crystallographic phases which do not exist in the bulk material, or exist only under very extreme conditions. An extensively studied case is that of iron, which is normally bcc but which can be stabilised in an fcc phase by epitactic growth on fcc (100) Cu. The magnetic properties of this phase are very different from those of bcc Fe, and in addition they are extremely sensitive to the value of the interatomic distance. Electronic structure calculations show that changes of a few % in the cell parameters of fcc Fe are enough to change the properties from being non-magnetic, to antiferromagnetic, and then to ferromagnetic with a strong moment [8]. It has been experimentally observed that different preparation techniques lead to fcc Fe with very different properties: when the iron is deposited onto Cu at low temperature, in general it is ferromagnetic [9] while it is antiferromagnetic when deposited at room temperature [10]. This difference comes from slight distortions in the crystalline arrangement of the fcc Fe film, which can induce big magnetic effects.

In addition to these purely structural effects, the choice of substrate can also directly influence the electronic structure of the deposited film. In fact, certain substrates have little or no direct electronic interactions with the deposited film, while the use of others leads to important hybridization effects between the electrons of the magnetic film and those of the substrate.

Substrates of the noble metals (Ag, Au) or of insulating materials (MgO, Al_2O_3) do not interact with the electrons of the deposited film. For example, thin films of bcc Fe on substrates of (001) Ag or (001) MgO have a magnetic moment very close to that observed for a 2-D film of Fe without a substrate.

On the contrary, transition metal substrates (V, Cr, Nb, Mo, Ru, Rh, Pd, W, Re) show significant hybridization effects with the electrons of the deposited film, resulting in novel properties. Consider the case of Pd which is a non-magnetic material that can be very easily polarised. If a thin film of Fe, Co or Ni is deposited on a substrate of Pd, the Pd acquires a magnetic moment from its surface down through a few atomic planes. Another example is that of iron deposited on (001) W. Band structure calculations predict that an atomic plane of Fe without a substrate has a moment of $3.1\ \mu_B$ / atom. A single atomic layer of Fe on (001) W is antiferromagnetic, with a moment of $0.93\ \mu_B$ / atom, due to strong hybridization with the W substrate. If a second atomic plane of Fe is deposited, the Fe becomes ferromagnetic [6].

Another manifestation of these effects is the appearance of dramatic interface anisotropy phenomena (see below). Thin films of Co sandwiched between films of Pd, Pt or Au have an easy direction of magnetization perpendicular to the plane of the film [11]. This behavior is very unusual for magnetic thin films, because the shape anisotropy (magnetostatic energy) of a thin film tends to align the magnetization in the plane of the film. This strong perpendicular magnetic anisotropy is often associated with significant magneto-optical effects, making these materials suitable candidates for magneto-optical recording media.

3.4. EFFECT OF REDUCED DIMENSIONALITY ON MAGNETIC PHASE TRANSITIONS

Consider again the free surface of a semi-infinite magnetic crystal. At sufficiently high temperature, the magnetic order always disappears, as the magnetic excitations become progressively more significant with increasing temperature. In the mean field approximation, the ordering temperature of ferromagnetic materials (the Curie temperature) is given by $T_C = J_0 S (S + 1) / 3k_B$ where S is the value of an individual spin, and J_0 is the sum of the exchange interactions with all neighbours. According to this expression, T_C is proportional to the number of neighbours (*via* J_0). Therefore, one can expect a reduction in the ordering temperature at the surface of a ferromagnetic material in comparison to the bulk value. This is true in a number of real cases, where one then refers to the creation of dead layers at interfaces and surfaces.

In addition to the modification of the ordering temperature, the critical exponents which characterise the variation in magnetization and susceptibility as a function of temperature are different from those of the bulk. For example, if one considers a ferromagnetic material in the mean field approximation, one can show that the magnetization of the bulk varies as $(T_C - T)^{1/2}$. The same model shows that, at the surface, magnetization varies as $(T_C - T)$ [12].

However, in certain cases, the dominant effect is not a reduction but an increase in the ordering temperature at the surface. For transition metals, this is again due to a narrowing of the d-band at the surface, which leads to an increase in the density of states at the Fermi level. This strengthens the magnetic stability at the surface according to the Stoner criterion. This increase in order at the surface in comparison with the bulk can also occur in rare earth metal thin films: it has been observed that the magnetic order in the uppermost layer of Gd persists until 310 K, that is 20 K above the ordering temperature of bulk Gd. This effect is due to the strengthening of the exchange interactions with the top two sub-surface layers, which is in turn due to a redistribution of charges near the surface [6].

In the same way, a reduction in the magnetic ordering temperature of ultra-thin magnetic layers, deposited on a substrate, is observed for thicknesses of less than a few atomic planes. This reduction is due to the reduced coordination of atoms at the surface in comparison to the bulk (see fig. 20.2).

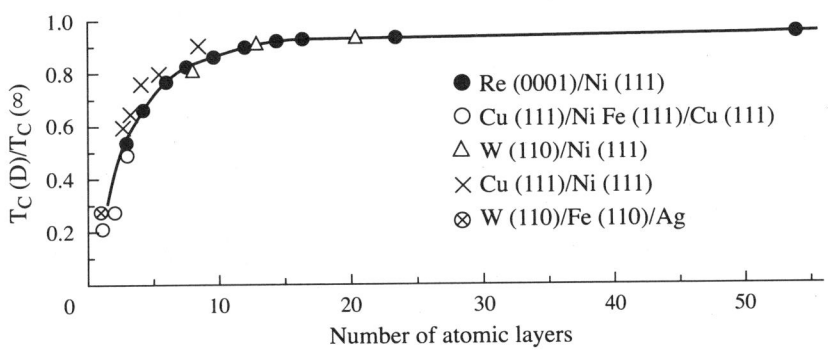

Figure 20.2 - Variation of the ordering temperature as a function of thickness for ultra-thin films of various transition metals, *after [13]*

The films are deposited on metal substrates: the substrate is indicated on the left, and the thin film material on the right.

3.5. *MAGNETIC ANISOTROPY OF THIN FILMS*

Magnetic anisotropy is the dependence of the magnetic energy of a system on the direction of magnetization within the sample. Thin films show very large anisotropy phenomena. The main reason for this is that their shape favours an orientation of magnetization within the plane in order to minimise the magnetostatic energy: this anisotropy energy is often described by a uniaxial anisotropy $E = -K \cos^2\theta$, where θ is the angle between the magnetization and the normal to the plane of the sample. By definition, a positive value of K implies an easy axis of magnetization perpendicular to the plane of the sample ($\theta = 0$). On the contrary, a negative value of K corresponds to an easy plane of magnetization ($\theta = \pi/2$). In the latter case, higher order energy terms may lower the symmetry, creating an in-plane easy axis of magnetization. Though in-plane anisotropy is sometimes observed, uniaxial anisotropy is more characteristic of thin films. Films showing perpendicular anisotropy, in particular transition metal films, are extensively studied because of their potential applications in the area of magneto-optical recording.

There are different sources of magnetic anisotropy in thin films. From the phenomenological point of view, the various sources are divided into two groups, those concerning the volume of the material, and those concerning its surface. The effective anisotropy of a thin film of thickness t is given by: $K = K^{eff} = K_v + K_s/t$, where K_v (in $J.m^{-3}$) represents the anisotropy terms throughout the volume, and K_s/t represents the difference in the energy of atoms at a surface or interface and those in the volume of the sample.

Experimentally, K_v and K_s are determined from the relationship $Kt = K_v t + K_s$ by plotting Kt as a function of film thickness, for a series of samples. Figure 20.3 shows such a curve for a multilayer of composition $(Co\ t_{Co}\ nm/Pd\ 1.1\ nm)_n$, for two deposition temperatures.

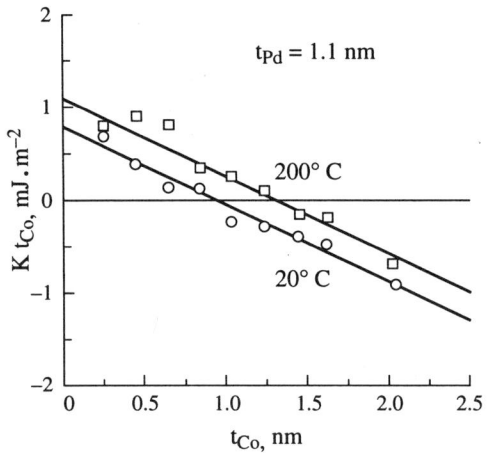

Figure 20.3 - *Magnetic anisotropy per unit area of Co layer for a (Co t nm / Pd 1.1 nm)$_n$ multilayer, after [14]*

The intersection with the K t axis represents the sum of the anisotropy energies of the two interfaces of each Co layer, while the slope is proportional to the volume anisotropy energy.

In this series of samples, a tilting of the easy axis of magnetization is observed for a Co thickness of 1.2 nm. For $t_{Co} < 1.2$ nm, the magnetization of the multilayer is spontaneously directed perpendicular to the interfaces while for $t_{Co} > 1.2$ nm, the magnetization is oriented in the plane of the multilayer.

Such reorientation of the easy axis of magnetization, as a function of the film thickness, has been observed in a large number of transition metal thin films and multilayers [15].

Some examples which have been studied include Co/Au (111) [11, 16], Co/Ni (111) [17], Co/Pd [18], Co/Pt [19], Co/Ru (001) [20], Fe (001)/noble metal (Ag [21], Au [22]), W (110)/Fe (110)/noble metal (Cu, Ag or Au) [23]. These reorientations manifest themselves as a change in the form of the hysteresis loop as measured using a SQUID or the polar or longitudinal magneto-optic Kerr effects. Figure 20.4 gives an example of the change in hysteresis loop for the case of an epitactic sandwich of composition Au/Co t_{Co}/Au.

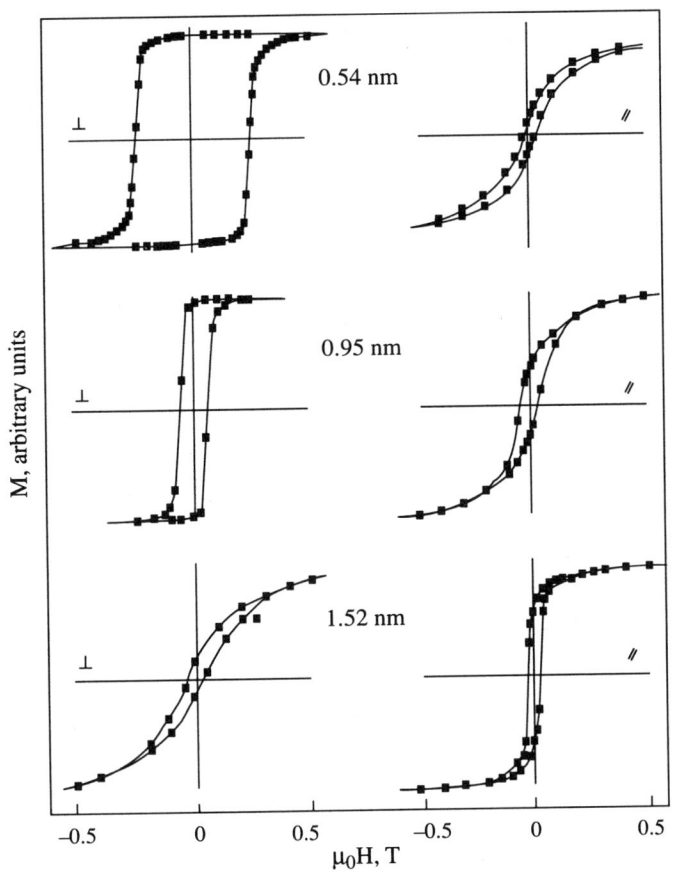

Figure 20.4 - Hysteresis loops of a series of samples of composition Au / Co t_{Co}/Au with $t_{Co} = 0.54$, 0.95 and 1.52 nm, measured on a SQUID at 10 K, after [24]

The surface anisotropy of thin films of transition metals can be significant. For example, at Co/Pd or Co/Pt interfaces, the surface anisotropy is of the order of $5 \text{ MJ}.m^{-3}$ ($300 \mu eV/$ atom of Co). In comparison [15], the anisotropy energy of bulk Co is only $0.85 \text{ MJ}.m^{-3}$ ($65 \mu eV/$ atom of Co), while that of Fe is $0.017 \text{ MJ}.m^{-3}$, and that of Ni is $0.042 \text{ MJ}.m^{-3}$. This surface anisotropy is comparable with the anisotropy of some permanent magnet materials, YCo_5: $5 \text{ MJ}.m^{-3}$, $Nd_2Fe_{14}B$: $12 \text{ MJ}.m^{-3}$.

At the microscopic level, there exist two principal sources of magnetic anisotropy. Firstly, there is the dipolar interaction between the atomic moments, which depends on the direction of the atomic spins and on the vector separating them. Secondly, there is the spin-orbit interaction which couples the atom's spin with its orbital moment. The orbital moment itself interacts with the crystallographic lattice through the crystalline electrostatic field, thus conferring preferential directions on the atomic orbitals. As the spin moment is coupled to the orbital moment by spin-orbit coupling, there is a resultant coupling between the magnetization and the crystallographic axes of the material.

Anisotropy effects in thin films are commonly described by adopting a more macroscopic approach, considering the following contributions: shape anisotropy, magnetocrystalline anisotropy, magnetoelastic anisotropy, and the effects of roughness and interdiffusion. Details of these contributions are given below.

3.5.1. Shape anisotropy

A thin film may be approximated by a very oblate ellipsoid of revolution. The demagnetising factor perpendicular to the plane is $N_\perp = 1$, so that the demagnetising field is practically zero as long as the magnetization stays in the plane of the sample, i.e. the demagnetising field $\mathbf{H}_d$ tends to maintain the magnetization in the plane of the sample. The shape anisotropy energy, per unit volume, of a magnetic thin film is given by $E_d = -\dfrac{\mu_0}{2V} \int \mathbf{M}.\mathbf{H}_d \, dv = \dfrac{1}{2}\mu_0 M_s^2 \cos^2\theta$, where M_s is the spontaneous magnetization of the material, and θ is the angle between the magnetization and the normal to the plane of the sample. If this term is the only anisotropy term, the applied field needed to saturate the magnetization of a thin film, perpendicular to its plane (out-of-plane), is given by minimising the total energy $E = E_d - \mu_0 M_s H \cos\theta$, and by calculating the value of H for which $\cos\theta$ equals 1. One finds that $H_{sat} = M_s$.

To saturate a film of iron out-of-plane, at room temperature, a field of $1,718 \text{ kA}.m^{-1}$ ($\mu_0 M_s = 2.16$ T) must be applied. In the case of cobalt, a field of $1,424 \text{ kA}.m^{-1}$ ($\mu_0 M_s = 1.79$ T) is needed, while for Ni a field of $484 \text{ kA}.m^{-1}$ ($\mu_0 M_s = 0.608$ T) is needed.

3.5.2. Magnetocrystalline anisotropy

This originates from spin-orbit coupling. As for bulk crystals, the symmetry in the bulk of the film determines the volume magnetocrystalline anisotropy. However, symmetry breaking at the surface, or at interfaces, gives rise to a supplementary surface magnetocrystalline anisotropy term. Néel was the first to predict, in a pair model, the existence of a surface anisotropy of the form $E_s = -K_s \cos^2 \theta$ [1].

For rare earth based films, the anisotropy is usually very significant because of the importance of spin-orbit coupling in these systems. The surface anisotropy can be relatively easily described within the point charge model. Considering the charge carried by each rare earth ion, one can calculate the electrostatic field at the surface, and the preferred orientation of the atomic orbitals in this field. One can thus deduce the surface anisotropy.

It is much more complicated to quantitatively interpret the results for transition metal films, because of the itinerant character of the magnetism of these materials, and because the spin-orbit coupling is much weaker than for the rare earth metals [25]. Nevertheless, very elaborate models based on *ab initio* band structure calculations allow a quantitative understanding of the observed surface anisotropies [26]. These approaches are based on a calculation of the perturbation to spin-orbit coupling by precisely identifying the orbitals which contribute most to magnetocrystalline anisotropy [27]. In this way, one can correlate the filling of spin ↑ or spin ↓ bands with the tendency towards planar or perpendicular anisotropy.

3.5.3. *Magnetoelastic anisotropy*

This is the inverse of magnetostriction (see chap. 12). The epitactic growth of a thin film onto a substrate with different cell parameters, or a multilayer of alternating layers of different cell parameters, produces enormous internal stresses which may exceed some tens of GPa. These stresses give rise to anisotropic terms *via* the magnetoelastic coupling.

Two situations are identified:

♦ If the cell mismatch between the two materials in contact is not too great, there is a *coherent deformation* of the unit cells of both materials, one contracts while the other dilates. Both materials adopt common cell parameters, which are intermediate in value between the original cell parameters of each material. This situation is observed for multilayers when the individual layers are sufficiently thin. The deformation is then homogeneous throughout the volume of the layers. This coherent deformation gives rise to a volume anisotropy term.

♦ For thicker layers, an elastic coherent deformation throughout the volume would cost too much elastic energy. Each material prefers to relax towards its bulk structure, creating a network of dislocations along the interface which allows the cell mis-match to be accommodated. This is known as an *incoherent* regime. As the deformation produced in this regime is localised at the interfaces, it gives rise to a surface anisotropy contribution.

Figure 20.5 illustrates the change from a coherent to an incoherent regime as a function of the film thickness for a film of Ni epitactically deposited on a Cu single crystal.

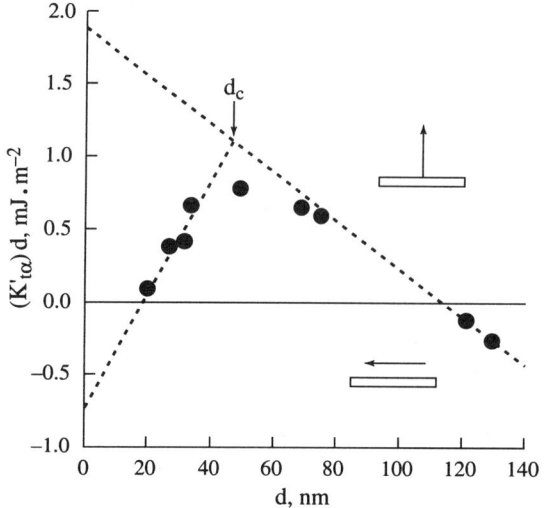

Figure 20.5 - Anisotropy of thin films of Ni epitactically grown on single crystals of Cu

The change in slope at about 4.5 nm indicates the transition from the coherent regime of the thinnest samples to the incoherent regime of the thicker samples, after [28].

3.5.4. Effect of roughness and interdiffusion

Thin films never grow in the ideal atomic layer by atomic layer fashion. The interfaces always have a certain roughness, characterised by the height and width of the terraces. This roughness produces a local demagnetising field that reduces the shape anisotropy compared with that of an ideal surface. In this sense, the roughness tends to favour an anisotropy perpendicular to the film plane. The resultant anisotropy varies as the reciprocal of the film thickness, and is thus a surface contribution [29]. The change in local environment resulting from the atoms being located in the middle of terraces, on terrace edges, terrace corners, crater bottom… also changes the local electronic properties. This can result in modifications to the surface anisotropy [30].

4. COUPLING MECHANISMS IN MAGNETIC MULTILAYERS

On the atomic scale, different types of coupling exist between magnetic moments in a bulk magnetic material. The most important are exchange interactions in metals, super-exchange interactions in insulators, and dipolar interactions. Similarly, in magnetic multilayers these energy terms give rise to coupling effects between magnetic layers, either directly across an interface or indirectly through a non-magnetic interlayer. Here we discuss three situations in which coupling gives rise to particular effects:

♦ direct coupling by *pinholes* or by dipolar interactions (orange-peel mechanism),

♦ coupling across a non-magnetic layer in a multilayer formed of alternating layers of a ferromagnetic transition metal and a non-magnetic noble or transition metal (example Co / Cu),

◆ coupling across an interface between a ferromagnetic transition metal layer and a rare earth layer (example Fe / Tb).

4.1. DIRECT FERROMAGNETIC COUPLING BY PINHOLES OR BY DIPOLAR INTERACTIONS

It is very difficult to obtain the ideal layer-by-layer epitactic growth mechanism when preparing multilayers. In many studies in which a careful control of the nanostructure of the sample is not needed, epitactic growth is not even sought. Films are often deposited on amorphous substrates (glass, Si covered by a silicon oxide layer...).

The resultant multilayer structure has many defects, in particular significant interface roughness, or even discontinuities in certain films. In a multilayer consisting of magnetic layers separated by non-magnetic layers, the presence of discontinuities in the non-magnetic layer (commonly called *pinholes*) gives rise to a direct contact between the magnetic layers that leads to ferromagnetic coupling between the layers (fig. 20.6). This situation is often encountered in sputtered multilayers with individual layer thicknesses less than about 2 nm. The presence or absence of these pinholes depend very much on the actual material and the growth conditions achieved.

Figure 20.6 - Ferromagnetic coupling through a pinhole

Two magnetic layers (F) are separated by a non-magnetic layer (NM). The typical thickness of the NM layer is of the order of 1 to 3 nm.

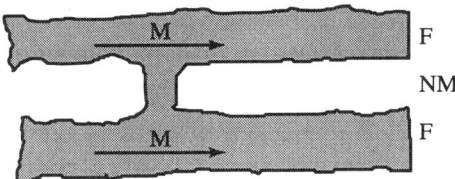

Another source of ferromagnetic coupling which is often encountered in multilayers of average crystallographic quality is the orange peel mechanism, initially described by Néel [1]. In multilayers, the roughness of the interfaces is often correlated from one interface to another by the very fact that the thickness of the layers is uniform (see fig. 20.7).

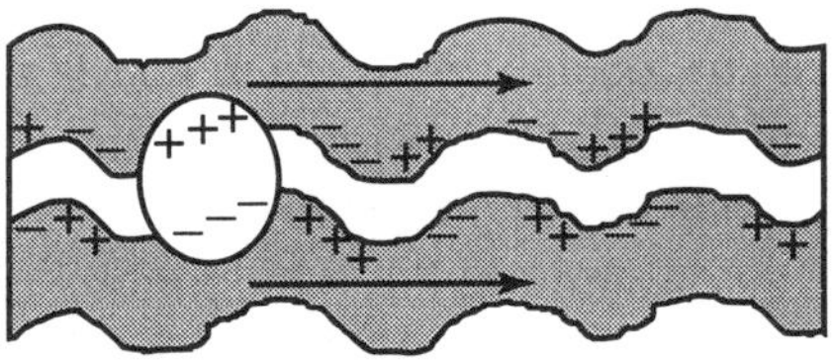

Figure 20.7 - "Orange peel" coupling mechanism

There exists a correlated roughness between the interfaces. When the magnetization vectors of two adjacent ferromagnetic layers are parallel, magnetostatic charges of opposite sign appear symmetrically on opposing interfaces (see for example the encircled area). The positive dipolar interactions between these opposing charges give rise to a ferromagnetic coupling between the magnetization vectors of the two magnetic layers.

Dipolar interactions between the magnetostatic charges which then appear on the interfaces give rise to a ferromagnetic coupling between the magnetic layers. Quantitatively, describing the interfacial roughness as a sinusoidal function of amplitude h (terrace height) and period L (double the width of the terraces), one can show, using the model of Néel, that the resultant dipolar coupling between the magnetic layers across a non-magnetic layer of thickness t_{NM} is given by [31]:

$J = \dfrac{\pi^2}{\sqrt{2}} \dfrac{h^2}{L} \mu_0 M_{sat}^2 \exp\left(-2\pi\sqrt{2}\,\dfrac{t_{NM}}{L}\right)$. In general, this coupling is weaker than

the direct coupling due to pinholes, and it usually manifests itself at greater thicknesses, i.e. at t_{NM} of the order of 2 to 10 nm.

4.2. COUPLING ACROSS A NON-MAGNETIC FILM IN MULTILAYERS CONSISTING OF ALTERNATING FERROMAGNETIC TRANSITION METAL AND NON-MAGNETIC TRANSITION OR NOBLE METAL LAYERS (EXAMPLE CO/CU)

Exchange type coupling between magnetic layers across a non-magnetic metal have been frequently observed in multilayers of the type $(F\,t_F\,/\,NM\,t_{NM})_n$ where F represents a ferromagnetic transition metal (Fe, Co, Ni, and most of their alloys) of thickness t_F, and NM is a noble metal (Cu, Ag, Au) or a non-magnetic transition metal (Cr, Mo, Ru, Re, Ir…) of thickness t_{NM}. When the crystallographic quality of the layers is sufficiently good, oscillations in the amplitude of this coupling is observed as a function of the thickness of the non-magnetic interlayer [32]. The period of these oscillations is typically of the order of a nanometre. For certain values of thickness t_{NM}, the coupling tends to favour a spontaneous parallel orientation of the magnetization in successive magnetic layers (ferromagnetic coupling). For other thicknesses, it tends to induce a spontaneous anti-parallel orientation of the magnetization (antiferromagnetic coupling). Oscillations in the amplitude of the coupling are also observed as a function of the thickness t_F of the magnetic layers but these are less pronounced, and in general do not change sign. An example of the oscillations in coupling as a function of t_{NM} is given in figure 20.8 for the trilayer $Ni_{80}Co_{20}\,/\,Ru\,/\,Ni_{80}Co_{20}$ [33].

This coupling originates in a polarization of the electrons of the non-magnetic layer induced by contact with the magnetic layers. It is known that, in bulk materials, a magnetic Mn impurity placed in a Cu matrix polarises the sea of conduction electrons of the Cu in the vicinity of the Mn. The densities of spin $\uparrow$ and spin $\downarrow$ electrons, instead of being equal, as in the case of pure Cu, are slightly different in the vicinity of the impurity, giving rise to a resultant polarization. This polarization oscillates with a wave vector $2\,k_F$ (where k_F is the Fermi wave vector), attenuating as $1/r^3$ where r is the distance from the impurity (see chap. 9). These are the so called RKKY (Ruderman-Kittel-Kasuya-Yosida) oscillations. If a second magnetic impurity is found at a certain distance from the first one, the reciprocal coupling effect between the conduction electrons of the matrix and this second impurity gives rise to an indirect coupling between the spins of the two impurities, which may be ferromagnetic or antiferromagnetic, depending on the distance between them.

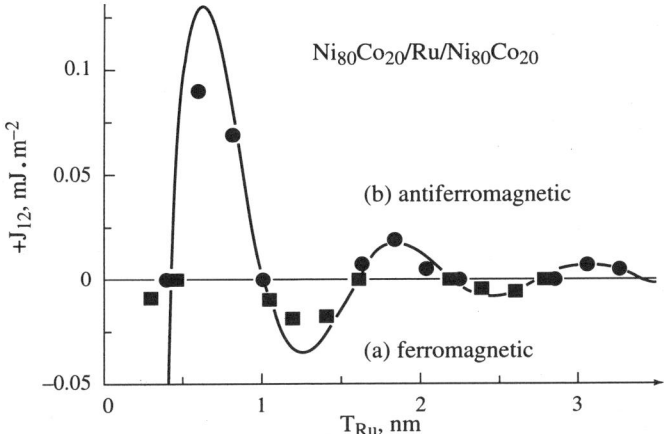

Figure 20.8 - Oscillations in coupling as a function of the thickness of the Ru interlayer, t_{Ru}, for the trilayer $Ni_{80}Co_{20}/Ru/Ni_{80}Co_{20}$, after [33]

The same coupling mechanism may occur in multilayers. In this case the oscillating polarization of the electrons of the non-magnetic interlayer are induced by magnetic atoms situated at the F/NM interface. The resultant coupling may be ferromagnetic or antiferromagnetic, depending on the thickness t_{NM}. Theoretical models allow us to relate the period of the oscillations with certain details of the Fermi surface [34]. In samples of very good structural quality, the superposition of a number of periods of oscillation may be observed.

4.3. INTERFACIAL COUPLING BETWEEN FERROMAGNETIC TRANSITION METAL FILMS AND RARE EARTH FILMS

Among bulk magnetic materials, rare earth-transition metal alloys are very important because of the exceptional properties which can be obtained by combining the high room temperature magnetization of the ferromagnetic transition metals with the high anisotropies of the rare earths (e.g. permanent magnet materials, materials with very high magnetostriction, materials with strong magneto-optical effects). In these alloys, the coupling between the rare earth moments, and the transition metal moments is ferromagnetic for the light rare earths (4f band less than half full), and antiferromagnetic for the heavy rare earths (4f band more than half full).

In a similar way, multilayers formed of alternating layers of ferromagnetic transition metals, and rare earths (in particular Gd or Tb) have been intensively studied. In Fe/Gd or Fe/Tb multilayers, an antiferromagnetic coupling exists across the interface between the magnetization of Fe, and that of the heavy rare earth. This coupling is of the same nature as that observed in homogeneous alloys of Fe, and heavy rare earths.

Therefore (Fe / Gd) multilayers show a ferrimagnetic macroscopic ordering: in zero field, each layer is ferromagnetic but its magnetization is antiparallel to that of its two neighbours. When a field is applied, a transition occurs at a certain threshold (the *spin-flop* transition) between a situation of antiparallel alignment of the magnetization of successive layers, and a situation in which the magnetization of successive layers is tilted at a certain angle on either side of the field direction. As the field intensity is increased, the tilt angle progressively increases as the magnetization approaches saturation in the field direction. Saturation is expected when the Zeeman energy associated with the field is sufficient to overcome the antiferromagnetic coupling between adjacent layers.

A trilayer formed of layers of amorphous alloys of $Y_{0.33}Co_{0.67}$, and $Gd_{0.33}Co_{0.67}$ is a model system which allows a detailed study of this type of magnetization process [35]. The first is a ferromagnetic material while the second is ferrimagnetic (the composition indices 0.33 and 0.67 will be omitted in the following discussion). Both these materials have anisotropies which are weak in comparison to the exchange energies at play. Below the compensation temperature of the ferrimagnetic material, the magnetization of Gd is greater than that of Co, so that the resultant magnetization of the alloy is parallel to that of Gd. The intensity of the ferromagnetic Co-Co exchange interactions is about ten times greater than that of the antiferromagnetic Gd-Co interactions, which is itself about ten times greater than the intensity of the ferromagnetic Gd-Gd interactions. In an YCo / GdCo / YCo sandwich, the magnetization of the Co is ferromagnetically coupled across the entire sandwich, and the magnetization of Gd is antiferromagnetically coupled to that of Co in the entire volume of the central layer. Figure 20.9 is a hysteresis loop of a trilayer of composition YCo 100 nm / GdCo 100 nm / YCo 100 nm at 4.2 K. The magnetization is expressed in units of magnetic moment per unit surface area, m / S, and is given in amperes.

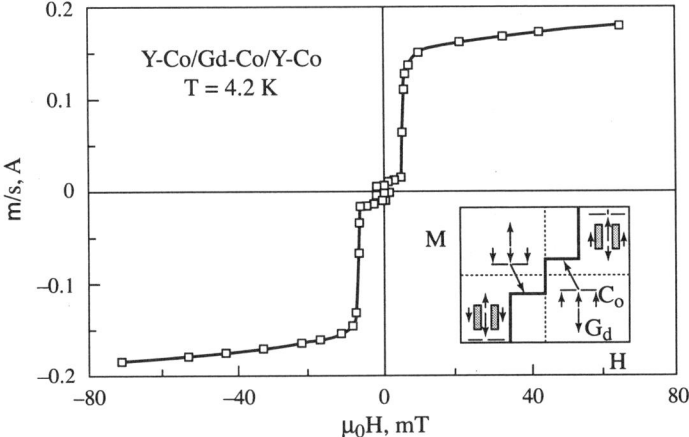

Figure 20.9 - Hysteresis loop of a trilayer sandwich measured at 4.2 K [35]
Trilayer composition: YCo 100 nm / GdCo 100 nm / YCo 100 nm.

In zero field, this system shows a ferrimagnetic macroscopic ordering similar to that previously described for Gd/Fe multilayers.

When a weak magnetic field is applied, the resultant moment of this ferrimagnetic system is oriented parallel to the applied field. Therefore the magnetization of the YCo films is parallel to the field direction while that of the GdCo film is antiparallel. When the field is increased, the magnetization of GdCo flips into the hemisphere of the field (spin-flop transition). This costs exchange energy because the flipping of the magnetization of the GdCo results in the formation of a Bloch wall in the magnetization of Co, as illustrated in figure 20.10.

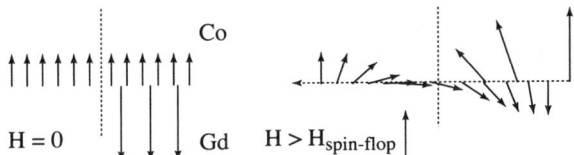

Figure 20.10 - Creation of a planar interfacial Bloch wall during the spin-flop transition in an YCo/GdCo/YCo sandwich [35]

For applied fields greater than $H_{spin-flop}$, the magnetization on both sides of the interface tends to saturate, in the direction of the field, driven by a progressive compression of the interface wall. Quantitatively, the wall profile is given by balancing the torque created by the exchange interactions within the wall $\Gamma_{ech} = -(A/2)(\partial^2\theta/\partial x^2)$, and the torque due to the local coupling of the magnetization with the applied field $\Gamma_H = \mu_0 M(x) H \sin\theta$, where $\theta(x)$ is the angle between the local magnetization, and the applied field, x is the axis perpendicular to the plane of the interface, and situated at $x = 0$, and A is the exchange constant (see eq. 5.1). For $x < 0$, $M(x) = M_1$, and for $x > 0$, $M(x) = M_2$, where M_1 is the magnetization of YCo and M_2 is the magnetization of GdCo. The equilibrium between these two torques $\Gamma_{ech} + \Gamma_H = 0$ leads to the Sine Gordon equation: $A \partial^2\theta/\partial x^2 = 2\mu_0 M_1 H \sin\theta$, and $A \partial^2\theta/\partial x^2 = 2\mu_0 M_2 H \sin\theta$, for $x < 0$, and $x > 0$, respectively.

This equation is very similar to that which gives the profile of a Bloch wall in a ferromagnetic material with the only difference being that the Zeeman energy $\mu_0 M H$ replaces the uniaxial anisotropy K. These equations can be easily solved analytically if the media on both sides of the interface are assumed to be semi-infinite; the domain wall width is then given by: $\delta = \pi\sqrt{A(M_1 + M_2)/2\mu_0 M_1 M_2 H}$, and its energy by: $\gamma = 2\sqrt{2\mu_0 AH}\left(\sqrt{M_1} + \sqrt{M_2} - \sqrt{M_1 + M_2}\right)$. It can also be shown that the approach to saturation is governed by the progressive compression of this macroscopic planar wall.

The magnetic moment per unit surface area approaches saturation according to the law: $m_{sat} - m(H) = 2\sqrt{2A/\mu_0 H}\left(\sqrt{M_1} + \sqrt{M_2} - \sqrt{M_1 + M_2}\right)$. Such an approach to saturation, varying as $(H)^{-1/2}$, has been experimentally observed in a very wide range of fields above the spin-flop transition in these systems [35].

5. TRANSPORT PROPERTIES OF THIN FILMS AND MULTILAYERS

5.1. ELECTRONIC TRANSPORT IN METALS

5.1.1. Classical image

The electrons of non-excited isolated atoms are bound to the nucleus, occupying states which are organised into electronic layers. These states correspond to well defined energy levels. A minimum energy, the ionization energy, is required to remove an electron from an atom.

When atoms approach each other, as in a crystal, they can no longer be considered isolated. An overlapping occurs between the trajectories of electrons (known as wavefunctions in the language of quantum mechanics) which occupy the outermost layers. This results in a broadening of the energy levels of these electrons which then form energy bands instead of energy levels.

The overlapping also results in a delocalization of the outermost electrons. These electrons, instead of remaining attached to a particular atom, become free to move from one atom to another in the crystal. These are called *free* or *conduction electrons*, because they are responsible for the circulation of an electric current when an electric field is applied to the metal. In the absence of an electric field, these electrons move randomly (Brownian motion). The modulus of the velocity of electrons (v_F, Fermi velocity) in random motion is extremely high, of the order of $c / 200$, where c is the speed of light. Nevertheless, as the velocity direction is random, the average velocity of all the electrons is zero so that no macroscopic electric current circulates in the metal. When an electric field is applied, the electrons experience a very weak acceleration in the direction of the electric field (the change in velocity due to the electric field remains infinitesimal in comparison with the random velocity in the electric field).

In a very short time (of the order of 10^{-12} s), a stationary regime is established in which the acceleration of the electrons in the direction of the electric field is counter-balanced by random scattering of these electrons. All points at which the translational symmetry of the crystal is broken may act as scattering centres: impurities, crystallographic defects, lattice excitations (phonons), magnetic excitations (magnons), interfaces, surfaces... The classical Drude model can be evoked to describe this stationary regime. In this model, the electrons are taken to be independent particles of mass m*, and of charge e. In the absence of an electric field, the electrons experience Brownian motion, similar to that of gas molecules. Their average velocity $<v_0>$ is zero. Consider the effect of an electric field $\mathbf{E}$ on any electron. Between two collisions, the electron experiences an acceleration $\gamma = -e\,\mathbf{E} / m^*$. If at the instant t_0 the electron moves with a velocity $\mathbf{v}_0$, at an instant t later, it has reached a velocity $\mathbf{v} = (-e\,\mathbf{E} / m^*)(t - t_0) + \mathbf{v}_0$. If we now average over all electrons, taking t_0 as the instant at which a scattering event occurs, and defining $\tau = <t - t_0>$ to be the average time between scattering events (equivalent to a time of flight), we obtain $<\mathbf{v}> = -e\,\mathbf{E}\,\tau/m^*$.

The quantity $\lambda = v_F\,\tau$, known as the *mean free path* (mfp), is the average distance travelled by an electron between two scattering events. Taking n to be the density of conduction electrons, the current density resulting from the application of an electric field is $\mathbf{j} = -n\,e\,<\mathbf{v}> = n\,e^2\,\mathbf{E}\,\tau/m^*$. Therefore the electrical conductivity σ, defined by $\mathbf{j} = \sigma\,\mathbf{E}$, is given by $\sigma = n\,e^2\,\tau/m^*$. This model has the advantage of being simple, but it does not consider the real electronic structure of material, in particular the existence of energy bands, and a Fermi surface which play a very important role in transport properties.

The Fermi velocity of electrons in normal metals is typically of the order of 1 to 2×10^6 m.s^{-1}, that is of the order of $c/200$, as stated above. On the other hand, the average drift velocity resulting from passing a current is very low. For example, in copper, supposing that there is one conduction electron per atom of copper, the drift velocity can be estimated, for a current density of 10^7 A.m^{-2}, to be 1 mm.s^{-1}, which is 9 orders of magnitude less than the Fermi velocity. At room temperature, the mfp in normal metals is typically of the order of a few tens of nanometres. In any case, at 300 K the mfp is limited by phonon scattering.

At low temperature, the density of phonons is very low. It is principally structural defects, and possibly surfaces, and interfaces for the case of thin films, and multilayers, which limit the mfp. Therefore the value of the mfp depends mostly on the crystallographic quality of the material. The ratio of room temperature to low temperature resistivity $\rho(300\ \text{K})/\rho(4.2\ \text{K})$ gives a good idea of the extent of structural scattering in a material. It can vary from about 2 in polycrystalline metallic films prepared by sputtering to about 10^6 in very good single crystals of Fe or Cu.

5.1.2. *Description of electrical conductivity in the framework of a band model, the free electron gas model*

Here we present the principal ideas of this description. For more details, the reader is referred to [36, 37].

In a crystal, the states which electrons can occupy are organised in energy levels (electrons occupying the inner electronic shells), and energy bands (less localised outer shell electrons). Due to the quantum nature of the electrons, and particularly because of their fermion character (spin 1/2), two electrons cannot occupy the same state (the Pauli exclusion principle).

Each material has a well defined number of electrons, $\mathcal{N}$, which is determined by the atomic number Z of its constitutive elements. These $\mathcal{N}$ electrons occupy all the energy levels between the lowest energy levels of the innermost electrons up to a maximum energy, ε_F, known as the Fermi energy. This energy is defined by $\mathcal{N} = n(\varepsilon_F)$ where the function $n(\varepsilon)$, determined by the electronic structure of the material, is the number of available levels of energy less than ε. Its derivative, $N(\varepsilon) = dn(\varepsilon)/d\varepsilon$, is the density of states (see § 3.2 of chap. 8).

In reciprocal space, the surface defined by all the wave vectors $\mathbf{k}$ corresponding to the Fermi energy is called the Fermi surface. At zero temperature, the occupation of electronic states switches from 1 to 0 at the Fermi energy. At finite temperature, this variation follows Fermi-Dirac statistics. The probability of occupation of a state of energy ε is given by the expression: $f(\varepsilon) = \{1 + \exp[(\varepsilon - \varepsilon_F)/k_B T]\}^{-1}$.

The transition from 1 to 0 spreads out over an energy band of the order of $k_B T$, generally much lower than ε_F, e.g. at 300 K, $k_B T$ is of the order of 1/40 eV while ε_F is of the order of a few eV in normal magnetic materials. Similarly, the electronic levels in crystals are all occupied except those of highest energy, which are partially full.

For ferromagnetic transition metals (Fe, Co, Ni, and most of their alloys), the 3d and 4sp bands are the partially full bands. For the rare earths, it is the 4f and 5sp bands which are partially full. Generally it is the electrons of these partially filled bands which are responsible for most physical, and chemical phenomena observed in materials.

In isolated atoms, the sp shells are the outermost shells, the d shells, and especially the f shells are deeper. As a consequence, in a crystal, the spreading of the electronic wave functions between neighbouring atoms is most significant for sp electrons, weaker for d electrons, and almost zero for f electrons. As a result the sp electrons are delocalised, and are very mobile in the crystal, the d electrons are more localised, and thus less mobile while the f electrons are very localised, and have no mobility. For this reason f electrons are not conduction electrons.

The electronic mobility is quantitatively given by the effective mass m* which is determined by the electronic structure of the material.

The sp electrons are generally assumed to be the principal contributors to conduction in metals as a result of their high mobility. However, the distinction between sp and d electrons is not clear when hybridization phenomena occur between bands. In addition, it has been proved, most notably in Fe, that certain electrons of d character are more mobile than electrons of sp character [38].

The free electron gas model is commonly used to describe sp electrons. In this model, the electrons are assumed to be independent, and their energy contains a kinetic energy term, $E = \hbar^2 k^2/2m^*$, only. The Fermi surface is defined by a series of wave vectors $\mathbf{k} = (k_x, k_y, k_z)$ satisfying $\hbar^2\left(k_x^2 + k_y^2 + k_z^2\right)/2m^* = \varepsilon_F$. Therefore the surface is a sphere centred on $\mathbf{k} = \mathbf{0}$.

When an electric field $\mathbf{E}$ is applied in the direction opposite to Ox, all the electrons are accelerated in the direction of $-\mathbf{E}$, so that the Fermi sphere is displaced along the k_x axis. This displacement liberates a thin layer of empty states of energy just below ε_F (left part of the Fermi sphere in figure 20.11).

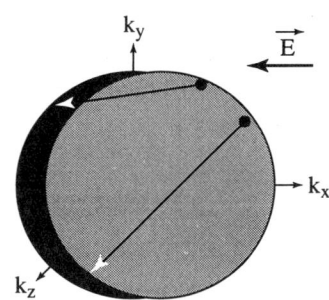

Figure 20.11 - Fermi surface before (black), and
after (grey) the application of an electric field

The white arrows represent the scattering between
occupied energy states of energy just above ε_F, and
empty states of energy just below ε_F.

Electrons with energy higher than ε_F (right side of the Fermi sphere) are scattered towards the empty states of energy just below ε_F (illustrated with white arrows in figure 20.11). In analogy with the Drude model, it can be shown that the displacement of the two Fermi surfaces, $\delta\mathbf{k}$, is given by $\hbar\,\delta\mathbf{k} = -e\,\mathbf{E}\,\tau$. However, this displacement depends directly on the average speed acquired by the electrons under the influence of the electric field: $m^* <\mathbf{v}> = \hbar\,\delta\mathbf{k}$. The same expression for conductivity is deduced as with the Drude model.

An important point to note is that all of the scattering events which are responsible for the resistivity of the material occur near the Fermi energy. Therefore the density of states at the Fermi level plays an important role in determining the transport properties of the system. More generally, Fermi's golden rule tells us that the greater the density of states at the Fermi level, the greater the probability of electron scattering.

5.1.3. The two current model, spin dependent scattering in magnetic transition metals

In transition metals the sp and d bands are the partially filled bands. In particular, the 3d and 4sp bands are the partially filled bands of the ferromagnetic 3d metals. The electrons have a spin 1/2, and thus two possible spin states: $\downarrow$ or $\uparrow$. The majority of scattering events do not affect the electron spin. The main sources of spin-flip in magnetic metal are the spin-orbit coupling, and magnon scattering. The spin-orbit coupling of 3d metals is weak. The density of magnons increases with temperature, from near zero at very low temperature to a relatively low value at room temperature because the Curie temperature of the most commonly-used 3d ferromagnetic materials such as Fe, Co, Ni and their alloys is well above 300 K. As a consequence, it may be considered that the two types of electrons, spin $\uparrow$ electrons, and spin $\downarrow$ electrons, conduct parallel currents which do not mix. The total conductivity is given by the sum of the conductivities of the two electron types: $\sigma = \sigma\uparrow + \sigma\downarrow$. This model is known as Mott's two current model.

The band structures of iron, cobalt and nickel can be represented in a schematic way as shown in figure 20.12. The 3d sub-bands of spin $\uparrow$ and spin $\downarrow$ electrons are shifted in energy by an amount equivalent to the exchange energy between the two spin states. Nickel and cobalt are called strong ferromagnets because the Fermi level intersects the minority 3d$\downarrow$ sub-band only. Iron is a weak ferromagnet because the Fermi level crosses both d sub-bands. As a result of the band displacement, the density of d states at the Fermi energy is very different for spin ↑ and spin ↓ electrons. Consequently the scattering probabilities are also very different for spin ↑ and spin ↓ electrons. The ratio of the probabilities depends on the ratio of the density of states at ε_F in the vicinity of the scattering centres for both electron types.

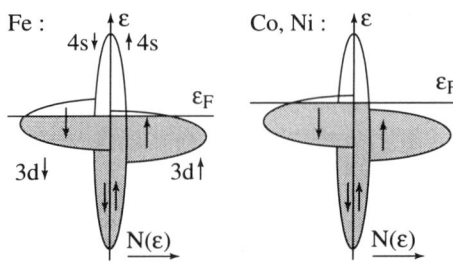

Figure 20.12 - Schematic representation of the band structures of the ferromagnetic transition metals (Fe, Co, Ni)

For example, in Permalloy (a magnetic alloy of composition $Ni_{80}Fe_{20}$ used for its high permeability), the scattering of spin ↓ electrons can be 5 to 10 times higher than spin ↑ electrons which implies, in terms of the mean free path: $\lambda_\uparrow / \lambda_\downarrow \approx 5$ to 10 (typically $\lambda_\uparrow = 6$ nm, and $\lambda_\downarrow = 1$ nm at 300 K). This spin dependent scattering leads to the giant magnetoresistance effect observed in transition metal-based magnetic multilayers (see § 5.3.2).

5.2. FINITE SIZE EFFECT ON THE CONDUCTIVITY OF METALLIC THIN FILMS

As for a viscous liquid, the electron drift velocity field cannot vary in a discontinuous manner. When the conductivity is locally perturbed, for example at a surface or an interface with another material, the conductivity relaxes towards its non-perturbed value over a characteristic distance which is equal to the mfp. The effect is described by Boltzmann's transport equation adapted to thin films in the Fuchs-Sondheimer theory.

Consider a multilayer (A/B) composed of alternating metallic layers (A and B) having different bulk conductivities, σ_A and σ_B. For a current circulating parallel to the plane of the multilayer, it would be incorrect to consider that the layers are connected in parallel, except in the limit where the thicknesses of the individual layers are much greater than the mfp values. The conductivity varies locally as shown in figure 20.13. In the interior of each individual layer, the conductivity tends to relax towards its bulk value for the given material, over a length scale equal to the mfp. If the thickness of the layers is comparable to the mfp, which is often the case (thicknesses of the order of some nanometres), the local conductivity may be very different from the bulk conductivity. The total conductivity of the structure is the integral of the local conductivities over the whole thickness of the multilayer.

The multilayer is composed of periodically alternating metallic layers (A and B) of bulk conductivities σ_A and σ_B in the direction perpendicular to the interfaces. The current is assumed to circulate parallel to the interfaces.

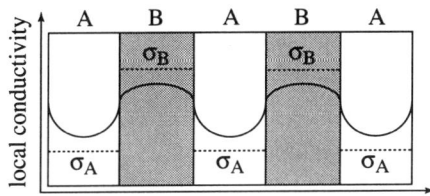

Figure 20.13 - Local variations in the conductivity of a multilayer

Similarly, in a thin film of thickness d limited by two planar surfaces, the local conductivity depends on the distance to the outer surfaces, and on the extent of scattering at these surfaces. Two extreme situations can occur at the exterior surfaces:

◆ if the surface is perfectly smooth, the electrons are specularly reflected in a similar way to a light beam reflected on a perfect mirror,

◆ if the surfaces are very rough, the electrons are reflected in a perfectly diffuse manner, i.e. with equal probability in all directions in the internal hemisphere of the layer. The optical analogy is the diffuse reflection of a light beam on an unpolished surface. All intermediate situations between these two extreme cases are possible, so that the electrons have a probability p to be reflected in a specular manner, and a probability (1 − p) to be reflected in a diffuse manner.

In the case of perfect specular reflection, the conductivity is uniform, and equal to the bulk conductivity.

In the case of diffuse reflection, the local conductivity is minimum in the immediate vicinity of the surface, and increases away from the interfaces. According to the Fuchs-Sondheimer theory, the average conductivity of the film is given by:

$$\frac{\sigma}{\sigma_0} = 1 - \frac{3}{2k} \int_1^\infty \left(\frac{1}{u^3} - \frac{1}{u^5} \right) [1 - \exp(-ku)] du,$$ where the variable of integration u is

the reciprocal of the cosine of the angle of incidence of the electrons with the surface, and $k = d / \lambda$, λ being the electronic mfp. This expression may be simplified in two limiting cases:

◆ if $\lambda \ll d$, $\rho = \dfrac{1}{\sigma} = \dfrac{m^* v_F}{ne^2} \left(\dfrac{1}{\lambda} + \dfrac{3}{8d} \right)$ and

◆ if $\lambda \gg d$, $\rho = \dfrac{1}{\sigma} = \dfrac{4 m^* v_F}{3 ne^2} \left[d \left(\ln \dfrac{\lambda}{d} + 0.423 \right) \right]^{-1}.$

Note that this last expression is not valid in the limit $\lambda \to \infty$ because it leads to a non-physical divergence of the conductivity of a film despite surface scattering. This divergence is due to electrons circulating perfectly parallel to the interfaces not being subject to interfacial scattering. The quantum theories of conductivity do not lead to a divergence because such electrons are forbidden by the Heisenberg uncertainty principle. Nevertheless, the semi-classical theories seem to correctly interpret the experimental results for thin films of thicknesses down to the order of $\lambda / 10$. Below this, quantum descriptions are necessary so as to not underestimate the scattering at surfaces and interfaces.

5.3. MAGNETORESISTANCE IN THIN FILMS AND METALLIC MAGNETIC MULTILAYERS

5.3.1. Anisotropy of the magnetoresistance in ferromagnetic transition metals in bulk and thin film form

Though discovered in 1857 by W. Thomson in Glasgow, the phenomenon of anisotropic magnetoresistance has only recently found an application in magnetic field sensors. Anisotropic magnetoresistance is the difference in local resistivity of a magnetic material depending on whether the current circulates parallel ($\rho_{//}$) or perpendicular ($\rho_\perp$) to the local direction of magnetization. The relative difference ($\rho_{//} - \rho_\perp$)/$\rho_{//}$ may be of the order of 5% at room temperature, and 20% at low temperature in NiFe or NiCo alloys [39]. In the majority of cases $\rho_{//} > \rho_\perp$. The local resistivity varies continuously as a function of the angle θ between the local magnetization, and the current according to the law:

$$\rho = (\rho_{//} + \rho_\perp)/2 + [\cos^2\theta - (1/2)](\rho_{//} - \rho_\perp).$$

In a demagnetised bulk sample, if one supposes an isotropic distribution of magnetic domains, the resistivity in zero field is given by a statistical averaging of the local resistivity calculated over all directions of the local magnetization with respect to the

current:
$$\bar{\rho} = \int_0^\pi \rho(\theta) \sin\theta d\theta \Big/ \int_0^\pi \sin\theta d\theta = \frac{\rho_{//} + 2\rho_\perp}{3}.$$

In a demagnetised thin film sample, the local magnetization usually remains in the plane of the sample due to shape anisotropy. The average over all directions of magnetization within the plane gives rise to a zero field resistivity given by:

$$\bar{\rho} = \frac{1}{\pi} \int_0^\pi \rho(\theta) d\theta = \frac{\rho_{//} + \rho_\perp}{2}.$$

The hysteresis loops of a ferromagnetic thin film possessing a well defined uniaxial anisotropy (for example a thin film of Permalloy $Ni_{80}Fe_{20}$), have the form given in figure 20.14, depending on whether the field is applied parallel or perpendicular to the plane of the sample:

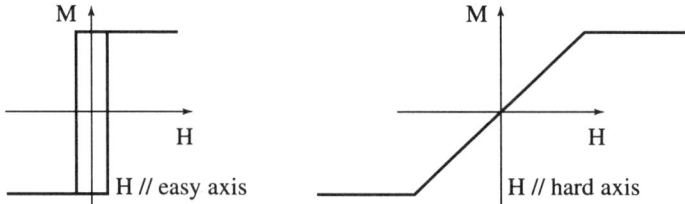

Figure 20.14 - Schematic diagram of the hysteresis loops of a ferromagnetic thin film having a well defined uniaxial anisotropy

When the field is applied parallel to the easy axis of magnetization, magnetization is reversed by the nucleation of reverse domains followed by the propagation of domain walls. Magnetization reversal at the coercive field may thus be very abrupt. When the field is applied parallel to the hard axis of magnetization, the magnetization is rotated, in a reversible fashion, away from the easy direction. Up until saturation is reached, for $H_{sat} = 2K_u/M_s$, the magnetization varies linearly with the field according to the law: $M = M_s^2 H/2K_u$. In this case, magnetization processes involve the continuous rotation of magnetization from positive to negative saturation.

Anisotropic magnetoresistance (AMR) has the form shown in figure 20.15, depending on the relative direction of the current and the field with respect to the easy and hard axes of magnetization. In thin films, the relative amplitude of the AMR decreases for thicknesses of less than about 50 nm due to the increased importance of the scattering of electrons by the external surfaces. The AMR phenomenon has been exploited from 1990 to 1998 in magnetoresistive read heads used to read high storage density computer hard disks (at densities between 0.2 and 2 Gbit.cm^{-2}), and is still used nowadays in other types of magnetic field sensors (see ref. [40] and § 6.6 of this chapter).

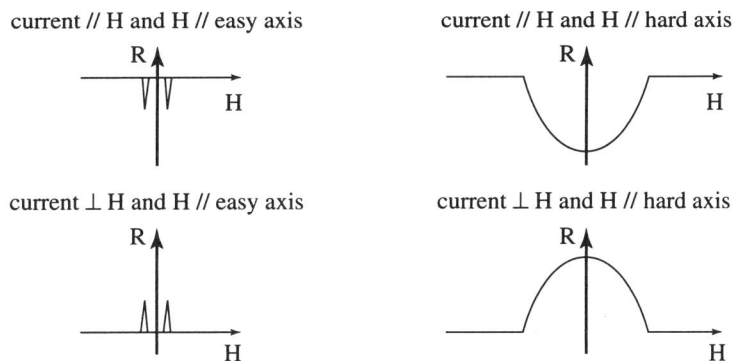

Figure 20.15 - Schematic representation of the anisotropic magnetoresistance of a ferromagnetic thin film having a well defined uniaxial anisotropy

5.3.2. Giant magnetoresistance (GMR)

The giant magnetoresistive effect was discovered in 1988 in (Fe 3 nm / Cr 0.9 nm)$_{60}$ multilayers [41]. In fact, two remarkable properties were observed in these systems.

The first is the existence of an antiferromagnetic coupling between the Fe layers across the Cr spacer layers (see § 4). For particular intervals of Cr thickness, this coupling tends to align the magnetization of successive Fe films in an antiparallel configuration in zero magnetic field. When a magnetic field is applied in the plane of the structure, the magnetic moments rotate towards the field direction until they become parallel at saturation.

The second remarkable observation made on these multilayers is that the change in the relative direction of the magnetization of the successive Fe films is accompanied by a significant decrease in the electrical resistivity of the structure as illustrated in figure 20.16.

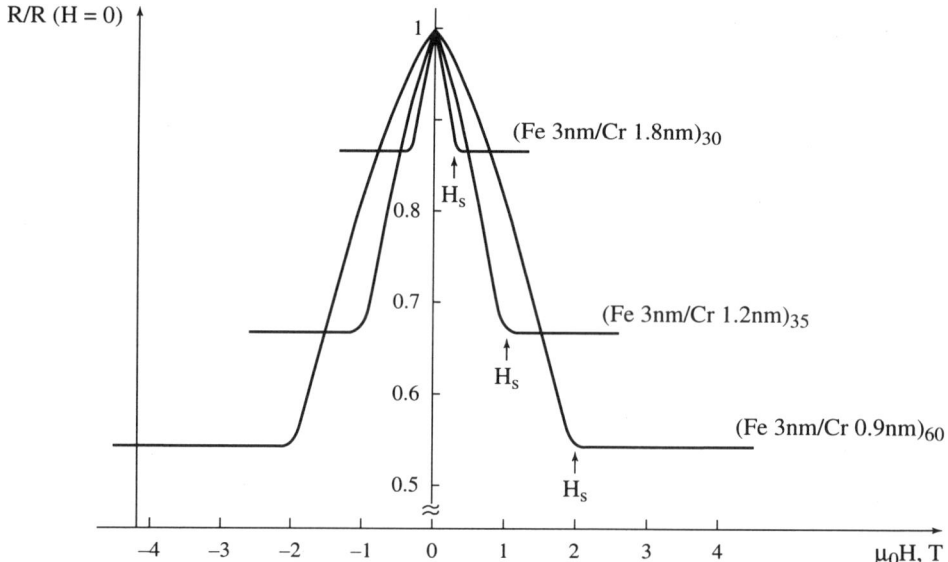

Figure 20.16 - Normalised resistance as a function of the magnetic field

Results observed at $T = 4\,K$ for various (Fe / Cr) multilayers coupled antiferro-magnetically, after [41]. The current and magnetic field are parallel to the plane of the film.

Since this first observation, giant magnetoresistive effects have been observed in a number of other systems of the form $B\,t_B/n*(F\,t_F/NM\,t_{NM})/C\,t_C$ where B is a buffer layer chosen to promote the growth of the given structure, n, is the repeat number of basic (F/NM) bilayers, F denotes a ferromagnetic transition metal (Fe, Co, Ni, and most of their alloys), NM a non-magnetic metal which is a good conductor (transition metals: V, Cr, Nb, Mo, Ru, Re, Os, Ir, or noble metals: Cu, Ag, Au…), and t_x represents the thickness of the film x (x = B, F or NM).

The amplitude of the giant magnetoresistance depends very much on the choice of materials used (F, NM), and on the thicknesses of the different layers. It varies from 0.1% in multilayers based on V or Mo to more than 100% in (Fe/Cr) [41, 42] or (Co/Cu) [43, 44] multilayers. In all these structures, it has been observed that the giant magnetoresistance is associated with a change in the relative orientation of the magnetization of successive magnetic layers [41, 45].

Two principal parameters are used to quantify giant magnetoresistance. The first is the GMR amplitude, often defined as $\Delta R / R = (R - R_{sat})/R_{sat}$, where R_{sat} is the resistance at saturation, i.e. the resistance measured when the magnetic moments are aligned parallel. The second is the variation in magnetic field ΔH needed to observe the full GMR value. For many applications, the figure of merit of a material is the ratio $(\Delta R / R)/H_{sat}$.

5.3.3. The physical origin of giant magnetoresistance

The GMR effect results from a combination of three factors:

♦ as the thickness of the individual layers (of the order of 10^{-9} m) is less than or comparable with the electronic mean free path, the electrons may pass from one magnetic layer to the next, crossing the non-magnetic separating layer;

♦ the scattering cross-section of electrons in the volume or at the interfaces of the magnetic layers depends on orientation of the spin of the electrons relative to the local magnetization (see § 5.1.3);

♦ the relative orientation of the magnetization in successive magnetic layers may be modified, through antiferromagnetic coupling across the non-magnetic interlayers or by magnetic pinning of certain layers by appropriate techniques.

Consider a multilayer consisting of layers of a magnetic metal separated by layers of a non-magnetic metal (for example $[Ni_{80}Fe_{20}\ 3\ nm / Cu\ 1\ nm]_n$). When the magnetization of all the NiFe layers are parallel, the spin up $\uparrow$ electrons have a long mean free path throughout the structure, and can thus conduct much current. The spin down $\downarrow$ electrons contribute little to the conduction. The short circuit effect due to the $\uparrow$ electrons gives rise to a low resistivity state. If the thicknesses of the individual layers is much smaller than the mean free path, the resistivity of the multilayer in the parallel magnetic configuration is: $\rho_p = \rho\uparrow\rho\downarrow / (\rho\uparrow + \rho\downarrow)$, where $\rho\uparrow$ and $\rho\downarrow$ are the resistivities of $\uparrow$ electrons and $\downarrow$ electrons in the magnetic metal, respectively. In the antiparallel magnetic configuration both electron types are strongly scattered, in alternate magnetic layers, giving rise to a higher electrical resistivity.

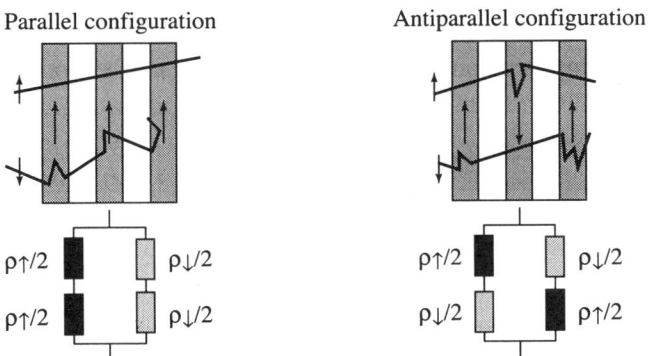

Figure 20.17 - Mechanism of giant magnetoresistance
The grey and white layers represent the NiFe and Cu layers, respectively. The electrons are weakly scattered if their spin is parallel to the local direction of magnetization while strongly scattered if antiparallel. Beneath, a schematic diagram of the equivalent resistances in both magnetic configurations.

The equivalent resistivity for this configuration is $\rho_{ap} = (\rho\uparrow + \rho\downarrow) / 4$. The amplitude of the giant magnetoresistance is then given by:

$$\frac{\Delta\rho}{\rho_{ap}} = \left(\frac{\rho\uparrow - \rho\downarrow}{\rho\uparrow + \rho\downarrow}\right)^2 = \left(\frac{\alpha - 1}{\alpha + 1}\right)^2 \text{ where } \alpha = \frac{\rho\uparrow}{\rho\downarrow}.$$

Qualitatively, three types of giant magnetoresistive multilayers can be distinguished, depending on how the magnetic configuration is controlled.

5.3.4. *Different types of giant magnetoresistive multilayers*

Antiferromagnetically coupled multilayers

An example of this type is (Fe / Cr) multilayers in which the GMR effect was first observed [41]. The saturation field of this type of system is, unfortunately, often too high for applications. When the thickness of the non-magnetic layer (here Cr) is varied, the magnetoresistance oscillates between a finite value and zero, due to the oscillation in coupling, changing from ferromagnetic to antiferromagnetic, depending on the thickness of the Cr layer. The magnetoresistance is significant when the coupling is antiferromagnetic (antiparallel configuration in zero field and parallel at saturation) but practically zero when the coupling is ferromagnetic (always a parallel configuration). Table 20.1 summarises the magnetoresistance characteristics of a number of antiferromagnetically coupled systems.

It is clear from this data that the GMR value (given in %) varies considerably from one system to another. The GMR value is a characteristic of each magnetic metal / non-magnetic metal couple, and not of each constituent considered separately. It is worth noting that of the various preparation techniques used, the sputtering process often gives the best results, which is an interesting observation from the applications point of view.

Table 20.1 - Giant magnetoresistance characteristics
of antiferromagnetically coupled systems

Multilayer	References	MR	T	$\mu_0 \Delta H_{max}^{MR}$	Prep.
(**Fe** 3 nm / **Cr** 0.9 nm)$_{40}$	[41]	92%	4.2	2	MBE
Cr 10 nm / (**Fe** 1.4 nm / **Cr** 0.8 nm)$_{50}$	[42]	150%	4.2	2	S
Fe 5 nm / (**Co** 0.8 nm / **Cu** 0.9 nm)$_{60}$ / Fe	[43, 44]	115%	4.2	1.3	S
(**NiFe / Cu**)	[46]	25%	4.2	1.5	S
(**NiFe** 2 nm / **Ag** 1 nm)	[47]	50%	4.2	0.1	S 77 K
(**Co** 0.6 nm / **Ag** 2.5 nm)	[48, 49]	41%	77	1	MBE
(**Co$_{70}$Fe$_{30}$** 0.4 nm / **Ag** 1.5 nm)	[50]	100%	4	0.3	S 77 K

T is the measurement temperature in kelvin, ΔH_{max}^{MR} represents the variation in field needed to observe the full GMR value, in tesla. Prep. indicates the sample preparation mode (MBE = molecular beam epitaxy, S = sputtering).

Among the various antiferromagnetically coupled multilayer systems, NiFe / Ag seems to be the most interesting for magnetic micro-electronic applications. This system has the highest GMR value for a sufficiently small saturation field. At room temperature, a (NiFe 2.5 nm / Ag 1.1 nm)$_{50}$ multilayer has a GMR value of 15% in a saturation field of the order of 12 kA.m^{-1} [46]. Moreover, these multilayers are very stable from the structural point of view, in contrast with the NiFe/Cu system in which the miscibility of NiFe in Cu leads to significant interdiffusion. The magnetoresistive properties of (NiFe/Ag) multilayers improve after annealing up to 250°C (increase in the GMR value, decrease in the saturation field, and thus an increase in sensitivity). Sensitivities of the order of 2.5%/kA.m^{-1} (0.2%/Oe), which are suitable for applications, can be achieved upon annealing.

Double coercivity multilayers

In the absence of antiferromagnetic coupling between the magnetic layers, it is possible to produce a change in the magnetic configuration by using two ferromagnetic materials having different coercivities (for example $Ni_{80}Fe_{20}$ and Co in a multilayer having $Ni_{80}Fe_{20}$/Cu/Co/Cu as the basic building block [51]). When the applied magnetic field is varied from a value which saturates the sample positively to a value which saturates it negatively, the magnetization of the softer magnetic material (i.e. with lowest coercivity) is reversed before that of the harder magnetic material. There exists, therefore, a range of magnetic field, between the two coercivity values, in which the magnetization of the different magnetic layers are aligned antiparallel. Room temperature GMR values of 16% in 4 kA.m^{-1} have been obtained in this type of system.

Spin valves

Spin valves are sandwich systems containing two magnetic layers (F_1 and F_2) separated by a non-magnetic metal layer (NM). The magnetization of the layer F_2 is pinned by exchange interactions with an adjacent antiferromagnetic layer (FeMn, $Ni_{1-x}Co_xO$, NiMn, PtMn, PtPdMn) [45, 52, 53]. The hysteresis loop, and the room temperature magnetoresistance of a spin valve of composition:

Si/Ta 5 nm/NiFe 6 nm/Cu 2.2 nm/NiFe 4 nm/FeMn 7 nm)/Ta 5 nm

are shown in figures 20.18-a and -b [45]. Figure 20.18-c shows the change in magnetoresistance as the applied magnetic field is varied between ±4 kA.m^{-1}.

The hysteresis loops of these systems are composed of two cycles. One cycle is centred around 0.48 kA.m^{-1}, having a coercivity of the order of 80 A.m^{-1}, and corresponds to the reversal of the non-pinned layer. The second cycle, shifted to about 32 kA.m^{-1}, and having a coercivity of the order of 8 kA.m^{-1}, corresponds to reversal of the layer coupled to FeMn. When the field is swept between ±80 kA.m^{-1}, the relative orientations of magnetization of the magnetic layers change: the magnetic moments are parallel below $\approx$ 320 A.m^{-1} and above $\approx$ 48 kA.m^{-1}, and are antiparallel between $\approx$ 640 A.m^{-1}m and $\approx$ 20 kA.m^{-1}. The shift by 480 A.m^{-1} of the non-pinned NiFe layer indicates the existence of a weak ferromagnetic coupling between the NiFe layers across the Cu interlayer, probably due to the "orange peel" mechanism (see § 4). The resistance of the structure varies very rapidly when the magnetization of the free (non-pinned) layer is reversed.

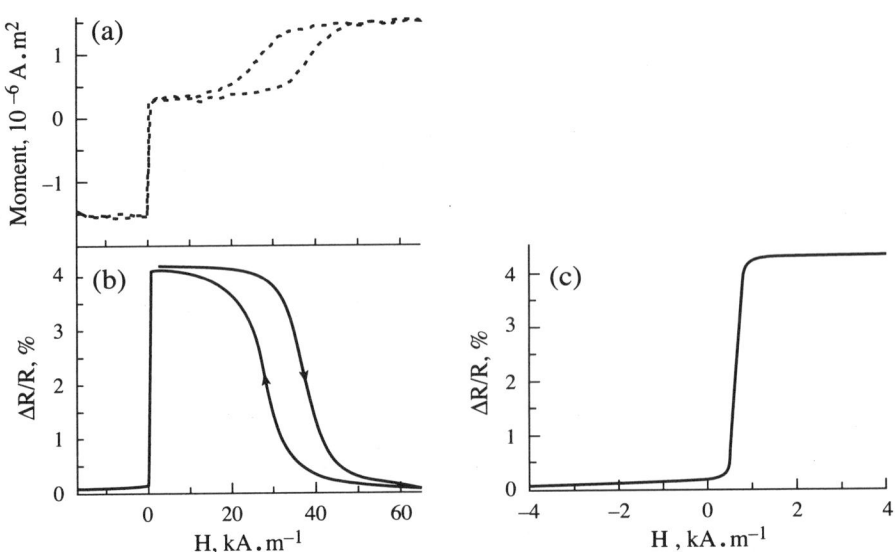

Figure 20.18 - Hysteresis loop (a) and magnetoresistance (b) at 300K of a spin valve of composition: Si / Ta 5 nm / NiFe 6 nm / Cu 2.2 nm / NiFe 4 nm / FeMn 7 nm / Ta 5 nm (c) magnetoresistance in low field [45]

Sensitivities $\Delta R / (R\Delta H)$ greater than $25\% / kA.m^{-1}$ have been obtained in this type of system. This very high sensitivity makes spin valves very appropriate field sensors for magnetoresistive read heads to be used for high density magnetic recording. Since 1998, spin valves have succeeded AMR thin films in read heads for storage densities higher than 1-2 $Gbit.cm^{-2}$.

Since spin-valves were invented in 1990, many improvements were brought to these systems to make them usable in recording technology. These improvements can be summarised as follow:

◆ A thin $Co_{90}Fe_{10}$ layer was introduced at the interfaces between NiFe and Cu. The purpose of this layer is twofold: NiFe has a moderately high Curie temperature (800 K) and is miscible with Cu. When spin-valves are annealed during the head manufacturing process, NiFe and Cu tend to mix, forming a NiFeCu alloy with even lower Curie temperature than NiFe. This results in a relatively low MR amplitude due to the presence of strong thermally activated magnetic fluctuations along the interfaces at room temperature. These fluctuations generate electron spin-flip, which means a loss of memory of the electrons as they pass from one ferromagnetic layer to the next, leading to a decreased GMR amplitude. In contrast, thin CoFe layers have a much higher Curie temperature than NiFe layers, and furthermore they are non miscible with Cu. They therefore constitute a barrier for interdiffusion between the ferromagnetic layers and the non-magnetic spacer, and magnetically strengthen the interfaces. This results in a much better structural stability of the structures upon annealing, and in about a twofold enhancement of the GMR amplitude.

- New antiferromagnetic materials with higher Néel temperature have been developed to ensure more stable pinning of the magnetization of the pinned layer over long periods of time (at least 10 years). The preferred material is nowadays PtMn which also has a good resistance to corrosion.

- The pinned layer has been replaced by a so-called synthetic pinned layer, typically with composition $Co_{90}Fe_{10}$ 1.5 nm / Ru 0.7 nm / $Co_{90}Fe_{10}$ 2 nm. For Ru thickness between 0.6 and 1 nm, a very strong antiparallel coupling between the two adjacent CoFe layers exists. Since these two layers are maintained in antiparallel configuration, their net moment is significantly reduced compared to the moment of a single layer of comparable thickness. Therefore the torque exerted by the applied field is much weaker than for a simple pinned layer, resulting in much more efficient pinning.

- Very effective materials (particularly NiFeCr alloys) have been developed for use as buffer layers: they highly improve the overall crystallographic structure of spin-valves.

- Very thin oxide layers are now introduced on the free and pinned layers to generate specular reflection of the conduction electrons within the part of the spin-valves where the GMR originates, i.e. within the free layer / NM / pinned layer core. The purpose is to channel the conduction electrons in this active part and reduce the parasitic diffuse scattering on the outer part of the spin-valves.

Thanks to all these improvements, magnetoresistance amplitudes of the order of 20% with pinning fields larger than 0.1 T at room temperature are nowadays achieved in spin-valves.

5.4. SPIN POLARISED ELECTRON TUNNELLING

Spin polarised electron tunnelling involves passing a current across a thin insulating layer (of thickness 2 to 3 nm) separating two magnetic electrodes. The difference in density of states in the region of the Fermi level, due to the magnetic character of the electrodes, results in a polarization of the tunnel electrons. Control of the relative alignment of the magnetization of the electrodes on either side of the junction allows one to vary the tunnel conductance of the junction. Tunnelling from a magnetic metal to a superconductor allows a measure of the percentage of polarization of the tunnel electrons [54, 55]. The percentage of polarization obtained for different magnetic metals is given in table 20.2.

Table 20.2 - Percentage of polarization of tunnel electrons in various magnetic metals

Element	Fe	Co	Ni	Gd
Percentage polarization	44%	+34%	+11%	+4%

These percentages are determined from tunnel effect experiments carried out between an electrode of the given magnetic metal, and a superconducting film of Al across a surface layer of aluminium oxide, after [54, 55].

Jullière was the first to study magnetoresistance in MOM (M = magnetic metal, O = oxide barrier) junctions [56]. He observed that the tunnel conductance was higher (by about 12% in these structures at 4 K) for a parallel alignment compared to an antiparallel alignment of the magnetic moments of the electrodes. This tunnel magnetoresistance can be explained as follows.

As for giant magnetoresistance, it is reasonable to assume that, at room temperature, the two channels of electrons (spin up and spin down electrons) carry the current in parallel. For each electron, the Fermi's golden rule states that the probability of tunnelling through the barrier is proportional to the density of states at the Fermi energy in the receiving electrode. Now the number of electrons which are candidates for a tunneling event is proportional to the density of states of the emitting electrode. Therefore, the current for each spin channel is proportional to the product of the densities of states on both sides of the barrier : In parallel magnetic configuration, the total current is therefore $j_{parallel} = D^\uparrow D^\uparrow + D^\downarrow D^\downarrow$ whereas in antiparallel configuration, it becomes $j_{antiparallel} = D^\uparrow D^\downarrow + D^\uparrow D^\downarrow$. In these expressions, $D^\uparrow$ and $D^\downarrow$ respectively represent the density of states of up and down electrons at the Fermi energy. The magnetoresistance is then given by the relative change in current between the parallel and antiparallel magnetic configurations. Introducing the polarization of the electrons given by $P = \dfrac{(D^\uparrow - D^\downarrow)}{(D^\uparrow + D^\downarrow)}$, the magnetoresistance can be expressed as

$$\frac{\Delta G}{G_{parallel}} = \frac{2P^2}{1 + P^2} \quad \text{or} \quad \frac{\Delta G}{G_{parallel}} = \frac{2P_1 P_2}{1 + P_1 P_2}$$ if the two electrodes have different

polarization. Recent experimental studies carried out with various types of oxide barriers and ferromagnetic electrodes have shown that the polarization of tunneling electrons is not characteristic of the magnetic material only, but actually of the couple insulator / magnetic material. For instance, it has been observed that the polarization of electrons emitted from Co through an alumina barrier is parallel to the Co magnetization, whereas the polarization of electrons emitted from the same ferromagnetic metal (Co) but through $SrTiO_3$ is opposite to the Co magnetization [57]. This points to the key role played by electron hybridization at the ferromagnetic / insulator interface. This hybridization determines the ease with which electrons can penetrate into the barrier.

The field of spin-polarized tunneling has undergone a very intense renewal in interest since 1995, following the observation of large tunnelling magnetoresistance (TMR) amplitude at room temperature in alumina based junctions [58, 59]. A large number of insulating and semiconducting materials have been tested (Al_2O_3, AlN, MgO, NiO, CoO, Ta_2O_5, SiO_2, Ge, BN, $SrTiO_3$). Today, the best results are obtained for Al_2O_3 barriers [60, 61, 62]. The preparation technique involves the initial deposition of the bottom metallic electrode, followed by an Al film about 1 nm thick. The Al layer is then fully oxidised, either with an oxygen plasma, or by bombardement with atomic oxygen, or more simply by natural oxidation in an oxygen or ozone atmosphere. Care is taken not to over-oxidize the Al layer, so as not to oxidise the underlying magnetic electrode. After oxidation of the aluminium barrier into alumina, the top magnetic electrode is deposited, as well as the antiferromagnetic pinning layer if any. The magnetoresistance curve of a tunnel junction is shown in figure 20.19.

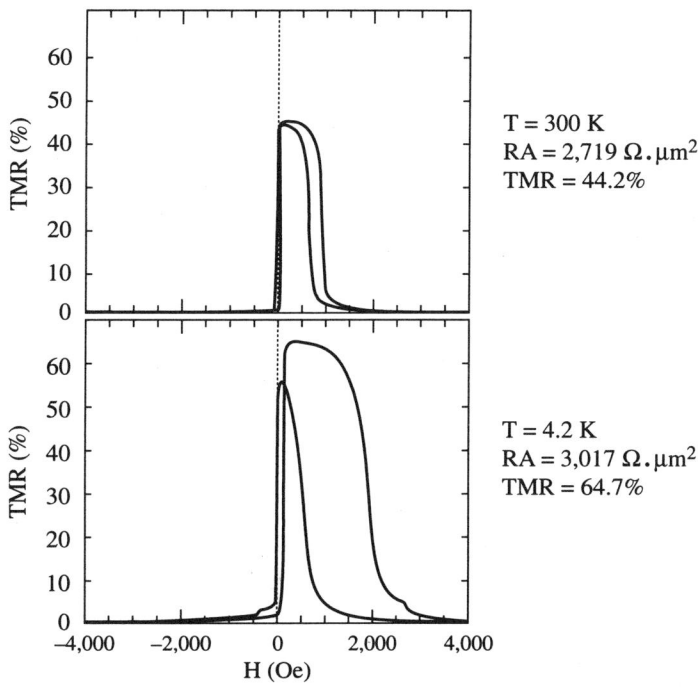

Figure 20.19 - Magnetoresistance of a tunnel junction of composition :
Buffer / IrMn 10nm / CoFe 4nm / Al 0.8nm-oxide / CoFe 4nm / NiFe 20nm / Ta 5nm
at room temperature and liquid helium temperature
The junction area is 5 ×5 μm² [after 61].

Variations in tunnel junction currents of 64% at low temperature and 45% at room temperature are now obtained in this type of structure. Regarding the control of the relative orientation of the magnetization in the two magnetic electrodes, the same techniques as those developed for spin-valves can be used, i.e. use of antiferromagnetic pinning layers and of synthetic pinned layers.

Although the MR transfer curves of spin-valves and tunnel junctions look rather similar, a number of differences must be pointed out:

♦ In GMR, the transport is ohmic as in any metal, which means that the voltage between two voltage probes varies linearly with the sensing current. In contrast, in tunnel junctions, the I(V) characteristics are non-linear. The conductivity through the barrier increases with the bias voltage. Furthermore, tunnel junctions cannot be exposed to large voltage (typically no more than 1.5 V), otherwise they undergo an electrical breakdown. The breakdown voltage in insulators is typically of the order of 1 V per nanometer of insulator.

- In metals, and even more in magnetic metals, the resistance increases with temperature due to the increasing contribution to the resistivity of phonon and magnon scattering. In tunnel junctions, the resistance of the barrier slightly decreases with temperature because the thermal excitation of the electrons is equivalent to a slight reduction (of the order of 25 meV at 300 K) of the tunnel barrier height. From quantum mechanics, it is known that the probability for an electron to tunnel through a barrier of height ΔE and width w is proportional to $\exp(-2kw)$, where k is the wave-vector of the evanescent wave in the barrier, given by $k = \sqrt{\dfrac{2m}{\hbar^2}\Delta E}$. With ΔE typically of the order of 2 eV, a reduction of ΔE by 25 meV leads to a decrease of resistance (inversely proportional to the probability of tunneling) of the order of 10%. The thermal variation of the resistance of oxide barriers is actually a good way to estimate their quality with respect to tunneling. A bad quality barrier usually shows either an increase in resistance with temperature in the presence of metallic pinholes, or a strong decrease of resistance with temperature in the presence of thermally activated conduction process within the oxide itself, as in doped semiconductors.

- In magnetic metallic multilayers, the GMR is independent of the voltage at which it is measured. In magnetic tunnel junctions, this is not true. In most cases, the TMR decreases as a function of bias voltage (see for instance [61]). Typically, with most commonly used alumina barriers, it drops to half its value at a low voltage of around 0.4 V. Several explanations have been proposed to explain this observation. One is that, as the voltage increases, more and more electrons from below the Fermi energy of the emitting electrode are allowed to tunnel. For an applied voltage V, the electrons which tunnel are those of energy between ε_F and $\varepsilon_F - V$. By considering the electronic structure of magnetic metals, it appears that the electronic polarization decreases as their energy decreases below the Fermi energy. As a result, the net polarization of tunnelling electrons decreases as the bias voltage increases, leading to a decrease in TMR amplitude. Another explanation has been proposed in terms of excitations of spin-waves by the tunneling electrons [63]: The tunneling electrons are accelerated by the electrical field which exists across the oxide barrier. They are therefore injected into the receiving electrode with an energy higher than the Fermi energy (hot electrons). These electrons then relax very quickly to the Fermi energy by generating phonons and magnons. This induces an excess density of magnetic fluctuations at the oxide / ferromagnetic interface, which in turn leads to spin-flip scattering, hence a decrease in TMR amplitude.

The very large magnetoresistance amplitude observed in magnetic tunnel junctions has stimulated a tremendous interest in these systems for sensor applications (especially in magnetoresistive read heads) as well as for non-volatile memory elements. It now seems that, because of the larger noise of tunnel junctions (shot noise) compared to spin-valves, tunnel junctions will not be superior to spin-valves for sensor applications. In contrast, their interest for memory applications remains considerable (see § 5.5.6).

5.5. APPLICATIONS OF MAGNETIC THIN FILM AND MULTILAYERS

Magnetic thin films have a wide range of applications, and thus a significant market. These applications have stimulated much research, most notably in the area of magnetic and magneto-optical recording technologies.

Among the principal applications for these materials, we consider here magnetic recording media, magneto-optical recording media, soft materials, magnetostrictive materials, and magnetoresistive materials.

5.5.1. Magnetic recording media

The storage information density on computer hard disks increased at a rate of 30% per annum during the two decades leading up to the 1990s. Since 1991, this growth rate exceeded 60% per annum. The magnetic *media* are thin films of a magnetic alloy of relatively high coercivity (at the time of writing, the quasistatic coercivity is about 0.3 T). These media are polycrystalline, with a crystallite size of the order of 10 to 20 nm. Today's highest performance planar media are thin films of ternary or quaternary alloys of the type $Co_{74}Cr_{17}Pt_5Ta_4$, deposited on thin films of Cr or a $Cr_{80}Mo_{20}$ alloy. These media have an in-plane magnetization. The CoCrPtTa alloys have a hexagonal structure, with a very large magnetocrystalline anisotropy along the c-axis. By proper choice of the buffer underlayer, these alloys can be textured in such a way that the c-axis of each crystallite lies in the plane of the film, thus leading to large in-plane coercivity. In 2000, demonstrations of magnetic storage at areal density of 65 Gbit.in^{-2} were achieved. At these densities, a *bit* of information consists of a rectangular domain with size of the order of 70 nm × 140 nm.

The criteria which define a good magnetic recording media suitable for high storage density recording are a high remanent magnetization and a high coercivity (though not so high as to prevent writing). Furthermore, the grains constituting the media must be very small, and well decoupled magnetically to allow sharp transitions between bits. Evaluation of the signal to noise ratio in magnetic recording media indicates that in continuous media, the number of grains within each bit of information must be at least 100. In CoCrPtTa alloys, the purpose of the Pt is to provide a large magnetocrystalline anisotropy, whereas the Cr and Ta are known to diffuse to the grain boundaries and therefore ensure good magnetic decoupling between adjacent grains.

The problem associated with the reduction in grain size is the so-called "superparamagnetic limit". For a given magnetic material, the barrier height that the magnetization experiences to switch from one direction to the opposite along its easy axis of magnetization is proportional to KV, K being the intrinsic magnetic anisotropy of the material and V the volume of the grain. Reducing the volume of the grain leads to a reduction in the barrier height, so that, at room temperature, the magnetization may become unstable with respect to thermal fluctuations. The characteristic time of fluctuations τ is given by an Arrhenius law: $\tau = \tau_0 \exp\left(\dfrac{KV}{k_B T}\right)$.

where τ_0 is the attempt time, typically of the order of 10^{-9} s. For the information to remain stable for at least 10 years, the KV product must therefore obey the relationship: $KV > 40\, k_B T$. For the materials used nowadays, $K \sim 2.2 \times 10^5$ J.m^{-3}, implying a minimum grain size of the order of 8 nm. The grain size used in the last demonstrations are not far from this minimum limit.

The present trends in research and development on magnetic recording media are the following:

♦ The simple in-plane CoCrPt magnetic storage layers are nowadays replaced by two antiferromagnetically coupled layers in a sandwich of the form underlayer / CoCrPt 10 nm / Ru 0.7 nm / CoCrPt 20 nm. The purpose of this structure is to increase the magnetic volume of the bit and therefore to push the superparamagnetic limit further. Since the two magnetic layers are very rigidly coupled through the Ru layer, their anisotropy energies add up, leading to a more stable magnetization.

♦ Perpendicular magnetic recording should soon supersede in-plane recording. At very high areal density, perpendicular recording becomes more favorable because the magnetostatic interactions between bits tend to decrease as their size decreases. Furthermore, very high perpendicular anisotropy can be achieved in perpendicular media, thanks to interfacial magnetic anisotropy which can be induced in magnetic multilayers. Examples of materials with good properties are the CoCrPt alloys but with c-axis oriented perpendicular to plane, (Co / Pt) multilayers and FePt L_{10} ordered alloys which consist in an alternation of atomic planes of Fe and Pt. Furthermore very high spatial resolution can be achieved in perpendicular recording by introducing a soft magnetic underlayer with in-plane magnetization under the perpendicular storage layer. With such media, a very efficient head for perpendicular recording can be designed, with very good flux closure between the head and the media, and therefore more efficient writing than with in-plane recording.

♦ To achieve areal density larger than 400 Gbit.in^{-2} it is believed that it will be necessary to use patterned media instead of continuous media. These patterned (also called discrete) media consist of an assembly of nanoscale magnetically independent dots, in which each dot represents one bit of information. The advantage of these media is that, since the transitions between bits are defined by the structuring of the media and not by the grain size, each dot may consist of a single magnetic grain. The superparamagnetic limit is then pushed towards much higher density. The difficulty in this approach is to find ways of producing these media rapidly and at low cost. Several approaches are investigated, based on self organization of nanoparticles [64], nano-imprint [65], or local ion irradiation [66].

5.5.2. Magneto-optical recording media

These thin films are based on materials which show an out-of-plane anisotropy, and a strong polar Kerr effect. The magnetization of each information bit points into or out of the plane of the disk. The information is read with a focused laser beam which is linearly polarised. During reflection at the surface of the media, the plane of polarization of the light is rotated, in one direction or the other, depending on the direction of local magnetization. The spot size of the laser is limited by diffraction, and therefore the storage density may be increased by reducing the wavelength of the laser light.

Until now, the most widely used materials are rare earth based amorphous alloys of the type $(Tb, Gd)_x(Fe, Co)_{1-x}$ with $0.2 < x < 0.3$. These materials show a good Kerr rotation (of the order of 0.2 to 0.4 degrees) for a wavelength of the order of 633 nm. However, the Kerr rotation is decreased at shorter wavelengths. Moreover, the technology which has been developed to prepare these materials is costly, so that, for the moment, magneto-optical recording does not compete with magnetic recording. New systems presently being studied include Co/Pt multilayers and ordered FePt alloys. These systems show a very strong perpendicular anisotropy which is associated with a strong Kerr rotation at blue wavelengths, and are therefore potential candidates for high density magneto-optical recording media.

5.5.3. Soft materials

Soft magnetic materials are frequently used as flux guides in magnetic micro-electronics. An example of such use is for magnetic pole pieces in read and write heads used in magnetic recording. These materials are required to have the following characteristics: a high saturation magnetization in order to produce a strong magnetic field in the air-gap for writing ($B > 1$ T), a high permeability ($\mu > 1000$) to make an efficient magnetic circuit, a low magnetostriction ($\lambda_s < 10^{-6}$) and a low coercivity ($H_c < 20$ A.m^{-1}) for low noise, a well defined anisotropy so as to have a good control of the domain structure, and a high resistivity ($\rho > 20$ $\mu\Omega$.cm) for reduced eddy current effects and therefore good high frequency performance. These materials must also be resistant to wear, corrosion and oxidation. The most frequently used materials are Permalloy ($Ni_{80}Fe_{20}$ type), iron or iron nitrides, CoFeCu and CoFeCr. These materials can be stratified to reduce their eddy current losses at high frequency. Amorphous materials of the CoZr type with additions of Nb or Re are sometimes used. These materials are deposited by various techniques: electrolysis, evaporation, or sputtering.

5.5.4. Magnetostrictive materials

Certain rare earth based alloys of the form R-Fe (R = rare earth), and in particular $Tb_{0.3}Dy_{0.7}Fe_2$ (known also as Terfenol-D), show considerable magnetostriction values of up to 2×10^{-3} at room temperature. This magnetostriction is due to the combination of strong spin-orbit interactions and the anisotropic charge distribution of the 4f electrons of the rare earth atoms. There are many studies underway to assess the suitability of these materials for applications such as micro-switches (linear translation drive): e.g. in the amorphous state, R-Co films show better magnetostrictive properties than R-Fe films (see § 4 of chapter 18 for more information).

5.5.5. *Materials for microwave applications*

These materials are used as wave guides (e.g. $Y_3Fe_5O_{12}$ in microwave devices), as microwave absorbers (e.g. in microwave ovens), and as radar absorbers (e.g. in defence applications).

5.5.6. *Magnetoresistive materials*

The first broad class of applications of magnetoresistive materials concerns magnetic field sensors. These materials allow a change in magnetic field to be detected as a change in electrical resistivity. The principal application is in magnetoresistive read heads for high density magnetic recording, but they are also used in linear or angular position sensors in the automobile industry, as well as for ticketing applications (ticket issuing). The basic principle being exploited is either the AMR effect (change in resistivity of a magnetic material as a function of the angle between the magnetization direction and the direction of the measuring current crossing the sample), or the GMR effect (change in the electrical resistance as a function of the relative angle between the directions of magnetization in neighbouring magnetic films). The most widely used materials for AMR sensors are Permalloys (of composition near $Ni_{80}Fe_{20}$, $Ni_{80}Co_{20}$). For detectors such as those used in the automobile industry, the working field value can be relatively high (10^{-2} tesla). For these applications multilayered materials, such as those with (NiFe 2 nm / Ag 1 nm) as the basic building block, show suitable characteristics.

In magnetoresistive read heads for computer disk drives, which detect fields of the order of a few millitesla, the *spin valves* have the most appropriate characteristics because of their large MR amplitude (up to 20% in spin-valves with nano-oxide layers) and high sensitivity.

In contrast, as mentioned in section 5.4, tunnel junctions do not look so promising for this application, despite their large MR amplitude, because of their large shot noise. Let us illustrate this point by comparing the signal to noise ratio in a metallic GMR element and in a magnetic tunnel junction.

♦ In a metal, the main source of noise is Johnson noise, due to the brownian motion of the electrons in the resistance. The noise amplitude is given by $V_{Johnson} = \sqrt{4k_BTR\Delta f}$ where k_B is Boltzmann's constant, T the temperature, R the resistance of the MR element (typically 20 Ω) and Δf the bandwidth (typically 400 MHz for a read-head). The signal is given by $\Delta V = \Delta R.I$ where I is the current flowing through the sensor and ΔR represents the change of resistance of the sensor due to its magnetoresistance. The signal to noise ratio (SNR) can then be expressed as $\dfrac{\Delta V}{V_{Johnson}} = \dfrac{\Delta R}{R} \dfrac{\sqrt{P}}{\sqrt{4k_BTR\Delta f}}$ where $P = R.I^2$ is the electrical power dissipated through the sensor. This shows that, in GMR element, the SNR can be improved by increasing $\Delta R / R$ as much as possible and working at the highest possible current density compatible with reasonable heating and electromigration in the MR element. In MR heads, current densities of the order of 4×10^7 A.cm^{-2} are nowadays used. Assuming a spin-valve total thickness of 35 nm, a length and height of the MR element of repectively 500 nm and 250 nm, a sheet resistance of 15 Ω and a relative MR amplitude of 10%, one finds a SNR = 690, i.e. 56 dB.

♦ In tunnel (TMR) junctions, besides Johnson noise which is also present, the main source of noise is shot noise which is due to the discrete nature of the electrical current tunneling through the barrier. The electrons tunnel one by one according to a Poisson process. This generates a noise given by $V_{shot} = \sqrt{2\,V\,R\,\Delta f\,\coth\dfrac{eV}{k_B T}}$

where e is the electron charge, V the voltage across the barrier. In the range of voltage used in a TMR head (typically 0.1 to 0.3 V), the cotanh term can be assumed equal to 1. Concerning the signal, one has to take into account that the TMR amplitude decreases with the bias voltage. This decrease can be expressed as $\dfrac{\Delta R}{R} = \left(\dfrac{\Delta R}{R}\right)_0\left(1 - \dfrac{V}{V_c}\right)$ with V_c typically of the order of 0.6 V. The signal to noise ratio is then given by :

$$\frac{\Delta V}{V_{shot}} = \left(\frac{\Delta R}{R}\right)_0 \frac{\left(1 - \dfrac{V}{V_c}\right)\sqrt{V}}{\sqrt{2e\,\Delta f}\,\sqrt{R}}.$$

It is maximum for $V = \dfrac{V_c}{3^{2/3}}$. Assuming a TMR amplitude of 30% at low voltage, a (resistance $\times$ area) product for the tunnel junction RA = 10 Ω.μm^2 which is already very low and difficult to achieve, a junction area A = 0.04 μm^2, one finds a maximum SNR equal to 316, i.e. 50 dB, which is not as good as for spin-valves in spite of the strong assumption made on the low value of the R $\times$ A product.

Tunnel junctions remain nevertheless extremely promising for non-volatile magnetic memories applications (Magnetic Random Access Memories). Several companies throughout the world are putting considerable R & D effort into developing this type of memories. They present several crucial advantages which are non-volatility, insensitivity to ionising radiation, very fast switching time (nanosecond), possibility of very high density, unlimited number of read/write cycles. The basic principle of MRAM is illustrated in figure 20.20 [67]:

In this design, each memory element is formed by the association of a tunnel junction and of a command transistor. The transistor is used as an electrical switch. By applying a voltage on the control line, the current can flow through the junction. If no voltage is applied to the gate, no current can flow. Each tunnel junction comprises a pinned magnetic layer and a soft magnetic layer, the magnetization of which can be set either parallel or antiparallel to the magnetization of the pinned layer. This defines the two states of the memory element.

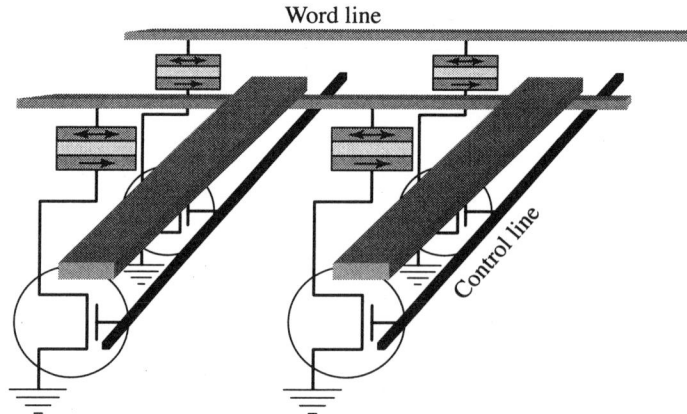

Figure 20.20 - Schematic representation of an array of 3 ×3 memory elements

Writing on a given memory element is then achieved by simultaneously sending two pulses of current, one along the corresponding word line, the other along the corresponding bit line. Assuming that the lines are cylindrical, these pulses of current generate circonferential magnetic fields around the lines. The memory element located at the intersection of the two lines feels the combined effect of two perpendicular in-plane fields, whereas the other elements along each lines only feel one field. The intensity and duration of the pulses are then adjusted so that the combined effect of the two perpendicular fields allows the magnetization of the addressed memory element to switch, whereas the other elements, which feel no field or only one field, do not switch.

For reading, a voltage is applied on the control line of the addressed memory element so as to make the transistor conducting. A current is then sent in the word line which now traverses the tunnel junction. The measurement of the voltage on the word line directly gives the magnetic state of the junction.

6. *SPIN ELECTRONICS AND MAGNETIC NANOSTRUCTURES*

The intense activity on the electronic transport properties in magnetic multilayers which was triggered by giant magnetoresistance and more recently tunnel magnetoresistance has given rise to a new and exciting new field in magnetism : spin electronics, also named magneto-electronics or spintronics [68, 69]. So far, in electronics, all microelectronic devices are based on the charge of the electrons. For instance, Digital Random Access Memories (DRAM) in computers are based on the level of charge of a capacitance which controls the voltage applied to the gate of a Field Effect Transistor. The specificity of spin electronics is to consider that the electrons also carry a spin in addition to their charge. The spin is used as an additional degree of freedom which allows to realise devices with new functionalities. These devices generally combine magnetic materials, which are used to polarise or analyse the electrons' spin, with semiconductors or oxide materials. MRAM are one example of such devices. Several other examples can be found in the literature, such as the spin rotation Field Effect Transistor [70] or the spin-valve transistor [71].

Furthermore, the increase in areal density of information stored on computer hard disks in magnetic recording technology has led researchers in magnetism to investigate the magnetic properties of submicron or even nanoscale magnetic objects. After a first evolution from the studies of bulk magnetic materials to magnetic multilayers with nanometer thick layers, a new evolution started a decade ago from magnetic multilayers to magnetic nanostructures. The lateral dimensions are now reduced to a few nanometers. These structures are produced by either natural or artificial means. The natural means consist for instance in taking advantage of the bad wetting properties of magnetic metals on oxides such as SiO_2 or Al_2O_3 . Thus a 2 nm Co layer deposited on a SiO_2 substrate spontaneously forms separated islands about 3 nm in diameter. Under certain circumstances, these islands can even exhibit a regular ordering on the surface. Such spontaneous ordering is called self-organization. Artificial means to prepare these nanostructures require the use of techniques developed in the microelectronics industry such as lithography, etching, lift-off, or ion milling. In agreement with Moore's law, which states that the number of electronic components per chip increases exponentially with time, these techniques have made tremendous progress over the last decades in terms of reliability, spatial resolution, chemical selectivity... In these fields of spin-electronics and nanomagnetism, the synergy between basic research and applications is very strong.

REFERENCES

[1] L. NEEL, Œuvres Scientifiques (1978) Editions du CNRS, Paris;
C.R. Acad. Sci. Paris **241** (1955) 533; *J. Phys. Rad.*, **29 suppl.** C2-87 (1968);
C.R. Acad. Sci. Paris **255** (1962) 1545 et 1676; *J. Appl. Phys.* **36** (1965) 944;
IEEE Trans. Magn. **1** (1965) 10; *J. Phys. Rad.* **15** (1954) 225; *Ann. Phys.* **2** (1967) 61.

[2] A. MARTY, S. ANDRIEU, *J. Physique IV* **C7-6** (1996) 3;
Epitaxial Growth, *Materials Science Series* (1975), J.W. Matthews Ed., Academic Press, New York-London.

[3] L.I. MAISSEL, M.H. FRANCOMBE, An introduction to thin films (1973) Gordon and Breach Science Publishers, New York.

[4] K.L. CHOPRA, Thin film phenomena (1969) McGraw-Hill, New York;
L.I. MAISSEL, R. GLANG, Handbook of thin film technology (1970) McGraw-Hill, New York;
J.L. VOSSEN, W. KERN, Thin film processes (1978) Academic Press, New York-London;
O'HANLON, User's guide to vacuum technology (1980) John Wiley & Sons, New York.

[5] U. GRADMANN, *J. Magn. Magn. Mater.* **100** (1991) 481.

[6] A.J. FREEMAN, RU-QIAN WU, *J. Magn. Magn. Mat.* **100** (1991) 497.

[7] R. WIESENDANGER, H.J. GUNTHERODT, *Surface science* **235** (1990) 1.

[8]   V.L. MORUZZI, P. M. MARCUS, *Phys. Rev. B* **39** (1989) 10.

[9]   D. PESCIA, M. STAMPANONI, G.L. BONA, A. VATERLAUS, R.F. WILLIS, D.F. MAIER,
       Phys. Rev. Lett. **58** (1987) 2126;
       R. ALLENSPACH, A. BISCHOF, *Phys. Rev. Lett.* **58** (1992)x 3385.

[10]   W.A.A. MACEDO, W. KEUNE, *Phys. Rev. Lett.* **61** (1988) 475;
       J. THOMASSEN, F. MAY, B. FELDMANN, M. WUTTIG, H. IBACH, *Phys. Rev. Lett.* **69**
       (1992) 3831.

[11]   C. CHAPPERT, P. BRUNO, *J. Appl. Phys.* **64** (1988) 5736;
       C. CHAPPERT, K. LE DANG, P. BEAUVILLAIN, H. HURDEQUINT, D. RENARD, *Phys.
       Rev. B* **34** (1986) 3192.

[12]   H.E. STANLEY, Introduction to Phase Transitions and Critical Phenomena (1971) Oxford
       Univ. Press, New York;
       P. KUMAR, *Phys. Rev. B* **10** (1974) 2928;
       T.C. LUBENSKY, M.H. RUBIN, *Phys. Rev. B* **12** (1975) 3885.

[13]   V. GRADMANN, Magnetism in ultrathin transition-metal films, *in* Handbook of Magnetic
       Materials vol. **7** (1993), K.H.J. Buschow Ed., North Holland.

[14]   F.J.A. DEN BROEDER, W. HOVING, P.H.J. BLOEMEN, *J. Magn. Magn. Mater.* **93** (1991)
       562.

[15]   W.J.M. DE JONGE, *in* Ultrathin magnetic structures I (1994), J.A.C. Bland & B. Heinrich
       Eds, Springer Verlag, Berlin-New York.

[16]   F.J.A. DEN BROEDER, D. KUIPER, A.P. VAN DE MOSSELAER, W. HOVING, *Phys. Rev.
       Lett.* **60** (1989) 2769.

[17]   G.H.O DAALDEROP, P.J. KELLY, F.J.A. DEN BROEDER, *Phys. Rev. Lett.* **68** (1992) 682.

[18]   F.J.A. DEN BROEDER, W. HOVING, P.J.H. BLOEMEN, *J. Magn. Magn. Mater.* **93** (1991)
       562;
       B.N. ENGEL, C.D. ENGLAND, R.A. VAN LEEUWEN, M.H. WIEDMAN, C.M. FALCO,
       Phys. Rev. Lett. **67** (1991) 1910.

[19]   C.J. LIN, G.L. GORMAN, C.H. LEE, R.F.C. FARROW, E. MARINERO, H.V. DO,
       H. NOTARYS, C.J. CHIEN, *J. Magn. Magn. Mater.* **93** (1991) 194;
       N.W.E. MCGEE, M.T. JOHNSON, J.J. DE VRIES, J. VAN DE STEGGE, *J. Appl. Phys.* **73**
       (1993) 3418.

[20]   A. DINIA, K. OUNADJELA, A. ARBAOUI, G. SURAN, D. MULLER, P. PANISSOD,
       J. Magn. Magn. Mater. **104-107** (1992) 1871.

[21]   B. HEINRICH, Z. CELINSKI, J.F. COCHRAN, A.S. ARROTT, K. MYRTLE, *J. Appl.Phys.*
       70 (1991) 5769;
       R. KRISHNAN, M. PORTE, M. TESSIER, *J. Magn. Magn. Mater.* **103** (1992) 47.

[22]   S. ARAKI, *Mater. Res. Soc. Proc.* **151** (1989) 123.

[23]   H.J. ELMERS, T. FURUBAYASHI, M. ALBRECHT, U. GRADMANN, *J. Appl. Phys.* **70**
       (1991) 5764.

[24]   C. CHAPPERT, P. BRUNO, *J. Appl. Phys.* **64** (1988) 5736.

[25]   G.H.O. DAALDEROP, Magnetic anisotropy from first principles, *in* Ultrathin magnetic
       structures I (1994), J.A.C. Bland & B. Heinrich Eds, Springer Verlag, Berlin-New York.

[26]   D.S. WANG, R. WU, A.J. FREEMAN, *Phys. Rev. B* **47** (1993) 14932.

[27]   D.S. WANG, R. WU, A.J. FREEMAN, *Phys. Rev. Lett.* **70** (1993) 869.

[28] R. JUNGBLUT, M.T. JOHNSON, J.A. DE STEGGE, F.J.A. DEN BROEDER, *J. Appl. Phys.* **75** (1994) 6424.

[29] P. BRUNO, *J. Appl. Phys.* **64** (1988) 3153.

[30] H.J.G. DRAISMA, F.J.A DEN BROEDER, W.J.M. DE JONGE, *J. Appl. Phys.* **63** (1988) 3479.

[31] J.C.S. KOOLS, *J. Appl. Phys.* **77** (1995) 2993.

[32] S.S.P. PARKIN, N. MORE, K.P. ROCHE, *Phys. Rev. Lett.* **64** (1990) 2304.

[33] S.S.P. PARKIN, D. MAURI, *Phys. Rev. B* **44** (1991) 7131.

[34] P. BRUNO, C. CHAPPERT, *Phys. Rev. Lett.* **67** (1991) 1602; *Phys. Rev. Lett.* (1991) **67**, 2592; *Phys. Rev. B* **46** (1992) 261.

[35] B. DIENY, D. GIVORD, J.M.B. NDJAKA, *J. Magn. Magn. Mater.* **93** (1991) 503.

[36] J.M. ZIMAN, Theory of solids (1972) Cambridge Univ. Press.

[37] N.W. ASHCROFT, N.D. MERMIN, Solid State Physics (1976) Saunders College Publishing, Philadelphia

[38] M.B. STEARNS, *J. Magn. Magn. Mater.* **104-107** (1992) 1745.

[39] T.R. MCGUIRE, R.I. POTTER, *IEEE Trans. Magn.* **11** (1975) 1018.

[40] D.A. THOMPSON, L.T. ROMANKIW, A.F. MAYADAS, *IEEE Trans. Magn.* **11** (1975) 1039.

[41] M.N. BAIBICH, J.M. BROTO, A. FERT, F. NGUYEN VAN DAU, F. PETROFF, P. ETIENNE, G. CREUZET, A. FRIEDERICH, J. CHAZELAS, *Phys. Rev. Lett.* **61** (1988) 2472.

[42] E. FULLERTON, M.J. CONOVER, J.E. MATTSON, C.H. SOWERS, S.D. BADER, *Appl. Phys. Lett.* **63** (1993) 1699.

[43] D.H. MOSCA, F. PETROFF, A. FERT, P.A. SCHROEDER, W.P. PRATT, R. LOLOE, *J. Magn. Magn. Mater.* **94** (1991) L-1.

[44] S.S.P. PARKIN, R. BHADRA, K.P. ROCHE, *Phys. Rev. Lett.* **66** (1991) 2152.

[45] B. DIENY, V.S. SPERIOSU, S.S.P. PARKIN, B.A. GURNEY, D.R. WILHOIT, D. MAURI, *Phys. Rev. B* **43** (1991) 1297.

[46] B. RODMACQ, P. MANGIN, C. VETTIER, *Europhys. Lett.* **15** (1991) 503; B. RODMACQ, G. PALUMBO, P. GERARD, *J. Magn. Magn. Mater.* **118** (1993) L-11.

[47] S.S.P. PARKIN, *Appl. Phys. Lett.* **60** (1992) 512; S.S.P. PARKIN, R. BHADRA, K.P. ROCHE, *Phys. Rev. Lett.* **66** (1991) 2152.

[48] S. ARAKI, Y. NARUMIYA, *J. Magn. Magn. Mater.* **126** (1993) 521.

[49] S.F. LEE, W.P. PRATT Jr, R. LOLOE, P.A. SCHROEDER, J. BASS, *Phys. Rev. B* **46** (1992) 548.

[50] O. REDON, J. PIERRE, B. RODMACQ, B. MEVEL, B. DIENY, *J. Magn. Magn. Mater.* **149** (1995) 398.

[51] T. SHINJO, H. YAMAMOTO, *J. Phys. Soc. Japan* **59** (1990) 3061.

[52] B. DIENY, V.S. SPERIOSU, J.P. NOZIERES, B.A. GURNEY, A. VEDYAYEV, N. RYZHANOVA, *in* Magnetism and structure in systems of reduced dimension, *NATO ASI Series B, Physics* **309** (1993), R. Farrow *et al.* Eds, Plenum Press, New York; B. DIENY, V.S. SPERIOSU, S. METIN, S.S.P. PARKIN, B.A. GURNEY, P. BAUMGART,

D. WILHOIT, *J. Appl. Phys.* **69** (1991) 4774;
B. DIENY, P. HUMBERT, V.S. SPERIOSU, B.A. GURNEY, *Phys. Rev. B* **45** (1992) 806.

[53]   J.C.S. KOOLS, *IEEE Trans. Magn.* **32** (1996) 3165.

[54]   P.M. TEDROW, R. MESERVEY, *Phys. Rev. Lett.* **26** (1971) 192;
P.M. TEDROW, R. MESERVEY, *Phys. Rev. B* **7** (1973) 318.

[55]   R. MESERVEY, D. PARASKEVOPOULOS, P.M. TEDROW, *Phys. Rev. Lett.* **37** (1976) 858;
J. Appl. Phys. **49** (1978) 1405 .

[56]   M. JULLIERE, *Phys. Lett.* **54A** (1975) 225 .

[57]   DE TERESA, A. BARTHELEMY, A. FERT, J.P. CONTOUR, F. MONTAIGNE, P. SENEOR,
Science **286** (1999) 507.

[58]   J.S. MOODERA, L.R. KINDER, T.M. WONG, R. MESERVEY, *Phys. Rev. Lett.* **74** (1995)
3273.

[59]   T. MIYAZAKI, N. TEZUKA, *J. Magn. Magn. Mater.* **139** (1995) L231.

[60]   W.J. GALLAGHER, S.S.P. PARKIN, Y. LU, X.P. BIAN, A. MARKLEY, R.A. ALTMAN,
S.A. RISHTON, K.P. ROCHE, C. JAHNES, T.M. SHAW, X. GANG, *J. Appl. Phys.* **81**
(1997) 3741.

[61]   X.F. HAN, A.C.C. YU, M. OOGANE, J. MURAI, T. DAIBOU, T. MIYAZAKI, *Phys. Rev.
B* **63** (2001) 224404.

[62]   R.C. SOUZA, J.J. SUN, V. SOARES, P.P. FREITAS, A. KLING, M.F. DA SILVA,
J.C. SOARES, *Appl. Phys. Lett.* **73** (1998) 3288.Motorola (40%), TDK (40%)

[63]   S. ZHANG, P.M. LEVY, A.C. MARLEY, S.S.P. PARKIN, *Phys. Rev. Lett.* **79** (1997)
3744.

[64]   S. SUN, C.B. MURRAY, D. WELLER, L. FOLKS, A. MOSER, *Science* **287** (2000) 1989.

[65]   S. CHOU, *Proceedings of the IEEE* **85** (1997) 652.

[66]   C. CHAPPERT, H. BERNAS, J. FERRE, V. KOTTLER, J.P. JAMET, Y. CHEN,
E. CAMBRIL, T. DEVOLDER, F. ROUSSEAUX, V. MATHET, H. LAUNOIS, *Science* **280**
(1998) 1919.

[67]   J.M. DAUGHTON, *J. Appl. Phys.* **81** (1997) 3758.

[68]   G.A. PRINZ, *Science* **282** (1998) 1660.

[69]   J. DE BOECK, G. BORGHS, *Physics World* **27** (1999).

[70]   S. DATTA, B. DAS, *Appl. Phys. Lett.* **56** (1990) 665.

[71]   D.J. MONSMA, J.C. LODDER, Th.J.A. POPMA, B. DIENY, *Phys. Rev. Lett.* **74** (1995)
5260.

CHAPTER 21

PRINCIPLES OF MAGNETIC RECORDING

Magnetic recording is based on remanence, i.e. on the possibility of writing stable or metastable magnetization configurations within a material. The information carrying medium is the heart of any recording (or storage) system. Of course it is complemented by write, erase, and readout devices.

The first part of this chapter describes the basic principles, and gives a short overview of the magnetic recording processes actually used today. They all rely on thin films for accessibility reasons.

The second part deals with the information supporting media. They can be particulate, or granular, media, in which information is written in the form of magnetised regions much larger than the grains, and not to be mixed up with domains. The medium can also be homogeneous, free of defects, and devoid of coercivity. In the latter case, domains in their equilibrium configurations are well suited to the storage of digital information, with the advantage that this information can be moved around within the medium. In this case the medium remains fixed (bubble memories), while the information on band or disk systems can be accessed only by moving the medium.

The third part is devoted to the writing processes. We describe the magnetic (or inductive) process, in which magnetization is written very locally by an applied field through a write head (which can generally also be used for reading). The thermomagnetic process involves localised heating of the medium through laser impact, with simultaneous application of a magnetic field. It is associated with the magneto-optical memories.

The fourth and last part is devoted to magnetic readout. In the inductive process, information is generally read by the write head. The magnetoresistive process involves a specialised head which cannot be used for writing, but which indirectly leads to a sizeable increase in the maximum storage density.

1. INTRODUCTION

The father of magnetic recording is the Danish engineer W. Poulsen. In 1898, he demonstrated an instrument which he called the *telegraphone*, the ancestor of our modern tape recorders. The telegraphone involved a small electromagnet, the forerunner of our present write and read heads, and the recording medium was just a hard steel wire (piano string).

The crucial progress which led to the industrial development of analogue recording, in its audio form from 1948 and its video form from 1951, involved two steps. One was the invention of the analogue recording process based on *AC-biasing,* by Carlson and Carpenter in 1921, and its rediscovery and improvement by German engineers during World War II. The other one was the development of various types of *magnetic tapes.**

On the other hand, the development of the first computers created, at the beginning of the 1950's, the need to store *digital* information in a manner that would combine capacity and speed of access. The first disk drive, the RAMAC, was produced by IBM in 1957. It already involved the main principles of the present disk units, although its performance may appear quite modest to us today. In particular its maximum information density was 2 Kbits.in^{-2} (300 bits.cm^{-2}), and its data transfer rate was 70 Kbits.s^{-1}. Its total capacity, 5 Mo, was reached by putting together no less than 50 disks with diameter 24 in (60 cm). In 1999, a typical hard disk unit (a commercial IBM product) features a maximum information density of 5.7 Gbits.in^{-2} (0.88 Gbits.cm^{-2}), a data transfer rate of 118 Mbits.s^{-1}, and it can store 6.5 Go on two 2.5 in (6.35 cm) diameter disks. The conservative theoretical limit [1] of 40 Gbits.in^{-2} (6.2 Gbits.cm^{-2}), the superparamagnetic limit, could already be reached by 2003-2004.

While the well-established technologies (longitudinal magnetic recording on disks and tapes) further progress towards their theoretical limits [1, 2], other principles are investigated in the laboratory, and could emerge in the next few years: perpendicular recording [3], "quantised" or "patterned" disks [4], and magnetic random access memories (MRAM) [5] using the magnetoresistance of a tunnel junction between two ferromagnetic metals [6].

In view of this particularly fast evolution, this chapter aims mainly at describing the principles, and the unchanging physical limitations of magnetic recording processes, while giving fairly little detailed informatoin on the technological aspects. The reader who wants to keep abreast of the state of the art should look up the literature.

---

*   Of course we should not forget the development of electronics, triggered by the invention of the triode in 1906.

It must also be mentioned that, within the general area of re-writable mass storage technologies, non-magnetic processes are also being investigated. Some of them were already convincingly tested, e.g. holographic [7] or microprobe storage [8].

It is natural to end this introduction by drawing the reader's attention on some recent books dealing with the most basic aspects of magnetic recording [9-13].

2. OVERVIEW OF THE VARIOUS MAGNETIC RECORDING PROCESSES

The most common recording process is so-called longitudinal recording. Its principle is shown in figure 21.1. The recording *medium* is a thin magnetic layer, with thickness h, which we will discuss later. It is supported either by a flexible plastic substrate (tape and floppy disks), or by a rigid substrate, usually made of aluminium (hard disk). The write head consists of a magnetic circuit involving a small gap, with thickness g, and an excitation coil with n windings.

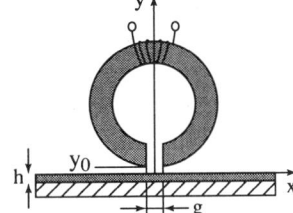

Figure 21.1 - Principle of longitudinal magnetic recording

The magnetic layer is submitted to the stray field, which is localised near the gap. We can thus consider that only a small region of the layer, with length roughly equal to the gap thickness g, and with width W >> g equal to that of the head, is submitted to the so-called write field produced by the head.

As we willl see, the essential component in this configuration is parallel to the axis Ox of the track.

2.1. ANALOG RECORDING

Let the tape run at constant speed v while a current I (t) proportional to the instantaneous value of the signal to be recorded is passed in the excitation coil of the head. Then the track, assumed to be initially demagnetised, will have along its length a magnetization distribution M (x) where x = v t, the space image of the time signal I (t).

We will see further how this space distribution of magnetization can, in turn, be changed into a time signal identical to the initial signal (readout). For the moment, we want to look more precisely at the write process.

Figure 21.2 shows the classical description of the static response of a rather hard ferromagnetic sample to an applied field. In particular, let us assume that the initially demagnetised sample is submitted to a field excursion with amplitude H (a field starting from zero to a maximum value H, and then returning to its initial zero value). There remains a so-called remanent magnetization M_r which is –at least insofar as H remains small enough– a growing function of H, with a striking non-linear character (fig. 21.2-b).

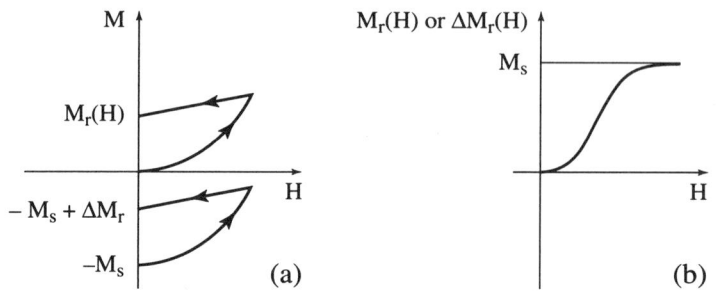

Figure 21.2 - Response of a coercive material to a field excursion

M_r *and* ΔM_r*: remanent magnetization and its variation;*
H: amplitude of the field excursion.

A qualitatively identical behavior is observed if we start, for example, from negative saturation, and plot the *variation* in remanent magnetization as a function of the amplitude of a positive field excursion (see lower loop in fig. 21.2-a).

These measurements are usually performed on size scales and under excitation configurations that are very different from those of recording on thin films (see chap. 2 and 26). However, in the latter case, we do not expect very different results, at least qualitatively and provided that:

♦ the thickness g of the gap and that, h, of the film, remain much larger than the lengths characteristic of the microstructure in the sample;

♦ the translation of the track at velocity v in the presence of a constant write current can be considered as equivalent to a field excursion with duration g / v, with the medium remaining fixed;

♦ the demagnetising effect associated in particular with the finite length of the region submitted to the write field is taken into account by correcting the applied field.

Then the response curve of the film, in terms of the remanent magnetization M_r(I) or of the variation in remanent magnetization that are locally induced on a moving track by a write current with instantaneous value I, retains the shape of the M_r(H) curve of figure 21.2-b, with marked non-linear character. Linearity can be improved by working around a point other than origin. Thus, from figure 21.2-a, it is tempting to work around the inflection point.

Actually, things are bit more complicated, but this DC biasing process has actually been used until a much more efficient process, AC biasing [13], was developed. The latter is based on the use of the *anhysteretic* curve, already introduced in § 2.5.1 of chapter 3 and in § 4.1.3 of chapter 6. We recall that this involves measuring the magnetization M_{an} created by a static field H when an *AC "unpinning" field is simultaneously applied*, with a slowly decreasing amplitude but starting from a value much larger than the coercive field. The anhysteretic magnetization M_{an} (H) is measured when the amplitude of the auxiliary AC field is reduced to zero, but actually the AC field has only a very small influence on the final value of M_{an} as soon as its amplitude is much smaller than the coercive field H_c. The procedure is repeated for each new value of the static field H. The M_{an} (H) curve thus obtained shows no hysteresis, and it remains linear up to a value of M_{an} typically around 0.4 M_s with a slope only limited by demagnetising effects. In other words, the initial *internal* anhysteretic susceptibility is infinite but, except in a toroidal geometry, the demagnetising field effect always leads to a finite *external* anhysteretic susceptibility χ_{an}.

Assume we work at a point of the anhysteretic curve defined by magnetization M_{an} and an applied external field H_0. After the AC unpinning field is suppressed, we now decrease H_0 to zero. Because the material again has coercivity, it is clear that there remains a remanent magnetization practically equal to M_{an} (H_0) = $\chi_{an}H_0$.

The recording head is fed with a *carrier current* with high frequency (typically 70 kHz for audio tapes), and with large amplitude. The write signal which is superimposed on the carrier has a frequency much lower than 70 kHz in audio recording, and it can be considered as static on the scale of the period of the carrier current. Consider a given region of the tape. As it passes across the head, it is submitted to an AC field with an amplitude that first grows to a nominal value H_{acn} much larger than the coercive field, and then tends to zero. Simultaneously, the write field grows to a nominal value H_{en}, then also goes to zero. This decrease is, in both cases, due to the fact that the region of interest moves away from the gap, and we will see later how it can be described.

We thus perform, in the region of interest of the tape, the experiment we described above, leading, at least to a first approximation, to the writing of a local magnetization proportional to the instantaneous write current.

This distribution of magnetization M and ΔM_r (x) thus written on the tape can later be recovered in the read step, which converts it into an electric signal in time V(t) through the reverse process: the recorded tape is passed at constant velocity v across a so-called read head, built in exactly the same way as a write head. Actually, the same head can be used for reading or writing. The principle involved in reading is induction. The voltage V(t) developed across the coil is proportional to $d\Phi / dt$, where Φ (t) is the instantaneous flux induced in the magnetic circuit of the head. We will see in § 5 that Φ (t) is proportional to M (x = vt), so that the signal picked up, V(t), is in fact the image of dM / dx. This leads to an enhancement of the high frequencies, called *harmonic distortion* by electronic engineers. This distorsion effect at reading, as well as another one related to recording, are corrected by filters in the amplification chain.

This section devoted to analog recording was brief. This reflects the tendency which will probably be confirmed in the coming years, viz the gradual replacement of all analog recording systems by digital systems, both in the audio (sound recording) and in the video (image recording) area.

The reader can find some complementary information in the recent books by R.M. White [14] and by P. Ciureanu and H. Gavrila [15]. The rest of this chapter only deals with *digital* recording.

2.2. DIGITAL RECORDING

Digital recording is, in its very principle, much simpler than analog recording. In the recording medium (tape or disk), it only aims at producing two values of magnetization, $+M_s$ or $-M_s$, M_s being the remanent magnetization that corresponds to the saturated loop. A track in a recorded digital tape or disk therefore consists in regions of alternating magnetization, with unequal lengths. These of course correspond neither to individual magnetic domains, nor to a *bit*, given the variety of *encoding systems* that are used.

Here we just mention this encoding problem. The interested reader should look up specialised books and papers [12, 14, 16]. A track in a tape or a disk is subdivided into small equal intervals (expressed indifferently in terms of length L or duration $T = L / v$), in which the binary data (*bits*) are arranged in a row.

The contents of the track can then be read in an understandable way only insofar as each of these intervals is recognised and identified through its rank in the sequence. Without going into details, this is obtained by synchronising the readout with the motion of the recorded track.

The simplest code consists in assigning one of the magnetization polarities the value 0 and the other the value 1. We must however remember that, as we willl see in detail later, the signal provided by a read head is not directly the written magnetization, but its *derivative* with respect to the coordinate x measured along the track. This means that only *transitions*, i.e. magnetization reversals, are in fact detected. It is thus clear that, if for some reason a transition is not detected in a sequence of binary data, the following bits will all have the wrong value (the value complementary to their real value). This is called error propagation.

A first sophistication possibility for the encoding scheme consists, in order to eliminate this problem, in using directly the transitions themselves. The convention is then that the presence of a transition, whatever its polarity, in a given readout interval corresponds to a binary "1", while the absence of such a transition corresponds to binary "0".

Other considerations (error detection, readout synchronization...) lead to further sophistication. Auxiliary transitions that do not correspond to data bits are then added. This implies, for a given information content, more cluttering of the track.

2.3. PERPENDICULAR RECORDING

In longitudinal magnetic recording, the demagnetising field associated with the finite length d of the uniformly magnetised region between two transitions grows with the h/d ratio. This demagnetising field tends to destroy the magnetization in the region of interest, and we understand how this leads to a lower boundary for distance d, hence to a limitation in the storage capacity.

One way around this difficulty is to magnetise the sample in the direction, not parallel to the plane of the film, but perpendicular to it. This is referred to as *perpendicular recording*. In this case, the demagnetising field decreases with the ratio d/h. On the other hand, it is a maximum for d >> h. This is why media with high uniaxial anisotropy, involving easy magnetization directions perpendicular to the film, must then be used.

Purely magnetic perpendicular recording on coercive media has been for years and is still at the laboratory stage. A strong revival of interest in this technique is, however, observed at present [2, 3]. Besides, magneto-optical storage as well as domain propagation memories are based on this "perpendicular" magnetization configuration and they have been commercially available for several years.

2.4. MAGNETO-OPTICAL RECORDING

These memories implement the perpendicular recording mode we just described, but writing and reading are performed optically [17, 18]. One of the main advantages of this technique is that, unlike in purely magnetic systems, the read or write head does not need to fly at a tiny height over the disk surface. The system is thus less sensitive to dust, which allows the disk to be removed and replaced at will.

Apart from a strong uniaxial anisotropy, with easy magnetization direction perpendicular to the plane of the film, the materials used, again in thin film form, must have, for reasons that will become clear soon, a large wall coercivity.

The writing process, called thermomagnetic, is based on a rapid decrease of the coercive field with temperature.

If a write field H smaller than the coercive field is applied to the whole film, along its normal Oz, at room temperature, there is by definition no alteration in the magnetization distribution of the medium. However, if a small region of the film is heated by a laser beam focused down to the diffraction limit, the coercive field can locally become smaller than H. The heated region then saturates in the direction of the applied field. This leads to the writing of stable bits, with size roughly equal to that of the laser's footprint on the film (hence of the order of the wavelength).

The readout process uses the polar Kerr effect, which is described in detail in chapter 13. The same laser is used for readout and writing, but of course at different output levels (typically 2 mW for readout and 10 mW for writing), at a wavelength λ in the near infrared for the first equipments ($\lambda \approx 0.8$ µm). The tendency is of course to go to shorter wavelengths to decrease the size of a bit.

2.5. DOMAIN PROPAGATION MEMORY

All the processes described above use the fact that the M (H) relation, as determined experimentally on samples with typically centimetre-range size, remains valid for micron-range sizes, possibly with corrections for the demagnetising field effect.

This is made possible in granular and particulate media, or even in continuous media involving defects, by the existence of characteristic sizes (the size of grains or particles, average distance between particles, size of defects) that always remain much smaller than the size of the regions submitted to the write field.

The situation is radically different if the medium is continuous, i.e. free of microstructure. The *wall coercive field* (the minimum field needed to unpin the wall) is then zero or very weak, and nothing prevents the film from taking on a domain structure strictly governed by magnetostatic equilibrium considerations. It is obvious that an *arbitrary* distribution of magnetization then cannot be imposed on the film, except with tricks. Memories can nevertheless be made with this type of material. They use in a very special way the free propagation property of walls. These devices are known as domain propagation memories, or more comonly *bubble memories*.

We will restrict their description to a brief discussion of their principle, because they did not encounter the expected success. They did however raise beautiful problems in magnetism, and the reader interested in the subject can look up the book by Eschenfelder [19].

Consider a thin film, with thickness h, spontaneous magnetization M_s, uniaxial anisotropy constant K, with easy axis Oz perpendicular to the plane of the film. The most stable equilibrium state in zero field for such a geometry depends on the ratio $K / \mu_0 M_s^2$. Two extreme situations can be predicted.

If $K / \mu_0 M_s^2 \ll 1$, the demagnetising field energy is dominant, the magnetization is parallel to the plane of the film, and the domain structure depends on the sample shape in the plane.

If $K / \mu_0 M_s^2 \gg 1$, the uniaxial anisotropy forces a magnetization direction parallel to Oz, and there appears a stripe domain structure (fig. 21.3-a). Its period d results from a compromise between the magnetostatic energy (which grows with the ratio d / h), and the wall energy (proportional to h / d).

Figure 21.3
Periodic stripe, maze
and cylindrical domains
(garnets for bubble memories)

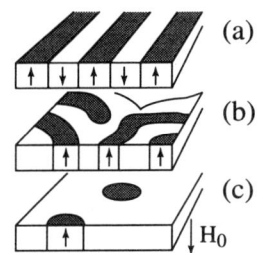

Actually, the naturally observed structures are rather of the two-dimensional maze type (fig. 21.3-b), but the band width remains very close to the prediction of a one-dimensional model. If a bias field H_0 parallel to Oz is applied and is large enough, the film of course saturates in the direction of H_0. It can however be shown that, if H_0 is not too large, stable cylindrical domains, with radius r, with magnetization opposite to H_0 (bubbles!) remain, isolated within a region magnetised along H_0.

These domains are stable if H_0 remains between two limits, H_b and $H_c > H_b$.

H_c is called the *collapse field*. When H_0 becomes larger than H_c, the bubble, which had finite radius r equal to r_c, collapses and disappears abruptly. If H becomes smaller than H_b, the cylindrical domain becomes unstable with respect to elliptic deformation: the bubbles change into stripes.

In the interval $\{H_b, H_c\}$, the bubble radius r is a decreasing function of H, continuous up to $r = r_c$. The bubble shape is stable with respect to small perturbations, and the bubbles can move very freely within the film, under the influence of small gradients in the bias field.

This property is used in so-called shift registers: the bubbles are shifted, in synchronism with a clock, along a track made of soft alloy (Permalloy) patterns deposited on the surface of the material.

These patterns are submitted to a propagation field which is rotated parallel to the film. The field gradients they create have the dual role of guiding the bubbles and shoving them along the track. The presence of a bubble in a sequence very naturally corresponds to a binary "1", whereas its absence correspond to "0".

The register includes a bubble generator / erasor which writes or modifies the data sequence during its circulation, and a detector which reads the information.

The device is equivalent to a more classical moving medium storage system (based on disk or tape), with the advantage of complete absence of mechanical motion, hence in particular of high insensitivity to shock and vibration.

On the other hand, the capacity remains very much smaller than that of disks, and the access times very much larger than those of semiconductor memories.

Bubble memories are only used today in very special applications (space and aeronautics), where their shock and radiation resistance, together with their non-volatility (the information is retained in case of power failure), are essential criteria.

3. RECORDING MEDIA

The various magnetic media used in recording belong to three categories. Particulate media consist of fine magnetic particles, spread in a polymer matrix. Granular media usually are actually ferromagnetic polycrystalline metals or alloys. Finally "continuous" and homogeneous media are thus called to distinguish them from granular media, and they can be either single crystalline, or amorphous.

More basically, one can distinguish between coercive and non-coercive media. The particulate and granular media are then in the first category, while the continuous homogeneous media are normally in the non-coercive range.

All media, as was already mentioned, are available in the form of thin films deposited on a substrate.

3.1. PARTICULATE MEDIA

The films are obtained by spreading on the substrate a polymerisable resin that contains a suspension of a fine powder, generally iron oxide (γ-Fe_2O_3).

The γ-Fe_2O_3 grains are typically shaped like prolate ellipsoids, with major axis a ~ 1 μm and minor axis b ~ 0.2 μm.

3.1.1. Stoner-Wohlfarth model

The Stoner and Wohlfarth model is the simplest description of such a composite. It is analysed in detail in chapter 5. It predicts a hysteresis loop characterised by a coercive field of the order of 1/2 H_a where H_a is the total anisotropy field (including the shape effect) of the particle. It is based on two strong assumptions.

One is that, within a particle, magnetization reverses through uniform rotation. The other is that interactions between particles are negligible.

Actually, the coercive field measured in these composites is two to three times lower than the predictions of this model (typically $\mu_0 H_C$ ~ 0.03 T instead of 0.09 T), which questions the validity of these assumptions.

Two non-uniform rotation mechanisms are analysed in detail in chapter 5. They do lead to a decrease in coercive field, provided however that the total anisotropy of the particle is mainly magnetostatic in origin (shape anisotropy).

It was established [20] that shape anisotropy contributes typically 2/3 of the total anisotropy in the γ-Fe_2O_3 particles used in recording, while 1/3 is due to magnetocrystalline anisotropy. Measurements performed on a single particle [21] seem to confirm that these non-uniform rotation mechanisms play a dominant role in magnetization reversal.

The effect of dipolar interaction between particles was not treated in chapter 5. This is a complex problem which can only be solved in a general way by numerical simulation.

Here we just discuss a *plausibility* argument showing that interaction decreases the coercive field. Consider a set of elongated particles, with their axes all parallel to a common direction Oz. Initially, all the particles are supposed to be magnetised in the same direction +Oz. A static magnetic field $H_z = -H\,(H > 0)$ is then applied. Assume that the magnetic moment in each particle is rotated by the same small angle θ: then there appears a restoring torque acting on each of these moments. What is the contribution just from the interaction to this torque?

This purely dipole interaction can be roughly estimated using an approach similar to that used in the theory of dielectrics (Lorentz field). Assume the "cavity" remaining when a given particle, assumed to be point-sized, is taken out of the composite is on average a sphere. The interaction field acting on a given particle is thus approximately the field that acts within a spherical cavity dug in a uniformly magnetised medium. The relevant magnetization here is that corresponding to the deviation θ, viz $c\,M_s \sin\theta$ (where c is the volume fraction of particles), and the interaction field is $(1/3)\,c\,M_s \sin\theta \sim (1/3)\,c\,M_s\,\theta$. Its direction is *perpendicular* to Oz and it is oriented so that the restoring torque, with modulus $(1/3)\,\mu_0\,c\,M_s^2\,\theta$, is *negative*. The restoring torque acting on the moment of a particle is thus:

$$\Gamma = [2\,K - H\,M_s - (1/3)\,\mu_0\,c\,M_s^2]\theta \qquad (21.1)$$

We see that the stiffness $d\Gamma/d\theta$ vanishes, hence that the position $\theta = 0$ becomes unstable, for a value of H smaller than $2\,K/M_s$. This value is, in the case under study, the coercive field of the isolated particle.

This simple model thus predicts a decrease in H_c proportional to the volume fraction of the particles in the composite, in agreement with experiment.

3.1.2. *Superparamagnetism in particulate media*

Around its stable state, with magnetization either along Oz ($\uparrow$) or along the opposite direction ($\downarrow$), the energy of the particle is a quadratic function of the direction cosines α and β of magnetization. The mean energy associated, at thermal equilibrium, with each of these degrees of freedom is, from the equipartition theorem of statistical physics, $(1/2)\,k_B T$, where k_B is Boltzmann's constant. For an isolated particle with volume V, we have: $V(<\alpha^2> + <\beta^2>)\,(1/2)\,(N_b - N_a)\,\mu_0\,M_s^2 = k_B T$, whence:

$$<\alpha^2> + <\beta^2> = 2\,k_B T\,/\,[(N_b - N_a)\mu_0\,M_s^2\,V] \qquad (21.2)$$

We here assumed that the anisotropy is solely due to the shape effect (see chap. 5). If the volume of the particle decreases, the amplitude of the thermal oscillation in the magnetization direction increases, and it is clear that the probability for spontaneous reversal of magnetization becomes sizeable.

According to Boltzmann statistics, the frequency f of spontaneous reversals is given by :

$$f = f_0 \exp\left[-(N_b - N_a)\mu_0 M_s^2 V / 2 k_B T\right] \qquad (21.3)$$

where the pre-exponential factor f_0 is the number of attempts per unit time. A good approximation for f_0 is the width of the natural gyromagnetic resonance line of the particle (see chap. 17). As an indication, f_0 is probably of the order of 50 to 500 MHz for a γ-Fe$_2$O$_3$ particle. The frequency of spontaneous reversals becomes of the order of f_0 for $(1/2)(N_b - N_a)\mu_0 M_s^2 V = k_B T$, and, if we take $V \sim ab^2$, with $b/a = 0.2$, we see that this occurs at room temperature for $b = 3$ nm hence $a = 15$ nm.

The spontaneous reversal frequency is then large on the scale of any quasi-static experiment. In particular, no remanent magnetization is then measured! This behavior is referred to as superparamagnetism, in analogy with the paramagnetism of atomic moments. This regime must absolutely be avoided in memories [1]. Fortunately, the variation of f with the ratio V/T is exponential, so that, for the γ-Fe$_2$O$_3$ classically used ($a \sim 1$ μm and $b \sim 0.2$ μm), the average time between spontaneous reversals is already much larger than the average lifetime of a generation of memories.

Superparamagnetism is treated in detail in a recent review paper [22], and briefly in chapters 4 (§ 2.3) and 22 (§ 3.1) of the present book.

3.2. GRANULAR MEDIA, METALLIC THIN FILMS

As we will see later, reducing the thickness of the magnetic film makes it possible, in the purely magnetic recording process, to increase the surface density of information, hence the storage capacity. However, the read signal amplitude is roughly proportional to the product of the thickness h by the spontaneous magnetization M_s, hence it is desirable to increase M_s when h is reduced.

The use of metal and alloy films is a considerable progress with respect to Fe$_2$O$_3$ based composites, because both the intrinsic value of magnetization and the filling factor are increased.

The metallic materials used in longitudinal magnetic recording are Co based alloys [23]. The films are textured and polycrystalline, with typical grain sizes of the order of 10 nm, and thickness around 20 nm. The local easy magnetization direction is oriented at random within the film plane. The magnetization processes are more complex than in single domain particle composites, because the interactions between grains are strong. The coercivity of these films is observed to be strongly correlated to their microstructure, which usually involves several characteristic scales (grains, subgrains, phase boundaries).

In several respects, however, the behavior of these granular media is not considered as very different from that of a collection of isolated particles. This is for example the current assumption for calculating their superparamagnetic limit [1, 2].

The materials for magneto-optical recording are usually amorphous R-FeCo type alloys (R being a rare earth metal Tb, Gd, Dy), with so-called *sperimagnetic* magnetic structure. The moments of the terbium and iron (and cobalt) atoms then make up partially disordered sublattices, with resulting moments in opposite directions (see fig. 4.21).

This structure, special to amorphous materials, is reminiscent of the ferrimagnetism of crystallised compounds such as ferrites. As in some ferrimagnets, these materials feature a compensation temperature, where the mesoscopic magnetization M_S goes to zero and changes sign. We note that the apparently physically sensible conclusion that, since the mesoscopic magnetization is zero, there must be no Kerr or Faraday rotation at the compensation temperature is wrong. The rare earth-metal and transition-metal "sublattices" contribute in a nearly additive way to the magneto-optical effect, and they have different contributions even when the absolute values of their magnetic moments are equal. The information can therefore be read at the compensation point too.

Finally, these materials feature, at least under some preparation conditions, both a strong perpendicular anisotropy and coercivity that is high at room temperature, and rapidly decreases with temperature. The anisotropy observed ($\mu_0 H_a$ is typically of the order of 1 to 2 T) is usually explained by a pair orientation order induced during the film deposition, and favored by the symmetry of this special forming process.

The origin of coercivity in these amorphous materials, assumed to be free of microstructure, is a question which deserves some more thought and comments.

A single domain sample of a *perfect* uniaxial material, with anisotropy field H_a larger than its demagnetising field NM_s, is *metastable* in the saturated state. Its magnetization reverses under a uniform inverse magnetic field only when the uniform rotation mode becomes unstable, which, in thin film geometry, requires an external field at least equal to $H_a - M_s$. In such a material, the coercive field is thus, to a first approximation, equal to $H_a - M_s$. This behavior is indeed observed in single-crystal or amorphous films used in bubble memories (see § 2.5). However, this is a very narrow view of coercivity. Consider now, not the saturated single domain state, but a multi-domain state, for example the stablest state in zero field (we saw in § 2.5 that this is a stripe domain structure, with zero mean magnetization). The curve of magnetization vs external field obtained starting from this state –which can be called an initial magnetization curve– features, as in soft materials, an initial part that is linear, and at any rate does not have an appreciable threshold effect. This means zero coercivity. Of course this results from the fact that the magnetization process operating in this case is wall displacement. In a perfect material, nothing opposes this displacement, and wall coercivity is zero.

In usual uniaxial materials, there are various kinds of defects, in other words a microstructure, more or less pronounced. As a consequence, the reversal field for the saturated film is smaller than the theoretical value $H_a - M_s$ because the presence of defects leads to weak points on which reversal starts before the uniform rotation mode becomes unstable. This process may be thermally activated, in which case the weak points are rather called nucleation sites.

The defects also have the opposite effect, as they are pinning points for the walls. This leads to non-zero wall coercivity, which can in particular stabilise states which would be unstable from the magnetostatic point of view (see § 2.5).

The result of these two effects is that the magnetization curve or the hysteresis loop now have a local meaning. In particular, the size of the analysed region has no influence insofar as it remains much larger than the length characteristic of the microstructure and than the film thickness.

In the R-FeCo alloys used in magneto-optical recording, the presence of spatial anisotropy fluctuations on a mesoscopic scale (whether the modulus of the anisotropy field or the orientation of the easy axis is involved) and the existence of a compensation point are considered as the essential ingredients to explain coercivity and its thermal variation [18]. The fluctuations lead directly to *coercive energy or pressure* terms independent of magnetization M_s. As a result, the coercive field H_C is inversely proportional to M_s, hence it diverges at the compensation point.

3.3. *CONTINUOUS MEDIA: EPITACTIC SINGLE CRYSTAL FILMS AND HOMOGENEOUS AMORPHOUS FILMS*

In domain propagation memories, the information supporting media must be free of microstructure and even of localised defects. Two types of materials satisfy this demand: the single crystal films obtained on a single crystal substrate, and homogeneous amorphous films.

In fact, while both types of materials have been developed in the laboratory, only epitactic films of magnetic garnets on a non-magnetic garnet have been used in commercial devices.

The magnetic garnets have the basic formula $R_3Fe_5O_{12}$, where R is a rare earth atom or yttrium. Innumerable substitutions are possible, both on the iron sites and on the rare earth sites, and the compositions used for bubble memories involve as many as ten components.

The non-magnetic substrate that is classically used is gadolinium gallium garnet $Gd_3Ga_5O_{12}$, commonly designated by the acronym GGG. The reader will find a lot of information on bubble memory materials in reference [19].

4. *THE WRITE PROCESS*

There are two main processes for creating or altering the magnetization of a small region in a magnetic thin film: the local application of a field using a write head, and thermomagnetic writing using a laser. Both were briefly described above. We return in detail to the magnetic process. The reader who wants to know more about thermomagnetic writing can look up references [18] and [24].

4.1. FIELD PRODUCED BY A MAGNETIC HEAD

The theory of magnetic circuits immediately provides the field H_e inside the gap, with thickness g of a head excited by current I:

$$H_g = (nI/g)/(1 + \ell/\mu g) \tag{21.4}$$

Here n is the number of windings in the excitation coil and ℓ the average perimeter** of the magnetic circuit, μ its permeability. The factor $\eta = 1/(1 + \ell/\mu g)$ is called the head efficiency. In the approximation $\mu g \gg \ell$, we get:

$$H_g = nI/g \tag{21.5}$$

but this formula of course does not describe the field to which the film is submitted.

In the simplified model of a head proposed by Karlqvist [25], the magnetic circuit is assumed to have infinite permeability, and to occupy the whole segment of space $0 < y < h_g$ ($h_g \gg g$) in the orthogonal reference frame Oxyz (fig. 21.4). Oz is perpendicular to the plane of the figure and the circuit also includes a gap bounded by planes $x = -g/2$ and $+g/2$.

Relations (21.4) and (21.5) express the field *within* the gap. Karlqvist [25] assumes that the x component of the field H_x remains equal to H_g down to $y = 0$. Since the permeability of the magnetic circuit is taken as infinite, this implies that $H_x = 0$ for $x < -g/2$ or $x > +g/2$. Determining the field in the whole half-space $y < 0$ then becomes a problem with specified boundary conditions for the tangential component of the field, hence for the magnetic potential, over the plane $y = 0$.

We know that the solution to this problem is unique, and it can be verified that a surface distribution of currents, with density $i_z = 2 H_g$ on a band with width g, localised on the plane $y = 0$ between $x = -g/2$ and $x = +g/2$, creates at height $y = -\delta$, with δ positive but arbitrarily small, a field H_x precisely equal to H_g for $-g/2 < x < +g/2$ and zero outside the same interval (fig. 21.4).

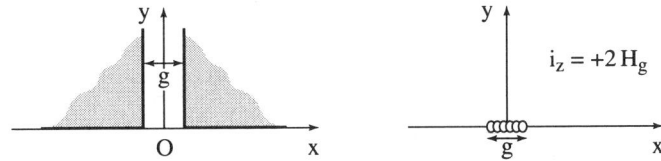

Figure 21.4 - Karlqvist's model

Left: schematic sketch of the region of the head gap near the recording media - Right: current distributions equivalent to the excited head for calculation of the field at y < 0

** The average is weighted by the ratio of the cross-section of the gap to the local cross-sectional area.

The field created by this fictitious distribution in the whole half-space $y < 0$ is thus necesssarily the solution to our problem. It is obtained by summing the elementary contributions of infinite rectilinear currents $i_z \, dx' = 2 \, H_g \, dx'$. The components obtained are $H_z = 0$ and:

$$H_x = -\frac{H_g}{\pi} \int_{-g/2}^{+g/2} \frac{1}{(x - x')^2 + y^2} \, y \, dx'$$

$$H_y = \frac{H_g}{\pi} \int_{-g/2}^{+g/2} \frac{1}{(x - x')^2 + y^2} \, (x - x') \, dx'$$

(21.6)

After integration, this provides:

$$H_x = -\frac{H_g}{\pi} \left(\tan^{-1} \frac{x + g/2}{y} - \tan^{-1} \frac{x - g/2}{y} \right)$$

$$H_y = \frac{H_g}{2\pi} \ln \frac{(x + g/2)^2 + y^2}{(x - g/2)^2 + y^2}$$

(21.7)

These functions are represented in figure 21.5 for various values of the ratio $|y| / g$.

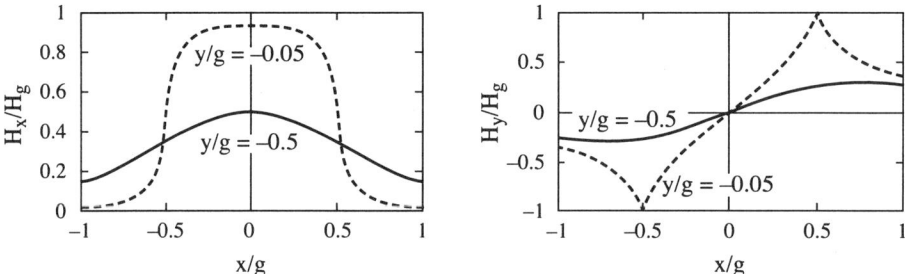

Figure 21.5 - Longitudinal (left) and vertical (right) fields
at various distances from the head

In present-day equipments (in 1998), the correct approximation is $g \gg \{h, y\}$, so that the profile of the field H_x is practically a *gate* function, equal to H_g between $-g/2$ and $+g/2$, and zero outside this interval. Near the edges of the gap, expression (21.7) can be simplified.

Thus, for x near $+g/2$, we have:

$$H_x \sim \frac{H_g}{\pi} \left(\frac{\pi}{2} - \tan^{-1} \frac{x - g/2}{y} \right)$$

(21.8)

In what follows, we will rather use the field gradient:

$$\frac{dH_x}{dx} = -\frac{H_g / \pi y}{1 + \left[(x - g/2)/y \right]^2}$$

(21.9)

Karlqvist's model is rather satisfactory for the most classical heads, involving a ferrite circuit in which the length of the part in contact with, or very near, the medium is indeed large with respect to the gap thickness, as in figure 21.4. This remains true in the most recent products, the *integrated planar* heads (fig. 21.6) [25].

In the so-called *vertical thin film* heads [14, 27], which appeared in the '80s, the magnetic circuit consists of soft magnetic films, which come near the medium in a plane perpendicular to the track, and whose thickness is not much larger than that of the gap (fig. 21.7). Karlqvist's model is clearly less adequate in this case. Calculations suited for this geometry were published as early as 1963 [28]. Numerical methods were also used (see for example ref. [26] for a well-documented review).

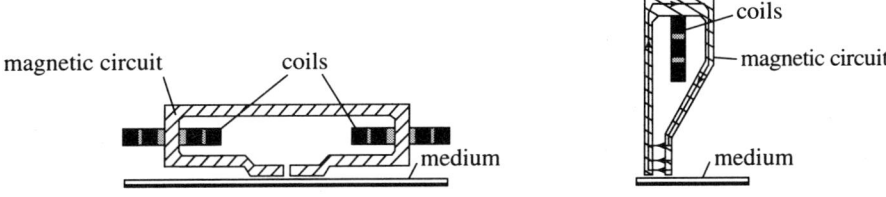

Figure 21.6   ***Figure 21.7***

Principle of an integrated planar head   ***Principle of a "vertical thin film" head***

Many models assume the permeability of the magnetic circuit to be infinite, or at least homogeneous and isotropic, and restrict the calculation to the static case. In more thorough models, where the ultimate limits are explored, in particular for thin film and integrated planar heads, sophistication is pushed far beyond the mere account of a finite homogeneous and isotropic permeability. This assumption is in fact not justified when the domain size is of the same order as the geometrical dimensions of the problem. The domain structure in the films must then be explicitly taken into account to determine (numerically) the response, which must furthermore be calculated in the dynamic regime. The reader can look up chapter 17 of the present book for more information on dynamic effects, and in particular the frequency dependence of response in a sinusoidal regime.

4.2. STABILITY OF WRITTEN MAGNETIZATION PATTERNS

The above section dealt with the shape of the field produced by the head, and we will later use these results to describe, at least in a semi-quantitative way, the write process. However, before tackling this problem, we investigate the conditions under which a *given* magnetization distribution in the film remains stable in the absence of a write field. This is one aspect of the basic problem of remanence stability, to be compared with the somewhat different view treated in section 3.1.2 of this chapter.

We define in a somewhat arbitrary way a standard distribution of magnetization, representative of those effectively encountered in written media:

$$M_x = M = (2/\pi) M_s \tan^{-1}(x/a) \qquad (21.10)$$

This *inverse tangent* distribution corresponds to an isolated transition from the saturated state with $M_x = -M_s$ to the saturated state with $M_x = +M_s$.

Here, Ox of course remains the axis parallel to the track, and we assume **M** to depend neither on coordinate z along the width of the track, nor on coordinate y along the medium thickness. We also neglect the perpendicular component M_y of **M**. The quantity 2a can be considered as the *length* of the transition.

This variation in magnetization produces a pole density $\rho = -\text{div}(\mathbf{M}) = -dM/dx$ which is, in turn, responsible for a demagnetising field $\mathbf{H_d}$.

We can assume the thickness h of the magnetic film to remain very small with respect to the transition length 2a. This approximation is not mandatory, but it simplifies the calculations, it is consistent with the assumption on the uniformity of M (x) vs thickness, and it remains realistic enough.

The magnetic film then reduces to the plane Oxz carrying a *surface* distribution of magnetic masses, with density $-h(dM_x/dx)$. We then have:

$$2\pi(x - x')dH_d = -\left(\frac{2M_s h}{a\pi}\right)\left(\frac{dx'}{1 + (x'/a)^2}\right) \tag{21.11}$$

Gauss's theorem is here used to express the elementary field produced at x by a line of magnetic masses, with linear density $-h(dM/dx')$, placed at x'. We note that this field has only one component, along Ox. We obtain:

$$H_d = -\frac{M_s h}{\pi^2 a} \int_{-\infty}^{+\infty} \frac{dx'}{(x - x')(1 + (x'/a)^2)} \tag{21.12}$$

which, after some simple transformations, gives:

$$H_d = -\left(\frac{M_s h}{\pi a}\right)\left(\frac{x/a}{1 + (x/a)^2}\right) \tag{21.13}$$

We see that the demagnetising field is zero at x = 0, i.e. at the middle of the transition. It is maximum, equal to $\pm(1/2)(M_s h/\pi a)$, for $x = \pm a$, respectively. If H_C is the coercive field of the material, the stability criterion is simply expressed as $(1/2)(M_s h/\pi a_0) = H_C$, hence:

$$2a_0 = M_s h/\pi H_C \tag{21.14}$$

The *minimum* length of a *stable* transition is thus proportional to the spontaneous magnetization in the material, to the film thickness and to the reciprocal of its coercivity. Increasing the maximum density of bits (which is of the order of $1/2a_0$) thus requires either decreasing $M_s h$ or increasing H_C. However, as we will see later, it is not advisable to decrease $M_s h$, because this leads to a decrease in the read signal. This is why the improvements now considered for the materials bear mainly on the increase of the coercive field. We recall that formula (21.14) is based on the approximation $2a_0 \gg h$, which implies $H_C \ll M_s/\pi$. If H_C becomes comparable to

M_s then a more exact calculation must be performed [14]. This leads, still under the assumption of a one-dimensional distribution of magnetization, to the conclusion that the transition can become infinitely steep provided $H_C \geq M_s$.

Another approach to evaluating the maximum density of stable bits in a medium starts from the assumption of a *sine-shaped* magnetization profile. The demagnetising field is again easy to calculate, as it was in the inverse tangent distribution we discussed above. The difference is that here we deal with the *magnetostatic* interaction *between bits*, and not just with the demagnetising effect of a single isolated transition. Let p be the period of the distribution (with p >> h), and $K = 2\pi/p$, so that:

$$M_x = M = M_s \sin(Kx) \tag{21.15}$$

We then find:

$$H_d = -(1/2) K M_s h \sin(Kx) = -(1/2) K h M \tag{21.16}$$

Applying the stability criterion $H_d = H_C$ leads to a minimum period p equal to:

$$p = 2\pi/K = M_s h / \pi H_C \tag{21.17}$$

This period should be compared to twice the length of the isolated transition, viz: $4a_0 = 2M_s h / \pi H_C$. We see that p is smaller by a factor two than $4a_0$, which practically means that a succession of transitions is more stable than a single isolated transition. This effect comes from the magnetostatic interaction between bits. A conservative value of the ultimate transition density will thus be: $1/2 a_0 = \pi H_C / M_s h$.

4.3. WRITING A TRANSITION WITH A KARLQVIST HEAD

We just investigated the stability of a transition without asking how it was written. This allowed us in particular to determine the minimum length of this transition.

In a way, this length sets an ultimate limit, which depends only on the coercivity of the recording medium. However, we may also suspect the existence of another limit, possibly a more restrictive one, resulting from the write process itself. We now analyse this write process by considering that the medium remains fixed and that the head moves (fig. 21.8): let x be the coordinate linked to the track, u that linked to the head. The head, constantly fed with the nominal write current I, is moved from left to right on the track, which is initially magnetised in the negative direction. The write field is assumed to be positive, it thus tends to reverse the existing magnetization.

If the write current is large enough, we understand that moving the head produces a magnetization reversal front, *stationary with respect to the head*, near the leading edge of the gap. Behind the head, magnetization has flipped over by 180°.

In a first, very crude approximation, we can neglect the demagnetising field, hence assume that the material is only submitted to the *field from the head*, given by equation (2.1.8) with a change of x for u. Knowing the hysteresis loop of the material, we can then deduce the magnetization profile M (u) in the transition, at least if dynamic effects are ignored (we assume that the material's response, as given by the hysteresis loop, is instantaneous).

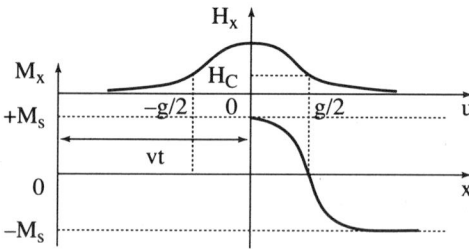

Figure 21.8 - Write process for a Karlqvist head

Actually, we are primarily interested not in a detailed knowledge of this profile, but only in obtaining a meaningful value for the length of the transition under the effect of the write field. We therefore just calculate the slope:

$$\frac{dM}{du} = \left(\frac{dM}{dH}\right)\left(\frac{dH}{du}\right) \approx \left(\frac{dM}{dH}\right)\left(\frac{H_g}{\pi y_0}\right)\left(\frac{1}{1 + [(u - g/2)/y_0]^2}\right) \quad (21.18)$$

where dM/dH is the slope of the rising branch of the hysteresis loop, and y_0 the distance between the head and the magnetic film, which is assumed to be thin with respect to y_0.

The parameter y_0 is commonly called the head spacing. We used the simplified expression (21.9) for the field gradient from the head. This gradient is a maximum for $u = g/2$, i.e. just below the leading edge of the gap. On the other hand, the slope of the loop dM/dH goes through a very sharp maximum, denoted by χ_{hy}, near the coercive field.

The maximum slope of the magnetization profile is thus obtained by setting the write current I so that H is equal to the coercive field precisely at the point where the field gradient of the head is a maximum. From (21.8), this implies $H_g = 2\,H_C$, hence $H_g / \pi y_0 = (2/\pi)\,H_C / y_0$.

An approximate value of the transition length $2a_1$ is then obtained by writing $M_s / a_1 = dM/du$, which leads to:

$$2a_1 = \frac{2M_s y_0}{\chi_{hy} H_g} = \frac{\pi}{2} y_0 \frac{\Delta H}{H_C} \quad (21.19)$$

where $\Delta H = 2\,M_s / \chi_{hy}$ is a parameter which measures the *squareness defect* of the loop. The length of the transition under the write field is thus proportional to $\Delta H / H_C$ and to the head spacing y_0. However, we have till now not taken into account the demagnetising field H_d.

In Williams and Comstock's model [29], H_d is introduced in an approximate manner by *assuming* that the transition again has an inverse tangent profile as used above:

$$M = \frac{2}{\pi} M_s \tan^{-1} \frac{(u - g/2)}{a_2} \quad (21.20)$$

Near the center of the transition, we have $M = (2/\pi) M_s (u - g/2)/a_2$. But, from equation (21.13):

$$H_d \approx -M_s h \frac{(u - g/2)}{\pi a_2^{~2}} = -\frac{h}{2 a_2} M \qquad (21.21)$$

This expression of the demagnetising field does not take into account the fact that the magnetic circuit of the head is nearby.

Equation (21.21) defines, near the middle of the transition, an *effective demagnetising field coefficient* $N = h/2 a_2$. The material's response can then be expressed as a function of the head's field alone, provided the classical demagnetising field (shearing) correction $+N M$ is applied to the *hysteresis loop* (see chap. 2 and 26). The main effect of this correction is that it increases the intrinsic squareness defect ΔH by an amount $2 N M_s$, so that the transition length $2 a_2$ becomes, using equation (21.19):

$$2 a_2 = \frac{\pi}{2} y_0 \frac{\Delta H + 2 N M_s}{H_C} = y_0 \left(\frac{\Delta H}{H_C} + \frac{M_s h}{H_C a_2} \right).$$

We thus obtain a self-consistency relation which takes the form of a second degree equation in a_2, with the physically satisfactory solution:

$$2 a_2 = \frac{\pi}{4} y_0 \frac{\Delta H}{H_C} \left\{ 1 + \left[1 + \frac{16}{\pi} \frac{h}{y_0} \frac{M_s}{H_C} \left(\frac{H_C}{\Delta H} \right)^2 \right]^{1/2} \right\} \qquad (21.22)$$

This relation evidences the importance of the head spacing y_0, of the shape of the loop (through parameter $\Delta H / H_C$), and of the medium thickness h. However, a_2 does not vanish with ΔH, and it thus remains finite for a medium with a perfectly square loop. We note that the vicinity of the high permeability material of the head's magnetic circuit leads to a decrease in the demagnetising field of the transition. This effect can be taken into account in a rather simple way by altering the expression for the effective demagnetising field coefficient we introduced above. The result is a slight decrease in the length of the transition.

The transition length $2 a_2$ is always larger than the ultimate minimum length $2 a_0$, which confirms that the density limit of the write process is set by the head-medium interaction and not by the sole medium.

Williams and Comstock's analytic model –which we slightly simplified in this presentation– has the advantage of clearly indicating the influence of the various parameters, and of even providing quantitative predictions for the transition lengths which are quite sensible.

Nevertheless, the present tendency is to use numerical simulation. Then, using the actual shape of the medium's hysteresis loop, the head's field and the demagnetising field are rigorously calculated.

A simple method consists in starting from a first magnetising distribution, for example that resulting from Williams and Comstock's approximation. The field acting on the material is then known, and from there a new distribution of magnetization is calculated via the hysteresis loop. Using successive iterations, the distribution is made to converge toward what is hoped to be the final and unique solution of the problem.

Values typical of the state of the art in 1997 are: $M_s = 1.1$ T; $\mu_0 H_C = 0.22$ T; $\Delta H / H_C = 0.2$; h = 20 nm; $y_0 = 20$ nm; g = 0.2 μm; (g >> h, y_0); such head spacings are actually obtained by quasi-direct contact through a solid lubricant film (carbon) shared between the medium (typically 15 nm) and the head (typically 5 nm). The ultimate transition length as calculated from equation (21.14) is then $2 a_0 = 0.03$ μm, and the write-limited transition length $2 a_2$ turns out to be 0.08 μm from equation (21.22).

It can be assumed that two successively written transitions must be at least $2 a_2$ apart to be correctly identified. In other words $2 a_2$ is also the writing resolution. The above data show that, in the present heads, this resolution is notably smaller than the gap g. On the other hand, we will see that the reading resolution is always of the order of g, so that the limiting process in terms of usable information density is the read process.

5. THE READ PROCESS

The presence of a transition on the track can be detected by induction. In this case, a single head usually serves both for writing and reading, and the signal is, as we now show, proportional to $v \, dM / dx$ where v is the medium velocity. So-called magnetoresistive heads have recently appeared. Their response is also proportional to dM / dx but independent of v. We describe and discuss these two readout processes.

5.1. INDUCTIVE READOUT

An inhomogeneous distribution of longitudinal magnetization M (x) necessarily entails a non-zero pole density and a demagnetising field, as discussed at length in the basic chapters of this book. The demagnetising field is not restricted to the magnetic film, it spills over into the neighbourhood. Detecting the transition is made possible by this *stray field*. In particular, the latter causes the induced flux variation in an *inductive head*.

To calculate this flux, we can use the very powerful *reciprocity theorem*, which was already used in chapter 17. A simple derivation is given in chapter 2.

Let **H** be the field produced at point P by the head when current I is passed through the coil. If the magnetic circuit of he head operates in the *linear* regime, **H** can be written in the form: $\mathbf{H} = \mathbf{C_H} I$, where the (vector) field coefficient $\mathbf{C_H}$ depends only on the point P considered. The reciprocity theorem then tells us that the flux Φ sent into the coil by a *point dipole* with magnetic moment **m** placed at P is given by $\Phi = \mu_0 \mathbf{C_H} \mathbf{m}$. We are interested in the flux variation produced by the passage of a transition under the head. This transition is characterised by a function $\mathbf{M}(x)$ of the coordinate x measured in a coordinate system moving with the track with velocity v. We assume again that $\mathbf{M}$ only has a longitudinal component:

$$M_x = M = (2M_s/\pi) \tan^{-1}(x/a_2).$$

The head's field coefficient profile is defined in the head's coordinate system, here assumed to be fixed, by a function $C_H(u)$, where C_H denotes the single longitudinal component of $\mathbf{C_H}$. The origin $u = 0$ is now chosen on the medium film, just below the middle of the gap. We decide that time zero corresponds to the two coordinate systems (u) and (x) having their origins in coincidence.

The element dx of the track, with width W, carries the moment $m = M(x) W h dx$. It induces in the readout coil the flux $d\Phi = \mu_0 C_H(u) M(x) W h dx$, where $u = x + vt$ is the instantaneous abscissa of this element in the coordinate system linked to the head. The induced voltage is: $dV = -d^2\Phi/dt = -v h W \mu_0 M(x) dx G(x + vt)$, where $G(x + vt) = dC_H/du|_{u = x+vt}$. Thus we have:

$$V(t) = -v\mu_0 W h \int_{-\infty}^{+\infty} M(x) G(x + vt) dx \qquad (21.23)$$

To make calculations easier, we assume that the head spacing y_0 is much smaller than the width of the gap g, in agreement with the orders of magnitude given above. Then the function $G(x, t)$ is approximately equal to:

$$G(x,t) = \frac{(C_H)_g}{\pi y_0} \left[\frac{1}{1 + \left(\frac{x + vt + (g/2)}{y_0}\right)^2} - \frac{1}{1 + \left(\frac{x + vt - (g/2)}{y_0}\right)^2} \right] \qquad (21.24)$$

Here $(C_H)_g$ is the head's field coefficient measured within the gap.

We can further simplify the calculation by assuming that a_2 is also much larger than y_0. Then the profile of G can be approximated by two Dirac peaks, each with content $\pi y_0 (C_H)_g / \pi y_0 = (C_H)_g$ centred respectively at $- vt - g/2$ and $- vt + g/2$.

The integral (21.23) reduces to:

$$V(t) = v W h \mu_0 (C_H)_g \frac{2}{\pi} M_s \left[\tan^{-1}\left(-\frac{vt}{a_2} - \frac{g}{2a_2}\right) - \tan^{-1}\left(-\frac{vt}{a_2} + \frac{g}{2a_2}\right) \right] \qquad (21.25)$$

It is worthwhile looking at two limiting cases. The first one is a bit academic, as it corresponds to $g \ll 2a_2$. The difference between the two inverse tangents is then practically a differential, and:

$$V(t) = \frac{1}{\pi} v W h \mu_0 (C_H)_g M_s \frac{g/a_2}{1 + (vt/a_2)^2} \qquad (21.26)$$

We thus have a *Lorentzian* voltage peak, with height proportional to $v W h (C_H)_g M_s g / a_2 = v W h \eta n M_s / a_2$ (remember that n is the number of windings and η the head's efficiency), with full width at half maximum, expressed in terms of the distance the tape has moved, equal to $2a_2$.

The other limiting case assumes that $g \gg a_2$. The schematic behavior of $V(t)$ is then close to a square pulse, with height proportional to $v\,W\,h\,\eta\,(n/g)\,M_s$, independent of a_2 and with width at half maximum equal to g.

In the general case where a_2 is of the same order of magnitude as g and where the head spacing and the medium thickness are no more neglected, it can be shown that the width at half maximum of the readout impulse corresponding to a transition (denoted as PW_{50} for *pulse width at 50%*) is given by:

$$PW_{50} = [g^2 + 4\,(y_0 + a_2)(y_0 + a_2 + h)]^{1/2} \qquad (21.27)$$

We note that this expression reduces to $[g^2 + 4\,a_2{}^2]^{1/2}$ if the head spacing y_0 and the medium thickness h are negligible with respect to g and $2\,a_2$. It is worth noticing that in the general formula, even if the transition is very steep ($a_2 = 0$), the readout pulse retains a finite width, at least equal to the gap thickness. For 1997 heads, the orders of magnitude given above lead to $PW_{50} \sim 0.25\ \mu m$. We note that $PW_{50} > 2\,a_2$ (see § 4.3). In other words, as already noted in § 4.3, the maximum linear density of useful bits in the track is limited by the read process.

Another way of identifying the theoretical limits of the read process is to look at the response, *in terms of induced flux*, to a sine-shaped magnetization distribution. If $M(x) = M_s \sin(K\,x)$, calculating $\Phi(t)$, in the approximation where y_0 and h are both very small compared to g and to the wavelength $\lambda = 2\pi/K$, yields:

$$\Phi(t) = -\frac{1}{2}\mu_0 W\,h\,g\,(C_H)_g\,M_s \sin(K\,v\,t)\,\frac{\sin(K\,g/2)}{K\,g/2} \qquad (21.28)$$

The readout flux is sine-shaped, with amplitude proportional to $(2/K\,g)\sin(K\,g/2)$. Therefore the *response involves zeros,* for characteristic wavelengths $\lambda_m = 2\pi/K_m = g/m$, with m an integer. This condition corresponds to the vanishing of the magnetic moment of the region of the film that is at any time submitted to the field of the gap.

In practice, a magnetic head is characterised [26] by using it first in the write mode, then in the read mode. First an AC square current, with variable amplitude and frequency, is fed into the head to write, at given speed, a periodic set of transitions on a *reference disk*. Then the voltage response to the recorded signal is recorded (fig. 21.9).

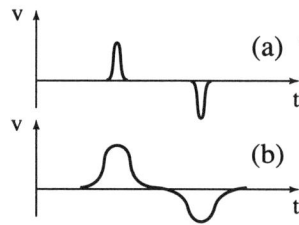

Figure 21.9 - Evolution of the shape of the read signal as a function of the written spatial frequency:
(a) low frequency
(b) high frequency

In the low frequency regime, the transitions are far from one another, and we measure voltage peaks with amplitude independent of the frequency, and width at half maximum equal, by definition, to that of an isolated transition (PW_{50}). With increasing frequency two phenomena occur. On the one hand, successive transitions start to overlap on writing, while the amplitude of the magnetization jump decreases. We thus approach a *sine-shaped* magnetization distribution, with *amplitude decreasing with f*. On the other hand, this effect superimposes on the read-mode response, as can be deduced from the transfer function given by equation (21.28).

For each value of the write current, the spectrum of the peak voltage vs spatial frequency, expressed as the number of flux reversals per mm of track length ($\mathrm{fr.mm}^{-1} = $ *flux reversal per mm*), is plotted. Such a spectrum is shown in figure 21.10 after J.M. Fedeli [26] the maximum operating frequency is usually taken to correspond to a decrease by 50% in the head sensitivity with respect to its low frequency value.

Figure 21.10 - Peak read voltage as a function of the written spatial frequency (after [26])

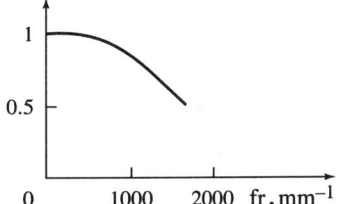

5.2. MAGNETORESISTIVE READOUT

The coupling between electric charge transport and magnetism is discussed in chapter 14, and *anisotropic magnetoresistance* in thin films, as well as *giant magnetoresistance*, which are of central interest here, are described in chapter 20.

Figure 21.11 shows the principle of a magnetoresistive sensor.

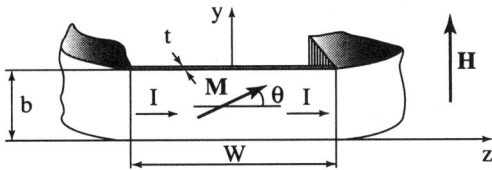

Figure 21.11 - Schematic description of the principle of a magnetoresistive sensor

The sensitive element is a soft magnetic thin film, typically a nickel iron alloy (Permalloy), with thickness t, width b (along Oy) and length W (along Oz). A current I is passed in the element (along the length W) and a voltage $V = \rho\,(W/b\,t)\,I$, proportional to the resistivity ρ of the alloy in the Oz direction of interest, is measured.

The film features a uniaxial anisotropy, with easy axis parallel to Oz (including in particular the shape contribution), and is characterised by the total anisotropy field H_A.

A first biasing field, small with respect to H_A, applied along Oz, is used to stabilise a single domain state. The field H to be measured is applied along Oy. It rotates the magnetization direction by an angle θ, leading to an induced component M_y. From chapter 20, we have $\sin\theta = H/H_A = M_y/M_0$ where M_0 is the spontaneous magnetization of the alloy. We also saw in chapter 20 that the *electrical resisitivity* of a ferromagnetic metal or alloy is different depending on whether it is measured parallel ($\rho_{//}$) or perpendicular ($\rho_\perp$) to magnetization. (The resistivity tensor has for its principal axes the magnetization direction and any two axes in the plane perpendicular to magnetization. Rigorously speaking, this is true only for amorphous or non-textured polycrystalline materials). Thus a uniaxial thin film will have resistivity ρ along the easy magnetization axis:

$$\rho = \rho_\perp + \Delta\rho\cos^2\theta = \rho_\perp + \Delta\rho\,(1 - H_y^2/H_A^2)$$
$$= \rho_{//} - \Delta\rho\,H_y^2/H_A^2 = \rho_{//} - \Delta\rho\,M_y^2/M_0^2 \qquad (21.29)$$

where the parameter $\Delta\rho = \rho_{//} - \rho_\perp$ is positive and typically a few percent of $\rho_{//}$ or $\rho_\perp$. This law leads to a parabolic variation of the resistance R of a magnetoresistive element submitted to a uniform field H_y (fig. 21.12-a), up to a saturation field equal to H_A. In practice, the variation is parabolic only for fields much smaller than the anisotropy field H_A due to the non-uniformity of the transverse demagnetising field (fig. 21.12-b). As a result, there appears an inflection point, around which the response to small field variations is linear. A simple method for linearization thus consists in biasing the element to the inflection point through a second small DC field along Oy.

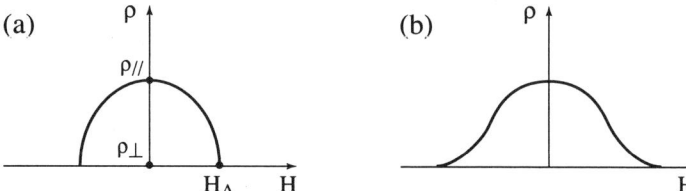

Figure 21.12 - Magnetoresistive effect in a thin film with uniaxial anisotropy
(a) theoretical behavior - (b) observed behavior

Another approach to linearization consists in passing the current at 45° to the easy magnetization direction, without biasing the film [30]. This technique is called "Barber pole" because it involves a pattern of highly conducting stripes (much better conductors than Permalloy!), which force the 45° direction for the current lines (fig. 21.13).

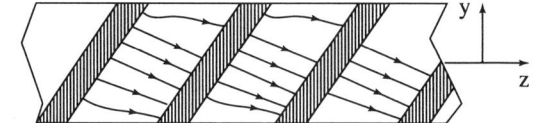

Figure 21.13 - "Barber pole" linearization technique

More information on biasing and linearization methods, and in particular those implemented in sensors for readout heads, is found in references [15, 31]. We now tackle the special application to readout heads. The field that must be detected is again the stray field of the recorded magnetic track. We place the magnetoresistive element as indicated in figure 21.14, the plane Oyz of the sensitive film being perpendicular to the axis Ox of the track.

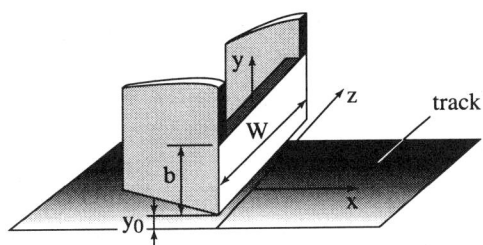

Figure 21.14 - Basic configuration of a magnetoresistive read head

When we discussed the inductive process, we saw that the stray field from the medium involves two components H_x and H_y. Component H_x is perpendicular to the sensitive field and produces only a very weak rotation of magnetization out of the Oyz plane. It can be neglected. We are thus interested only in measuring the field H_y parallel to the width b of the element. However, there appears an important difference with respect to the field sensor discussed above. Here H_y varies rapidly with the height y above the medium, and the element is therefore submitted to a highly inhomogeneous field.

Each elementary strip of the sensitive film, with width dy and length W, features a different variation in resistivity $D\rho(y)$. Due to the small values of the relative changes in resisitivity involved, it can be shown that the relative variation in resistance of the magnetoresistive element DR/R is given by:

$$\frac{DR}{R} = \frac{1}{b} \int_{y_0}^{y_0+b} \frac{D\rho(y)}{\rho} \, dy \qquad (21.30)$$

Here y_0 is again the head spacing, but the origin of the coordinates is taken on the magnetic track. We assume that the response of the magnetoresistive element is suitably linearised, so that we can write $D\rho(y)/\rho = \rho'M_y$. On the other hand, the sensitive film is also a *soft* magnetic layer ($H_A \ll M_s$). It therefore has high susceptibility, so that the *induction* B_y is practically $\mu_0 M_y$. Thus equation (21.30) can be written as:

$$\frac{DR}{R} = \frac{\rho'}{b\mu_0} \int_{y_0}^{y_0+b} B_y \, dy \qquad (21.31)$$

Imagine a coil, with axis Oy, involving n' windings per metre, with length b, is tightly wound around the magnetoresistive element. The flux Φ through the solenoid thus produced would be:

$$\Phi = n't \, W \int_{y_0}^{y_0+b} B_y \, dy \qquad (21.32)$$

Thus the relative variation in resistance of the magnetoresistive element can be expressed in the form:

$$DR/R = \rho'\Phi/(b\,W\,t\,\mu_0\,n') \tag{21.33}$$

Although it appears a bit artificial, this equation is valuable because it allows us once again to use the reciprocity theorem. Let $K_X(x, 0)$ be the x component of the field that would be created on the track if a current of 1 A were passed through the fictitious solenoid, with the soft yoke effect of the magnetoresistive film taken into account.

The flux induced in this fictive solenoid by the magnetization $M_x(x)$ of the written film (assumed to have small thickness h) is then:

$$\Phi = n'\mu_0 \int_{-\infty}^{+\infty} W\,h\,M(x)K_x(x,0)\,dx.$$

Whence:

$$DR/R = \rho'(h/bt)\int_{-\infty}^{+\infty} M(x)K_x(x)\,dx \tag{21.34}$$

This relation can be written in the equivalent form:

$$DR/R = \rho'(h/bt)\int_{-\infty}^{+\infty}\frac{dM_x}{dx}\,\Psi(x,0)\,dx \tag{21.35}$$

where $\Psi(x)$ is now the magnetic potential created by unit current density in the fictive solenoid. Note that here Ψ is, dimensionally speaking, a length.

In the simple configuration of figure 21.14, the field $K_x(x, 0)$ created by the fictive solenoid on the recording medium reduces, at least approximately, to that of two infinite lines of magnetic mass, with densities per unit length respectively $m' \sim -M_y\,t \sim -(M_0/H_A)n't$ at $y = y_0$, and $+(M_0/H_A)n't$ at $y = y_0 + b$, which leads to:

$$\Psi(x,0) = -\frac{M_0\,t}{4\pi H_A}\ln\left(\frac{x^2 + y_0^2}{x^2 + (y_0 + b)^2}\right) \tag{21.36}$$

In this formula, we recall that M_0 and H_A characterise the magnetoresistive film, while in equation (21.34), $M(x)$ is the longitudinal magnetization of the track. The signal produced by the passage of an abrupt transition from $-M_s$ to $+M_s$ (which gives a Dirac peak with content $2\,M_s$ for dM_x/dx) is simply (with $b \gg y_0$):

$$\frac{DR(x)}{R} \approx \frac{h}{b}\frac{\rho'M_0}{2\pi H_A}M_s\ln\left(\frac{x^2 + y_0^2}{x^2 + b^2}\right) \tag{21.37}$$

In this relation, x is the position of the transition. We check that the read signal, proportional to DR/R, is a peak centred on coordinate $x = 0$, with width at half maximum $2\,(b\,y_0)^{1/2}$. This width characterises the limitation in resolution due strictly to the magnetoresistive head. We note than the broadening due to the read process vanishes with the height, but, with the orders of magnitude for the state of the art in 1997 ($b = 0.6\ \mu m$ and $y_0 = 0.05\ \mu m$), it turns out to be $0.35\ \mu m$. This value remains above that characterising the best inductive readout heads (see preceding section).

However the situation changes radically if *magnetic shields* are associated with the magnetoresistive element.

The implemented configuration is similar to that of vertical thin film inductive heads, at least very near the medium (fig. 21.15-a). The two soft shielding films are perpendicular to the track axis, and define a very thin gap where the magnetoresistive element is placed. The sizes indicated on the figure are given as an indication of the state of the art in 1997, and it must be kept in mind that evolution is very fast in this area.

The likeness to an inductive head disappears as soon as we move further from the recording medium. The shields have limited height, it is not necessary to close the magnetic circuit, and of course even less necessary to include a coil!

The response of such a head can again be calculated using the reciprocity theorem. The magnetic potential $\Psi(x, 0)$ produced on the recording medium by a fictive solenoid with the dimensions of the sensitive element and carrying unit current, is used.

In this approach, we can use the same assumption as Karlqvist, viz that the distribution of magnetic potential between the plates, which is linear in the center of the structure, is conserved as far as the end of the gap, i.e. to the plane $y = y_0$. The potential Ψ_t on this plane thus has the shape indicated in figure 21.15-b. It is a trapezoid with height $\overline{\Psi_t}$ and width at half maximum $(1/2)(g + t)$ which can be decomposed into two Karlqvist potentials. Then the potential *at arbitrary height y* can be found analytically, using already established results (see § 4.1). The problem is thus reduced to determining $\overline{\Psi_t}$. This is done using the model of a line with distributed reluctance, similar to a tree-plate resistive line. More information can be found in the very detailed book by P. Ciureanu and H. Gavrila [15]. Here we just give some indications.

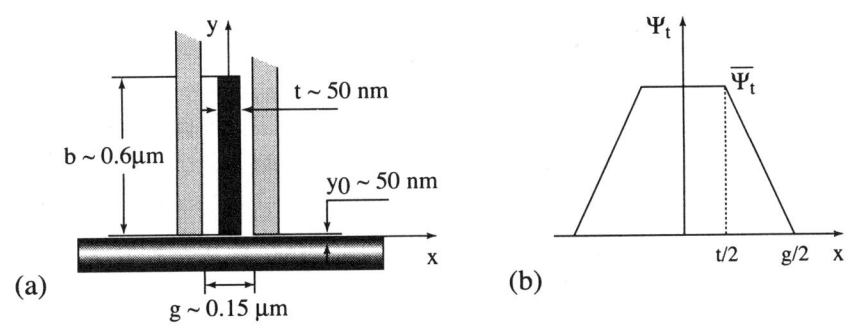

Figure 21.15 - Shielded magnetoresistive head

(a) geometry and axis definition
(b) shape of the magnetic potential induced by the fictive solenoid on plane $y = y_0$

In the distributed reluctance line model, the potential Ψ of the central plate (the sensitive film) and the flux Φ through it are functions of the single variable y, solutions to the differential equation system:

$$d\Phi / dy = -\Psi \Lambda' \; ; \;\; d\Psi / dy = n'I - \mathcal{R}'\Phi \qquad (21.38)$$

where $\mathcal{R}'$ is the reluctance per unit length of the central plane (the magnetoresistive film) and Λ' the permeance per unit length of the two half-gaps in parallel. n'I is the magnetomotive force per unit length associated with the fictive solenoid. It can be checked that $(\Lambda' \mathcal{R}')^{-1/2}$ is dimensionally a length, denoted as L_c.

If μ is the *intrinsic* permeability of the magnetoresistive film, then:

$$\mathcal{R}' = \frac{1}{\mu_0 \mu W t} \; ; \;\; \Lambda' = \frac{4\mu_0 W}{g - t} \; ; \;\; L_c = \frac{1}{2}[\mu t(g - t)]^{1/2} \qquad (21.39)$$

There exists a simple solution if the height b of the magnetoresistive element is small with respect to the characteristic length L_c. The variations of the potential and of the flux with y are then linear, and from symmetry we necessarily have $\Psi = 0$ at the centre of the element, i.e. at $y = b/2$. Then we can check that: $\overline{\Psi_t} = n'Ib/2 = n'b/2$ (since, from the very definition of Ψ, I = 1 A.

The potential on the exit plane of the gap is then a trapezoid with height n'b/2, with width at half maximum $(g + t)/2$, which can be used as the boundary condition for a combination of the two Karlqvist solutions, as we already explained. The result to be remembered is that, by adding shields, we recover readout pulses with width comparable to those of inductive heads.

But the significant advantage of magnetoresistive heads over their inductive counterparts is the *amplitude* of the output signal, and the fact that it is *independent of the head-track velocity*. Calculations show, and it is simple to check, that the readout signal corresponds practically to the saturation of magnetoresistance, in other words $DR/R \sim (\rho_{//} - \rho_\perp)/2\rho \sim 1\%$ to 2%.

The voltage picked up, $DV = I_0 DR = V_0 DR/R$ then depends only on the biasing voltage V_0 of the element. The latter is restricted by *thermal dissipation*. If P'_m is the maximum power that can be dissipated per unit area of the magnetoresistive film, then the optimum bias is: $V_0/W = (\rho P'_m)^{1/2} t^{-1/2}$.

An acceptable order of magnitude is $P'_m \sim 30 \, \mu W.\mu m^{-2}$ (we note that this corresponds to 30 MW.m^{-2}!). As ρ is of the order of 20 $\mu\Omega.cm$ in the alloys used, we have: $(\rho P'_m)^{1/2} = (2.5 \, V.m^{-1})^{1/2}$. With $t = 0.05 \, \mu m = 5 \times 10^{-8}$ m and $DR/R = 1\%$, we get $DV/W \sim 100 \, \mu V.\mu m^{-1}$.

This value is roughly ten times higher than for an inductive head. For given amplitude, we can thus in particular *sharply decrease the track width* W. For given bit length, this is another way of increasing the density of information per unit area of a disk or tape.

These heads also have considerable potential for progress through a decrease in the thickness of classical materials, and the implementation of new materials featuring giant magnetoresistance effects (see chap. 20).

6. CONCLUDING REMARK

We did not mention, in this chapter, one aspect of disk or tape storage that deserves a lot of attention from the designer although it is not magnetic. This is the problem of the mechanical interaction between the medium and the head, which move at relative speeds ranging typically from 0.5 to 5 $m.s^{-1}$.

The present trend is to work practically at contact. This was always the case for tapes, but not for hard disk storage, where the head was made to literally fly at a fraction of a micrometer above the disk surface. The friction problems now become more serious than the aerodynamical problems. The interested reader can in particular look up reference [32].

REFERENCES

[1]   S.H. CHARAP, PU-LING LU, YANJUN HE, Thermal stability of recorded information at high densities, *IEEE Trans. Magn.* **Mag-33** (1997) 978.

[2]   D.A. THOMSON, J.S. BEST, The future of magnetic storage technology, *IBM Journal of Research and Development* **44** (2000) 311.

[3]   Y. SONOBE, Y. IKEDA, Y. TAGASHIRA, Composite perpendicular recording media consisting of CoCrPt with large Hk and CoCr with positive inter particle interaction, *IEEE Trans. Magn.* **Mag-35** (1999) 2769.

[4]   S.Y. CHOU, Patterned Magnetic Nanostructures and Quantised Magnetic Disks, *Proc. IEEE* **85** (1997) 652.

[5]   Z.G. WANG, Y. NAKAMURA, Spin tunneling random access memory (STRAM), *IEEE Trans. Magn.* **32** (1996) 4022.

[6]   M. JULLIERE, *Phys. Lett.* **54A** (1975) 225.

[7]   J. ASHLEY, M.P. BERNAL, G.W. BURR, H. COUFAL, H. GUENTHER, J.A. HOFFFNAGLE, C.M. JEFFERSON, B. MARCUS, R.M. MCFARLANE, R.M. SHELBY, G.T. SINCERBOX, Holographic storage technology, *IBM Journal of Research and Development* **44** (2000) 341.

[8]   P. VETTIGER, M. DESPONT, U. DRECHSLER, U. DüRIG, W. HÄBERLE, M.I. LUTWYCHE, H.E. ROTHUISEN, R. SCHUTZ, R. WIDMER, G.K. BINNIG, The "Millipede" - More than thousand tips for future AFM storage, *IBM Journal of Research and Development* **44** (2000) 323.

[9]   J.C. MALLINSON, The foundation of magnetic recording (1987) Academic Press, New York-London.

[10]   C.D. MEE, E.D. DANIEL, Magnetic recording (1990) McGraw-Hill, New York.

[11]   J.J.M. RUIGROK, Short wavelength magnetic recording (1990) Elsevier, New York.

[12]   H.N. BERTRAM, Theory of magnetic recording (1994) Cambridge Univ. Press.

[13]   W.K. WESTMIJZE, *Philips Res. Rep.* **8** (1953) 148, article reproduit dans [2].

[14]   R.M. WHITE, Introduction to magnetic recording (1986) IEEE Press, Piscataway, NJ.

[15]   P. CIUREANU, H. GAVRILA, Magnetic heads for digital recording, *Studies in electrical engineering* **39** (1990), Elsevier, New York.

[16]   S.B. LUITJENS, *in* High density digital recording, *NATO ASI Series E, Applied Sciences* **229** (1993) chap. 8, 217, K.H.J. Buschow, G. Long & F. Grandjean Eds, Kluwers Academic Publishers, Dordrecht.

[17]   T. SUZUKI, *MRS Bulletin* **21** (1996) 42.

[18]   M. MANSURIPUR, The physical principles of magneto-optical recording (1995) Cambridge Univ. Press.

[19]   A.H. ESCHENFELDER, Magnetic bubble technology, *Springer Series in Solid State Sciences*, Vol. 14 (1980), Springer Verlag, Berlin-New York.

[20]   D.F. EAGLE, J.C. MALLINSON, *J. Appl. Phys.* **38** (1967) 995.

[21]   J.E. KNOWLES, *IEEE Trans. Magn.* **Mag-16** (1980) 62.

[22]   J.L. DORMANN, D. FIORANI, E. TRONC, *Adv. Chem. Phys.* **98** (1997) 283.

[23]   T.C. ARNOLDUSSEN, Film media, *in* Magnetic Recording Technology, 2nd ed. (1995) chap. 4, C.D. Mee & E.D. Daniel Eds, McGraw-Hill, New York.

[24]   K.H.J. BUSCHOW, *in* High density digital recording, *NATO ASI Series E, Applied Sciences*, **229** (1993) chap. 12, 355, K.H.J. Buschow, G. Long & F. Grandjean Eds, Kluwers Academic Publishers, Dordrecht.

[25]   O. KARLQVIST, *Trans. Royal Inst. Techn. Stockholm* **86** (1954), article reproduit dans [2].

[26]   J.M. FEDELI, *in* High density digital recording, *NATO ASI Series E, Applied Sciences*, **229** (1993) chap. 9, 251, K.H.J. Buschow, G. Long & F. Grandjean Eds, Kluwers Academic Publishers, Dordrecht.

[27]   R.E. JONES, W. NYSTROM, *U.S. Patent* 4 190 872 du 26 février 1980.

[28]   I. ELABD, *IEEE Trans. Audio* **Au-11** (1963) 21.

[29]   M. WILLIAMS, R. COMSTOCK, *AIP Conf. Proc.* **1,** No 5 (1971) 738, article reproduit dans [2].

[30]   K.E. KUIJK, W.J. VAN GESTEL, F.W. GORTER, *IEEE Trans. Magn.* **Mag-11** (1975) 1215.

[31]   J.C. MALLINSON, Magnetoresistive heads. Fundamentals and Applications (1996) Academic Press, New York-London.

[32]   BHARAT BHUSHAN, *in* High density digital recording, *NATO ASI Series E, Applied Sciences*, **229** (1993) chap. 10, 281, K.H.J. Buschow, G. Long & F. Grandjean Eds, Kluwers Academic Publishers, Dordrecht.

[33]   C.D. MEE, E.D. DANIEL, Magnetic recording (1987) McGraw-Hill, New York.

[34]   A.S. HOAGLAND, J.E. MONSON, Digital magnetic recording, 2nd ed. (1991) John Wiley & Sons, New York.

[35]   J.J.M. RUIGROK, Short wavelength magnetic recording (1990) Elsevier, New York.

[36]   J.C. MALLINSON, The foundation of magnetic recording (1987) Academic Press, New York-London.

CHAPTER 22

FERROFLUIDS

A ferrofluid is a suspension of small magnetic particles in a carrier liquid. We begin by discussing the characteristics of a ferrofluid, in particular the criteria for the stability of such a suspension, as well as methods of preparation. We then go on to describe the properties and the effects of a magnetic field: superparamagnetism, dipolar interactions leading to the formation of chains, birefringence, viscosity....We also present a certain number of applications which, essentially, bring into play the effects of a magnetic field: confinement in field gradients, variations in permeability of the material through deformation of the fluid, variation of birefringence or of viscosity. We conclude by presenting a spectacular effect specific to ferrofluids which arises from the liquid state of this material: surface instabilities in the presence of a magnetic field.

1. INTRODUCTION

A ferrofluid can be called a magnetic liquid. In reality, however, it is actually a colloidal suspension of small magnetic particles in a carrier liquid.

In general, magnetic materials have a Curie temperature very much lower than their melting temperature. Helium 3 can, of course, be polarised at less than 2.7 mK , and might be considered to be a magnetic liquid, but at this temperature it does not lend itself to practical applications. Liquid oxygen, which is paramagnetic, might also be considered to be a magnetic liquid, and has been used to reveal magnetic domains at low temperatures. More recently, a long range magnetic order has been evidenced in alloys of cobalt-palladium, around the composition $Co_{80}Pd_{20}$, in the under-cooled liquid state [1, 2]. In this case the material is brought to a metastable liquid state below its melting temperature by quenching.

In practice, for a material to share both liquid and magnetic properties, magnetic particles are dispersed in a liquid. The idea is an old one: it is the *Bitter powder* used since the 1930s to reveal magnetic domain walls. When the liquid is placed on the polished surface of a magnetic sample subdivided into domains, the particles are attracted by the fringing magnetic fields and decorate the domain walls. In the 1940s magnetic fluids for use in brakes and clutches were prepared using powdered iron in oil, the particles being of the order of μm or more. Such liquids are, however, unstable (the particles either sink or agglomerate), and when a magnetic field is applied they solidify. Only in the 1960s did the knowledge of how to make what are called stable ferrofluids, using particles 3 to 15 nm in size, become available. They remain liquid even when subjected to intense magnetic fields, and display a magnetic susceptibility sufficiently strong for them to behave like magnetic liquids.

This chapter offers a general outline of these materials, their properties, and their applications. For further information the reader may refer to a very thorough work of reference by R.E. Rosensweig [3], as well as to an article that appeared in French in the journal *La Recherche* [4], which offers an excellent presentation of ferrofluids and contains a more specialised bibliography. Finally, every three years since 1983 the proceedings of the "International Conference on Ferrofluids" [5] have given a very comprehensive list of publications on ferrofluids, as much on the fundamental as on applied aspects, and patents. The volumes for the years 1993 and 1995, in particular, contain review articles concerned with applications.

2. CHARACTERISTICS OF A FERROFLUID

As has been stated, a ferrofluid is composed of small magnetic particles suspended in a carrier liquid.

2.1. STABILITY

One of the characteristics of a good ferrofluid is its stability:
- stability vis-à-vis gravitational forces: the particles must not settle,
- stability vis-à-vis magnetic field gradients: the particles must not cluster in regions where the field is intense,
- stability vis-à-vis the agglomeration of particles under the effet of dipolar forces or Van der Waals type interactions.

The necessary conditions for this stability lead first to a criterion regarding the size of the particles. They must be sufficiently small for the thermal agitation, the brownian motion of the particles, to oppose settling or concentration in a magnetic field gradient. One can obtain an order of magnitude for the acceptable size of particles by comparing the energy terms in play: thermal energy: $k_B T$, gravitational energy: $\Delta\rho \, V \, g \, l$, magnetic energy: $\mu_0 \, M_p \, H \, V$, where k_B is the Boltzmann constant, T the absolute temperature, $\Delta\rho$ the difference in density between the particles and the liquid, V the volume of the particles, g the acceleration of gravity, l the height of liquid in the gravitational field, μ_0 the permeability of free space, M_p the magnetization of the particles and H the magnetic field.

The criterion for stability with regard to gravitational forces is obtained by writing:

$$k_B T / (\Delta \rho \, V \, g \, l) \geq 1 \qquad (22.1)$$

Assuming spherical particles of diameter d, a $\Delta \rho$ of $4,300 \ \text{kg.m}^{-3}$ (typical of magnetite Fe_3O_4), a container 0.05 m in height, and a temperature of 300 K, one obtains $d \leq 15$ nm.

To estimate the size of the particles that ensure stability in the presence of a field gradient, it is assumed that the magnetic energy $\mu_0 M_p H V$ corresponds to the work performed to move a particle of magnetization M_p in the fluid, from a region where the field has a value H to a region where the field is zero. One then has:

$$k_B T / (\mu_0 M_p H V) \geq 1 \qquad (22.2)$$

Taking a magnetization of $4.46 \times 10^5 \ \text{A.m}^{-1}$ (5,600 G, the value for magnetite), and a maximum field of $8 \times 10^4 \ \text{A.m}^{-1}$ (0.1 T), one obtains $d \leq 6$ nm.

The stability in field gradients thus seems to be the more demanding factor, leading to particles of size less than 10 nm. These criteria for stability assume that the particles remain small, in other words that they do not agglomerate. But these are small dipoles, and the dipolar interactions tend to cause them to agglomerate. In the same way, at very short distances the Van der Waals force between particles is attractive. The thermal energy needed to oppose the agglomeration of dipolar origin has the same order of magnitude as that which opposes sedimentation. However, agglomeration of Van der Waals origin is irreversible since the energy required to separate two particles, once agglomerated, is very large. Consequently it is necessary to find a way of preventing the particles from getting too close to each other. This can be done:

♦ either be coating the particles with a polymer layer to isolate one from the other. These are surfaced ferrofluids, the polymer in question being a surfactant.

♦ or by electrically charging the particles, which will then repel because of the Coulomb interaction: these are the ionic ferrofluids.

2.2. TYPES OF FERROFLUIDS AND THEIR PRODUCTION

2.2.1. Surfaced ferrofluids

The surfactant is made up of polymer chains analagous to soap molecules, one end of which adsorbs on the surface of the magnetic particles while the other end has an affinity with the carrier liquid. The particles are thus coated with a layer of polymer which keeps them a certain distance apart. This type of ferrofluid is obtained by milling a coarse powder, generally of magnetite (the grains being of the order of μm in size), in the presence of the surfactant. This operation can take a very long time, up to 1,000 hours. It is the presence of the surfactant during the milling process that makes such a large reduction in size (down to 10 nm) possible, and which results in each grain being covered by a single polymer layer. Using this method it is possible to use different solvents (carrier liquids) such as water, or hydrocarbons such as kerosene. This type of ferrofluid is the most common, and is marketed by Ferrofluidics®.

2.2.2. Ionic ferrofluids

In this type of ferrofluid the magnetic particles carry an electrical charge. These are, in a way, large ions which maintain a certain distance from one another by electrostatic repulsion. In reality this repulsion is partially screened by the ions of opposite signs present in the solution to ensure its neutrality. Nevertheless the magnetic particles repel each other at short distance. These ferrofluids are obtained by a process of precipitation from dissolved iron salts, in general ferrous chloride and ferric chloride, the proportion of which determines the size of the particles. This method of preparation is the simplest, the quickest and works better than reducing the size by milling. By adding surfactants as well it is possible to use a variety of carrier liquids: water, oil, organic solvents… This type of ferrofluids correspond to the French line, as developed by Massart [6].

3. PROPERTIES OF FERROFLUIDS

3.1. SUPERPARAMAGNETISM

The magnetic particles in a ferrofluid have a typical size of the order of 10 nm, and are therefore single domain particles. For the materials currently used to make ferrofluids this size is less than, or of the order of, the thickness of the domain wall, and consequently there is no space in a grain for such a wall. Even for those materials with a very strong anisotropy, resulting in domain walls some nm thick, a subdivision into domains would be too costly in energy terms. The grains are, therefore, small magnets suspended in the carrier liquid.

Due to thermal agitation, the grains undergo a brownian motion which moves them in all directions (which –as we have seen– prevents sedimentation of the ferrofluid), and which also continuously disorients them: their orientation fluctuates in all directions, and in zero field an average resultant magnetization is not observed. The ferrofluid behaves like a paramagnetic material: it is called a superparamagnet because the particles are, in a way, carriers of super-large magnetic moments. In a magnetic field the moments still fluctuate, but on average the net magnetization in the direction of the field is no longer zero. The ferrofluid can be magnetized, and in a sufficiently high field it can reach saturation when all the particles are aligned (see § 2.3 in chap. 4). Assuming that the particles are noninteracting, one can describe the variation of magnetization of a ferrofluid as a function of the field by a Langevin law (eq. 4.20).

If the magnetization of the particle material is $\mathbf{M_p}$, the magnetic moment of a particle of volume V is: $m = \mathbf{M_p} V$. When the magnetic moment makes an angle Θ with the applied field the particle is subjected to a torque: $\Gamma = \mu_0 \, m \, H \sin \Theta$, and the magnetic energy of the particle in the field is given by:

$$E_H = -\mu_0 \, m \, H \cos \Theta \qquad (22.3)$$

In zero field, the orientation of the particles is uniformly distributed, and the probability density of finding a particle the magnetic moment of which makes an angle Θ with a given direction is equal to:

$$p(\Theta) = (1/2) \sin \Theta \qquad (22.4)$$

When a magnetic field is applied, the probability density for the moment to make an angle Θ with the direction of the field is, in addition, proportional to the Boltzmann factor $\exp(-E_H / k_B T)$. Only the average of the component of the moment in the direction of the field is not zero, and is given by (see eq. 4.18):

$$\langle m \cos \Theta \rangle = \frac{\int_0^\pi m \cos \Theta \exp\left(\dfrac{\mu_0 \, m H \cos \Theta}{k_B T} \right) \dfrac{\sin \Theta}{2} \, d\Theta}{\int_0^\pi \exp\left(\dfrac{\mu_0 \, m H \cos \Theta}{k_B T} \right) \dfrac{\sin \Theta}{2} \, d\Theta} \qquad (22.5)$$

By setting $x = \mu_0 \, m H / k_B T$, one obtains:

$$<m \cos \Theta> = m \, (\coth x - 1/x) \qquad (22.6)$$

which is Langevin's function. The magnetization of the ferrofluid is equal to the average moment per unit volume:

$$\mathbf{M_f} = <m \cos \Theta> \Phi / V = \Phi \, \mathbf{M_p} \, (\coth x - 1/x) \qquad (22.7)$$

with: $\qquad\qquad x = \mu_0 \, m \, \mathbf{H} / k_B T = \mu_0 \, \mathbf{M_p} \, V H / k_B T \qquad (22.8)$

$\mathbf{M_p}$ is the saturation magnetization of the material constituting the magnetic particles, V the volume of the particles, and Φ is the fraction of the volume occupied by the grains. This is Langevin's law which describes the magnetization of a paramagnetic substance made up of non-interacting particles as a function of the field H and of the temperature. The saturation magnetization of a ferrofluid is, therefore, equal to $\Phi \mathbf{M_p}$, and its initial susceptibility, obtained by the expansion of the Langevin function for small x is:

$$\chi_i = \lim_{H \to 0} (M_f / H) = (\Phi \mu_0 M_p^2 \, V) / (3 \, k_B T) \qquad (22.9)$$

As has been stated, the orientation of magnetic moments fluctuates because of the brownian motion of the particles. In fact there is an additional source of fluctuation in this orientation, this is the rotation of the moment within the particle. In the case of a particle with uniaxial anisotropy, the moment can fluctuate between the two opposite easy magnetization directions by crossing the energy barrier linked to the anisotropy. This is a thermally activated phenomenon. This fluctuation mechanism, called Néel's mechanism, can co-exist with the fluctuation in the orientation of the particles associated with the brownian motion. The time constants associated with these two mechanisms depend on the size of the grains, the anisotropy constant, the viscosity of the carrier liquid, and the temperature. The characteristic times which describe, for example, the relaxation of the magnetization when the field is switched off abruptly are given by:

$$\tau_B = 3\,V\eta/k_BT \qquad (22.10)$$

for the brownian mechanism, where V is the volume of the particles, η is the viscosity of the carrier liquid, and:

$$\tau_N = f_0^{-1}\exp(KV/k_BT) \qquad (22.11)$$

is the relaxation time for the Néel mechanism, where K is the anisotropy constant, and f_0 the attempt frequency for the particle to cross the energy barrier, typically of the order of 10^9 Hz. KV corresponds to the height of the energy barrier to be overcome for the magnetization of a particle to reverse. When $\tau_N \ll \tau_B$, one speaks of intrinsic superparamagnetism, and when $\tau_B \ll \tau_N$, of extrinsic superparamagnetism. Typically τ_N is of the order of 10^{-9} s and τ_B of the order of 10^{-7} s. By altering the parameters one can change the regime, and by lowering the temperature one will meet the blocking temperatures of the two mechanisms successively. At low temperature the system is magnetically frozen, and will exhibit glassy behavior, in the sense of spin glasses, with all the associated relaxation phenomena.

3.2. INTERACTIONS BETWEEN PARTICLES: FORMATION OF CHAINS

The single domain magnetic particles of a ferrofluid act on one another through the dipolar interaction. There is no exchange interaction (it is of very short range) between particles, and so, here, the dipolar interaction is not negligible if the ferrofluid is not too dilute. The interaction energy between two dipoles m_1 and m_2 is, according to equations (2.24) and (2.81), given by:

$$E_{dip} = \frac{\mu_0}{4\pi}\left[\frac{m_1 m_2}{r^3} - 3\frac{(m_1\,\mathbf{r})(m_2\,\mathbf{r})}{r^5}\right] \qquad (22.12)$$

where $\mathbf{r}$ is the vector which connects the two dipoles. The energy of a system consisting of two dipoles is minimum when they are in contact and are oriented parallel to $\mathbf{r}$. In this case, for spherical particles of volume $V = \pi d^3/6$ and of moment $m = M_p V$, the energy is equal to:

$$E_{dip} = -\mu_0 M_p^2 V/12 \qquad (22.13)$$

The ratio between thermal energy and dipolar energy per particle, $24\,k_BT/\mu_0 M_p^2 V$, defines, therefore, the size of the particle, for a given magnetization, such that thermal agitation opposes agglomeration due to dipolar interaction. This criterion regarding particle size is fulfilled in zero field for conventional ferrofluids, where the particles repel one another at short distance. The thermal fluctuations are all the more efficient when the moment fluctuates within the particle, and the orientation of the particle also fluctuates.

When a field is applied, the moments have a tendency to align with the field, the thermal fluctuations are less efficient, and the particles form chains in the direction of the field, the higher the field, the longer the chains (in fact, even in zero field the particles can form chains, but they are short and of random orientation). This effect of chain formation in a field is reversible, i.e. the chains break up when the field is brought to zero.

The magnetization of a ferrofluid in which particles do not interact has been described by a Langevin function, and an initial susceptibility varying as $1/T$. One way of taking the dipolar interactions into account is to express them by a mean field that acts on a given particle. This field, which is produced by the rest of the particles in the fluid, is added to the applied field. This approach, applied to the case of frozen ferrofluids, and more generally to small particles undergoing dipolar interaction in a solid matrix, results in a Curie-Weiss type behavior, that is to say in an initial susceptibility of the form $1/(T - T_0)$ (and no longer $1/T$) where T_0 is an ordering temperature. The nature of this order is still controversial.

In the case of a ferrofluid in a liquid state it is also necessary to take into acount the spatial correlations, due to the fact that particles change position, and form chains. The mean fields representing the interactions should no longer be those estimated for the isolated particles, monodisperse (all of the same size), and uniformly distributed. Another approach, which tries to take these effects into account, is numeric simulation of the Monte-Carlo type [7, 8]. The energy of the system is written as:

$$E = E_r + E_{dip} + E_H \tag{22.14}$$

where E_r is a term representing short-range repulsion, E_{dip} is the dipolar energy of a particle interacting with its neighbours, and E_H the energy term related to the magnetic field. At a given temperature T the particles change position, and orient themselves in such a way as to minimise the energy while at the same time obeying the Boltzmann distribution, $\exp(-E/k_B T)$. This type of simulation demonstrates the presence of short and disoriented chains in zero field, the formation of longer chains in the direction of the applied field when it is no longer zero, and an initial susceptibility of the Curie-Weiss type varying as $1/(T - T_0)$.

3.3. VISCOSITY

As we have already mentioned, in contrast to the original magnetic liquids which were composed of micron size iron particles in oil, a ferrofluid remains liquid in the presence of a magnetic field, even when it is magnetized to saturation. On the other hand, its rheologic properties are modified by the field, and its viscosity is, therefore, an important characteristic.

The viscosity of a ferrofluid is governed by that of the carrier liquid. The possibility of choosing different solvents (water, kerosene, oil…) permits the choice of a range of viscosities. The presence of the particles in the liquid increases the initial viscosity, even in the absence of a magnetic field.

When a magnetic field is applied and the ferrofluid is subjected to a shearing force, the particles tend to remain aligned with the field. The velocity gradients around the particles in the fluid, and hence the viscosity, are then increased. When the *vorticity* of the fluid is parallel to the applied field, the particles can rotate freely, and the field has no effect on viscosity. By contrast, if the field and the vorticity are perpendicular to each other, the increase in viscosity due to the field is maximal.

3.4. OPTICAL BIREFRINGENCE

The magnetic particles of a ferrofluid have a linear birefringence (rectilinear in the notation of chapter 13), that is to say they show a different optical index according to whether the light is polarised parallel or perpendicular to the easy direction of magnetization. In other words, the propagation velocities parallel or perpendicular to the easy axis of magnetization are not equal. A weak dichroism, that is a difference in the absorption coefficient between a light polarised parallel or perpendicular to the easy magnetization axis, may also be present. However this effect is very weak, and it is essentially the birefringence that is operative.

In zero field the orientations of the particles are distributed uniformly within the ferrofluid, and hence the average birefringence is zero. When a field is applied in a given direction a certain degree of orientation of the particles is obtained, and the birefringence is no longer zero on average.

In the case of zero birefringence ($n_{//} - n_\perp = 0$), a linearly polarised light will remain so after crossing the material. By using an analyser crossed with respect tot the polariser one can stop transmission of the light. In the case of non-zero birefringence induced by a magnetic field, light linearly polarised along one of the principal directions of the birefringence (parallel or perpendicular to the field) will also remain so polarised. By contrast, light polarised linearly in another direction will become elliptical after crossing the sample, and, under the preceding conditions of crossed analyser and polariser, the transmitted light will not be extinguished. The effect of a magnetic field can thus be made visible, since the ferrofluid allows light to pass through it when it becomes magnetized while it does not in zero field.

A ferrofluid is, in general, very absorbing for visible light, and, to display this optical property, it is necessary to work either with thin layers of ferrofluids or with highly dilute materials. Recently ferrofluids based on magnetic garnets (YIG) have been produced [9]: these garnets are insulators and, when sufficiently thin, are transparent to visible light, one might therefore expect interesting optical properties from them.

4. APPLICATIONS

One of the benefits of ferrofluids lies in their applications, actual or potential, arising from their liquid and magnetic character. These applications take advantage of the fact that the material is deformable, and hence can adopt the desired shape. It may be maintained in place, or moved about, by means of magnetic field gradients. The properties are dependent upon a magnetic field which, furthermore, acts at a distance. Being made up of small particles in a variety of solvents, the material can equally be incorporated into various types of substances or materials. Some of these applications have reached the industrial stage but many remain at the prototype or even theoretical level. Clearly, research and development into new uses of ferrofluids will continue to be a very active field.

4.1. LONG-LIFE SEALS

The most common, and really commercial, use of ferrofluids, is in the realization of seals for rotating shaft systems, in particular those operating at high speed. The ferrofluid is maintained in place around an axis by ring-shaped permanent magnets. The carrier liquid is oil based, and is, therefore, a lubricant. The fact that there is no wear and tear, and hence no dust particles in the system, confers upon these seals a very long life, more than 10 years. These seals are also very effective.

4.2. LUBRICATION - HEAT TRANSFER

Certain applications exploit the fact that a coating of ferrofluid can be held, in a tube for instance, by permanent magnets. This film of ferrofluid can ensure lubrication or improve the flow of a fluid, and consequently the transport of heat. The latter can also be modified by controlling convection by a magnetic field, since a field acts on the viscosity, and a field gradient and temperature affect the magnetization, hence the forces involved in the process of convection.

4.3. PRINTING

A printer using a jet of ferrofluid has been made: using an array of permanent magnets a kind of extrusion is produced in the ferrofluid. An electric field pulse then tears off a droplet of ferrofluid, which attaches to paper placed between the ferrofluid and the electrode. This printer is comparable in performance to a traditional laser printer.

4.4. ACCELEROMETERS AND INCLINOMETERS

Certain devices are based on the principle that a film of ferrofluid can be held fixed by magnets, but remains deformable under the influence of acceleration or of gravity.

These are accelerometers and inclinometers. The deformation of the fluid leads to a modification of the permeability of the medium between the measuring coils, which can be detected by the difference in inductance between the coils, as shown in figure 22.1.

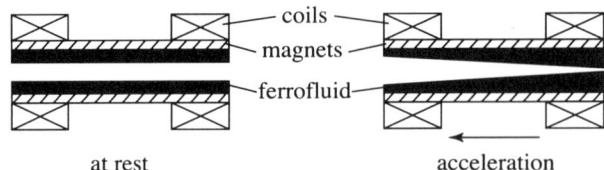

at rest   acceleration

Figure 22.1 - Schematic diagram describing the principles of an accelerometer

4.5. POLISHING

The polishing process is an area to which ferrofluids can bring improvement. The abrasive particles incorporated into a ferrofluid float freely in it, thus allowing the active region to be fed continously with fresh particles. Furthermore, the polishing tool itself floats on the ferrofluid, resulting in a compression force between the tool and the material to be polished. This force may be adjusted by use of a magnetic field which gives control over the quantity of material to be removed by the polishing process.

4.6. DAMPERS AND SHOCK ABSORBERS

As in the case of the polishing process, inertial shock absorbers use the lift or levitation exerted by a ferrofluid on a body, magnetic or non-magnetic. This effect is controlled by the magnetic field. For loudspeakers of the moving coil type, for example, the ferrofluid stabilises the movement of the coil thereby reducing sound distortion. Other shock absorbers make use of the fact that the viscosity of the ferrofluid is modified by a magnetic field: their damping ability can be adjusted to the load or the roughness of the terrain by controlling viscosity using the field.

4.7. APPLICATIONS OF THE OPTICAL PROPERTIES OF FERROFLUIDS

As we have seen, when a field is applied, the ferrofluid is magnetized and becomes birefringent. When the polariser and analyser are cross positioned light may pass through them, while in zero field it cannot. This phenomenon can be exploited in magnetic field detectors or in light modulators, and may also be used in a more indirect way to produce a viscosimeter. The principle is to dilute ferrofluid in a liquid whose viscosity is to be measured, and to determine the law governing the variation of the transmitted light over time after the magnetic field has been abruptly cut off. This law is exponential, and its time constant is directly related to the viscosity of the fluid.

4.8. BIOMEDICAL APPLICATIONS

Currently a whole range of applications in medical and biomedical fields is in the process of development (see also chap. 25). Firstly there is the possibility of improving the contrast in medical imaging, for NMR or other systems. Not only can the signal be improved but also the contrast between tissues. To do this a knowledge of how to prepare non-toxic ferrofluids, and how to direct them to the relevant sites is required.

The transport of biologically active molecules is, thanks to encapsulated ferrofluids, also being explored, even though specific biological receptor systems, for tumors for example, already exist and are in current use. The marking of cells by magnetic particles, in order to establish their location or to separate them, still causes problems, particularly because of the particle size distribution in the ferrofluids, which gives rise to uncertainty about the quality of marking. The difficulty encountered in placing the ferrofluid very selectively on the cells to be marked has not yet been fully solved, but work is progressing in this field.

This possibility of achieving the separation of cells or particles is also being explored in fields other than biology. The effect of a magnetic field gradient associated with the gravimetric method increases the ease and efficiency of separation of small metallic chips.

Very specific applications of ferrofluids are already in use; for example in the localised heat treatment of tumors, whereby the ferrofluid is directed onto and fixed to the tumor, and where an alternating field can be used to provide localised heating by a process of dissipation inside the ferrofluid.

4.9. FORESEEABLE DEVELOPMENTS

In general terms new applications are expected to appear because it is possible to incorporate ferrofluids into numerous materials, those of a biological nature, as has already been stated, but also liquid crystals, polymers, micro-capsules of phospholipids (the properties of which are close to those of biological membranes), etc.

Limitations, as for example a weak susceptibility or a weak saturation magnetization, tend to disappear. For example ferrofluids have been produced using Fe_3N particles with a saturation magnetization of 1.86×10^5 A.m^{-1} (2,330 G), and an initial specific susceptibility of 160, which is high and much larger than that expected from Langevin's law. The interactions between particles must play a role in this increased susceptibility [10].

The lower limits of particle size have also been pushed back: it is now known how to produce particles of maghemite (γ-Fe_2O_3) less than 3 nm in diameter [11].

5. SURFACE INSTABILITIES

Among the phenomena observed in ferrofluids, one of the most spectacular and most widely studied is related to their surface instabilities. This phenomenon can cause problems in certain applications. For example, a seal may be destroyed if it is the site of an instability triggered by the vibration of the shaft. For this reason, in addition to its fundamental interest, this surface instability is the subject of detailed studies in order to overcome the problem.

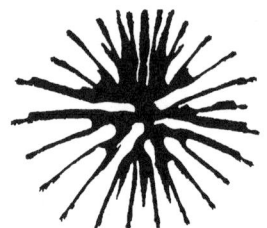

Figure 22.2 - Digitation of a droplet of ferrofluid in a Hele-Shaw cell, subjected to a field normal to the cell, from [4]

It shows up initially on a drop of ferrofluid held with a non-miscible fluid between two plates (a Hele-Shaw cell), and put into a magnetic field perpendicular to the plates. Depending upon the amplitude of the suddenly applied field the droplet takes on various shapes, going from a simple deformation to highly branched shapes reminiscent of the *digitation* obtained when one viscous liquid is pushed into another.

The problem is to understand the formation of these branches, their number and their length. This is interesting because these complex forms observed in very diverse systems, present a certain generality, but it is difficult because at the basic level these shapes are sensitive to the initial conditions.

The instability also manifests itself on the free surface of a volume of ferrofluid subjected to a field normal to this surface. Above a critical field the surface bristles into spikes which, apart from a few defects, form a triangular lattice whose period depends on the field, and is smaller the higher the field. A thin film of ferrofluid will exhibit the same phenomenon as a free surface, but instead of surface spikes one observes modulations in the thickness of the film. These modulations also organise into a lattice whose period depends on the applied field.

It is very interesting to see in these systems, as in the case of magnetic domain structures, how the presence of long range interactions and of competing interactions drives a homogeneous system, subjected to a uniform magnetic field, to spatially structure, with a characteristic length determined by a ratio of competing energies.

All these instabilities are the result of competition between the energy terms in the system: stabilizing terms such as gravity or surface tension, and destabilizing terms related to the demagnetizing field, and hence to the magnetization of the ferrofluid in the applied field. Beyond a threshold field the destabilizing terms dominate, and the most unstable mode is selected. A linear analysis of the stability of the initial shape (circular for the droplet, flat for the surface or the film) allows the prediction of this most unstable mode for a given field, and hence the number of branches in the case of the droplet, or the period of the lattice in the case of the surface or the film. More precise predictions about the shape or the geometry adopted cannot be obtained analytically because of the non-linear character of the equations involved. Numerical approaches, however, remain possible.

There is a strong analogy between these systems and magnetic domain structures. The energy associated with surface tension and gravity plays the role of the domain wall energy. The characteristic length, the period of the spikes or the modulation of the film thickness, is analogous to the domain size, which results from the compromise between wall energy and magnetostatic energy.

The degrees of freedom of the two systems are, however, different. In the case of domains it is their size and shape that allows the energy to be minimized, taking into account the initial conditions of the system. In the case of the spikes or the modulations in the thickness of a film of ferrofluid, it is the value of the magnetization (the fluid is superparamagnetic) and the shape of the sample that can adjust themselves.

For example, the modulations in thickness locally reduce the demagnetizing field, allowing the magnetization of the ferrofluid, and hence the absolute value of the magnetic energy term in an applied field, $-M_f H$, to be increased. As in the case of magnetic domains, despite the fact that the system is fluid, one observes hysteresis in the evolution of the spatial structures when the magnetic field, which is the controlling parameter, is modified. The system can be trapped in a period of the peaks or modulations corresponding to the initial field. If the field is increased suddenly, the system will not be able to adopt the smallest period corresponding to the new value of the field. It will then resolve the energy compromise by invoking another parameter, such as the height of the modulations, or in creating disorder and distributions in the size of modulations. For certain values of the field, the system can form an inhomogeneous lattice having the same period as the initial lattice, with a superstructure where the initial modulations are decorated by much smaller modulations [12], see figure 22.3. Thus, as in the case of domains, this trapping leads to a great wealth of observed structures.

Recently, experiments with a ferrofluid confined in a Hele-Shaw cell with a non-miscible liquid have been carried out [13]. All the geometries observed for magnetic domains in an epitaxic garnet film (bubble domain material) were encountered: parallel stripes, mazes, bubble lattices, cellular structures analogous to the soap froth in 2D, and so on.

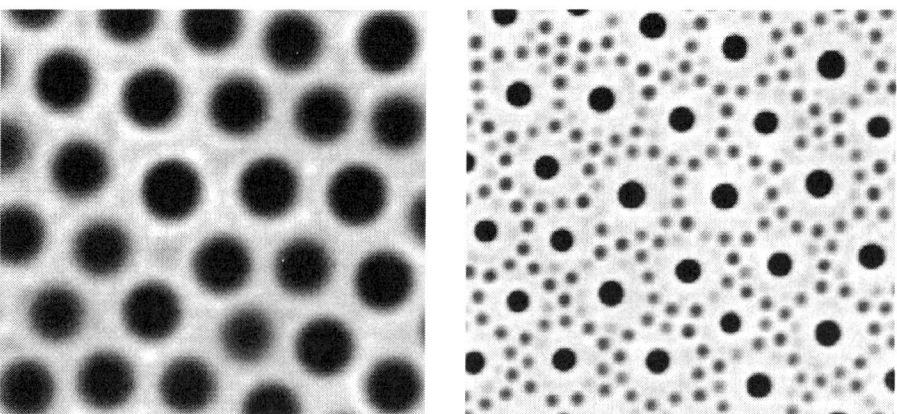

*Figure 22.3 - Homogeneous and inhomogenous structure of thickness
modulations in a ferrofluid in a field applied normal to the film*

It is no longer simply the instability leading to these structures that is being
investigated. Interest now focuses on the metastable equilibrium states, which depend
on the initial conditions, and result in sharing the space between the magnetic and
nonmagnetic fluids in such a way as to find the best compromise between the surface
energy (here the surface tension between the two fluids) and the dipolar energy.

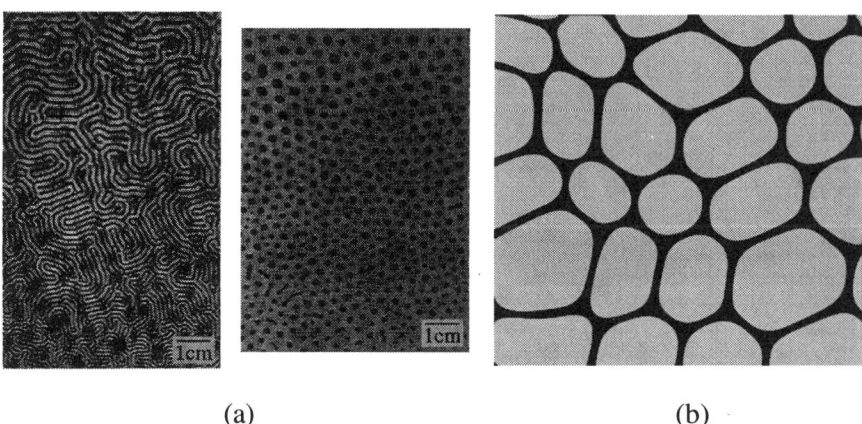

(a)   (b)

*Figure 22.4 - Spatial structures of a ferrofluid confined in a Hele-Shaw cell
with a non-miscible fluid and subjected to a magnetic field normal to the cell:
maze, bubble lattices and cellular structure.* Photos: from [13] (a) and F. Elias (b)

These observations reinforce the notion of an analogy between structures in
ferrofluids and magnetic domains, even if the variations in a field are opposite: in fact,
in the case of domains, the state with no wall is the equilibrium state in a strong field,
while for the ferrofluid the equilibrium state with minimum surface energy is obtained
in zero field.

REFERENCES

[1] D. PLATZEK, C. NOTTHOF, D.M. HERLACH, G. JACOBS, D. HERLACH, K. MAIER, *Appl. Phys. Lett.* **65** (1994) 1723.

[2] J. RESKE, D.M. HERLACH, F. KEUSER, M. MAIER, D. PLATZEK, *Phys. Rev. Lett.* **75** (1995) 737.

[3] R.E. ROSENSWEIG, Ferrohydrodynamics (1998) Dover Publications.

[4] J.C. BACRI, R. PERZYNSKI, D. SALIN, *La Recherche* **192** (1987) 1152.

[5] Proc. Intern. Conf. on Magnetic Fluids, *J. Magn. Magn. Mater.* **39**, No 1-2 (1983); **65**, No 2-3 (1987); **85**, No 1-3 (1990); **122**, No 1-3 (1993); **149**, No 1-2 (1995).

[6] R. MASSART, *IEEE Trans. Mag. Mag. Mater.* **17** (1981) 1247.

[7] R.W. CHANTRELL, A. BRADBURY, J. POPPLEWELL, S.W. CHARLES, *J. Appl. Phys.* **53** (1982) 2742.

[8] K. O'GRADY, A. BRADBURY, S.W. CHARLES, S. MENEAR, J. POPPLEWELL, *J. Magn. Magn. Mater.* **31-34** (1983) 958.

[9] S. TAKETOMI, Y. OZAKI, K. KAWASAKI, S. YUASA, H. MIYAJIMA, *J. Magn. Magn. Mater.* **122** (1993) 6.

[10] I. NAKATANI, M. HIJIKATA, K. OZAWA, *J. Magn. Magn. Mater.* **122** (1993) 10.

[11] A. BEE, R. MASSART, S. NEVEU, *J. Magn. Magn. Mater.* **149** (1995) 6.

[12] P.A. PETIT, M.P. DE ALBUQUERQUE, V. CABUIL, P. MOLHO, *J. Magn. Magn. Mater.* **122** (1993) 271.

[13] F. ELIAS, C. FLAMENT, J.C. BACRI, S. NEVEU, *J. Phys. 1 France* **7** (1997) 711.

OTHER ASPECTS OF MAGNETISM

CHAPTER 23

MAGNETIC RESONANCE IMAGING

The phenomenon of Nuclear Magnetic Resonance (NMR) was discovered independently by Bloch, Hansen and Packard [1], and by Purcell, Torrey and Pound [2]. In 1952 Bloch and Purcell were awarded a Nobel prize for their work.

*NMR has gradually shown itself to be essential in **physics** where it constitutes a probe for the study of materials, in **chemistry** where it has become a powerful analytical tool in studies of macromolecular structures and molecular dynamics, in **biology** where its impact is important in determining the structure of proteins and in the in vivo exploration of metabolism, and in **medicine** where, with the development of imaging, it offers a means of non traumatic routine exploration. The award of a further Nobel prize to Richard Ernst has served to emphasize the importance of the last fifty years of continued progress. It is to him that we owe the development of the two-dimensional NMR methods which form the foundations of the study of molecular structures and the Fourier transform based NMR imaging techniques.*

This chapter is devoted to the Magnetic Resonance Imaging (MRI) techniques that have secured very wide interest in this application of magnetism. It is comprised of three parts: the first two deal with the presentation of the basic principles of NMR and MRI respectively. The third and last part presents a fascinating application of NMR, namely the imaging of cerebral activity.

1. THE PHYSICAL BASIS
OF NUCLEAR MAGNETIC RESONANCE

In this first part we will briefly present the basic characteristics of an NMR experiment. A more detailed description of the physical bases of NMR can be found in a number of specialised works [3-7].

1.1. ENERGY LEVELS IN A MAGNETIC FIELD

Nuclear Magnetic Resonance has its origin in the magnetic properties of nuclei. Certain nuclei have an angular momentum, and consequently a magnetic moment (fig. 23.1). The phenomenon of NMR is intimately linked to the coexistence of these two physical characteristics. As with half of the known nuclei, ^1H, ^{13}C, ^{19}F, ^{23}Na, ^{31}P, ^{35}K... have a non-zero nuclear spin I. In this introductory chapter to NMR methods we shall restrict ourselves to the discussion of the properties of nuclei of spin I = 1/2.

When a nucleus of this type is placed in a magnetic field $\mathbf{B_0}$, the interaction energy is given by:

$$\mathscr{E}_m = -\gamma \hbar B_0 m \tag{23.1}$$

where m = ± 1/2, h (= $2\pi\hbar$) is Planck's constant, and γ the gyromagnetic factor, a constant, characteristic of the type of nucleus (tab. 23.1). For spin 1/2 nuclei there are, therefore, two energy levels, and the separation between these two levels is:

$$\Delta\mathscr{E} = \gamma \hbar B_0 \tag{23.2}$$

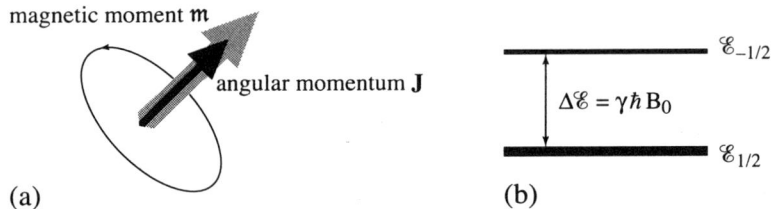

(a) (b)

Figure 23.1 - (a) a nuclear spin - (b) energy levels

By introducing the transition frequency F_0 ($\Delta E = h F_0$), one obtains the Larmor equation which expresses the proportionality between this frequency and the magnetic field:

$$F_0 = \gamma B_0 / 2\pi \tag{23.3}$$

Table 23.1 - Resonance frequencies and the natural abundance
of some of the nuclei of interest in the fields of biology and medicine

Nucleus	Natural abundance (%)	I (spin)	NMR Frequency (MHz) (B_0 = 1 T)
^1H	99.98	1/2	42.6
^{13}C	1.1	1/2	10.7
^{19}F	100	1/2	40.1
^{23}Na	100	3/2	11.3
^{31}P	100	1/2	17.2

In general the phenomenon of NMR necessitates a quantum mechanical description. However, many characteristics of Magnetic Resonance Imaging (MRI) can be understood within the classical formalism, and we will confine ourselves to this aspect.

1.2. A COLLECTION OF NUCLEI IN A MAGNETIC FIELD

At thermal equilibrium, the spin states are distributed between the two energy levels in the proportions given by Boltzmann statistics:

$$n_{1/2} / n_{-1/2} = \exp(\Delta \mathcal{E} / k_B T) \tag{23.4}$$

where $n_{1/2}$ is the number of nuclei in the low energy state ($m = 1/2$) and $n_{-1/2}$ the number of nuclei in the high energy state ($m = -1/2$), T is the absolute temperature, and k_B Boltzmann's constant. As is clear from table 23.1, the transition frequencies in magnetic fields of a few teslas lie within the radio frequency range. It follows that, at ordinary temperatures, $\Delta \mathcal{E}$ is very much smaller than $k_B T$, and the excess population in the lower energy level is very small (fig. 23.2).

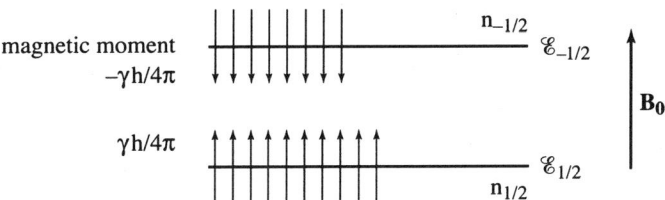

Figure 23.2 - Populations of the two energy levels

The magnetic moment of the spins with quantum number $m = 1/2$ (low energy) is equal to $\gamma h / 4\pi$, and that of the spins of quantum number $m = -1/2$ (high energy) is equal to $-\gamma h / 4\pi$.

The existence of a non-zero macroscopic magnetic moment m_0 arises from the excess population in the low energy state, such that:

$$m_0 = (\gamma h / 4\pi)\{n_{1/2} - n_{-1/2}\} \tag{23.5}$$

At thermal equilibrium this macroscopic moment is aligned parallel to the polarising field B_0. As we shall see, an NMR experiment consists of moving m_0 away from its equilibrium direction along B_0, and thus creating a transverse component in the *macroscopic nuclear magnetization* ($M_0 = m_0 / V$, see § 1.2.1 of Chap. 2)

1.3. RADIOFREQUENCY PULSES

An NMR experiment aims at inducing and observing transitions between energy levels. This can be achieved using a photon source in the form of a magnetic field B_1, perpendicular to B_0, and rotating around B_0 with a frequency equal or close to the transition frequency:

$$F_{rf} \approx F_0 = \gamma B_0 / 2\pi \tag{23.6}$$

The sense of rotation is defined by the direction of the $-\gamma B_0$ axis. The field B_1 is often produced by a coil with its axis normal to B_0, in which a sinusoidal current of frequency F_{rf} flows (fig. 23.3).

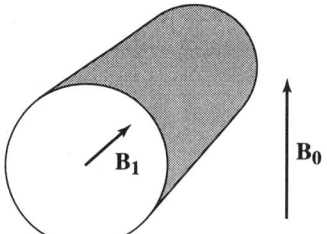

Figure 23.3 - Geometry of the coil creating the radiofrequency field

The linearly polarized field produced by the coil may be decomposed into two components rotating in opposite directions. Only one of these is effective. The field $\mathbf{B_1}$ is applied for a limited time, and one often speaks of a radiofrequency (rf) pulse. In practice one employs radiofrequency pulses whose spectral width is sufficiently large as to cover the entire NMR spectrum. A pulse of duration Δt covers a band of frequencies approximately equal to $\Delta F \approx 1 / \Delta t$ centred around the main frequency F_{rf} (see § 2.1). Depending on the type of application, the width of an rf pulse may vary from a few microseconds when the irradiation of a very large frequency band is required, to several milliseconds when it is desired to selectively irradiate a very narrow frequency range. An rf pulse induces transitions between energy levels, and moves the macroscopic magnetization $\mathbf{M_0}$ away from its thermal equilibrium direction along $\mathbf{B_0}$. Immediately following the excitation of the spin system, the magnetization $\mathbf{M}$ makes an angle θ with $\mathbf{B_0}$ (if the field $\mathbf{B_1}$ is of constant amplitude and is applied during a time τ, then $\theta = \gamma B_1 \tau$). Pulses for which $\theta = 90°$ are frequently used. They bring the magnetization to a position perpendicular to $\mathbf{B_0}$. $\theta = 180°$ pulses are also often used, these invert both the magnetization and the populations, but do not create a transverse component of magnetization.

1.4. SPIN LATTICE RELAXATION

After any kind of perturbation of the thermal equilibrium state (for example after a perturbation associated with a $90°$ pulse), the return to equilibrium of the longitudinal component of the magnetization takes place in exponential fashion with a time constant T_1. This mechanism corresponds to an exchange of energy between the system of nuclear spins and its environment (the lattice). In tissues the value of T_1 varies between about 100 ms and a few seconds. The return to equilibrium is governed by:

$$M_z(t) - M_0 = [M_z(0) - M_0] \exp(- t / T_1) \qquad (23.7)$$

where M_0 is the value of the longitudinal magnetization at thermal equilibrium, and $M_z(t)$ is its value at time t after perturbation of the equilibrium (fig. 23.4). Thus, after a $90°$ pulse (usually intended to permit signal acquisition), it is necessary to wait for some time before repeating the experiment (with a view, for instance, to increasing the signal to noise ratio).

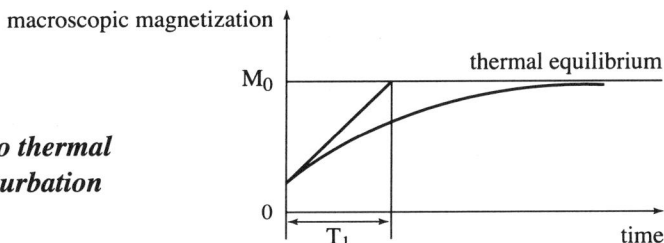

Figure 23.4 - Return to thermal equilibrium after perturbation

1.5. FREE PRECESSION SIGNAL

An rf pulse perturbs both the distribution of the spin states and the orientation of the nuclear magnetization. Immediately following the rf pulse one can observe the free precession, at the Larmor frequency F_0, of the macroscopic magnetization **M** about the director field **B$_0$**.

The rotation of this magnetic moment actually induces an electromotive force (emf) at the frequency F_0 in a coil enclosing the sample (the axis of this coil must be perpendicular to **B$_0$**):

$$e \sim M_0 \cos 2\pi F_0 t \tag{23.8}$$

In practice the detection of the signal calls for a frequency change, from the high frequency F_0 towards a low frequency f_0 ($f_0 = F_0 - f_{rf}$). For this, one carries out a signal multiplication operation using a reference voltage at the excitation frequency f_{rf}.

After a filtering process, the signal received at the end of the chain by the data processing computer may be written as:

$$e \sim M_0 \exp j (2\pi f_0 t + \Phi) \tag{23.9}$$

where, in order to obtain a complex signal, one uses two channels, shifted in phase by $\pi / 2$ of a lock-in amplifier.

Figure 23.5 shows an NMR signal following the application of an rf pulse, as it emerges from the detection chain.

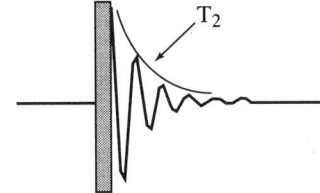

Figure 23.5 - The free precession signal as it emerges from the detection chain (after the change of frequency)

Spin-spin relaxation - Figure 23.5 shows that the signal decreases as a function of time. This decrease, due to a loss of coherence between spins, is exponential, and is characterized by the time constant T_2. In tissues, and for water protons T_2 is usually of the order of a few hundred milliseconds or less. This relaxation mechanism is known as spin-spin or transverse relaxation. The process of spin-spin relaxation is irreversible. It does not involve any exchange of energy with the environment. The induced emf should, therefore, be rewritten in the form:

$$e \sim M_0 \exp \hat{\jmath} \, (2\pi f_0 t + \Phi) \exp(-t/T_2) \qquad (23.10)$$

1.6. CHEMICAL SHIFT

The Larmor frequency of a particular nucleus (^{31}P for example), placed in a given magnetic field, varies slightly with the electronic environment. This phenomenon is known as the *chemical shift* since it produces a shift in the resonance frequencies. It lies at the heart of the value of NMR in the fields of chemistry and biochemistry, and has its origin in the demagnetizing field associated with the electronic environment of the nucleus, which modifies the local field:

$$\mathbf{B_{loc}} = \mathbf{B_0}(1 - \sigma) \qquad (23.11)$$

where σ is the screening constant.

The effective transition frequency is, therefore, written as:

$$F = \gamma \mathbf{B_0}(1 - \sigma)/2\pi \qquad (23.12)$$

The screening constant (a dimensionless quantity) lies in the range 10^{-6}-10^{-3}, its value depending on the structure of the electronic environment. The NMR signal in the time domain may thus be written as:

$$s(t) = \sum_i s_i(0) \exp(\hat{\jmath} 2\pi f_1 t) \exp\left(-t/T_2^{\,i}\right) \qquad (23.13)$$

where i characterises each of the various electronic environments in which the nucleus can be found in the sample under consideration ($f_i = F_0^i - F_0^{ref}$). The spectrum is obtained by performing the Fourier transform of s(t):

$$S(f) \; \Leftarrow \; TF \; \Rightarrow \; s(t) \qquad (23.14)$$

Thus the NMR spectra display several resonance lines (fig. 23.6), each at a frequency characteristic of the chemical environment.

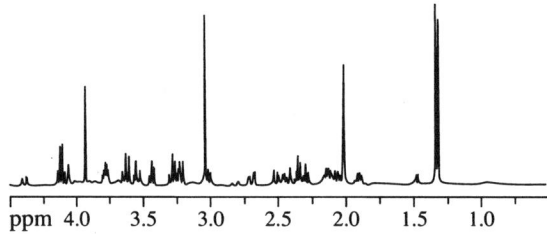

Figure 23.6 - Typical ^{1}H (proton) NMR spectrum from a biological sample

The horizontal scale represents the chemical shift. The chemical shift δ_1 of a nucleus of interst within chemical environment i is defined as $\delta_1 = (F_0^i - F_0^{ref})/F_0^{ref}$, where F_0^{ref} corresponds to a standard chemical environment.

1.7. SPIN ECHOES

As was stated above, the transverse magnetization, and hence the NMR signal, decays with the time constant T_2. In practice, as sketched in figure 23.7, the microscopic and macroscopic heterogeneities of the magnetic field lead to a faster decline in the transverse magnetization. The sample is no longer subject to a uniform field, and this results in a dispersion of the resonance frequencies within it. The decrease in the signal is thus better described by a time constant T_2^* which reflects the inhomogeneities of the field:

$$1/T_2^* \ = \ 1/T_2 + \gamma \Delta B / 2 \tag{23.15}$$

where ΔB is the width of the field distribution in the sample.

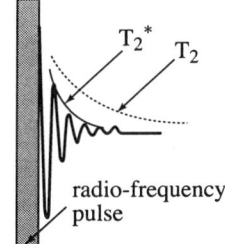

***Figure 23.7 - Decrease of magnetization associated
with inhomogeneities of the magnetic field***

However, the decrease in signal associated with the presence of inhomogeneities in the field is not an irreversible phenomenon since it can be reversed by using a 180° pulse, called the refocussing pulse. As was indicated earlier, a 180° pulse of inverts the populations of both energy levels. An additional characteristic of this type of pulse is that it also affects the signal phase. Under appropriate conditions, the phase is reversed.

Let us consider a signal emanating from a nucleus in a clearly determined environment. In an inhomogeneous field the frequency depends on the position $\mathbf{r}$ in space. Let the resonance frequency be $f(\mathbf{r})$.

The sample is subjected to the pulse sequence shown in figure 23.8 (spin echo sequence: $(90° - T_E/2 - 180° - T_E/2)$. Immediately prior to the 180° pulse [time $t = (T_E/2)_-$] the elementary signal coming from the point at position vector $\mathbf{r}$ is:

$$s\,[(T_E/2)_-, \mathbf{r}] \ \sim \ \exp[j\,2\pi f(\mathbf{r})\,T_E/2]\,\exp(-T_E/2\,T_2) \tag{23.16}$$

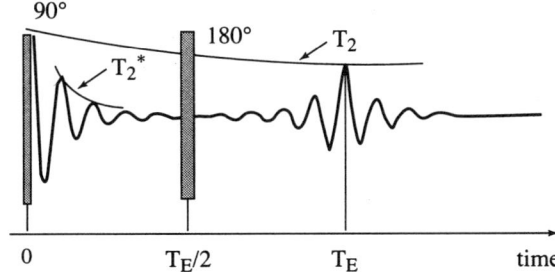

Figure 23.8 - Spin echo sequence

In principle one may calculate the signal received by the detection coil by integrating the elementary signal over the volume of the sample. As the 180° pulse reverses the phase, the signal immediately after this pulse (time $t = (T_E/2)_+$), is given by:

$$s[(T_E/2)_+] \sim \exp(-j\, 2\pi f\, T_E/2) \exp(-T_E/2\, T_2) \qquad (23.17)$$

At a time $t > (T_E/2)_+$ one obtains:

$$s(t) \sim \exp(-j\, 2\pi f\, T_E/2) \exp[j\, 2\pi f\, (t - T_E/2)] \exp(-t/T_2) \qquad (23.18)$$

Thus, at time $t = T_E$:

$$s(t) \sim \exp(-T_E/T_2) \qquad (23.19)$$

A 180° pulse, therefore, refocuses the dephasing that appears in time as a result of the chemical shift or of field inhomogeneities.

1.8. THE NMR EXPERIMENT

In summary an NMR experiment consists of the following stages:
♦ evolution towards thermal equilibrium,
♦ perturbation of thermal equilibrium using an rf field produced by a coil,
♦ setting of the electronic conditions (amplification, change of frequency, filtering, digitalization) and analysis (Fourier transform) of the free precession signal induced across the receiving coil (the same coil may be used both to produce the rf field and to receive the signal).

Figure 23.9 shows the block-diagram of an NMR spectrometer.

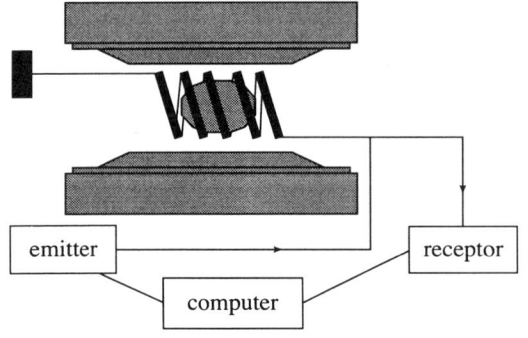

Figure 23.9 - Block diagram of an NMR spectrometer

2. MAGNETIC RESONANCE IMAGING

MRI can provide information in many areas: solid-state physics, chemistry (analysis), studies of three-dimensional molecular structures, etc. Here we shall limit ourselves to presenting the use of NMR methods in obtaining images of samples containing nuclear spins. The principles of NMR imaging were put forward simultaneously, and independently in 1973 by P. Lauterbur [8] on the one hand, and by P. Mansfield and

P.K. Granner on the other. These initial demonstrations were followed by immense efforts research and development, which resulted in imaging by NMR becoming a very widespread diagnostic medical tool.

Below we describe the general principles of imaging methods. The first stage is generally a procedure to select a slice of the object which one wishes to image. The principle by which a slice is chosen rests on the notion of a selective pulse; this will be the subject of the first part of this presentation. In the second part we shall describe the characteristics of the basic tool of imagery: uniform field gradients. The slice-selection process, using selective pulses in the presence of gradient, will be described in the third part. The fourth part describes the general principles by which space is encoded to allow the reconstitution of the cross-sectional image.

The concept of an image immediately brings to mind a question: what is the physical quantity being imaged? In NMR imaging the reply is complex. The first element is that an NMR image concerns only one type of spin, most frequently that of protons. Moreover it most often also concerns protons in a well defined molecular environment, water protons as a rule, bearing in mind their abundant presence in tissue. One can now refine the answer by specifying that the physical quantity imaged is the intensity of the transverse component of the macroscopic magnetization of water protons at the moment of observation. This magnetization is, of course, always proportional to the density of spins at every point considered, but its magnitude may be weighted by a number of physical parameters: relaxation time, translational diffusion, effects of magnetic susceptibility, etc. One can influence the "weight" taken by each parameter by using a suitable preparation period.

2.1. SELECTIVE PULSES: THE LINEAR RESPONSE APPROXIMATION

We have seen that an NMR experiment consists of subjecting a sample immersed in a field $\mathbf{B_0}$ to a rotating field $\mathbf{B_1}(t)$, at right angle to $\mathbf{B_0}$. The rotation frequency F_{rf} of $\mathbf{B_1}$ around $\mathbf{B_0}$ is close to the Larmor frequency $F_0 = \gamma B_0 / 2\pi$ of the nuclear spins under consideration. The field $\mathbf{B_1}$ is applied for a period which is often short compared to the relaxation times T_1, T_2, and $T_2{}^*$. One can then ignore the effect of relaxation over the duration of the pulse (less than a few milliseconds). For instrumental reasons, the transverse magnetization $\mathbf{M_\perp}(F_0)$ produced by the perturbation of the system is generally observed only after the excitation has ended.

The response of a spin system to a radiofrequency pulse is, in fact, fundamentally non linear. Thus, if a 90° pulse acting on a spin system produces a certain transverse magnetization $M_\perp$, a 180° pulse will not produce a signal $2\,M_\perp$, as would occur in a linear system, but a zero transverse magnetization. However, if a pulse is of short duration (and / or has a weak amplitude), such as to perturb the spin system only slightly, then one can show that the system behaves in a linear fashion. Under these conditions the transverse magnetization, produced by the radiofrequency pulse, and observed immediately after the end of the pulse (time $t = 0$), is proportional to the value of the Fourier transform $b_1(F)$ of $B_1(t)$ at the Larmor frequency F:

$$M_\perp(F_0, t = 0) \sim M_0(F_0)\, b_1(F_0) \qquad (23.20)$$

where $M_0(F_0)$ is the longitudinal magnetization of the spins with resonance frequency F_0 and

$$b_1(F) = \int_{-\infty}^{+\infty} B_1(t) \exp(-j2\pi F t)\, dt \qquad (23.21)$$

is the Fourier transform of $B_1(t)$.

The rotating field $\mathbf{B_1}(t)$ is expressed here in complex notation:

$$B_1(t) = B_1{}^X(t) + j B_1{}^Y(t) = B_1{}^m(t) \exp(j2\pi F_{rf} t) \qquad (23.22)$$

where $B_1{}^m(t)$ is the amplitude of the rotating field, and X, Y are two orthogonal axes of the plane perpendicular t_0 $\mathbf{B_0}$.

The subsequent evolution of the magnetization then takes place freely under the effect of the Zeeman term (free precession) and of relaxation:

$$M_\perp(F_0, t) = M_\perp(F_0, 0) \exp(j2\pi F_0 t) \exp\left(\frac{-t}{T_2^*}\right) \qquad (23.23)$$

Pulses with a rectangular envelope are often used (fig. 23.10). In this case the rotating field is written in complex notation:

$$B_1(t) = B_1{}^m \exp(j2\pi F_{rf} t) \qquad (23.24)$$

where $B_1{}^m$ is a constant.

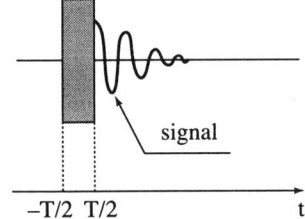

Figure 23.10 - Excitation of a system of spins by a rectangular pulse

The signal received is proportional to the transverse magnetization produced by the pulse.

It is easy to verify that the Fourier transform of a rectangular pulse of duration T and of amplitude $B_1{}^m$, centred at $t = 0$, is equal to:

$$b_1(\Delta f) = B_1{}^m T \frac{\sin(2\pi \Delta f\, T/2)}{2\pi \Delta f\, T/2} \qquad (23.25)$$

where $\Delta f = F - F_{rf}$.

If one is only interested in the amplitude of the signal but not its phase, then one has:

$$M_\perp(\Delta f) \sim M_0(\Delta f) \frac{\sin(2\pi \Delta f\, T/2)}{2\pi \Delta f\, T/2} \qquad (23.26)$$

The function $M_\perp(\Delta f)$ is zero for $\Delta f = n/T$, where n is an integer different from zero (fig. 23.11). Thus, the width of the response to a rectangular pulse is equal to $2/T$ if one considers only the principal lobe.

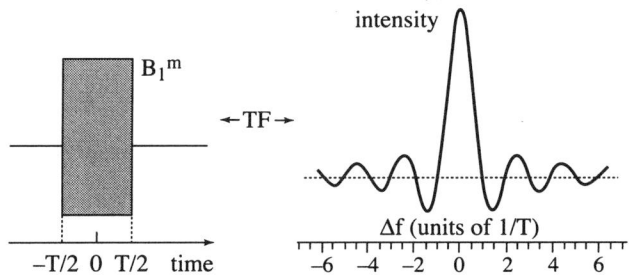

Figure 23.11 - The Fourier transform of a rectangular pulse

If $\Delta f = 1/2T$ then $[\sin 2\pi\,\Delta f]/2\pi\,\Delta f = 0.63$ which corresponds to an already important reduction in the signal with respect to its amplitude at $f = 0$, that is to say the resonance.

The frequency response of a rectangular pulse clearly demonstrates the known characteristics of frequency selectivity, but the presence of relatively intense side lobes distinguishes it from an ideal selective excitation which must, in the frequency domain, take the form of a gate. In using the characteristics established above (the frequency distribution of magnetization produced by a pulse is given by the Fourier transform of the pulse), one immediately thinks of using amplitude modulated pulses having the form of a *sinc* (function $\sin x/x$).

Unfortunately a sinc function, extending in the time domain from minus infinity to plus infinity is not feasible. The pulse must, therefore, be truncated, usually to the second or third zero on each side of the pulse. This truncation is responsible for the oscillations in the frequency profile (convolution with the Fourier transform of the truncated window) but the result continues to be of good quality (fig. 23.12).

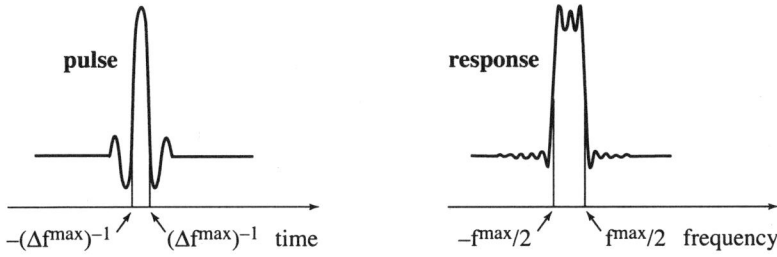

Figure 23.12 - The sinc pulse truncated to its third zero, and its Fourier transform

Pulses of Gaussian form are also frequently used. They too are truncated (2 or 5% of their maximum value). The Fourier transform of a gaussian $g(t) = \exp(-\pi t^2)$ is a gaussian $G(f) = \exp(-\pi f^2)$.

More generally, it can be verified that a gaussian $g(t) = \exp(-a\,t^2)$ has a Fourier transform:

$$G(f) = \sqrt{\frac{\pi}{a}}\,\exp\left(-\frac{\pi^2 f^2}{a}\right).$$

One can, by way of an exercise, establish the relationship between the width at half-height of a gaussian in the time domain ($t_{1/2}$), and the corresponding width ($f_{1/2}$) of the frequency window. One can verify that:

$$f_{1/2} = 4\,\frac{\ln 2}{\pi\,t_{1/2}} = \frac{0.882}{t_{1/2}} \tag{23.27}$$

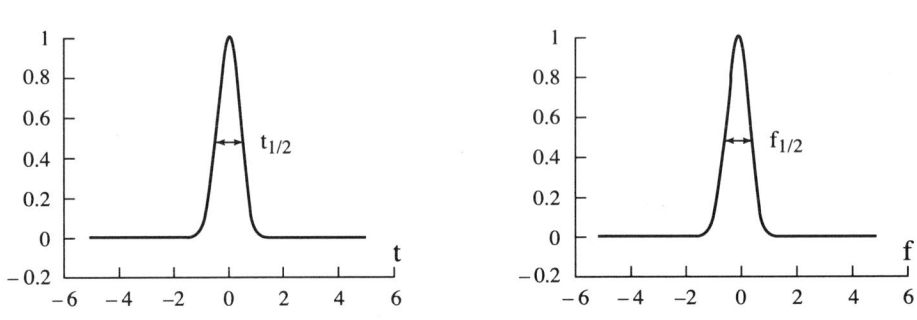

Figure 23.13 - A Gaussian pulse, and the corresponding response in the frequency domain

2.2. FIELD GRADIENTS

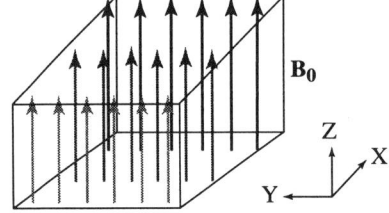

**Figure 23.14
Field gradient in the X direction**

The selection of a slice, and all NMR imaging, rests on the presence of a system of coils capable of producing a magnetic field variable in space and time. In general one seeks to have at one's disposal a field that can vary linearly in one or other of three directions X, Y, and Z of space. One then speaks of a field gradient.

The field at the point of position vector **r** is then given by:

$$\mathbf{B(r)} = (B_0 + \mathbf{G \cdot r})\,\mathbf{1_Z} \tag{23.28}$$

where **G** is a constant vector in space with components G_X, G_Y, and G_Z.

One therefore has:

$$G_X = \partial B/\partial X,\; G_Y = \partial B/\partial Y,\; G_Z = \partial B/\partial Z \tag{23.29}$$

The expression (23.28) now takes the form:

$$\mathbf{B(r)} = (B_0 + G_X X + G_Y Y + G_Z Z)\,\mathbf{1_Z} \qquad (23.30)$$

The amplitude of the gradients G_X, G_Y, G_Z, must be adjustable and amenable to modulation in time. A gradient is, of course, expressed in $T.m^{-1}$, but one can also use the unit $Hz.m^{-1}$ if one recalls the relation $F = \gamma B / 2\pi$. In practice the order of magnitude of the gradients used in clinical imaging is of the order of $10\ mT.m^{-1}$. The relation (23.28) shows that the precession frequencies vary spatially according to the law:

$$f(\mathbf{r}) = f_0 + \gamma \mathbf{G.r} / 2\pi \qquad (23.31)$$

2.3. EXCITATION OF A SYSTEM OF SPINS IN THE PRESENCE OF A GRADIENT: SLICE SELECTION

Here we shall consider samples containing one type of molecule "visible" by NMR (water for example). The process of slice selection is aimed at producing a transverse magnetization in a well defined region of the space. Ideally this region should be limited by two planes, normal to a given direction **U**, of the laboratory frame of reference (see fig. 23.15). The distance e between the two planes defines the thickness of the slice. The excitation of the spins contained in a slice of the material normal to **U**, and with a thickness e, is effected naturally with the help of selective pulses.

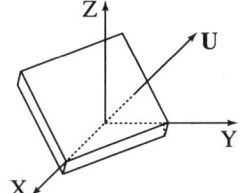

Figure 23.15 - Slice selection

Let us consider a homogeneous material, placed in a field gradient G_X, and let us apply a selective pulse (a sinc or a gaussian for example). Let ΔF be the frequency width over which the pulse is effective (the width at half height of the frequency response $M_\perp(F)$). Since $F = F_0 + \gamma G_X X / 2\pi$, the thickness of the slice e is related to the intensity of the gradient by the relation:

$$e = 2\pi \Delta F / \gamma G_X \qquad (23.32)$$

One also has:
$$F = F_0 + \gamma G_X X / 2\pi \qquad (23.33)$$

The centre of the excited region (i.e. the position of the centre of the slice), X_t, is at $F = F_{rf}$, thus

$$X_t = (F_{rf} - F_0) / 2\pi \gamma G_x \qquad (23.34)$$

The adjustment procedure for the slice selection is contained in the relations (23.32) and (23.34) and illustrated in figure 23.16. The frequency width ΔF is fixed from the moment the selective pulse is decided upon (its form, amplitude, duration). *The choice* *of the slice thickness e determines the intensity of the gradient (relation 23.32), while the position of the slice X_t is adjusted by manipulating the frequency of the rotating field F_{rf} (relation 23.34).* Due to variations of the signal phase within the body of the slice, it is necessary to reverse the gradient for a time of the order of the half-width of the pulse, after its application.

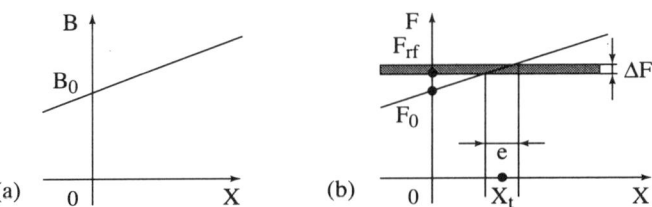

Figure 23.16 - Gradient X: Spatial variation of (a) the magnetic field (b) the Larmor frequency in the laboratory frame of reference

2.4 IMAGING: THE RECIPROCAL SPACE

When the longitudinal magnetization in a three dimensional sample is perturbed by a spatially selective pulse, which is assumed to be applied in the presence of a field gradient G_X (fig. 23.17), the transverse magnetization produced by the selective pulse comes from a Y, Z "plane". The coordinate X_t of this plane depends on the frequency of the pulse. An imaging method should allow determination of the intensity of this magnetization at each point of the plane.

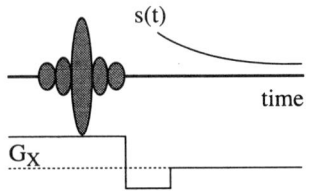

***Figure 23.17 - Relative timing of the radiofrequency pulse and the inversion of the gradient G_x during the slice selective excitation** s(t) represents the NMR signal from the slice.*

If the influence of the transverse relaxation is neglected, then the signal at time t_0, in a homogeneous static field determined by the slice-selection process (see fig. 23.17), is given by:

$$s(t_0) \sim \exp(j\,2\pi\,f_0 t_0)\int \rho(Y,Z)\,dYdZ \qquad (23.35)$$

where $\rho(Y, Z)$ is the density of spins at the point X_t, Y, Z, and f_0 the Larmor frequency (in the absence of a gradient, and after a change of frequency). As the magnetic field is assumed to be homogeneous within the slice, f_0 does not depend on the position.

Let us now introduce a gradient pulse in the Y direction, before the acquisition of the signal (fig. 23.18). During this pulse the field is given by $B = B_0 + G_Y Y$. From this one deduces:

$$F = F_0 + \gamma G_Y\,Y/2\pi \qquad (23.36)$$

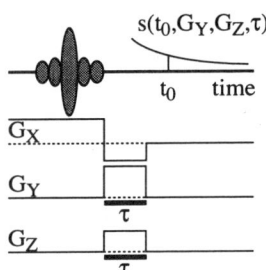

**Figure 23.18 - Phase encoding
in the Y and Z directions**

Consequently, this pulse of duration τ introduces an additional dephasing:

$$\Phi = \gamma G_Y Y \tau \tag{23.37}$$

which can also be written as $\Phi = K_Y Y$, with:

$$k_Y = \gamma G_Y \tau \tag{23.38}$$

The signal at time t_0 after application of the gradient pulse is given by:

$$s(t_0, k_Y) \sim \int \rho(Y, Z) \exp(jk_Y Y) dY dZ \tag{23.39}$$

In the same way one can introduce a gradient pulse along the Z direction. This pulse introduces a phase difference $\Phi = k_Z Z$ (where $k_Z = \gamma G_Z \tau$). The signal is then given by:

$$s(t_0, k_Y, k_Z) \sim \int \rho(Y, Z) \exp(jk_Y Y) \exp(jk_Z Z) dY dZ \tag{23.40}$$

One notes that $s(t_0, k_Y, k_Z)$ is proportional to the Fourier transform of $\rho(Y, Z)$. The reconstruction of $\rho(Y, Z)$ may, therefore, be effected by simply carrying out the inverse Fourier transform:

$$\rho(Y, Z) \sim \int s(t_0, k_Y, k_Z) \exp(-jk_Y Y) \exp(-jk_Z Z) dY dZ \tag{23.41}$$

However, this can only be done if the signal $s(k_Y, k_Z)$ is known for each coordinate of the plane (k_Y, k_Z) (fig. 23.19), or at least for a sufficiently large number of points of the (k_Y, k_Z) plane, also known as the Fourier plane or reciprocal space.

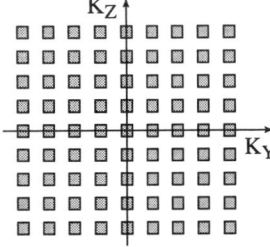

**Figure 23.19
Sampling of the Fourier plane**

Sampling of the reciprocal space may be carried out by repeating the experiment with different values of k_Y and k_Z. However this procedure takes a lot of time since for each experiment one obtains one point only in the reciprocal space.

The sampling of reciprocal space over 256×256 points requires 256×256 experiments. The time between experiments must be sufficiently long as to permit the longitudinal magnetization to return to a value close to its thermal equilibrium value; one often needs to allow a time delay of 1 to 2 seconds between the pulses, which leads to a duration of 20 to 40 minutes for each experiment.

One can sample Fourier space more efficiently by sampling the free precession signal in the presence of a gradient, called the *readout gradient*. The method is illustrated in figure 23.20. A gradient pulse G_Y of negative intensity is switched to a positive value at time t_A. The gradient waveform is adjusted so that at time T_E:

$$\int_0^{T_E} G_Y dt = 0 \qquad (23.42)$$

The NMR signal may then be written as:

$$s(t, G_Y, k_Z) \propto \int \rho(Y, Z) \exp[j\gamma G_Y Y(t - T_E)] \exp(jk_Z Z) dYdZ \qquad (23.43)$$

With the change of variable $k_Y = \gamma G_Y(t - T_E)$, Equation (23.43) becomes:

$$s(k_Y, k_Z) \propto \int \rho(Y, Z) \exp(jk_Y Y) \exp(jk_Z Z) dYdZ \qquad (23.44)$$

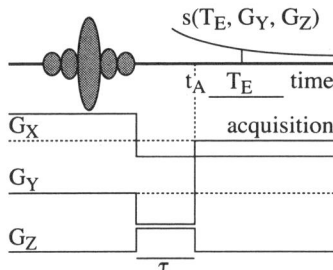

Figure 23.20
Gradient echo technique

Consequently, when a free precession signal is sampled in the presence of a gradient G_Y, the samples sit on a line of Fourier space. The procedure may be repeated line by line for all the values of k_Z. This method is known as the gradient echo technique (the dephasings introduced by the gradient G_Y are refocused at time T_E, hence the term echo). The gradient G_Y is called the readout or measurement gradient (measurement takes place in the presence of this gradient), and the encoding process in this direction of the space is called frequency encoding (the frequency of the signal, $\gamma G_Y Y$, is position dependent). The gradient G_Z is called the phase encoding gradient (the gradient pulse introduces a phase factor $\Phi = \gamma G_Z Z \tau$ which depends on position). Numerous imaging techniques exist, imaging by spin echo for example. All of these techniques consist of sampling reciprocal space.

2.5. CONTRAST

So far we have assumed that transverse relaxation was very slow, and did not affect the NMR signal. In practice, with an echo gradient technique, the signal decreases with the time constant T_2^*. The degree of T_2^* weighting may be adjusted by acting on the echo time T_E. A spin echo technique is weighted by the relaxation time T_2. In the same way, one can introduce a T_1 weighting by using a repetition time T_R smaller than T_1 (fig. 23.21).

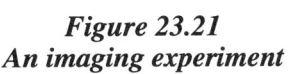
Figure 23.21
An imaging experiment

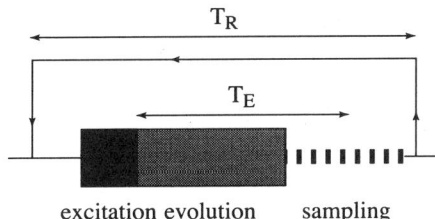

excitation evolution   sampling

If $T_R \gg T_1$ and $T_E \ll T_2$, the image is essentially an image of spin density. The use of a repetition time such that $T_R \approx T_1$ and $T_E \ll T_2$ introduces a T_1 contrast. In the same way, if $T_R \gg T_1$ and $T_E \approx T_2$ one obtains a T_2 contrast. Many other physical, physico-chemical or bio-physical parameters may be similarly imaged by MRI. For example, NMR is a powerful technique for imaging coherent and incoherent displacement (translational diffusion).

3. AN EXAMPLE OF MRI APPLICATION: CEREBRAL ACTIVATION IMAGING

The applications of NMR imaging are numerous. In clinical studies it has become a major tool of morphological investigation. Moreover, since the early nineties MRI has lent itself to new applications allowing access to the functioning of the brain. This access to cerebral function by MRI is based on the modulation of the signal by the haemodynamic parameters (cerebral blood volume, cerebral blood flow), and by the degree of blood oxygenation. Taking into account the central role played by the haemodynamic parameters and oxygenation, the characteristics of cerebral activation detected by MRI present a certain similarity to those observed by positron emission tomography (PET) or by single photon emission computed tomography (SPECT). In fact, the most widely used method exploits the intrinsic magnetic properties of blood.

3.1. HAEMODYNAMIC AND BLOOD OXYGENATION CHANGES INDUCED BY NEURONAL ACTIVATION

Numerous PET studies have shown the association of neuronal activity with an increase in the local cerebral blood flow (CBF) of the order of several tens % [10]. The increase in CBF itself follows an increase in local cerebral blood volume (CBV), which leads to a decrease in arterial resistance. The mechanisms governing the relationship between the metabolism and CBF remain, as yet, imperfectly understood. A relative increase in oxygen supply corresponds to the increase in CBF. Nevertheless this increase in the oxygen supply remains well above the increase in the consumption of oxygen. During neural activity, the fraction of oxygen extracted therefore falls; this leads to an increase in the oxygen partial pessure, and hence to an increase in the saturation of haemoglobin with oxygen.

Thus the disproportion between the increases in blood flow on the one hand, and oxygen consumption on the other, results in an over-oxygenation of the venous side of the capillaries, and in the venous system. Nevertheless, using functional MRI techniques one often observes a certain shift between, on the one hand, the increase in oxygen consumption induced by the increase of metabolic activity, and, on the other, the increase in CBF, which comes about rather belatedly [11]. Thus, in the first moments following the onset of neuronal activity, the degree of blood oxygenation is lowered with respect to its resting value, then increases with the increase in local CBF until it largely exceeds the resting value.

Functional MRI (fMRI) exploits the signal variations induced by modifications of the CBV, the CBF, and the oxygen saturation of haemoglobin. Depending on the method of imaging used, the signal weighting may favor one or another of these three parameters. The first demonstration of the potential of MRI for the observation of cerebral activity was carried out by using a method appropriate to the local variation of CBV [12]. Since this initial demonstration, the methods employed to obtain functional images of the brain by MRI have, nevertheless, generally been based on the use of BOLD (Blood Oxygenation Level Dependent) contrast [13 to 18], that is to say, on the exploitation of the variations in signal induced by modifications of the degree of blood oxygenation as a function of neuronal activity. Other, more recent work, is concerned with methods sensitive principally to CBF [19 to 21]. Although techniques centred on the observation of the local modification of CBV or CBF are promising, they have not (yet?) reached the stage of development of those based on the production of the BOLD contrast. We shall therefore restrict ourselves here to the presentation of the latter.

The physiological parameters that intervene in the biophysical mechanisms underlying the BOLD effect are the following:

♦ the haemotocrit (Hct):

$$Hct = \frac{\text{volume of red blood cells}}{\text{blood volume}} \tag{23.45}$$

♦ the rate of blood oxygenation (Y), linked to the partial pressure in oxygen [22]:

$$Y = \frac{\text{concentration of oxygenated haemoglobin}}{\text{concentration in haemoglobin}} \tag{23.46}$$

♦ the blood volume fraction (f):

$$f = \frac{\text{blood volume}}{\text{tissue volume}} \tag{23.47}$$

3.2. *BOLD CONTRAST: BIOPHYSICAL MECHANISMS*

BOLD contrast is concerned with variations in the NMR signal induced by modifications in the rate of blood oxygenation. It originates in the difference in magnetic behavior of the intra- and extra-vascular sectors. The importance of these differences depends on the state of oxygenation of the haemoglobin. In the following section we present the biophysical mechanisms which lie at the origin of BOLD contrast.

The basic characteristic of the NMR methods is the linear relationship between the resonance frequency and the applied magnetic field:

$$F_0 = \gamma B_0 / 2\pi \tag{23.48}$$

where γ is the gyromagnetic ratio. Any temporal or spatial variation in the field brings about a temporal or spatial variation in the resonance frequency.

When a sample is placed in a magnet, the value of the field in the sample does not depend solely on the strength of the magnet but also, of course, on the nature of the sample. In biological tissues, the difference between the field $\mathbf{B_0}$ which may be measured in vacuum (or in air), and that of $\mathbf{B}$ which establishes itself inside the sample, is generally very small, but NMR is sensitive to very small variations in frequency (hence the field).

The *materials* which play a role in the applications of MRI to biology and medicine can, from the point of view of their magnetic properties, be classified into three broad categories: diamagnetic, paramagnetic, and ferromagnetic materials. Two of these in particular interest us: the diamagnetic and paramagnetic materials (see chap. 4). In these two types of materials a linear relationship exists between the induced magnetization $\mathbf{M}$ and the field $\mathbf{H}$ ($\mathbf{M} = \chi\mathbf{H}$), with the result that the magnetic induction $\mathbf{B}$ in the sample differs by a quantity $\Delta\mathbf{B}$ from the induction $\mathbf{B_0}$ in vacuum.

$$\mathbf{B} = \mu_0(\mathbf{H} + \mathbf{M}) = \mu_0(1 + \chi)\mathbf{H}$$
$$= \mathbf{B_0}(1 + \chi) = \mathbf{B_0} + \Delta\mathbf{B} \tag{23.49}$$

where χ is the magnetic susceptibility of the sample, and μ_0 is the permeability of free space. The quantities $\Delta\mathbf{B}$ and $\mathbf{B_0}$ are linked by the relation:

$$\Delta\mathbf{B} = \chi\mathbf{B_0} \tag{23.50}$$

$\Delta\mathbf{B}$ is, therefore, proportional to $\mathbf{B_0}$, but it is much smaller. In diamagnetic materials χ is negative, and is of the order of -10^{-5} to -10^{-6} (the field inside the material is in fact slightly smaller than would be measured in vacuum). The susceptibility of water is equal to -8.6×10^{-6}. In fact, the diamagnetic contribution is a property that concerns all biological tissues. Nevertheless in certain cases other contributions may add to, or even dominate, those associated with the orbital magnetic moment. This is the case, for instance, for materials containing molecules characterized by the presence of unpaired electrons. In this case χ is positive, and the material is said to be paramagnetic. The field inside the material is then higher than would be measured in the absence of the sample. For a pure paramagnetic body the susceptibility is of the order of 10^{-4} to 10^{-5}. The free radicals (copper chloride $CuCl_2$, oxygen O_2...) are paramagnetic molecules. It is however rare in biology to find pure materials made up of paramagnetic molecules. One most often finds liquid mixtures or solutions in which paramagnetic molecules may be present. Among the paramagnetic molecules that play a role in biology, it is interesting to mention NO. The presence, in a normally diamagnetic liquid (e.g. water), of paramagnetic molecules (deoxyhaemoglobin for instance) increases the algebraic value of the magnetic suceptibility. When a solution is made up of a number of different constituents of molar concentration by volume c_i, the susceptibility of the solution is given by the expression:

$$\chi = \sum c_i \chi_i^{m} \qquad (23.51)$$

where χ_i^{m} is the molar susceptibility by volume of the constituent i (the susceptibility of a constituent comprising one mole of the material per unit volume).

Ferromagnetic materials will not be considered here since they do not play a role in the current methods of fMRI (however, these materials present interesting properties when used in the form of microcrystals at very low concentrations in certain contrast agents called superparamagnetics).

When a homogeneous material, dia or paramagnetic, is placed in a uniform magnetic field $\mathbf{B_0}$, then ΔB is itself uniform (except in the vicinity of the walls). It is not the same when the material is spatially heterogeneous, and is made up of sections having different magnetic susceptibilities, which is indeed the case for biological tissues.

3.2.1. Effects of magnetic susceptibility in tissues

The extra-vascular environment as well as the blood plasma are diamagnetic environments of virtually identical susceptibility. The blood contains red cells, which are themselves laden with haemoglobin.

Deoxygenated haemoglobin is a paramagnetic molecule, while oxygenated haemo-globin is diamagnetic. In fact, since the work of Linus Pauling on the subject [23], the correlation between the oxygenation state of the haemoglobin and its magnetic susceptibility has been known. This gives blood a magnetic susceptibility higher (in numerical value) than that of the extra-vascular medium. This increase in magnetic susceptibility is more or less pronounced, depending upon the degree of blood oxygenation. Taking into account the relation (23.51), the magnetic susceptibility of the blood may be written in the form [24]:

$$\chi_{blood} = Hct\, Y \chi_{oxy} + Hct\,(1 - Y)\chi_{deoxy} + (1 - Hct)\chi_{plas} \qquad (23.52)$$

where Hct is the haematocrit, Y is the rate of blood oxygenation (Eq. 23.46), χ_{oxy}, χ_{deoxy}, and χ_{plas} the susceptibilities of the oxygenated and deoxygenated red blood cells, and of the plasma.

The order of magnitude of the difference of susceptibilities between deoxygenated blood and oxygenated blood [25] is about $+ 9,5 \times 10^{-7}$ (haematocrit 0.42, SI units). Brain tissue, like all other vascularized tissues, is magnetically heterogeneous. It is traversed by a collection of vessels filled with blood, the magnetic susceptibility of which may differ from that of its surroundings. The magnetic field is perturbed locally by the difference in magnetic susceptibility $\Delta\chi = \chi_1 - \chi_2$ between two sectors of magnetic susceptibility χ_1, and χ_2, and ceases to be uniform. Let us consider for example an object which somewhat resembles a blood vessel, a cylinder of radius a, of susceptibility χ_1, whose axis is placed arbitrarily in a direction normal to the magnetic field (fig. 23.22).

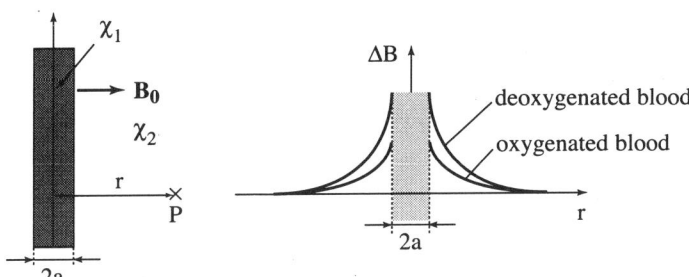

Figure 23.22 - Inhomogeneities of the magnetic field induced by the presence of a cylinder of blood with susceptibility χ_1 immersed in a medium of susceptibility χ_2

When the cylinder is immersed in a medium of susceptibility $\chi_2 \neq \chi_1$, the field in its vicinity is perturbed. For a given point, this perturbation depends on the distance of the point from the axis of the cylinder. For example, one can show that, in the plane defined by $\mathbf{B_0}$ and the axis of the cylinder, ΔB at a point P is proportional to a^2 / r^2, and to $\Delta\chi$ (r is the distance of P to the axis of the vessel). The perturbation decreases very rapidly, therefore, as one moves away from the cylinder. However, it increases with the difference in magnetic susceptibilities of the two regions, and with the magnetic field intensity. One has therefore:

$$\Delta B_{int} = \frac{2\pi}{3} \Delta M \left(3\cos^2 \theta - 1 \right) \tag{23.53}$$

$$\Delta B_{ext} = 2\pi \Delta M \, (a/r)^2 \cos 2\phi \, \sin^2 \theta \tag{23.54}$$

It should be noted that we have limited ourselves to discussing the form of the spatial variations of $\Delta B(r)$ in a particular plane, defined by the axis of the cylinder and by $\mathbf{B_0}$. One finds, for instance in reference [26], the general expressions of the spatial perturbation of the magnetic field, both inside and outside the cylinder.

The cortical tissue is crossed by a large number of vessels: arteries, arterioles, capillaries, venules, and veins. Arterial blood is saturated in oxygen, and carries only very little deoxygenated haemoglobin. Its susceptibility is, therefore, close to that of the surrounding tissues, and does not vary with neuronal activity. This compartmentalization of blood does not, therefore, introduce significant (magnetic) heterogeneity. In contrast, the degree of blood oxygenation in the capillaries, venules, and veins, is normally very much weaker. Venous blood presents a susceptibility that is much stronger than that of the tissues, the difference in susceptibility of the two mediums depends on the degree of saturation of the haemoglobin, which itself depends on the intensity of neuronal activity. The BOLD method exploits the perturbations of the magnetic field thus introduced into the extravascular sector in the vicinity of the blood vessels on the venous side.

3.2.2. Susceptibility effects, and the NMR signal: BOLD contrast

Differences in magnetic susceptibility between vascular and extra-vascular sectors introduce magnetic field inhomogeneities even outside the vessels. The NMR signal of water protons situated in the vicinity of the capillaries, venules or veins is sensitive to these inhomogeneities. Therefore, the concern here is not in the direct observation of haemoglobin or of oxygen, rather, it is in the observation of the influence of these molecules on the signal of the water protons. As in classical imaging, functional images are constructed from the signal coming from the water protons. In fMRI, the intensity of the signal is modulated by the magnetic susceptibility effects, described above.

Two types of mechanisms contribute to this modulation: the inhomogeneity of the field, which introduces a dispersion in resonance frequencies (static effect), and the diffusion of water molecules within the *susceptibility gradients* characteristic of these inhomogeneities, which introduces a temporal modulation of the resonance frequency [27] (dynamic effect).

Static effects

The cortical tissue may be modelled, in a very simplified manner, by a complex vascular network immersed in a diamagnetic medium. The susceptibility of this network may be stronger or weaker depending on the degree of neuronal activity. The blood volume fraction, f, being of the order of 4%, the largest part of the signal comes from the extravascular sector. The differences in magnetic susceptibility between the vascular and extravascular regions introduce magnetic field gradients in the neighbourhood of the vessels (see fig. 23.22), and hence a distribution of the resonance frequencies of the water protons. The width of this distribution increases with the concentration of deoxyhaemoglobin, and also with cerebral blood volume (vasodilatation or recruitment of capillaries).

Associated with cerebral activation are:

♦ a decrease in the concentration of deoxyhaemoglobin, and hence in the width of the frequency distribution (fig. 23.23),

♦ an increase in CBV, which tends to increase the width of the frequency distributions.

Figure 23.23 - Distribution
of resonance frequencies
in a voxel of cortical tissue

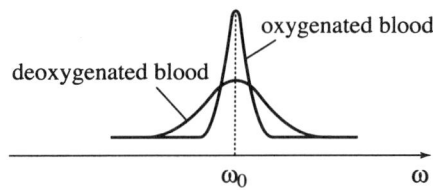

There are, therefore, two opposing effects, but the first is dominant in that, on the whole, the width of the frequency distribution decreases when one moves from a state of rest to one of cortical activation. This double influence, of modifications of CBV, and of the rate of blood oxygenation, is a first sign that illustrates the difficulties of making a quantitative analysis in fMRI.

It is known that the NMR signal is the response of the spin system (here the spins of the water protons) to an excitation (radiofrequency pulse). The frequency of this signal is the mean resonance frequency of these protons in an inhomogeneous field. The amplitude of the free precession signal decreases with the time constant T_2^* which partially reflects the degree of homogeneity of the magnetic field (see § 1.1.7). The more homogeneous the field, the longer is T_2^*. In a perfectly homogeneous field T_2^* tends towards T_2.

The passage from a state of rest to one of activity decreases the width of the distribution of resonance frequencies (the field becomes more homogeneous), and hence increases the parameter T_2^* (the signal lengthens). Figure 23.24 illustrates this point.

Figure 23.24 - Decrease of
the free precession signal
in a voxel of cortical tissue

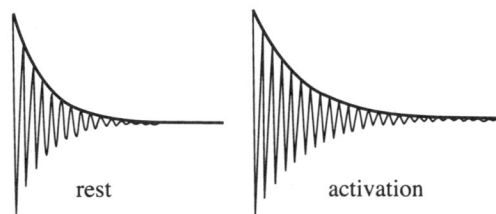

One can now confront these results with the properties of the two important classes of MRI pulse sequences: spin echo, and gradient echo pulse sequences. In a gradient echo experiment the signal is weighted by T_2^*. At time $t = T_E$, it has the form:

$$S(T_E) = S(0) \exp(-T_E/T_2^*) \tag{23.55}$$

By contrast, in a spin echo experiment, no T_2^* weighting is introduced. The signal is weighted by T_2: the amplitude at time T_E does not contain, if one limits oneself to this *static* analysis (taking no account of the mobility of water molecules), any information about the field inhomogeneities, and hence on the state of cerebral activation. Things are, in fact, less clear cut if one considers the influence of molecular diffusion.

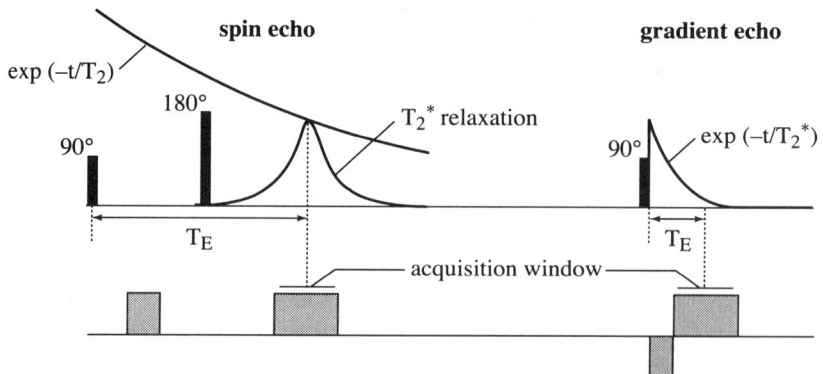

Figure 23.25 - Signal weighting in a spin echo and in a gradient echo experiment

The waveforms of the readout gradients in the corresponding imaging experiments have been specified for the sake of clarity, but the effect of these gradients on the signal has not been illustrated.

Dynamic effects

In an fMRI experiment the duration of observation of the signal is of the order of 30 to 100 ms. During this time of data acquisition, the water molecules diffuse more or less freely within the extra-vascular space. One generally considers that during this time the probability of a molecule crossing from the vascular to the extra-vascular region is negligible. In the extra-vascular region these molecules diffuse in the field gradients associated with the differences in susceptibility. On examining the expression (23.54), one notices that ΔB in the vicinity of the wall (r = a) does not depend on the radius of the vessel being considered. The amplitude of ΔB depends, in fact, only on $\Delta \chi$. On the other hand the extent of the region of inhomogeneous field scales down with the radius of the vessel.

When a molecule diffuses in the vicinity of a small vessel, the experiment time (i.e. the echo time is long enough for the molecule to "see" the whole range of local fields. In contrast, molecular diffusion in the vicinity of a large vessel does not permit a given molecule to experience the whole range of these field variations within the experiment time (fig. 23.26).

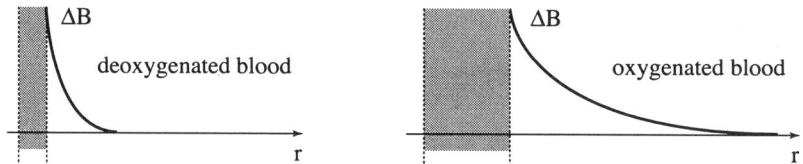

**Figure 23.26 - Field perturbations around a vessel of
small diameter d extend over smaller distances than around a larger vessel**

The amplitude of the field variations does not depend, however, on the radius of the vessel. The field gradients (dΔB/dr) are, therefore, much stronger around small vessels.

This has two consequences:

♦ The movement tends to reduce the resonance frequency dispersion. In the case of small vessels (capillaries) each water molecule has, for the duration of the experiment, travelled across the entirety of the distribution frequencies. Because of the movement, which introduces an averaging effect, the mean resonance frequency differs little from one molecule to another, which is not the case in the absence of movement. The loss of coherence due to the spatial dispersion of frequencies is thus reduced. This is the phenomenon of motional narrowing [7]. Hence, during a rest period, T_2^* is longer than it would be in the absence of diffusion: for small vessels, the rest-activity difference is reduced. For large vessels, the situation is less changed in the presence of diffusion (the average distance travelled by a molecule over the duration of the experiment does not allow it to experience all the resonance frequencies). At this point, it is useful to introduce the variation ΔR_2^* of the speed of relaxation between rest and activation:

$$\Delta R_2^* = (1/T_2^*)_{rest} - (1/T_2^*)_{act} \qquad (23.56)$$

The variation ΔR_2^*, that is to say the sensitivity of a gradient echo experiment, is, therefore, stronger for the venules and veins than for the capillaries for a given volume fraction of blood (fig. 23.27).

♦ In a spin echo experiment, the situation is reversed. In the absence of diffusion, this type of experiment would be insensitive to the dispersion of the resonance frequencies (refocalization). Molecular diffusion breaks the symmetry of a spin echo experiment, and, in the presence of susceptibility gradients, introduces a variation ΔR_2 of the speed of transverse relaxation which does not exist in the absence of diffusion. Thus diffusion makes a spin echo experiment sensitive to the activation state of the brain, whilst having the tendency, at least for vessels of small diameter, to make the gradient echo experiment less sensitive to cerebral activation. This sensitization of spin echo techniques is, nevertheless, very selective since it depends heavily on the radius of the vessels (fig. 23.27). It is to be noted, however, that the capillaries (diameter of the order of 5 µm) contribute in a preferential manner to the signal, but this favourable aspect is counterbalanced by the much weaker overall sensitivity of this type of experiment. This aspect must again give rise to caution if one wishes to pursue a quantitative analysis of the results of an fMRI procedure.

The BOLD mechanism introduces a modulation of the NMR signal as a function of the degree of blood oxygenation. This modulation is complex since it may depend on the size of the vessels in the volume concerned, and on the fact that other parameters may contribute to the signal modulation as, for example, the variations of the CBV or even, depending on the type of experiment, the CBF. The BOLD effect is based on the existence of field inhomogeneities created by the heterogeneity of magnetic susceptibility. One notes on looking at the expression (23.54), that these inhomogeneities increase with the intensity of the field. The sensity of fMRI, therefore, increases with the intensity of the field. From this point of view, it is interesting to work in the highest fields, a field of three teslas being considered to be a good compromise between sensitivity and flexibility of use.

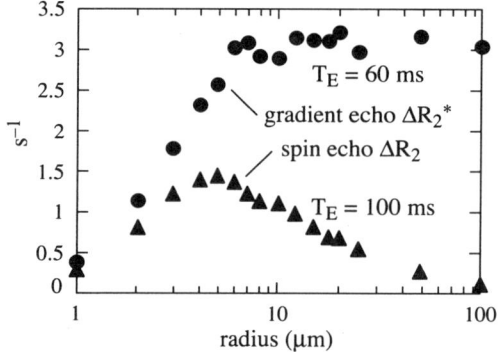

Figure 23.27 - The sensitivity of spin echo (ΔR_2) and gradient echo ($\Delta R_2{}^*$) experiments to cerebral activation

This sensitivity is shown as a function of the radius of the capillaries (Monte Carlo modelling, for a constant blood volume fraction). Courtesy S. Grimault, INSERM unit 438.

3.3. THE fMRI SEQUENCES

The most widely used fMRI sequences aim at exploiting the BOLD effect. They are generally constructed in such a way as to give a strong sensitivity to the parameter $T_2{}^*$. Inevitably, therefore, they are methods based on the acquisition of a gradient echo signal.

Two major categories of methods may be identified:

◆ the *rapid gradient echo* methods based on an excitation of the spin system with the help of a small-angle pulse. The arrangement of gradients is such that each excitation permits the acquisition of one line of reciprocal space (fig. 23.28). In principle an image comprising 64 lines necessitates 64 successive acquisitions (however, to speed up the experiment, one can make a partial exploration of the reciprocal space). The echo time is often of the order of 30 ms, which allows the optimization of sensitivity to the variations in $T_2{}^*$. The repetition time is of the order of 50 to 100 ms, which leads to an acquisition duration for a single slice of less than 5 s. The acquisition of N slices increases the acquisition time by a factor of N.

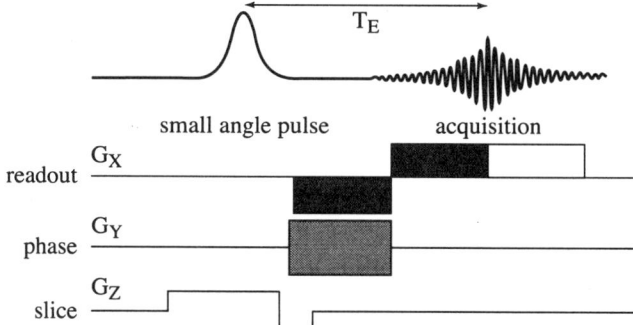

Figure 23.28 - Gradient echo type sequence. Line by line acquisition of Fourier space

◆ *Ultra rapid gradient echo* methods of type EPI (Echo Planar Imaging), where the arrangement of the gradients is such that the whole of Fourier space is covered during the acquisition of the signal which follows excitation by a single 90° pulse (fig. 23.29). The duration of acquisition is of the order of 100 ms so that some ten slices can be obtained in 1 second.

Note - *The term "gradient echo" is often associated with the methods of the first class. The acronym EPI designates the methods of the second class. It is important to note that the EPI methods used in fMRI are in general also gradient echo methods.*

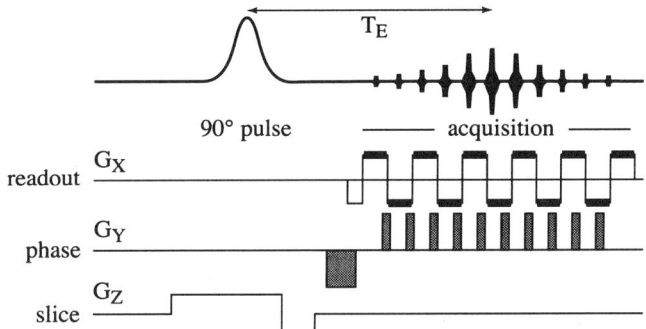

Figure 23.29 - Gradient echo type sequence, EPI mode.
***Complete acquisition of Fourier space** (11 lines in the case of this simplified figure)*

The time resolution of an EPI experiment is, therefore, much better than that which can be reached with rapid gradient echo type methods, notably when multi-slice exploration is necessary. The sensitivity per unit time of EPI methods is also superior, notably in the case of multi-slice exploration. Nevertheless, the setting up of this type of experiment necessitates very efficient gradient systems which are available only in the most recent generation of machines.

3.4. THE fMRI PROTOCOL
AND THE PROCESSING OF THE fMR IMAGES

fMRI is fundamentally a difference technique. The absolute value of the intensity of a pixel of the image cannot, in fact, be associated in a quantitative way to a parameter describing the state of cortical activity. Whatever the protocol and the method of analysis, the concern will always be to compare measurements performed under different conditions.

The subject is asked to carry out a certain number of tasks sequentially. This often involves alternating periods of "rest" (can one speak of cerebral rest?), and periods during which a sensitive or sensory stimulation, or a clearly defined motor or cognitive activity, is carried out. This succession of tasks is often designated by the term paradigm. For the entire duration of the experiment, images of a slice or of a collection of slices are obtained with as good a time resolution as possible, for example 1 second in EPI mode.

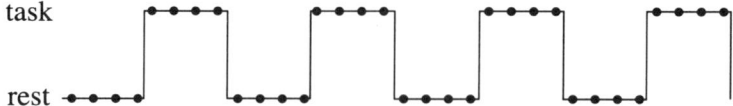

**Figure 23.30 - Example of sampling during a paradigm
consisting of rest-task alternations**

Analysis of the images is then carried out pixel by pixel. A pixel of an activated region is characterised by an increase in intensity when one passes from a period of rest to a period of activity. A pixel of an area not involved in the task or by the stimulation retains a constant intensity over the course of the different periods. In fact, the variations of the signals associated with neuronal activity are extremely weak (frequently of the order of 1 to 2%), and it is not always easy to extract information from the noise.

3.5. EXAMPLES OF APPLICATIONS

The following few examples show the functional maps obtained during sensory (visual) stimulation, motor activity or cognitive (language) tasks. The fMRI experiments that produced these results were all carried out following the same protocol.

They all begin by a rapid, conventional MRI "scout" sequence, generally in the sagittal plane, with a view to positioning the volume to be explored by functional MRI. Then, the functional MRI sequences are applied during performance of a given paradigm. In these functional sequences, spatial resolution, and the contrasts between different soft tissues (white and grey substances, cerebro-spinal fluid) are generally sacrificed in favor of the time resolution and "BOLD" type contrast. Finally, a conventional MRI sequence, of high contrast and high spatial resolution, is obtained. The images provided by this last sequence serve as anatomical references to the functional charts.

The latter are obtained by comparing, pixel by pixel, the intensities of the signals measured both under *control* conditions, and during conditions of *stimulation* or of *task*. In the following examples this comparison is made by estimating the intercorrelation existing between the temporal evolution of the pixels and a model function representing the paradigm applied. Those pixels presenting a sufficiently high intercorrelation factor will be considered as being representative of an "activated" voxel.

3.5.1. Vision

The experiment described here had, as its objective, the identification of the visual cortical areas specifically involved in the perception of motion (areas called MT or V5). Hence the paradigm used alternate control and stimulation periods, the latter differing principally through a motion component of a visual stimulus.

The latter is transmitted to the subject by a system comprising computer, video projector, projection screen, and some mirrors.

The visual stimuli used in this experiment consisted of concentric rings, stationary during *control* periods, and in continuous expansion during periods of *stimulation* (see fig. 23.31).

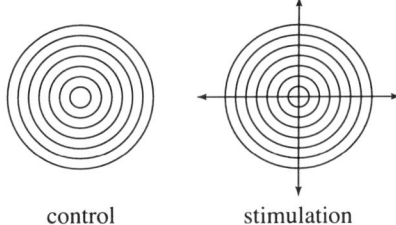

control   stimulation

Figure 23.31 - Visual stimuli

The visual areas being localised in the occipital area of the brain, the signals were detected with the aid of a small circular, flat radiofrequency coil, centred on the cerebral region of interest. This coil presents optimal detection sensitivity for the regions close by.

Figure 23.32 shows a typical result. The two cortical areas presenting functional responses correspond to V5. These results illustrate that different functional areas of the visual cortex (V5 in this case) can be identified by modulating the appropriate parameters of the visual stimuli (motion in the present case).

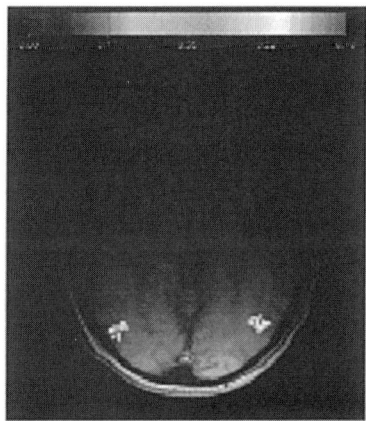

Figure 23.32 - Activity detected in the visual areas (V5 or MT) which present a particular sensitivity to movement

3.5.2. Cognition

The objective of the following experiment was to evaluate the hemispheric dominance for language. The determination of hemispheric dominance for language constitutes a question of major clinical interest, for example when considering patients suffering from temporal epilepsy resistant to treatment. For such patients, surgical resection can be envisaged only in the non-dominant hemisphere with respect to language. The determination of the hemispheric dominance of language in patients is normally carried out with the help of the *Wada test*. During this test, a barbiturate (amobarbital) is injected at the level of the internal carotids, with a view to anaesthetising a hemisphere transiently (for some minutes).

A comparison of linguistic performance before and just after the injection permits the evaluation of the involvement of the corresponding hemisphere in language. The Wada test is invasive, not without clinical risks, and often presents difficulties of interpretation. fMRI, therefore, offers an alternative, or at least a complementary non-invasive, route. In fMRI, the determination of the hemispheric dominance for language clearly requires the application of a paradigm giving rise to functional responses in the language areas. A number of approaches are followed with a view to obtaining these responses. They can be classified into two categories. In the first, the subject is asked to judge the connection between certain characteristics of words (or possibly of "non-words", consisting of a series of characters not representing known words) which are presented to him or her. These characteristics are for instance of orthographic, phonologic, or semantic type. In the other, the subject is led to produce words. This is done preferably silently, as articulation generally creates motion artefacts in the functional images.

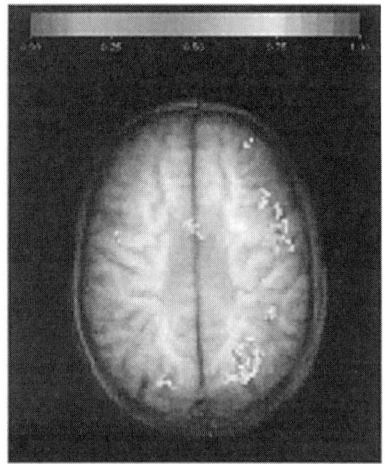

Figure 23.33 - Activity detected in the language areas and in the associative visual areas during a cognitive task presenting a component of visual imagery and of language

The functional image shown here represents functional responses obtained using the second approach. The paradigm consisted of the alternation of control periods, during which the subject avoided speaking (even in an internal fashion), and of task periods when the subject silently named objects identified while mentally picturing familiar scenes. The image represents the projection of functional and anatomical images acquired over a volume of 35 mm thickness. A left-right asymmetry of responses is evident, with a predominance of the left (the subject was right handed; in accordance with radiological convention, the left hemisphere is represented on the right, and *vice versa*).

This result illustrates that it is possible to obtain functional responses by fMRI during a cognitive activity (language). In particular, it demonstrates that the hemispheric dominance of language can be easily determined by fMRI.

3.5.3. Motor function

The voluntary execution of a motor act requires the involvement of a number of cortical areas which translate the intention of making a voluntary movement into motor action. These are comprised of the supplementary motor area (SMA), the premotor area (PM), the primary motor area (M1), and the primary somatosensory area (S1).

Traditionally it has been considered that the initial stages of voluntary movement, such as its planning and programming, involve the SMA and PM areas, with M1 being solely responsible for its execution. This view has been challenged by the results of functional imaging (PET and fMRI) obtained when subjects mentally simulate movements without actually executing them (*mental imagery* of movement). Interest in mental imagery experiments resides in the fact that they provide a "key" to the motor representations present even before execution of the movement. The functional imaging experiments indicate, contrary to traditional opinion, that the primary motor area is involved during the mental simulation of motor action. The two fMRI images shown below show examples of the results obtained during a motor action of the right hand actually carried out (fig. 23.34), and mentally simulated (fig. 23.35). The plane of the images has been centred over the motor region of the hand.

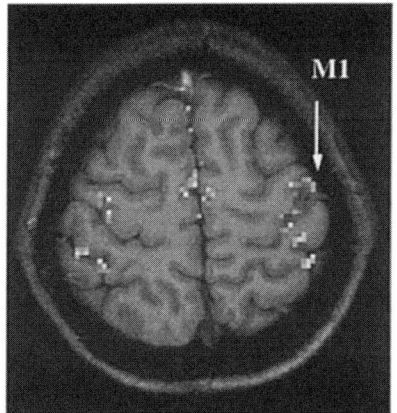

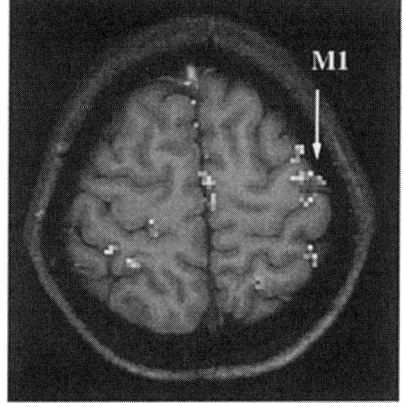

Figure 23.34 - Actual execution   **Figure 23.35 - Mental simulation**

The paradigm applied here alternated between control periods, during which the subject did not carry out any motor task (period of *rest*), and periods during which the subject placed his thumb against the other fingers of the right hand, with a self-regulated frequency of about 1 Hz. Figures 23.34, and 23.35 show that the sensory-motor responses are principally localised within the controlateral hemisphere with regard to the limb in motion. Moreover they demonstrate that the primary motor area M1 is active during the mental simulation of a motor act. It appears, therefore, that there is a strong similarity between the neuronal processes involved in a motor action actually executed, and one that is mentally simulated.

REFERENCES

[1]   F. BLOCH, V.W. HANSEN, M. PACKARD, *Phys. Rev.* **69** (1946) 127.

[2]   E.M. PURCELL, H.C. TORREY, R.V. POUND, *Phys. Rev.* **69** (1946) 37.

[3]   P.T. CALLAGHAN, Principles of nuclear magnetic resonance microscopy (1993), Oxford University Press.

[4]   P. MANSFIELD, P.G. MORRIS, NMR imaging in biomedicine (1982) Academic Press, New York-London.

[5]   R.R. ERNST, G. BODENHAUSEN, A. WOKAUN, Principles of Nuclear Magnetic Resonance in One and Two Dimensions (1987) Oxford Univ. Press, Oxford.

[6]   A. ABRAGAM, Principles of Nuclear Magnetism (1961) Oxford Univ. Press, Oxford.

[7]   C.P. SLICHTER, Principles of Magnetic Resonance (1978) Springer Verlag, Berlin-New York

[8]   P. LAUTERBUR, *Nature* **242** (1973) 190.

[9]   P. MANSFIELD, P.K. GRANNER, *J. Solid State Phys.* **8** (1973) L-422.

[10]   P.T. FOX, M.E. RAICHLE, *Proc. Nat. Acad. Sci. USA* **83** (1986) 1140.

[11]   R.S. MENON, S. OGAWA, X. HU, J. STRUPP, P. ANDERSON, K. UGURBIL, *Magn. Reson. Med.* **33** (1995) 453.

[12]   J.W. BELLIVEAU, D.N. KENNEDY, R.C. MCKINSTRY, B.R. BUCHBINDER, R.M. WEISSKOPF, M.S. COHEN, J.M. VEVEA, T.J. BRADY, B.R. ROSEN, *Science* **254** (1991) 716.

[13]   S. OGAWA, T.M. LEE, A.R. KAY, D.W. TANK, *Proc. Nat. Acad. Sci. USA* **87** (1990) 9867.

[14]   S. OGAWA, T.M. LEE, A.S. NAYAK, P. GLYNN, *Magn. Reson. Med.* **14** (1990) 68.

[15]   K.K. KWONG, J.W. BELLIVEAU, D.A. CHESLER, I.E. GOLDBERG, R.M. WEISSKOPF, B.P. PONCELET, D.N. KENNEDY, B.E. HOPPEL, M.S. COHEN, R. TURNER, H.M. CHENG, T.J. BRADY, B.R. ROSEN, *Proc. Nat. Acad. Sci. USA* **89** (1992) 5675.

[16]   S. OGAWA, D.W. TANK, R. MENON, J.M. ELLERMAN, S.G. KIM, H. MERKLE, K. UGURBIL, *Proc. Nat. Acad. Sci. USA* **89** (1992) 5951.

[17]   P.A. BANDETTINI, E.C. WONG, R.S. HINKS, R.S. TIKOFSKY, J.S. HYDE, *Magn. Reson. Med.* **25** (1992) 390.

[18]   D. LE BIHAN, R. TURNER, T.A. ZEFFIRO, C. CUENOD, P. JEZZARD, V. BONNEROT, *Proc. Nat. Acad. Sci. USA* **90** (1993) 11802.

[19]   S.G. KIM, *Magn. Reson. Med.* **34** (1995) 293.

[20]   R.R. EDELMAN, B. SIEVERT, D.G. DARBY, V. THANGARAJ, A.C. NOBRE, M.M. MESULAM, S. WARACH, *Radiology* **192** (1994) 513.

[21]   K.K. KWONG, D.A. CHESLER, R.M. WEISSKOFF, K.M. DONAHUE, T.L. DAVIS, L. ÖSTERGAARD, T.A. CAMPBELL, B.R. ROSEN, *Magn. Reson. Med.* **34** (1995) 878.

[22]   L. STRYER, *Biochemistry* (1981) W.H. Freeman & Co, New York.

[23]   L. PAULING, C. CORYELL, *Proc. Nat. Acad. Sci. USA* **22** (1936) 210.

[24]   R. WEISSKOFF, S. KIIHNE, *Magn. Reson. Med.* **24** (1992) 375.

[25]   P.A. BANDETTINI, E.C. WONG, *Int. J. Imag. Syst. Tech.* **6** (1995) 133.

[26]   S. OGAWA, R.S. MENON, D.W. TANK, S.G. KIM, H. MERKLE, J.M. ELLERMAN, K. UGURBIL, *Biophys. J.* **64** (1993) 803.

[27]   J.L. BOXERMAN, L.M. HAMBERG, B.R. ROSEN, R.M. WEISSKOFF, *Magn. Reson. Med.* **34** (1995) 555.

CHAPTER 24

MAGNETISM OF EARTH MATERIALS
AND GEOMAGNETISM

The discovery of the earth's magnetic field belongs to the origins of physics. Although less well known than the beginnings of astronomy and celestial mechanics associated with Copernicus, Galileo and Newton, its history is nevertheless exemplary of the development of science. It was the existence of a real need by society (precise navigation using the magnetic compass), the experimental deduction by Gilbert, and the problem of the rationalization of action at a distance (the only scientific example for a long time being the force between magnets) that were responsible for this early progress. The demonstration by Gilbert in 1600 of the internal origin of the earth's magnetic field based on the similarity of the lines of force around the earth and those around a naturally magnetic sphere of lodestone, is considered to be the first publication of modern physics, even before the solution by Galileo of a body falling under gravity. This internal origin along with the mystery of action at a distance probably contributed to the "diabolization" of magnetism during the following centuries.

The study of the earth's magnetic field and of the magnetism of natural materials now constitutes two fully fledged disciplines: Geomagnetism and Rock Magnetism. The former is today particularly concerned with the origin of the earth's magnetic field and the latter is an offshoot of physical magnetism and has benefited from the pioneering theories of Louis Néel as applied to palaeomagnetism. These two disciplines have since developed their own methodology, leading to many applications, essentially in the field of the Earth Sciences, but also in Extraterrestrial, Life and Environmental Sciences.

1. INTRODUCTION

After a short presentation of some experimental techniques, this chapter is concerned with the magnetic properties of natural materials, essentially terrestrial rocks. These can be defined as coherent aggregates of different solid phases usually crystallised (minerals). Also discussed are the properties of free mineral particles found in the air, water, soils and sediments, living beings, as well as those of extra-terrestrial materials and synthetic analogues of these natural materials likely to be present in the environment: ashes, concrete sewage sludge…

It is useful to distinguish three types of magnetic properties:

♦ the intrinsic magnetic properties (susceptibility, spontaneous magnetization, hysteresis, ordering temperature, etc.) which give information on the nature of the material, its chemical composition in magnetic elements, and how these elements are assembled into minerals. At this level all materials mentioned can be studied, whereas in the following only consolidated materials are concerned [1, 2, 3].

♦ the anisotropy of these properties, due to the preferential orientation or petrofabric of the anisotropic magnetic minerals. This magnetic fabric gives an insight into the structure of materials [4, 5].

♦ the natural remanent magnetization (NRM), which is the recording by a rock of the magnetic fields it has experienced since its formation, is the basis of palaeomagnetism. This allows one to trace the history of the earth's magnetic field during geological times, and also the changes undergone by a rock, such as displacement, rotation, reheating, etc. [1, 6, 7, 8].

As palaeomagnetism is practically unique in giving geophysical information about the past state of the earth, the subject has undergone a rapid development in the last few decades. In effect the geologist is most often a detective reconstituting past events from the evidence observed in rocks, whilst the geophysicist observes the actual state of our planet [9, 10, 11, 12].

Geomagnetism will be treated within the perspective of palaeomagnetism as the several centuries of direct measurements of the field are very little compared to the 3.9 Ga record in the geological archives [13, 14, 15, 16]. (*In this chapter ka is a millennium, Ma a million years and Ga a billion years, etc.*).

2. EXPERIMENTAL TECHNIQUES

2.1. GENERALITIES

The instruments particular to palaeomagnetism need to meet the following requirements:

♦ the magnetic grains are highly diluted (concentrations practically always less than 10^{-2} and often in the range 10-100 ppm, that is 10^{-5} to 10^{-4}) which implies the need for a sensitivity that is unusually high for physical magnetism.

♦ because of this dilution, the variable granulometry (typically in the range 0.01 μm to 1 mm), the presence of small-scale heterogeneities, and the need for a precise specimen orientation, the samples are quite large (at least 10 cm^3). The field sampling techniques as well as the sample holders have been rapidly standardised. The norm for rocks being a cylinder 25 mm in diameter and 22 mm long, whilst for unconsolidated materials a 20 mm cube is inserted into a protective plastic box. The cylindrical rock cores are obtained in the field using a portable two-stroke petrol coring drill equipped with a water-cooled diamond tipped coring tube. The cores 10-20 cm in length are later sliced in the laboratory into specimens 22 mm long using a non-magnetic saw blade. The orientation of the rock sample is defined in the field by two angles given by an orientation device which combines a clinometer and a magnetic or solar compass with an angular precision approaching one or two degrees. The geographical reference frame is defined by the trihedron, North, East and Down. A given direction is described by the declination, which is the angle between the geographical north and the projection of the magnetic vector in the horizontal plane, and the inclination defined as the angle between the magnetic field direction and the horizontal plane.

♦ the inherent variability of natural materials, the need to extrapolate the results to a kilometric or continental scale, as well as the study of the temporal variation has led to the study of a great number of samples which compensate for this variability by using a statistical approach. In general a regional study concerns a few tens of sites or outcrops (an area of around 10 m in size to average out any inhomogeneities in the recording processes in the rock), and each site is made up of a dozen cores with each core giving several specimens. In the case of a detailed study of the temporal variation within a sedimentary sequence, it may be necessary to take a sample every two centimetres over a core length of several metres. The thousands of samples, which have to be measured each year in a laboratory, require analytical equipment, which is simple to use, where the specimens stay at room temperature and pressure, and need only a few minutes of measurement time per specimen.

Finally the units commonly used in palaeomagnetism are volumetric as most of the specimens are of a standard volume. Specific units are reserved for quantitative studies, for example in rock magnetism.

2.2. MEASUREMENT OF REMANENT MAGNETIZATION (NRM)

The range of NRM observed in rocks is large, from 10^{-5} to 10^2 A.m^{-1}, and meaningful results can be obtained throughout all of this range. This implies that the measuring equipment must have a very large dynamic range and sensitivity better than 10^{-6}, which at the scale of the standard 10 cm^3 specimen, corresponds to the detection of an induction of the order 10^{-14} T. A system, which corresponds to these specifications, is a magnetometer using the Josephson effect (SQUID device) specially designed for palaeomagnetism, and often called a cryogenic magnetometer. Systems based on new high temperature superconducting materials and operating with liquid nitrogen are being developed but liquid helium systems (developed at the end of the 70's) will probably remain the norm due to the substantial progress made in cryogenic techniques: a fill of helium can last up to two years. The usual configuration of the apparatus consists of a horizontal cylindrical helium reservoir of around 100 L completely traversed by a central tube allowing the insertion into the central measuring zone of single samples or in the case of sedimentary cores continuous tubes 1.5 m long. This tube or hot finger is surrounded by three superconducting coils linked to Josephson junctions (flux transformers) giving simultaneously the three components of the NRM vector in an acquisition time of around one second. Under optimum conditions one can attain a precision of 0.1%. Magnetometers based on classical induction but amplifying the signal by a fast and prolonged rotation of the sample (spinner magnetometer) are also widely used because they are less expensive, and are more robust. Their sensitivity lies in the range 10^{-5} to 10^{-4} A.m^{-1} with a precision approaching 1% at the best. Whatever the measuring system, the NRM vector is calculated from a combination of several successive measurements with the sample in different orientations which allow one to estimate the noise, the signal from the sample holder, the instrumental drift, an eventual induced magnetization, a poor centring of the magnetic moment of the sample and sample positioning errors.

The measurement of remanence implies the absence of any induced magnetization that is in the absence of the ambient magnetic field, which is normally around a few tens of μT. The measurement system should be shielded using a set of Helmholtz coils (little used today) or by means of cylindrical MuMetal tubes. The use of 3 to 5 layers of MuMetal 1 mm thick reduces the field at the centre of the tubes to a few nT. As the absence of field variations is essential to the working of a Josephson junction a superconducting shield is added inside the MuMetal ones. Non-magnetic rooms have also been developed, whose walls are covered with silicon-iron sheet or MuMetal so that all the experimental work can be carried out in a space where the field is of the order of 0.1 to 1 μT, so minimising the risk of the samples picking up a parasitic magnetization.

Stepwise demagnetization treatments, obtained (i) by heating or (ii) by alternating magnetic field are essential for the analysis of the complex NRM signal, see § 5.3. In (i) the samples are heated to temperatures ranging from 100 to 700°C in a non-magnetic oven placed inside MuMetal shields, and then cooled in zero field. In the second case (ii) the oven is replaced by a coil fed by an alternating current whose maximum value corresponds at the centre of the coil to the chosen field (up to 0.1 T or 0.3 T depending on the equipment) then slowly decreasing regularly to zero. Between each step of increasing temperature or demagnetising field the residual remanence is remeasured.

2.3. *SUSCEPTIBILITY AND ANISOTROPY MEASUREMENTS*

The magnetic susceptibility that is commonly measured is the reversible initial susceptibility χ_0 obtained in a weak alternating field (< 1 mT) at room temperature.

The choice of experimental conditions follows several objectives:

- ◆ the susceptibility measured should be independent of the applied magnetic field to be able to
  - – compare measurements made in different weak fields, and
  - – for anisotropic material, use the tensor relation (2.53) $\mathbf{M} = \chi \mathbf{H}$ between $\mathbf{M}$ and $\mathbf{H}$; from χ_0, an estimation of the induced magnetization of the basement in the geomagnetic field is useful for the interpretation of magnetic anomalies (§ 6);
- ◆ ease of measurement of χ (measurement time typically 2 s) at the highest sensitivity.

The measurement apparatus (impedance bridge type) have a sensitivity of 10^{-15} to 10^{-16} in the field or in diagraphy (borehole logging), and 5×10^{-8} in the laboratory, the values encountered range from -10^{-5} to 1. The measured susceptibility is always the *apparent susceptibility* whose maximum value is 3 for a sphere, according to the equation (2.80), $\mathbf{M} = [\chi / (1 + N\chi)] \mathbf{H}_0$, the intrinsic susceptibility requiring corrections for the shape of the grains. The frequencies used vary from 10^2 to 10^4 Hz, the variations of χ_0 as a function of frequency in this range are small because of the very low electrical conductivity of minerals. However, in the case of very fine grains (transition superparamagnetic-single domain) a relative decrease of χ_0 is observed, up to 20% for an increase of frequency of a factor of 10. One can also determine χ_0 by measuring the induced magnetization in a weak field (for example in a cryogenic magnetometer), the remanent magnetization having been subtracted from measurement after inversion of the sample. However, the problem of viscosity limits the general use of this method.

Another interesting susceptibility is that measured in a high field $\chi_{hf,}$ which is with a field strong enough to saturate the ferromagnetic minerals. Once saturation is achieved the magnetization varies in a linear manner with the field. χ_{hf} is a tensor, which gives information on the non-ferromagnetic minerals (matrix) generally making up the essential part of the volume of the rock. It should be noted that $\chi_0 = \chi_{wf} + \chi_{hf}$ where χ_{wf} is the weak field susceptibility of the ferromagnetic grains.

The susceptibility is anisotropic, and is a tensor quantity: it is defined by six independent parameters (three principal values, χ_1, χ_2, χ_3, and three directions which give the orientation of the trihedron of the eigenvectors). One determines the susceptibility anisotropy either by measuring the susceptibility in carefully chosen directions, or by determining the susceptibility differences in three perpendicular planes of rotation. For the latter, one uses a torsion balance or a system of induction with an applied alternating field acting on a rotating sample.

2.4. CHARACTERIZATION OF MAGNETIC MINERALS

This characterization is done by measurements of induced or remanent magnetization as a function of the temperature or the applied field. The apparatus used are similar to those used in physical magnetism. The study of the behavior of artificially induced remanences is frequently used to identify the carriers of the NRM. Amongst the most frequently used are the Isothermal Remanent Magnetization (IRM) acquired by exposing the sample to a continuous or pulsed field for a short time, and the Anhysteretic Remanent Magnetization (ARM) produced by subjecting the sample to a weak direct field along with a strong alternating field which decreases slowly to zero. To give a sample an ARM one uses a modification of the apparatus used for alternating field demagnetization.

Physical methods other than magnetic ones are also used, such as X-ray diffraction, Mössbauer spectroscopy and microanalysis... They should be adapted to the low concentration, the size of the magnetic grains, the systematic presence of a mixture of different phases and of solid solutions, etc. A magnetic separation and/or separation by density can overcome these difficulties.

The study of thin sections under the optical microscope either in the transmission or reflection mode is an indispensable first step in the characterization of the mineral phases in a rock. A reflected light study is particularly useful for the oxides and sulphides responsible for the magnetic properties. As the magnetic carriers, especially in sediments, are often smaller in size than 10 μm, electron microscopy of the magnetic extract complements the optical methods.

2.5. MEASUREMENT OF MAGNETIC FIELDS

Naturally occurring magnetic fields are weak and can be observed over a large frequency range. In the case of continuous fields or when their time variations exceed a second, one can use magnetometers based on nuclear or paramagnetic resonance that combine rapidity with ease of use (precision approaching 10^{-5} and a sensitivity up to 0.1 nT). When the vector components of the field are required, a magnetometer based on the flux gate sensor is most frequently used. Many different configurations of this sensor exist but it typically consists of two parallel soft magnetic cores driven to saturation by an alternating current in two primary coils but with opposite phase. A secondary winding around both cores measures the asymmetry produced by any direct magnetic field parallel to the axis of the cores. For alternating magnetic fields various induction systems are used with or without an iron core.

3. INTRINSIC MAGNETIC PROPERTIES
OF EARTH MATERIALS

3.1. INTRODUCTION

Earth materials, principally rocks, are essentially made up of crystalline phases, *minerals* whose properties will be described in § 3.2, and much more rarely of amorphous phases (glasses produced by the rapid cooling of a silicate magma or certain phases formed in the presence of water at room temperature).

Minerals are formed from the available store of natural chemical elements, consisting of 91 different elements. The specific crystal chemistry of each element or class of elements is such that they are segregated into different phases. However, each phase is often a solid solution with various pure end members generally involving 4 to 10 different principal elements: for example, for the phase $(Ca, Mg, Fe)CO_3$, these are: $CaCO_3$, $MgCO_3$ and $FeCO_3$. In addition one finds traces of a few tens of other elements in the crystalline lattice, which substitute for the normal constitutive elements or exist as inclusions.

This complexity of natural minerals is moderated by the relative abundance of the elements: in terms of mass percentage only ten elements have an average concentration in the earth's crust larger than 0.1%, viz in decreasing order: O, Si, Al, Fe, Ca, Na, Mg, K, Ti, Mn. To this list of the so-called major elements one could add those light elements, which although their average mass concentration is less than 0.1%, are major components of certain minerals: H, C, S, P, Cl, F. All other elements are present as traces (< 400 ppm), and only in exceptional cases form a specific phase.

Out of the above list only two elements are "magnetic", that is to say capable of exhibiting a spin moment: Fe and Mn, with average crustal concentrations of 4 and 0.1% respectively. Magnetic mineralogy is essentially reduced to that of iron and accessorily of manganese minerals.

The other magnetic elements Cr, Ni, Sm, Co, Nd, U, to mention only the most abundant, have average concentrations of 122, 99, 39, 29, 7 and 3 ppm, respectively. This table of the abundance of the elements in the crust is quite different from that in the solar system, where iron is the most abundant element (Fe / Si = 0.6), and from the magnetic point of view nickel is close behind (Ni / Si = 0.03). This is due to the density segregation which began at the very beginning of the history of the earth: 86% of Fe and 99% of Ni are stored in the metallic core of the earth below a depth of 2,900 km.

Minerals are classed according to their principal elements:
- silicates, minerals formed from SiO_4 groups with varying substitution of Si by Al. As these 3 elements constitute 80% of the mass of the earth's crust, one can understand that this class is quantitatively essential,
- oxides, associate the metals with oxygen,
- sulphides, associate the metals with sulphur,
- carbonates (CO_3), sulphates (SO_4), phosphates (PO_4) and oxyhydroxides.

From the point of view of rock magnetism one distinguishes three classes of minerals:

a. diamagnetic minerals which do not contain any magnetic elements, and are therefore characterised by a negative susceptibility, for example quartz (crystallised SiO_2, $\chi = -14.5 \times 10^{-6}$) and calcite ($CaCO_3$, $\chi = -12.3 \times 10^{-6}$).

b. paramagnetic minerals which contain iron (or manganese), but at concentrations too low to be magnetically ordered at room temperature. Their susceptibility is independent of field up to several Tesla, and follows a Curie-Weiss law: $\chi = \mathscr{C} / (T - \Theta_p)$. These minerals being almost always solid solutions, the value of the Curie constant $\mathscr{C}$ and the paramagnetic Curie temperature Θ_p vary according to the degree of substitution.

c. ferromagnetic minerals in the broadest sense of the term, that is to say magnetically ordered at room temperature, are found only amongst the oxides, sulphides and oxyhydroxides. They are characterised by their ordering temperature (T_C or T_N), weak field susceptibility and hysteresis cycle: spontaneous magnetization M_s and remanent magnetization after saturation M_{rs}, coercive force H_c and remanent coercive force H_{cr}. It is ferromagnetism in a broad sense because this term groups together real ferromagnetism (which does not exist in terrestrial materials), with ferrimagnetism, and an antiferromagnetism associated with a parasitic ferromagnetism. Considering the intensity of the remanent magnetization measured and the absence in nature of pure and perfectly crystallised phases, the antiferromagnetic minerals always exhibit a parasitic magnetization.

The following section essentially describes group (c) which is present in all rock magnetic applications. The paramagnetic minerals, which only contribute their paramagnetic susceptibility, will be described more briefly.

3.2. MAGNETIC MINERALS

Table 24.1 summarises the properties described in this chapter [1, 2]. As one is dealing with fine grains, the effects of size will greatly modify the properties such as remanence, coercivity and susceptibility. In particular one is interested in grains that cover the superparamagnetic-single domain (SP-SD), and single domain-multidomain (SD-MD) transitions at room temperature. For elongated grains of magnetite, these critical sizes are respectively 50 and 500 nm. The other minerals have much larger critical sizes. In the case of grains whose size is slightly larger than the second transition, their properties (remanence and coercivity) are closer to those of SD than those of MD grains, due to blocking of domain walls. This is known as pseudo-single domain behavior (PSD). The transition between this type of grain and the MD grains with good wall mobility occurs at a size of up to 10 times that corresponding to the transition SD-MD.

3.2.1. Oxides

In the system Fe-O, three phases exist in nature: Fe_3O_4 (magnetite), $\gamma\text{-}Fe_2O_3$ (maghemite), $\alpha\text{-}Fe_2O_3$ (hematite). These phases can be either pure or substituted, with Ti being the principal element substituting for Fe, for example in the case of titanomagnetite $Fe_{3-x}Ti_xO_4$, with $x = 1$. Other elements such as Al, Mg, Mn, Cr, Ni. can also replace Fe. This substitution, by diluting the Fe ions, has the effect of reducing T_C and M_s without modifying in a fundamental manner other properties. For the sake of simplicity, only the pure phases are described below.

Magnetite is the best-known magnetic mineral as it is the stable form of iron for low oxygen partial pressures, and also because it is responsible for the strongest magnetizations observed. Black in colour, it crystallises in the cubic system and belongs to the spinel family, already described in § 6.1 of chapter 17. It exhibits a crystallographic transition, that of Verwey, at 118 K.

Table 24.1 - Characteristics of some natural magnetically ordered compounds at room temperature

Mineral	Formula	System	M_s $(A.m^2.kg^{-1})$	T_C, T_N (°C)	$\chi_{app.}$ $\times 10^3$
Iron	Fe	cubic	218	765	3000
Magnetite	α-Fe_3O_4	cubic	92	580	3000
Titanomagnetite	$Fe_{2.4}Ti_{0.6}O_4$	cubic	25	150	3000
Maghemite	γ-Fe_2O_3	cubic	84	675	3000
Hematite	α-Fe_2O_3	rhombohedral	0.2 to 0.5	*675*	2-50
Titanohematite	α-$Fe_{1.4}Ti_{0.6}O_3$	rhombohedral	20	110	-
Ferroxyhite	δ-FeOOH	hexagonal	7 to 20	187	-
Goethite	α-FeOOH	orthorhombic	0.001 to 1	*90-120*	1-5
Greigite	Fe_3S_4	cubic	31	330	-
Pyrrhotite	Fe_7S_8	monoclinic	17	325	100-300
Pyrrhotite	Fe_9S_{10}	hexagonal	(15 to T>200°C)	*290*	1-2

These compounds are ferrimagnetic (except metallic iron) or antiferromagnetic with a weak or parasitic ferromagnetism (in italics).

Note - χ_{app} *is the apparent susceptibility, which takes into account the demagnetising field. For a soft magnetic material made up of spherical grains the demagnetising factor is N = 1/3 (see chap. 2).*

Maghemite is also ferrimagnetic with the same crystalline structure as magnetite, but with the Fe^{2+} cations replaced by Fe^{3+} and lattice vacancies. This defect lattice is metastable, with an irreversible transformation to hematite when the temperature exceeds 350°C. This brownish-red mineral is relatively rare as it is formed under non-equilibrium conditions, either by slow oxidation of magnetite, or by low temperature processes which have a biological or organic aspect. Its magnetic properties are similar to those of magnetite.

Hematite, along with the oxyhydroxides described below, is the stable form of iron in an oxidising environment. This hexagonal mineral, which as a powder has a characteristic blood-red colour, is antiferromagnetic with the spins oriented perpendicular to the c-axis. Above the Morin temperature (260 K), an asymmetric, otherwise known as Dzyaloshinsky, coupling produces a weak spontaneous magnetization perpendicular to the antiferromagnetic axis. Below this temperature massive hematite should not have a spontaneous magnetization.

In addition to the intrinsic anisotropic moment perpendicular to the c-axis, in fine-grained hematite there is also an isotropic parasitic moment parallel to the c-axis *in the basal plane* due to an imperfect compensation of the two antiferromagnetic sublattices. In consequence, the observed remanent magnetization for fine grains is little or not at all reduced below the Morin temperature. Hematite has a very strong anisotropy with an easy plane of magnetization perpendicular to the c-axis, and a uniaxial and triaxial anisotropy in this plane.

3.2.2. Sulphides

Sulphides are formed at the surface in an anaerobic environment containing organic matter due to the action of bacteria taking oxygen from sulphate ions, and at depth from hydrothermal fluids rich in H_2S. The majority of metallic ore deposits are associated with sulphides, which explains their economic importance. The iron sulphides are the most abundant, but the principal one, pyrite FeS_2, is non magnetic. In effect, the covalent bonds Fe-S are such that the electrons are in a low spin state. Pyrite does not have a Langevin type of paramagnetism, only a weak paramagnetism independent of temperature ($\chi = 10^{-5}$) of the Van Vleck type.

Two other iron sulphides are ferrimagnetic: greigite (Fe_3S_4), which crystallises in the cubic system, and monoclinic pseudohexagonal pyrrhotite (Fe_7S_8). The magnetic structure of the former is similar to that of magnetite, but the mineral is metastable and decomposes at temperatures above 300-350°C. Its Curie temperature is equal or greater than the transformation temperature. In the case of Fe_7S_8, the ferrimagnetism is due to an ordering of vacancies in planes perpendicular to the c-axis. The magnetization is strictly confined to this plane, as is the intrinsic magnetization of hematite, but accompanied here by a large biaxial and triaxial anisotropy. Fe_7S_8 also has a low temperature magnetic and electronic transition at 34 K.

Hexagonal pyrrhotite Fe_9S_{10}, is antiferromagnetic (disordered vacancies) but a reversible disorder-order transition occurs at 200°C, and above this temperature this composition is ferrimagnetic up to its T_C of 290°C. The iron sulphides found in nature are often a mixture of these different phases (and others not yet discovered) on a micron scale. It should be noted that troïlite FeS and its equivalent oxide, wustite FeO, practically do not exist in nature.

3.2.3. Other magnetic minerals

There exist several iron oxyhydroxides, FeOOH, which are formed in association with or in place of hematite in surface processes. The most important of these is the α isomer gœthite, which dehydrates at around 350°C. This yellow-orange orthorhombic mineral is antiferromagnetic and is usually in the form of needle shaped crystallites (along the antiferromagnetic c-axis), of micron length and width of the order of 0.01 μm. The poor crystallinity and the large number of defects could explain the existence of an important parasitic moment even for synthetic samples. This magnetization is remarkable for its very high coercivity.

Native iron, even though being practically absent from terrestrial rocks, merits being mentioned as it is abundant in lunar rocks and in meteorites. It occurs in certain rocks from the mantle and also as Fe-Ni alloys in meteorites.

3.2.4. *Paramagnetic minerals*

The great majority of minerals are either diamagnetic or paramagnetic at room temperature. Amongst the latter one can mention those oxides poor in Fe, Mn or Cr, for example ilmenite $FeTiO_3$ ($T_N = 60$ K), siderite $FeCO_3$ ($T_N = 35\text{-}40$ K), as well as all the ferromagnesian silicates. The principal examples are: olivine $(Fe, Mg)_2SiO_4$, the pyroxenes –chain silicates whose formula is $(Fe, Mg)SiO_3$ or $(Fe, Mg)Ca(SiO_3)_2$, the amphiboles similar to the pyroxenes but hydrated, and the phyllosilicates with a hydrated sheet structure (e.g. micas and clays). The ordering temperatures are at the most 60 K (iron end member of the olivines) and most often less than 10 K; varying with the Fe content and the ratio Fe^{2+}/Fe^{3+}. The paramagnetic Curie temperature Θ_p is greater or equal to the ordering temperature in absolute terms but can be negative. In many silicates containing iron the amount of this element is too low for any order to be observed down to 2 K, and Θ_p is close to 0 K.

3.3. *APPLICATIONS OF MAGNETIC MINERALOGY TO THE EARTH SCIENCES*

Iron is one of the most abundant elements in natural materials. It is present in different oxidation states and in a multitude of minerals with contrasting magnetic properties that give information on the physico-chemical conditions of their formation. The identification and dosing of these phases by magnetic measurements, known as magnetic mineralogy, is therefore a prime technique for studying geological processes, due in particular to the sensitivity and rapidity of the measurements: for example, a susceptibility measurement takes one second, without preparation of the sample, for a sensitivity equivalent to 0.1 ppm of magnetite.

The applications are very numerous and correspond either to the mineralogical transformations of iron or to the variation in the amount of a magnetic phase. In this way one can determine the temperature reached by a rock containing pyrite: above 400°C, this mineral is transformed into pyrrhotite, which multiplies the susceptibility by 10^4. In the same way, the oxidation of a rock (for example by infiltration of surface water) will be detected by the transformation of magnetite into hematite.

In sedimentary sequences one often observes a variation of susceptibility, which is perfectly correlated with the climatic cycles revealed by other means (fig. 24.1). These cycles, which are now well established, arise from the control of the insolation by the earth's rotational and orbital parameters with periods of 97, 40 and 28 ka. Magnetic measurements can be used, either in the laboratory or *in situ* inside a borehole, to determine the climatic cycles and so date the different sedimentary levels. In the present example –a sequence of Quaternary aeolian deposits (< 2 Ma) in China– the warm periods correspond to an increase in susceptibility due to the formation of maghemite in the soil; whilst during glaciations, the biochemical activity in the soil is non-existent and no maghemite is formed. In figure 24.1, the cold periods are shaded. In oceanic sediments, on the contrary one observes a decrease in the concentration of magnetite (coming from continental erosion) during warm periods, due to the dilution of the detrital input by diamagnetic carbonates synthesised by plankton in larger quantities than during cold periods.

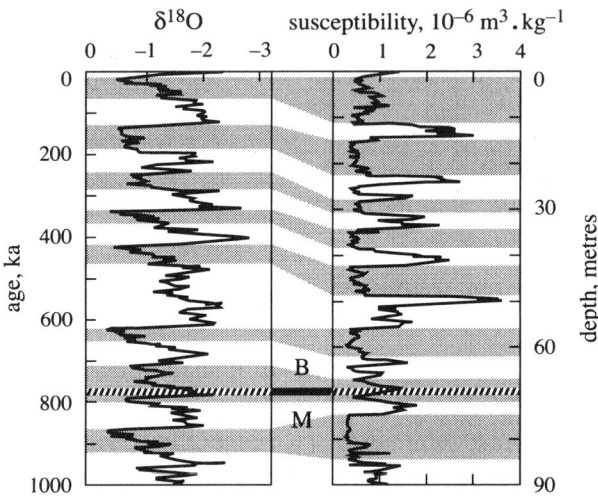

**Figure 24.1 - Magnetic Susceptibility as a function of depth
in the lœss series (Aeolian deposit) of Xifeng in China**

χ (right side) is correlated with the climatic cycles recorded by the oxygen isotope ratio expressed as the parameter $\delta^{18}O$ in the calcite of marine microorganisms as a function of age (left side). B/M corresponds to the recording of the last field reversal (Brunhes/Matuyama, see fig. 24.10). The cold periods are shaded in grey and correspond to an increase in ^{18}O in the seawater in which the microorganisms grow [3].

An example of another application is shown in figure 24.2 where one observes a good correlation between the remanence after saturation (M_{rs}) of different surface samples and their concentration in cadmium, a highly toxic metal discharged into the environment by anthropogenic activities. The iron content of the materials studied varies little, but this iron is found in weakly magnetic materials (paramagnetic or antiferromagnetic) in those particles which have a natural origin, whilst it is in the form of magnetite (from combustion processes or metallurgical activity) in the anthropogenic deposits which are also loaded with heavy metals. On the other hand, a magnetic extraction that would concentrate the particles of magnetite would also concentrate the heavy metals (Cd, Pb, Zn, Cu...). Therefore one could use magnetic separation, already extensively used for the purification of ores or the sorting of garbage, as a process for the depollution of sewage, of muds or of soils contaminated with heavy metals [3].

The ratio Cd/Al is plotted against the ratio M_{rs}/Al for preindustrial (■) and actual (□) sediments, atmospheric dusts (○) and plane tree leaves (▲) sampled in or around the "Etang de Berre" (a lake near Marseille).
Normalization with respect to aluminium (an element occurring naturally in the clays) makes the analyses independent of the mass of the sample and the dilution by organic matter.

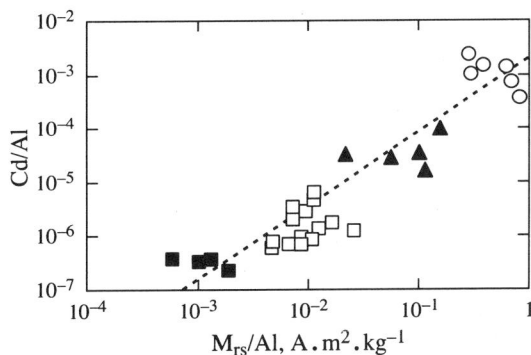

Figure 24.2 - Correlation between Saturation remanence and Cadmium content

4. MAGNETIC ANISOTROPY: APPLICATION TO THE DETERMINATION OF ROCK FABRIC

The determination of the preferential orientation of crystals in rocks, or petrofabric, is an essential step in the study of the mode of formation of natural materials, and of the deformations that they have subsequently undergone during their long history. A knowledge of the petrofabric allows one to look back at the modes of flow, at the deformations and the tectonic constraints, and therefore at numerous dynamic aspects of geological processes.

The petrofabric can be determined visually in a qualitative manner on those rocks, which show a highly developed fabric, and by using a microscope for fine-grained material. This observation enables one to define a plane of preferential orientation (foliation) and a direction of alignment (lineation) of the shape axes or of the network of grains. For the sake of simplicity one can model the shape of the grains by an ellipsoid, the most often of revolution (plate or needle-like grains). The crystal lattice is usually coaxial with the shape axes. For example, the phyllosilicates are in the form of plates perpendicular to the pseudo-hexagonal c-axis. Rheological models allow one to quantitatively estimate the petrofabric as a function of the shape of the grains, of the relative viscosity of the matrix around these grains, of the type of deformation (coaxial or rotational) and of the intensity of this deformation. Foliation, lineation and degree of preferential orientation therefore allow in certain conditions the determination of the plane of flattening or of flow, the direction of extension or of shear, as well as the degrees of deformation.

The **anisotropy of magnetic susceptibility** in a weak field (**AMS**) of a grain will be determined by its crystalline lattice, except in the case of minerals of high susceptibility (like that of magnetite) where the form effect will dominate. For phyllosilicates the susceptibility ellipsoid will be a flattened ellipsoid of revolution ($\chi_1 = \chi_2$) about the c-axis, with a degree of anisotropy $P = \chi_1 / \chi_3$ of the order of 1.2 to 1.3. For hematite or pyrrhotite, the relationship between the crystallographic and magnetic axes is the same, but the values for P can exceed 100. Finally for an ellipsoidal grain of magnetite of dimensions a > b > c, the susceptibility ellipsoid will be practically identical to the inverse of the demagnetising field ellipsoid N whose principal axes are functions of the ratios a/b and b/c. The measurement of the AMS of a rock containing anisotropic magnetic minerals allows the determination of the preferential orientation of these minerals. The fabric that is determined is that of the lattice (and indirectly of shape through the relationship between crystalline lattice and the shape of a grain) except in the case of magnetite where one directly determines the shape fabric.

The development of AMS was an important step forward in the Earth Sciences [4, 5] because of the numerous advantages in comparison with other petrofabric techniques:

♦ rapidity, as the measurement of a sample takes about five minutes, compared to several hours using an X-ray texture goniometer or image analysis of oriented thin sections,

♦ result is directly representative of the volume of the sample, whilst the above mentioned techniques apply only to planes. The extrapolation to three dimensions requires several hypotheses and measurements in three mutually perpendicular planes,

♦ independent of the size of the grains,

♦ sensitivity that allows the determination of rocks reputed to be "isotropic", for example magmatic rocks, sediments and soils.

However, the transition from the susceptibility ellipsoid to the petrofabric requires in each case knowledge of the nature of the magnetic minerals present. A quantative application, for example translating the degree of anisotropy P into the amount of flattening is in principle possible if one knows the parameter P of the magnetic grains and a relevant rheological model. These conditions are rarely fulfilled and one often has to be satisfied with a semi-quantitative application within a given mineralogical composition and a mechanism of homogeneous deformation.

Figure 24.3, showing the measurements of AMS in some strongly deformed metamorphic rocks, illustrates these limitations. In figure 24.3-a the directions of χ_1 and χ_3 of layers rich in phyllosilicates are plotted. The plane of horizontal flattening, and the direction of tectonic stretching E-W are respectively perpendicular to χ_3 and parallel to χ_1, as one would expect knowing the AMS of the phyllosilicates. The degrees of planar ($F = \chi_1 / \chi_2$) and linear ($L = \chi_2 / \chi_3$) anisotropy are high with L < F, in agreement with the intense flattening that these rocks have undergone. On the other hand the results obtained on the decimetric carbonate beds which are intercalated with the previously mentioned beds (fig. 24.3-b), show an inverse directional relationship, along with a lower degree of anisotropy and L = F.

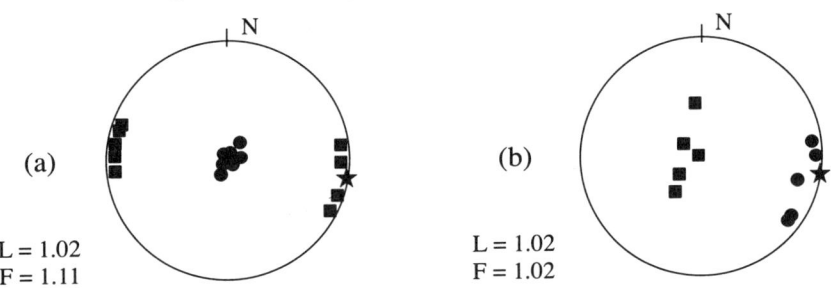

L = 1.02
F = 1.11

L = 1.02
F = 1.02

Figure 24.3 - Stereographic projection of the directions of maximum (χ_1, squares) and minimum (χ_3, circles) susceptibilities of the Dauphine schists *(a) Phyllosilicate beds - (b) Carbonate beds*

The projection of the hemisphere on the circle gives the diameter for a horizontal plane (which is here the plane of flattening), and the centre is the vertical direction. Star: stretching direction.

This is due to the fact that under the effect of intercrystalline deformation, the ferriferous carbonates develop to a moderate degree an alignment of their c-axes parallel to the direction of flattening. The c-axis of the carbonates turns out to correspond to the magnetic axis χ_1. The difference in the AMS of these two kinds of rock, although having undergone the same tectonic conditions, can be explained by the magnetic properties and a quite different deformation mechanism for those minerals responsible for the magnetic susceptibility.

Figure 24.4 shows the AMS observed in a phyllosilicate rich sediment deposited in two days by a flood of the River Isère at Grenoble (France).

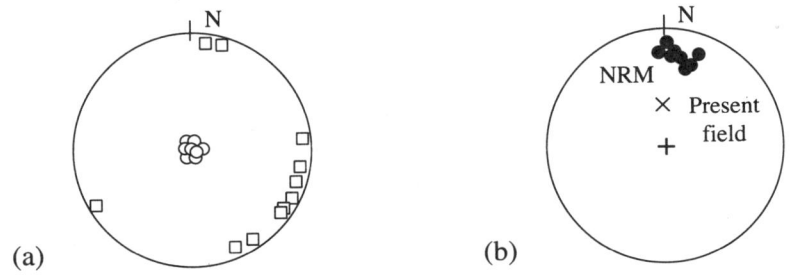

Figure 24.4 - Stereograms of the directions measured in a sediment deposited in 1989 by a flood of the River Isère at Grenoble *(a) Stereogram of the anisotropy of susceptibility (see fig. 24.3) (b) stereogram of the natural remanent magnetization*

Cross: direction of present day field.

A strong magnetic foliation (plane perpendicular to the mean direction of χ_3), which is horizontal, corresponds to the plane of sedimentation, whilst the magnetic lineation (average direction of χ_1) turns out to be parallel to that of the current during deposition. The determination of palaeocurrents is seen as one of the interesting applications of AMS.

A deep-sea sediment, sampled by coring in the middle of the Pacific Ocean, is characterised by a low sedimentation rate (1 mm.ka^{-1}), and has quite a different AMS (fig. 24.5, see also fig. 24.10 concerning the same site). The first few metres of sediment appear to be isotropic (random χ_3 directions and a degree of anisotropy F < 1.002), whilst lower down a slight magnetic foliation appears, with a progressive increase in F. These observations, which are quite different from the previous case, are explained by the fact that the deposition of the sediment was completely destroyed by the action of burrowing organisms (bioturbation) which are found in the first 20 cm of sediment. This interval corresponds to a lapse of time of 0.2 Ma, which is largely sufficient to produce the observed complete isotropy. Lower down compaction will progressively reduce the water content of the sediment and re-orientate the grains into the plane of horizontal flattening so producing a progressive increase in F.

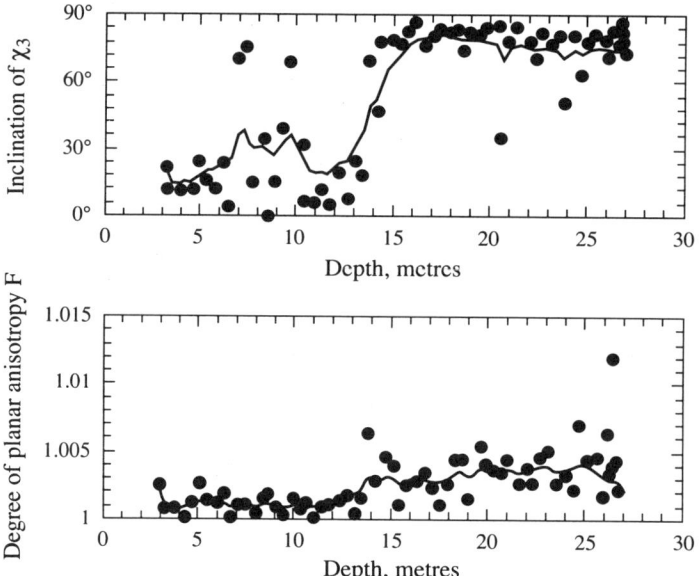

Figure 24.5 - Degree of planar anisotropy and inclination of the axis of minimum susceptibility as a function of depth in a sediment core from the Indian Ocean

AMS is a technique for characterising rock fabric, and is increasingly used to solve many problems in geology. In order to separate the different contributions to the low field susceptibility one can use the high field anisotropy, which gives information on the fabric of the paramagnetic minerals, and also the anisotropy of remanence, which is related only to the ferromagnetic.grains.

5. NATURAL REMANENT MAGNETIZATION

5.1. PRINCIPLES OF PALAEOMAGNETISM

The **natural remanent magnetization (NRM)** of a material can give information on the magnetic fields recorded by the ferromagnetic grains during the history of the material [1, 6]. Palaeomagnetism is based on the following assumptions linking the **NRM** vectors and the earth's magnetic field $\mathbf{B_T}$:

♦ **NRM** is acquired at a known instant of time t_0 in the history of the rock; and if this at the moment of formation of the rock (determined by radiochronology or palaeontology, for example) this is called a primary **NRM**;

♦ **NRM** $= c\,\mathbf{B_T}$ (where c is a positive scalar), that is the **NRM** is parallel to and in the same sense as $\mathbf{B_T}$. This measurement of the direction of ancient field can be completed by an estimation of the palaeointensity, either absolute if one is capable of determining c by reproducing in the laboratory the NRM acquisition mechanism, or relative if c is just taken as some constant. This constant is a function of the process of acquisition of the magnetization, of the intrinsic properties of the carriers of the magnetization, and of their concentration. If one considers the first two conditions to be constant for a particular series of rocks, one obtains a relative palaeointensity by normalising the NRM using the susceptibility or an artificial remanent magnetization.

Palaeomagnetism can therefore be used to determine:

♦ the intensity and the direction of the field at the time t_0 $[B_T(t_0)]$, if the rock has not been disturbed since t_0,

♦ the age of the rock if one knows function $B_T(t)$,

♦ the movement undergone by the rock since t_0, based on a knowledge of t_0 and the direction of $B_T(t_0)$.

These applications will be dealt with in § 7 and 9. A description of the different processes of NRM acquisition (§ 5.2) enables one to see how and under what conditions the above assumptions are valid. Concerning the first of these, one sees that the total NRM is the sum of multiple components acquired at different epochs, in particular after the formation of the rock (secondary NRM). The techniques used to detect and separate these different components will be described later (§ 5.3), as well as the criteria used to determine the age of each component. For the second assumption, the parallelism between the NRM and B_T implies a negligible anisotropy effect. In the case of the strongly anisotropic sediment in figure 24.4 this is not the case as the NRM is deflected by 30° towards the horizontal plane of easy magnetization (perpendicular to χ_3). The possibility that the two vectors could be antiparallel exists. Known as auto-inversion, a phenomenon predicted by Louis Néel, this process is limited to the case of a TRM acquired by grains with interlaced phases of different Curie temperature.

5.2. *NRM ACQUISITION PROCESSES*

The earth's magnetic field (several tens of μT) is significantly less than the coercive force of the fine magnetic grains found in nature. Therefore a NRM cannot be acquired by the simple application of a field $B_T > B_{cr}$ which would produce an IRM (see § 2.4). However an exception exists, for the large currents (10^4 to 10^6 A) produced by a lightning strike can remagnetise rocks over a surface of several square meters in a concentric fashion around the point of impact.

The first three types of natural magnetization that are described below require the notion of a relaxation time τ. For monodomain grains, which carry a strong remanence, Néel showed that $\tau = c\,e^{-vK/k_B T}$, where $c = 10^{-9}$, v is the grain volume, and K the magnetic anisotropy constant.

When τ is much longer than the measurement time t_m, the moment of the grain is blocked; if this is not the case the moment fluctuates spontaneously between the different directions of easy magnetization. This latter state is called *superparamagnetism* (see § 2.3 in chap. 4 and § 3.1 in chap. 22).

When a superparamagnetic grain changes its temperature or volume, one can define a blocking temperature T_b below which its time of relaxation satisfies the inequality $\tau \gg t_m$, or a blocking volume v_b above which $\tau \gg t_m$. In the case of multi-domain grains, the same concepts of relaxation time and temperature can also be applied.

5.2.1. *Thermoremanent magnetization (TRM)*

Thermoremanent magnetization is produced by the cooling down of a rock in an ambient magnetic field. When a grain cools below its blocking temperature ($T_b < T_C$) there is a greater probability that its moment will be blocked in the direction of easy magnetization that is closest to that of B_T, so that a population of randomly oriented grains will acquire a TRM which is parallel and proportional to B_T for weak fields.

Thus, for non-interacting magnetite grains the susceptibility χ_{TRM} ($\chi_{TRM} = \mu_0 A T R / B$) lies between 1 and 100 (maximum value for single domains). By creating a TRM using a known field in the laboratory one can determine the palaeointensity of B_T using the so-called Thellier method. TRM is the primary magnetization of magmatic rocks.

5.2.2. *Crystallization remanent magnetization (CRM)*

Crystallization or chemical remanent magnetization is produced by the growth of a superparamagnetic grain past its blocking volume v_b at constant temperature. A CRM can also be a primary magnetization in the case of a rock whose magnetic grains have crystallised during formation of the rock. For example in certain sediments or in soils magnetic oxide or sulphide grains crystallise just below the surface due to bacterial processes. CRM can also be secondary in origin due to a remobilization of iron in the rock by a change in physico-chemical conditions. Thus when a deep rock is exposed at the surface due to erosion it is in direct contact with oxygen and water, and primary magnetite can be replaced by hematite or goethite that will carry a secondary CRM.

5.2.3. Viscous remanent magnetization (VRM)

Even at constant temperature and volume the property of relaxation time can allow the formation of a magnetization over a geological time scale (e.g. the last 1 Ma). This is known as a *viscous remanent magnetization*, which would appear to be stable on a laboratory time scale, and is a recording of the average field during a time $< \tau$.

5.2.4. Detrital remanent magnetization (DRM)

The NRM acquisition processes that have been described so far do not involve any movement of the magnetic grains. Only the magnetic spins follow the direction of the magnetic field. However, in the case of a sediment formed by deposition at the base of a water column of particles of remanent moment m, these can turn freely under the effect of a couple $m \times \mathbf{B_T}$. When they arrive at the water-sediment interface, the ferromagnetic particles will statistically conserve their orientation, acquired during their settling, which is parallel to $\mathbf{B_T}$. Known as *detrital remanent magnetization*, it is often deflected towards the horizontal plane. This is called inclination error (fig. 24.4). In a very recent sediment, the water content can exceed 60%. The cohesion of the aggregate of particles is very weak and the magnetic grains can eventually re-orientate in the ambient field, particularly if bioturbation occurs and if the granulometry is fine. One then has a post-depositional magnetization (PDRM), as shown in figure 24.5. Although a PDRM does not exhibit an inclination error, it does have a time lag with respect to the age of deposition, corresponding to the time required to deposit 10 cm of sediment. Far from any continental influx, this could correspond to more than 10 ka. Then compaction begins to occur, eventually producing an inclination error, then the sediment will start to cement by partial solution-crystallization of certain phases. This cementation can be accompanied by the acquisition of a CRM. It can be seen that sediments, on which a large part of palaeomagnetic applications repose, have NRM acquisition processes which are both multiple and complex.

5.2.5. Piezoremanent magnetization (PRM)

Finally one should mention the possibility for a material to acquire a remanent magnetization under the effect of a directed stress: this is piezoremanent magnetization. Natural directed stresses are often quite small (a few tens of bars) and PRM would appear to be inexistent in common rocks. However, it can be produced during coring of a rock. The strong magnetization of certain meteorites could be explained by the presence of a PRM.

5.3. NRM ANALYSIS TECHNIQUES

The central problem of the separation and identification (remanence carrier, magnetization mechanism, age) of the different components of the NRM is approached using the techniques of progressive demagnetization by heating or applying alternating magnetic fields in a field-free enclosure. By following the evolution of the NRM vector during its destruction by ever increasing temperatures or alternating magnetic fields, one can determine the superposed directions if they are carried by grain populations having different blocking temperatures or coercive field spectra. When the direction no longer changes during the demagnetization, one has isolated the characteristic magnetization (ChRM). One generally expects the secondary magnetizations to disappear first. This is the case for a VRM as T_b or H_{cr} decrease with the relaxation time. However, the correct interpretation of the demagnetization should be linked with the identification of the magnetic minerals. In the example of a primary remanence of a sediment (DRM) carried by magnetite to which is superposed a secondary magnetization due to alteration (ChRM) and carried by goethite, an alternating magnetic field will first remove the primary DRM, whilst a thermal demagnetization will make the secondary ChRM disappear above 120°C. In the case of a CRM due to hematite, it is the secondary magnetization that will better resist both alternating fields and thermal demagnetization, and the CRM risks being wrongly considered to be the primary magnetization.

In order to evaluate the coherence of ChRM directions measured on N samples from the same exposure, one uses Fisher spherical statistics. It is assumed that the measurements have a normal isotropic distribution about the mean direction, so that the probability p that a given direction has an angle θ with the true mean direction is given by, $p(\theta) = c \, e^{K\cos\theta} \sin\theta$, where $c = K / 2 \, \text{sh}(K)$. K is the precision parameter: the larger is K, the smaller is the dispersion of the directions. (It should be noted that here K is not an anisotropy constant). The value of K of a finite population of N directions of ChRM can be estimated from the mean vector $\mathbf{R} = \Sigma \, \mathbf{ARN}_{ci} / | \mathbf{ARN}_{ci} |$. One can show that $K = (N - 1) / (N - |\mathbf{R}|)$. In order to represent the uncertainty in the mean direction $\mathbf{R}$, one defines an angle of confidence α_{95}, such that the probability for the true mean direction to make an angle $< \alpha_{95}$ with $\mathbf{R}$ is 0.95. The angle $\alpha_{95} = 140 / \sqrt{KN}$ in degrees defines a cone of confidence of semi-angle α_{95}, around $\mathbf{R}$ (example fig. 24.6).

This angle, which tends to 0 when N approaches infinity, is much less than the circular standard deviation of the distribution: $\delta = 81 / \sqrt{K}$. Typically for a site of 10 samples showing a well-defined ChRM, one can expect values for K between 100 and 1,000 and with $\alpha_{95} < 5°$.

The age of the ChRM can also be deduced from the relation of the palaeomagnetic directions with the geological phenomena. Two examples are given in figure 24.6, which allow a test of the primary character of the magnetization:

♦ in folded sedimentary rocks, if the ChRM was acquired before the folding, one should find different directions according to the tilting of the beds. If one puts the beds back into their original horizontal position, the corrected NRM directions should regroup together (fig. 24.6-a and -b).

♦ in the assembly, known as a conglomerate, which is formed from blocks resulting from the erosion of an older rock, should the directions of ChRM of these blocks be greatly dispersed, then the original NRM acquired at formation of the rock has resisted the erosion, and the transport of the blocks (fig. 24.6-c).

The coherence of a palaeomagnetic interpretation may be reinforced by the agreement between the obtained direction and that which would be expected from palaeomagnetic studies carried out in the same region or the same continent (see § 7 and 9).

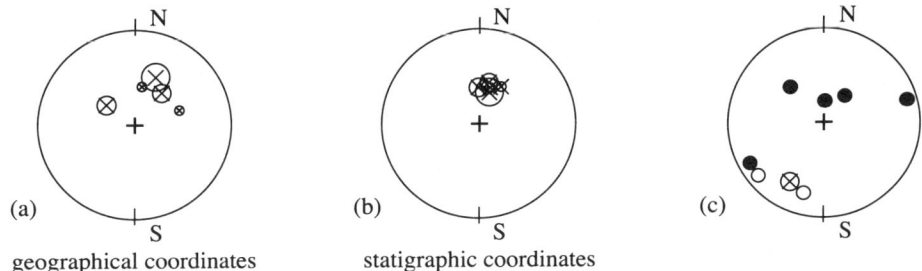

Figure 24.6 - Three examples of stereographic projections of ChRM

Fold test comparing the mean directions of ChRM (represented with their circle of confidence of radius α_{95}): (a) before unfolding of the beds (geographical coordinates) and (b) after (stratigraphical coordinates) for Jurassic subalpine calcareous sediments (190 Ma) - (c) Conglomerate test on Permian volcanic pebbles (280 Ma) from the Estérel range (south of France), in an interstratified layer between two volcanic flows. The mean direction of the regional flows is indicated by a cross together with its circle of confidence. The stereogram shows both hemispheres: open circles (filled) for vectors pointing up (down).

6. PRESENT GEOMAGNETIC FIELD

6.1. GENERALITIES

The ambient magnetic field is continuously measured in many magnetic observatories around the world (in France at Chambon-la-Forêt) in order to monitor its time variation, whilst magnetometers carried by satellites, aircraft, boats or on the ground allow one to follow its spatial variation at all possible scales.

The temporal analysis is done either by filtering or processing the signal using a Fourier transform in order to separate the different frequencies, whilst the most important tool in the spatial analysis of the field is its decomposition into spherical harmonics. This was first carried out by Gauss in 1838 using observatory measurements to completely define the earth's magnetic potential $\mathcal{V}$ in spherical coordinates, where r is the distance to the centre of the earth, θ the colatitude (90° - latitude) and ϕ the longitude, with the help of a set of functions of decreasing wavelength (spherical harmonics) satisfying Laplace's equation: $\Delta\mathcal{V} = 0$ (that is **rot B** = 0). This condition implies that no electric current crosses the reference surface. To a first approximation this is the case for the surface of the earth. The potential $\mathcal{V}$ can be written:

$$\mathcal{V} = \sum_{\ell=0}^{\infty} \sum_{m=0}^{\ell} \left(A_\ell^m r^\ell + B_\ell^m r^{-(1+\ell)} \right) Y_\ell^m (\theta, \phi) \qquad (24.1)$$

where A_ℓ^m and B_ℓ^m are the two sets of coefficients of the decomposition, measuring respectively the external and internal sources of the field, whilst Y_ℓ^m are the spherical harmonics, derived from Legendre polynomials P_ℓ^m (app. 6, in volume I).

The external sources of the earth's field are the electric currents in the ionised part of the atmosphere. It can be shown that if one uses the field values averaged over one year the terms A_ℓ^m cannot be detected. The external field intervenes only in the temporal variations around a zero annual mean value (< 1 nT). One can simplify the expression for the mean potential, measured in amperes, to give:

$$\mathcal{V} = (a/\mu_0) \sum_\ell \sum_m (a/r)^{1+\ell} P_\ell^m (\cos\theta)\left(g_\ell^m \cos m\phi + h_\ell^m \sin m\phi \right) \quad (24.2)$$

where a = 6,371 km.

The sources of the internal field could lie in the magnetization of rocks or internal electric currents. These electric currents are to be found essentially in the metallic core of the earth, between 2,800 and 6,371 km depth, whilst magnetised rocks occur only at less than 30 km depth (Curie isotherm of magnetite). These quite different source depths mean that the core field can only be detected at the surface in harmonics up to the order $\ell = 8$ (wave length around 5,000 km), whereas rock magnetization, which varies in an incoherent manner as a function of the distance, only contributes to harmonics of a much higher order [13, 14].

6.2. CORE FIELD

The variations with θ of the Legendre polynomials of order 1, 2, and 3 corresponding to dipolar, quadrupolar and octopolar fields are given in the appendix 6 in volume I (fig. A6.2). Table 24.2 gives the coefficients g_{lm} and h_{lm} in μT for these terms.

Table 24.2 - Coefficients for the expansion of the mean field
into spherical harmonics, expressed in μT

{ℓ, m}	1, 0	1, 1	2, 0	2, 1	2, 2	3, 0	3, 1	3, 2	3, 3
g_ℓ^m	−30.0	−1.95	−2.04	3.03	1.65	1.29	−2.15	1.24	0.85
h_ℓ^m	–	5.63	–	−2.13	−0.18	–	−0.04	0.26	−0.24

Fields which are symmetric with respect to the axis of rotation of the earth, called zonal, correspond to terms with m = 0 and are uniquely defined by the coefficient g. The coefficient g_{10} corresponds to an axial dipolar field (along z, its sign indicating a dipole pointing towards the S pole), whilst g_{11} and h_{11} give the components along x and y of the dipole field. It can be seen that approximately 90% of the earth's magnetic field can be represented by a dipole inclined at 11°
$$\left(= \arctan\left[\sqrt{{g_{11}}^2 + {h_{11}}^2} \Big/ g_{10} \right] \right)$$ with respect to the rotation axis of the earth.

This inclination justifies the notion of geomagnetic poles: places where the field is vertical, and distinct from the geographical poles. These poles are found on maps (fig. 24.7) of the intensity and of the declination (D is the angle between the geographic north and the horizontal component) of the mean field at the surface of the earth. The geomagnetic poles appear on a declination map as two singular points (where is D is undefined as the field is vertical), and correspond to the intensity maxima in the Canadian Arctic and in the Antarctic near Adélie Land.

It is clear from these maps that the field is not uniquely dipolar: the lines of force are not symmetrical with respect to the magnetic equator, a third maximum in intensity is observed in Siberia, etc. Table 24.2 shows the quadrupole and octupole terms to be of the order of 1 to 10% of g_{10}. Beyond order 3, the coefficients are always less than 1 μT, but never fall below 0.1 μT out to order 7.

Precise measurements of the mean field show that it varies continuously, and the maps shown here are only valid for 1990. The annual variation of the different coefficients is of the order of 10 nT.a^{-1} (tab. 24.2). More prosaically, this slow displacement of the lines of force, otherwise known as secular variation, is more readily observed in the declination: the isolines D = 0 both in Europe and North America drift to the west at approximately 100 km.a^{-1}.

The precise use of the magnetic compass requires a good knowledge of the secular variation: for example the declination will be exactly zero in Paris during the year 2010. Observatory measurements of the direction of the field, which have been carried out in Paris since 1650 (fig. 24.8), confirm the nature of this secular variation. For the temporal variations of the field before this date, one has to rely on **archaeomagnetism** and palaeomagnetism (§ 7).

6.3. SHORT WAVELENGTH SPATIAL VARIATIONS: CRUSTAL MAGNETIZATION

The difference between the mean field and the core field as described by the spherical harmonics (1 < 10), is called the anomaly field. On average it is zero but it can fluctuate by up to ±1 μT. However, the order of magnitude of these anomalies is 10 to 100 nT. The spatial variations of the field having a wavelength < 1,000 km do not have their source in the core, and are therefore produced by contrasts in crustal magnetization.

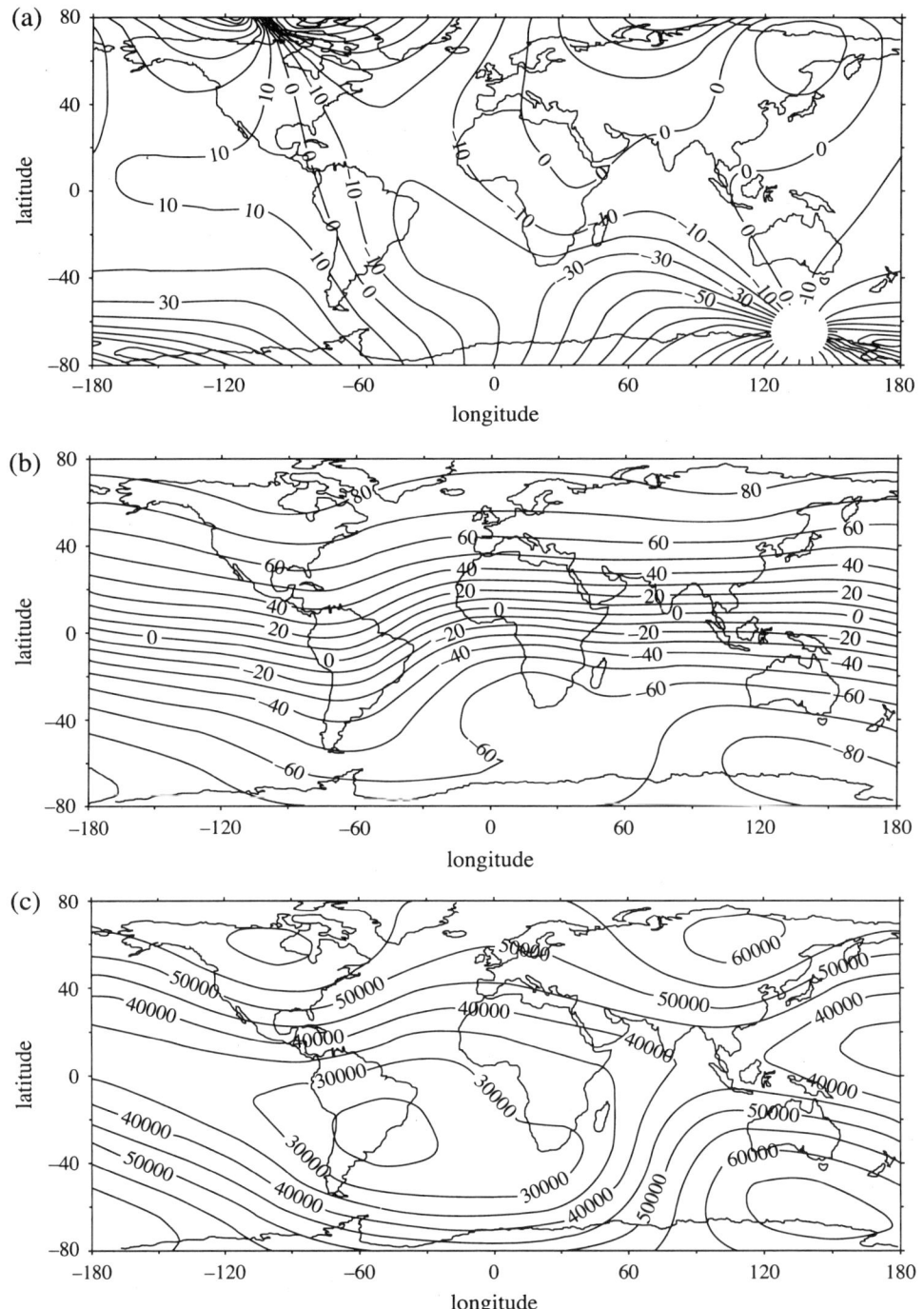

***Figure 24.7 - World maps of the declination,
inclination and intensity of the mean field for 1990 [14]***

*The isolines of intensity and angle are respectively in nT and in degrees
(positive to the east for the declination and downwards for the inclination).*

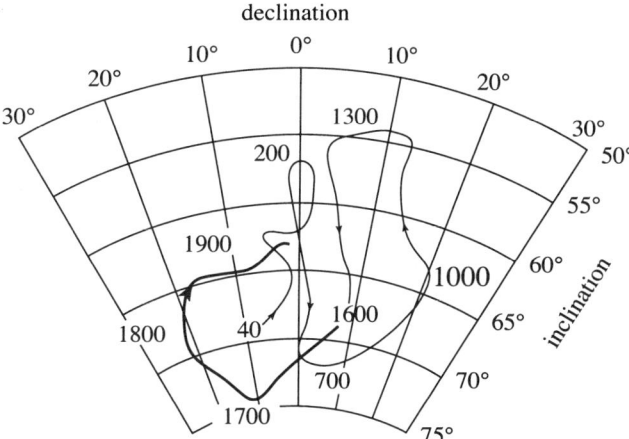

Figure 24.8 - Secular variation of the geomagnetic field at Paris during the last 2000 years

Stereographic projection of the variation of the declination and inclination, as measured directly since 1650 (thick line), or deduced from archaeomagnetic studies (thin line, see § 7) between 40 and 1,600 A.D. [15].

This magnetization can be induced by the ambient magnetic field. Any contrasts are solely due to the susceptibility, which depends on the concentration of magnetic minerals. This type of anomaly is used to detect the contrasts of susceptibility in the ground, and discover metallic ore bodies (very often associated with pyrrhotite or magnetite), or to determine the depth or geometry of a rock body more magnetic than the surrounding rocks. The sensitivity of this method can be evaluated from the yardstick that $1 m^3$ of magnetite buried at a depth of 30 m will produce an anomaly with an amplitude of 1 nT, that is a little more than the detection limit of a magnetometer.

Above a layer of rock which appears homogeneous one does not observe a magnetic anomaly (constant χ), except when the remanent magnetization varies and is of the same order of magnitude as, or greater than χB_T. In this way magnetic anomalies enable a palaeomagnetic study to be carried out at a distance. This is the case for the anomalies recorded in the oceanic crust (see § 7).

6.4. RAPID VARIATIONS: EXTERNAL FIELD

Apart from the very localised phenomenon of lightning, an ionised atmosphere, that is electrically conducting, is only found from an altitude of about 50 km until 1,500 km. This ionosphere undergoes cyclic heating by the sun's rays, producing transient electrical currents mainly with a daily period, which generate a field of a few tens of nT [13].

On the other hand the "solar wind", a flux of charged particles given out by the sun (protons, electrons, helium nuclei) at a velocity relative to the earth of $400 \, km.s^{-1}$, is slowed down and deflected when it meets the earth's magnetic lines of force. This encounter creates a supersonic shock wave; with the particles that envelop the earth "combing" the dipolar lines of force to form a sheath called the magnetosphere whose dimensions are several earth radii (fig. 24.9). The field lines in this magnetosphere vary due to the rotation of the earth's dipole (inclined with respect to the earth's rotation axis), and also because of the very great variability of the intensity of the solar wind. The interactions between the plasma of solar particles, the magnetospheric field and the ionospheric currents produce many very complex electromagnetic phenomena. The most prominent is the magnetic storm, which results in a rapid variation (on the scale of one minute to one hour) of the field at the surface of the earth, and which can reach 1 μT. The initial phase is generated by the sudden arrival of a cloud of particles that are linked to the instability of sunspots. The resulting displacement of the lines of force induces strong ionospheric currents, in particular at low latitudes, the equatorial electrojet. The particles emit electromagnetic waves by synchrotron radiation over a large frequency band and with an energy sufficiently great to seriously disrupt radio telecommunications. These phenomena are made worse at low latitudes, where the emission can develop in the middle of the day even without a magnetic storm. Finally it can be seen in figure 24.9 that the poles are favoured regions for the penetration of solar particles into the lower atmosphere. The collisions between the particles and the oxygen in the air cause the emission of light, giving rise to the aurora borealis around the north pole, and its counterpart the aurora australis in the southern hemisphere.

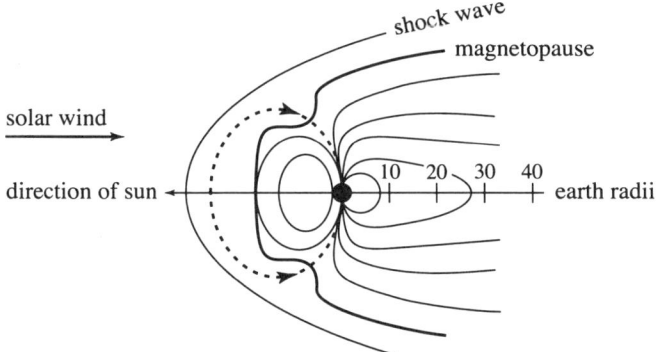

Figure 24.9 - Magnetic environment of the earth

The magnetopause (thick line) separates the magnetosphere (zone filled by the earth's magnetic field whose lines of force are shown by thin lines) from the conducting medium formed by the particles of the solar wind. The dotted line shows the easy path for particles penetrating the magnetosphere.

Low frequency waves (1 to 10,000 Hz) detected at the surface are linked to lightning strikes in atmospheric storms. A network of receiving antennas allows the localization and an estimation of the storm intensity for meteorological studies. These waves are capable of inducing electric (telluric) currents in the earth's upper crust whose resistivity varies from 10 to 10^4 Ω. m. The determination of the resistivity at depth by the study of these currents is known as the magnetotelluric method.

The environment can be affected by artificial magnetic fields. On the ground beneath high-tension lines the alternating magnetic field (50 Hz) is of the same order as the local geomagnetic field (10 μT). The biological effects of such alternating fields remain controversial. The electrical current of a lightning strike causes as much damage by the field impulse that it produces, and the associated induction in electrical circuits, as the direct passage of the lightning current into facilities. The enormous amplification of these phenomena in the "fire-ball" of an atmospheric nuclear explosion, capable of destroying by overvoltage the electrical installations over a very extended area, is a very potent weapon that is part of electromagnetic warfare.

7. *THE ANCIENT FIELD RECORDED BY PALAEOMAGNETISM*

A knowledge of the past geomagnetic field is possible using palaeomagnetism, provided that the reliability criteria mentioned previously in § 5.1 are fulfilled, that the age of the acquisition of the NRM is known, and that the magnetised object has not been subsequently displaced (in the case of a directional study). The secular variation curve from the Roman period onwards, established by E. Thellier, and shown in figure 24.8, was constructed by measuring the TRM of well dated archaeological baked clays (oven walls, *in situ* fired bricks, etc.) which were collected from all over France. It can be seen that the secular variation during the last two millennia is not the cyclic phenomenon suggested by the historical measurements, which give the impression of a secular variation due to the precession of an inclined dipole. Apart from this chaotic aspect, the average field can be seen to be statistically indistinguishable from the theoretical field calculated for a central axial dipole.

This agreement, which could be fortuitous, is seen to be true for each part of the globe where there are sufficient palaeomagnetic data from volcanic rocks of the last few Ma. Volcanic rocks are used because their ChRM is acquired in less than a year, and therefore it does not risk being filtered by the secular variation, as is the case with sediments. If the field at our scale is non-dipolar with its complex variations, it appears at the scale of 10^3-10^6 years as essentially composed of a central axial dipole together with a random noise causing an angular dispersion which varies according to the latitude from 15 to 20° [16]. This stable state characterises the major part of geological time, but sometimes instabilities occur whose amplitudes are much greater than the secular variation. This is called an excursion, as occurred approximately 40 ka ago and lasted of the order 1 ka, when rocks recorded field directions that were practically opposed to those of the actual dipole field.

However, the major instability of the geomagnetic field is its reversal. Palaeomagnetic studies show that the earth's dipole has reversed very many times during the geological past, and the last occurred 0.78 Ma ago (fig. 24.10). The observation of reversed magnetization in ancient rocks, reported for the first time by Bruhnes at the beginning of the twentieth century in lava flows from Cantal, in the French Massif Central, can only exceptionally be explained by the phenomenon of self reversal invoked by L. Néel (see § 5.1). The process of reversal itself lasts only a few thousand years. On the other hand the average duration of periods where the axial dipole remains pointing to the south (present day state, called normal polarity) or to the north (reversed polarity) is 0.3 Ma for the last 5 Ma.

This process is not cyclic: it obeys Poisson statistics. The elaboration of the geomagnetic reversal time scale (fig. 24.10-b) is based on the compilation of the polarities found in well-dated rocks, and on the use of the magnetic anomalies of the oceanic crust. The reversal sequence can be recovered as a function of the distance from the dorsal axis where the oceanic crust is formed (fig. 24.10-c), or as a function of depth in a sedimentary sequence (fig. 24.10-a).

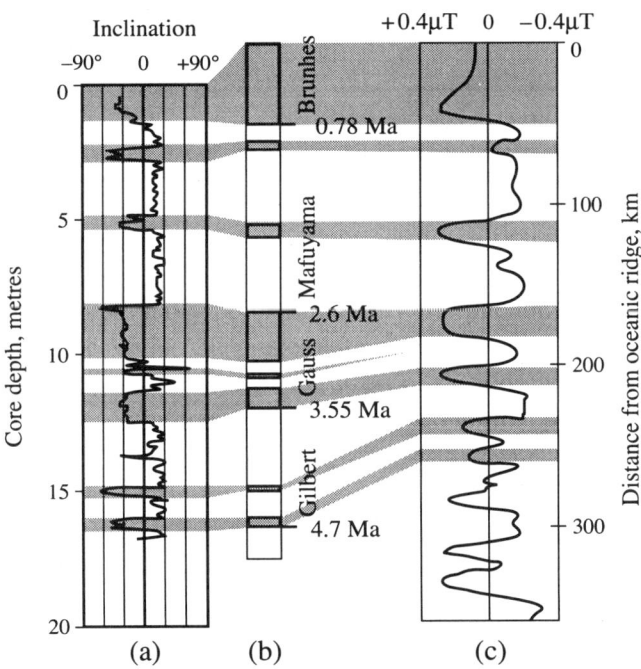

Figure 24.10 - Reversals of the Geomagnetic field

(a) *Record of the inclination of ChRM as a function of the depth in a sediment core from the southern Indian Ocean (inclination of –30° for normal polarity, present day) correlated to the scale in (b);*

(b) *Geomagnetic reversal time scale for the last 5 Ma showing the succession of normal periods (in grey as the present day) and reversed (white);*

(c) *Geomagnetic field anomalies measured on the surface of the Indian Ocean as a function of the distance from the oceanic ridge and correlated with the reference scale shown in (b).*

This polarity scale is reliable up to 180 Ma (age of the oldest oceanic crust). Palaeomagnetic studies of continental rocks indicate that before this time the geomagnetic field had similar characteristics. However, there were long periods without reversals known as quiet intervals or superchrons (Permian reversed period between 312 and 262 Ma, and the Cretaceous normal from 118 to 83 Ma).

From the point of view of its intensity the geomagnetic field also shows large variations, both during stable periods (for example g_{10} was 30% greater 3 ka ago) as well as during excursions or reversals (reduction by a factor of 5 to 10). This fact as well as the recordings of the same event at different points around the globe, seems to prove that during excursions and reversals there is a temporary disappearance of the dipole field, and that it does not undergo a progressive rotation by 180°.

8. ORIGIN OF THE CORE FIELD: DYNAMO THEORY

The possibility of maintaining a permanent magnetic field by the transformation of mechanical energy into a self-excited field-current system is termed the dynamo effect. A very simple laboratory device, the disc dynamo, can show these properties: it consists of a conducting disc rotating in a field **B** so inducing a current I that is fed into a conducting loop. This creates a field that reinforces the original field. Above a certain critical rotational velocity, one can remove the initial external field, and so produce a self-excited field. Moreover, for the same set up and sense of rotation two polarities of the field are possible (fig. 24.11). This is quite similar to the behavior of the geomagnetic field, which can exhibit two opposite "stationary" states.

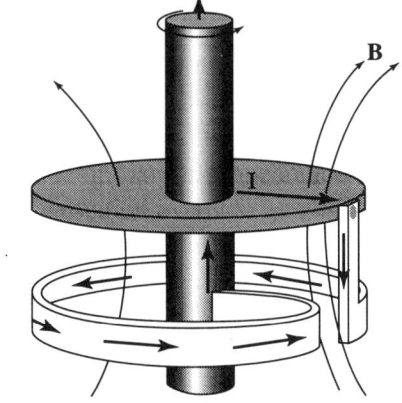

Figure 24.11
Example of a machine capable of producing a self-excited magnetic field

*A conducting disc (shaded) rotates in a magnetic field **B** with an angular velocity ω. The induced current I in the disc flows in a fixed circuit (unshaded) in the form of a loop which produces a magnetic field in the same sense as the initial field **B**.*

Obviously the processes occurring in the outer core (conducting liquid, with a temperature in its upper part of around 3,000 K and a viscosity similar to that of water, figure 24.12-a) are eminently more complex than that of a simple disc dynamo, and a serious description of the actual theory of the earth's dynamo is well beyond the scope of this book [10, 14, 16]. In order to compensate for ohmic dissipation, energy has to be supplied and estimates vary according to the dynamo model from 10^{10} to 10^{12} W. This energy comes from the conversion of a small part of the energy produced in the core and which escapes by thermal convection, and then by heat transfer to the mantle. There are many possible sources of energy: heat from the disintegration of radioactive isotopes such as K^{40}, latent heat of crystallization and gravitational potential energy associated with the crystallization of liquid at the core-mantle boundary and subsequent sedimentation at the top of the solid inner core.

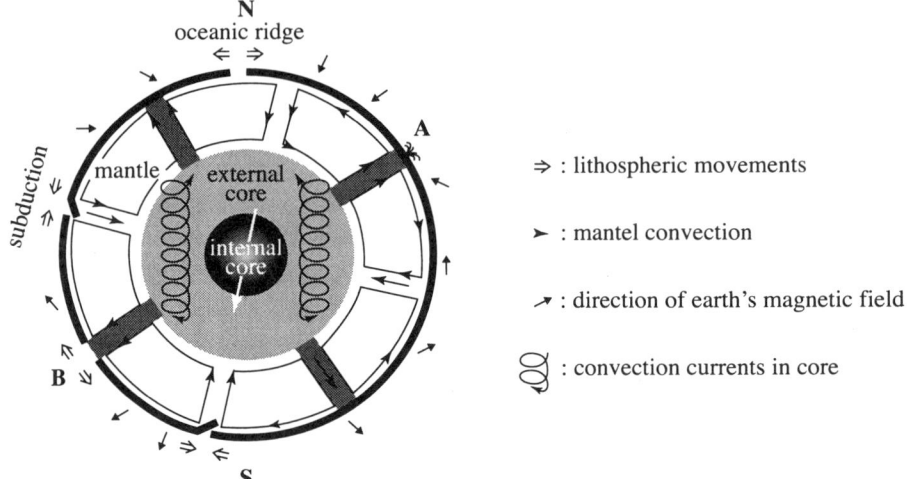

a - Schematic cross-section of the earth, showing the type of convection in the mantle and in the outer core (light grey) and flow directions. The rigid lithosphere (black) is divided up into plates, which have different relative movements (the movements at the plate boundaries are indicated by arrows with a double line). The virtual dipole in the inner core is shown by a white arrow, which is equivalent to the geomagnetic field observed at the surface (small arrows). The plumes (dark grey) can be correlated (B, Iceland) or not (A, hot-spots of Reunion and Hawaii) to the oceanic ridges.

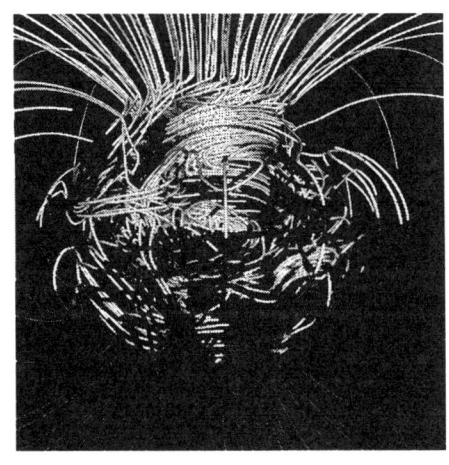

b - Result of a numerical dynamo model in the core showing the lines of force at a given instant. The core mantle boundary corresponds to the bending of the lines of force [9, 10, 14].

Figure 24.12
Origin of the earth's magnetic field

It can be shown that the movements of material required by the geodynamo should have a component with respect to the rotation axis of the earth. Only those movements produced by thermal convection can contribute to the dynamo mechanism, and the earth's rotation intervenes only through the Coriolis force, which deflects the movements induced by the convection. Another important theorem, the "antidynamo theorem" of Cowling, states that a field that is perfectly symmetrical cannot be self-sustaining. The inclination of the central dipole is therefore an essential element, with the Coriolis force only intervening to keep this inclination low. In the so-called "frozen-flux" hypothesis the lines of force in the core are "frozen" into the core fluid and follow its movement (over a short time scale), which is in good agreement with the magnetohydrodynamic conditions in the core. Thanks to this hypothesis, one can transform the secular variation maps into a map of the velocity at the core-mantle boundary. These velocities are typically of the order of $10 \, km.a^{-1}$ and are organised following zonal and cyclonic patterns similar to that of the atmospheric circulation.

Only modelling can provide information regarding processes below the core-mantle boundary. Convection in the core is probably turbulent but could be in the form of columns which are parallel to the earth's rotation axis and which touch the inner core. The solid inner core, apart from providing a geometrical constraint, stabilises the field by an induction effect. However, the existence of a core is not essential for the working of the geodynamo: as the inner core is thought to have been formed some 1.7 Ga ago, whilst palaeomagnetism indicates that there was already a geomagnetic field in existence from 3.5 Ga. It will be seen that dynamos are found on other planets whose cores are quite different from that of the earth (§ 10). As an illustration of the actual state of research in this field figure 24.12-b shows a good example of the instantaneous lines of force inside and outside the core as calculated by a realistic numerical model of a convective terrestrial dynamo. Modelling of reversals of the field is still in its infancy. However, it is interesting to note that the frequency of reversals in the long term (200 Ma) varies in a synchronous manner with the average velocity of the drift of the continents which is itself a function of the strength of the thermal convection in the mantle. The key to the reversal mechanism is probably related to the fluctuations of the transfer of energy at the core-mantle boundary.

9. PALAEOMAGNETIC APPLICATIONS

9.1. TECTONICS AND CONTINENTAL DRIFT

The observation that geomagnetic field is an axial dipole over the long term (§ 7) is essential for the tectonic applications of palaeomagnetism, for if one can calculate the average value of the field and so eliminate the secular variation, then the direction of the ChRM can be perfectly predicted: D = 0 (that is the NRM vector points towards the geographical pole), and $I = \arctan(2 \cot \theta)$ (the angle of inclination of the field with respect to the horizontal is simply a function of the colatitude). Palaeomagnetism can therefore provide the proof that a particular formation has undergone a rotation about a vertical axis and/or a shift in latitude, subject to the condition that one can determine by other means an eventual rotation about a horizontal axis. This is simple for sediments or lava flows where the palaeo-horizontal is visible. On the other hand the palaeo-longitude remains undetermined.

The normal method for the interpretation of a ChRM direction is to calculate the geographical position of the virtual geomagnetic pole (VGP) corresponding to the pair (D, I) under the hypothesis of a centred axial dipole. For rocks of the same age sampled at different sites on the same **large** continental block (having evolved as a rigid block since this age), it can be seen that if D and I vary as a function of the actual geographical position of the sites studied, they all give the same VGP. This confirms that the hypothesis of the centred axial dipole is also valid beyond the last few Ma mentioned in § 7.

One observes that with increasing age of the rocks, the more the VGP of a continent diverges from the present pole (fig. 24.13), and with velocities of the order of 1 to 10 cm.a^{-1}. This "apparent drift" could be interpreted as a global displacement of the lithosphere by slipping on deep deformable layers: this is known as true polar wander. This phenomenon is possible at certain epochs in connection with a mass imbalance or orbital instabilities due to planetary interactions. Astronomical modelling over a time scale of a billion years shows that the moon stabilises the axis of rotation of the earth. An eventual true polar wander can be explained only if the apparent polar wander paths of Europe and North America are greatly different. If one closes the Atlantic Ocean using the coastal contours, the curves can be brought into coincidence between 450 Ma and 180 Ma. This proves that these two continents drifted together during this time period, and then split apart to form the actual Atlantic Ocean.

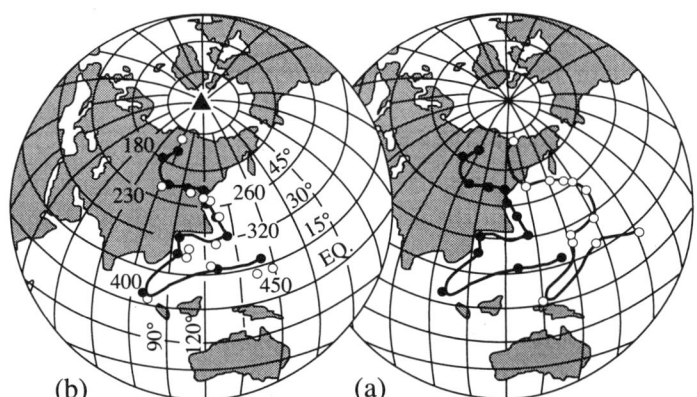

Figure 24.13 - Apparent virtual geomagnetic pole wander paths [8]

These paths are given as a function of age (in Ma) for North America (full circles) and for Eurasia (open circles), in their actual position (a) or after closing the Atlantic Ocean by a rotation of 38° around the rotation pole marked by a triangle (b).

This proof of continental drift given by palaeomagnetism during the 1950's, in opposition to the "fixist" dogma of the period, preceded the general acceptation of the theory known as plate tectonics. It was only in the middle of the 1960's that plate tectonics really imposed itself, following the exploration of oceanic ridges and trenches (zones of expansion and destruction of the plates), and the interpretation of ocean crust magnetic anomalies (fig. 24.10). The determination of the VGP wander paths of all of the major continental blocks allows the retracing of their movements during the geological past. In order to reverse the movement in time the continent is placed so that its VGP coincides with the actual pole. The most spectacular movement concerns the Indian sub-continent, situated some 80 Ma ago to the east of Madagascar, which broke away from the African continent to arrive at the Eurasian continent 30 Ma later. On the way it joined up with the equatorial island that corresponds to present day Tibet, and so formed the Himalayan mountain chain. On its path northwards India passed above the hot-spot responsible for the present day Reunion volcano and which produced the catastrophic eruption of millions of km^3 of magma in the Deccan province in northwest India. Lasting less than 1 Ma (see § 9.2) and synchronous with the extinction of the dinosaurs, this volcanism could have played a decisive role in the important climatic changes which characterised the Cretaceous-Tertiary boundary some 65 Ma ago. The Deccan lavas have a normal inclination of –40° identical to that of the present day inclination of the Reunion lavas, which shows that hot-spots can have a stable position over geological time, due to their deep roots in the mantle over which the plates slide (fig. 24.12).

In deformed zones palaeomagnetism can detect the rotations and latitude shifts of exotic terranes, by a comparison of their VGP with the VGP of an adjacent stable continent of the same age. For example, parts of the Californian coast have been shown to come from beyond the equator, drifting like oceanic islands over 5,000 km before docking against the North American continent. The Mediterranean is also a region with a very complex tectonic history, and palaeomagnetism can provide key information for the understanding of the *"Mare Nostrum"* mosaic. For example rocks of 280 Ma from Spain, Corsica-Sardinia and Italy have declinations which are respectively lower by 40, 60 and 50° with respect to declinations recorded in rocks of the same age to the north in stable Europe. These anticlockwise rotations produced respectively, the opening of the Bay of Biscay and the formation of the Pyrenees, the opening of the Gulf of Lion, and the formation of the Alps.

Throughout this chapter we have considered the radius of the earth to be constant. However, certain authors have proposed an expansion of the earth to explain the opening of the oceans and the increase in the length of the day (only 19 hours 1 Ga ago). The coherence of the VGP's from the same large continent would exclude any significant variation of the radius. It is now known that the opening of the oceans is compensated by an equivalent closing due to subduction (fig. 24.12-a), and that the slowing down of the earth's rotation is due to a transfer of angular momentum to the moon (which recedes from the earth) by tidal friction.

9.2. DATING

The geomagnetic field at a particular place varies with time on different scales [16]:
♦ 0.1-1 ka for the secular variation,
♦ 0.1-10 Ma for periods of stable normal or reverse polarities separating two reversals,
♦ 10-100 Ma for the effect of continental drift.

If one knows *a priori* the evolution of the direction of the field as a function of time at a given place, one can use this palaeomagnetic information to date the acquisition of the NRM, that is to say the date of formation of the rock in the case of a primary NRM. This type of dating, in which a comparison is made with a reference curve which has been calibrated using other methods of dating (most often isotopic), implies that the geological formation to be dated covers a sufficient interval of time to cover any uncertainties. In a similar manner the reference curve of the secular variation in France for the last two millennia enables the dating of *in situ* archaeological baked clays (oven walls, hearths, bricks, etc). However, if one finds a site which gives a direction of $D = 2°$, $I = 57°$, the date given by this method is either 1,350, or 200 A.D. (fig. 24.8). This incertitude can be resolved by the archaeological context. Archaeomagnetic dating can also use the estimation of the palaeointensity, which enables the method to be applied to displaced objects such as potsherds and tiles.

This incertitude is even greater in the case of the polarity, for if one finds a site where the polarity is reversed all one can conclude is that it must be older than 0.78 Ma. On the other hand if one has a sedimentary sequence with many reversals, and one has several reference points –for example that the top of the core is younger than 0.78 Ma– then the dating, using a depth-time curve, is straightforward (fig. 24.10). Known as magnetostratigraphy this method is an essential element for the dating of sedimentary sequences, and it is the only method applicable for those sediments without characteristic fossils or volcanic layers which can be isotopically dated. This is very often the case for continental deposits containing hominid remains. The oldest of the genus Homo found in Europe (site of Atapuerca in Spain) has been dated at more than 0.78 Ma by magnetostratigraphy. Thanks to a compilation of all the studies of the magnetic anomalies, and those of magnetostratigraphy one can now make a correlation with absolute ages, and the reference magnetostratigraphic scale is now established with a great precision: 0.01 Ma for the last few 10 Ma and 0.1 Ma beyond this [7]. It can be used to date or evaluate the time of deposition of sedimentary or volcanic series. For example the duration of the emplacement of the Deccan traps (§ 9.1) is estimated to at least 1 Ma by this method.

Finally the third time scale, that of continental drift, allows one to use the VGP from a given stable continental site, and to compare it with the reference curve (fig. 24.13) of the VGP's of the continent. Although imprecise this method could be very interesting for those formations where no other method is available or has been used to date the acquisition of the NRM. In this way one could reveal secondary magnetizations that could easily pass unnoticed.

Geomagnetism also intervenes indirectly in radiocarbon dating that uses the isotope carbon 14 and covers the periods before 45 ka. This method determines the age of fossil organic material using the difference between the ratios C^{14}/C^{12} of the fossils and that of the present day atmosphere which remains constant. This ratio results from equilibrium between the radioactive disintegration of C^{14} (half-life 5.6 ka), and its production by collision between a N^{14} nucleus present in the upper atmosphere and a slow secondary electron produced by energetic particles from the solar wind. The atmospheric ratio depends directly on the average flux of solar particles. This flux is modulated by the intensity of the dipole field, as the charged particles are deviated more or less efficiently according to the thickness / depth of the magnetosphere (fig. 24.9). A knowledge of the temporal variations of the dipole intensity can be used to predict the correction necessary for the C^{14} dates (based on the hypothesis of a constant atmospheric ratio) to give the true ages. Such a prediction based on palaeomagnetic data, would appear to be in good agreement with the direct calibration of C^{14} dates by dendrochronology (tree-ring counting) or U/Th (fig. 24.14).

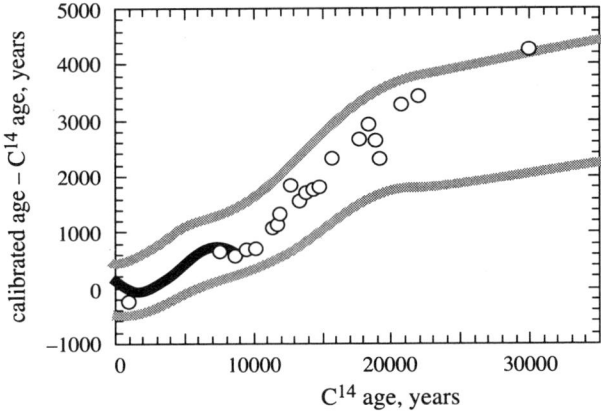

Figure 24.14 - Correction of Carbon 14 dates

This correction is obtained directly by counting tree rings (thick continuous line), and by the simultaneous dating of corals by C^{14} and U/Th (circles), or indirectly by a model calculated from the fluctuations of the dipole field as determined from palaeomagnetism. The two shaded curves define the error limits [17].

10. EXTRATERRESTRIAL MAGNETISM

Observations made from the earth or using magnetometers carried on board satellites show that the magnetic phenomena observed on earth are also seen on other planets. A field of internal origin created by a dynamo effect is supposed to exist for the majority of planets as well as for the satellites of Jupiter, Io and Ganymede. The observation of a logarithmic law between magnetic moment and angular momentum would tend to prove that an extremely simplified scaling law such as that of Busse (fig. 24.15) appears to operate and enables a prediction of the order of magnitude of the magnetic moment of the planets..

These magnetic moments, which are always close to the rotation axis of the planet, come from a dynamo effect due to thermal convection in the electrically conducting core. In the case of the giant planets, the conducting material is probably a high-pressure metallic form of hydrogen rather than an iron-nickel alloy. The magnetic field at the surface of Jupiter is ten times stronger than that observed on the earth.

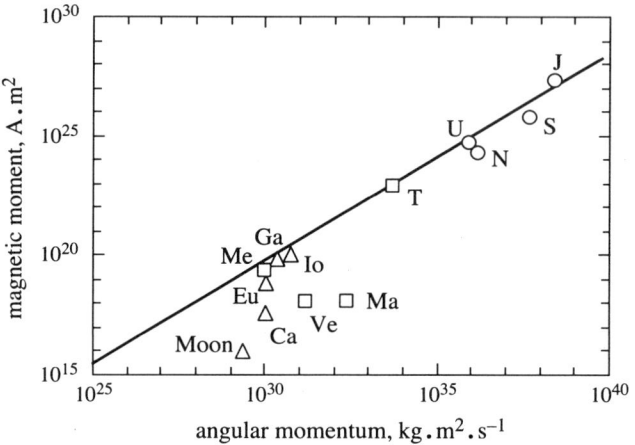

Figure 24.15 - Busse's law

Logarithmic relation between the magnetic moment m and the angular momentum £ (Busse's law) for the outer or giant planets (circles), the inner or terrestrial planets (squares), and the moons of Jupiter and of the earth (triangle), after [18].

On the other hand, Venus, Mars and the Moon exhibit very weak fields probably due uniquely to the magnetization of crustal rocks. The absence of an internal dynamo in these bodies can be interpreted differently:

♦ Venus, which is similar in size to the earth, has a very long rotation period (243 days) so that the Coriolis force would be too weak to guide convection as on the earth and produce a coherent dipole field;

♦ Mars and the Moon are much smaller than the earth, and the size of their cores and the heat flux from the cores are insufficient to initiate a dynamo effect. Io and Ganymede, which are similar in size have an additional energy source: tidal friction.

The sun generates a strong dipole field with a surface intensity of the order of 100 μT, which reverses roughly every 11 years. The dense plasma that makes up the interior of the sun, and the nuclear fusion reactions in the core provide the necessary conditions for a solar dynamo. Locally fields up to 100 mT are observed at sunspots on the sun's surface, related to strong turbulence in the solar convection.

The existence of much stronger fields is deduced from the electromagnetic waves emitted by other types of stars. The record being held by neutron stars with fields of the order of 10^8 T which explains the "pulsar" effect. At the other end of the scale the interplanetary field is of the order of 1 to 10 nT whilst the interstellar field is estimated at 0.1 nT.

Palaeomagnetism, which requires that rock samples be brought back to the laboratory, has up till now only provided information on the ancient fields of the Moon and of meteorites [14].

The magnetic anomalies observed on the Moon and Mars produced by rocks magnetised by a significant magnetic field are evidence for an ancient magnetic field that has since disappeared. Lunar rocks, formed by the cooling of a magma some 3 Ga ago, have a strong magnetization. If this is a TRM one would deduce that these rocks cooled down at this time in a field of the order of 10 to 100 μT whereas the present ambient field due to the crustal magnetization does not exceed 0.3 μT. The thermoremanent origin of this NRM is questionable; in particular one could explain the strong magnetization of the lunar rocks by meteoritic impact in the presence of a weak field. The violent shock produces simultaneously a remobilization of the magnetic domain walls by stress, and a plasma that induces a magnetic field. If one retains the hypothesis of a TRM, the origin of a strong lunar field 4 Ga ago remains controversial: was it an internal field due to an early dynamo (as is suggested by similarity between the actual Io and the Moon), or was it an external field, either terrestrial or solar? A strong solar field would lead to the hypothesis, put forward for other reasons, of an early T-Tauri phase of the sun at the very beginning of the solar system associated with a solar wind and a very strong field. Such a field would have the merit of giving the initial impulse to start the planetary dynamos. The confirmation of the existence of a strong field throughout the solar system some 4.5 Ga ago can begin to be found in the strong NRM found in certain meteorites. However, it is difficult to exclude the possibility that this magnetization was not entirely due to the multiple shocks suffered by these wanderers of the solar system. Rock samples from Mars or from drilling on the Moon are awaited with great impatience before further progress can be made in this field!

REFERENCES

[1]   M. WESTPHAL, Paléomagnétisme et magnétisme des roches (1986) Doin, Paris.

[2]   D. DUNLOP, Rock Magnetism: fundamentals and frontiers (1997) Cambridge Univ. Press.

[3]   Quaternary Climate and Magnetism (1999) B. Maher & R. Thompson Eds, Cambridge Univ. Press.

[4]   P. ROCHETTE, M. JACKSON, C. AUBOURG, *Rev.. Geophys.* **30** (1992) 209.

[5]   D.H. TARLING, F. HROUDA, The Magnetic Anisotropy of Rocks (1993) Chapman & Hall, London.

[6]   R. BUTLER, Paleomagnetism: Magnetic Domains to Geologic Terranes (1992) Blackwell, Oxford.

[7]   N. OPDYKE, J.E.T. CHANNEL, Magnetic Stratigraphy (1996) Academic Press, New York-London.

[8]   R. VAN DER VOO, Paleomagnetism of the Atlantic Thetys and Iapetus Oceans (1993) Cambridge Univ. Press.

[9]   L. LLIBOUTRY, Géophysique et Géologie (1998) Masson, Paris.

[10]   J.P. POIRIER, Les profondeurs de la Terre (1996) Masson, Paris.

[11]   A. COX, R.B. HART, Plate tectonics: how it works (1986) Blackwell, Oxford.

[12]   W. LOWRIE, Fundamentals of Geophysics (1997) Cambridge Univ. Press.

[13]   J.J. DELCOURT, Magnétisme Terrestre - Introduction (1990) Masson, Paris.

[14]   R. MERRILL, M.W. MCELHINNY, P. MCFADDEN, The magnetic field of the Earth (1996) Academic Press, New York-London.

[15]   E. THELLIER, *Phys. Earth Planet. Inter.* **24** (1981) 89.

[16]   V. COURTILLOT, J.P. VALET, *C.R. Acad. Sci. Paris II* **320** (1995) 903.

[17]   E. BARD, *Science* **277** (1997) 532.

[18]   P. ROCHETTE, *La Recherche* **308** (1998) 36.

CHAPTER 25

MAGNETISM AND THE LIFE SCIENCES

*The curiosity of "magneticians" has recently extended to magnetic substances synthesised by some living organisms: bacteria, bees, pigeons, dolphins... The list gets longer every year. This is one aspect of **magnetobiology**, which also studies the influence of magnetic fields on the growth of plants or on animal metabolism. The magnetic properties of organic matter, whether living or inert, will be treated in the first part of this chapter.*

*On the contrary, the term **biomagnetism** is reserved for the study of magnetic fields generated within living beings by muscular movements or by a cerebral activity, and also embraces those magnetic techniques used to explore living beings. These will all be presented in the second part of this chapter.*

Medicine *does not restrict itself to the exploration of the living, it also intervenes in vivo, and a description of some magnetic intervention techniques on living beings will be given in the third part.*

1. MAGNETIC PROPERTIES OF ORGANIC MATTER

1.1. INERT ORGANIC MATERIALS

Organic matter is based on the *chemistry of carbon*, associated with H, O, N and various other elements. Pure carbon is diamagnetic, as are hydrogen and nitrogen, whilst oxygen is antiferromagnetic at very low temperature. Some interesting magnetic properties were observed on organic compounds containing only C, H, O and N. Such is the case of tanol suberate $(C_{13}H_{23}O_2NO)_2$ every molecule of which contains two free radicals N-O. An electron is localized on every N-O bond, and this substance behaves antiferromagnetically below 0.38 K, which shows how weak the interactions are in this case [1]. For an induction of only 10 mT, a metamagnetic transition occurs, and at less than 60 mT the material is practically saturated with a moment of the order of 10 A.m^2 per mole.

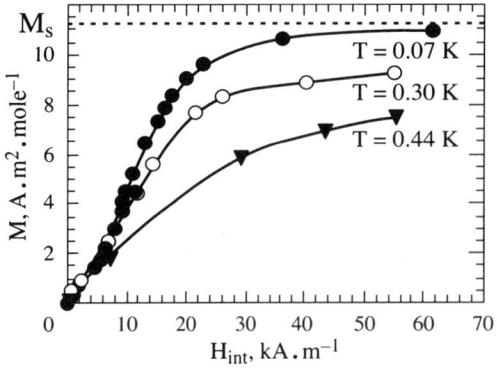

Figure 25.1 - Magnetization of tanol suberate powder as a function of the internal field

Note the S-shape of the isotherms measured at temperatures lower than the critical temperature $T_N = 0.38$ K.

In 1996, the world record for the Curie temperature of a purely organic substance $T_C = 1.46$ K, was established by the dupeyredioxyl molecule (1,3,5,7-tetramethyl-2,6-diazaadamantane-N,N'-dioxyl) [2], whose ferromagnetism originates from two spins 1/2 situated on two N-O groups.

Molecular magnetism, as this very young science is called, is concerned with the magnetic properties of organic molecules, including the search for molecules that might have an ordering temperature above room temperature. The potential candidates are to be found amongst the organo-metallic compounds where the magnetic centers are atoms of the 3d metals, for example iron or chromium. There are a great variety of more or less complex organic molecules that include one or several metallic atoms. Their magnetic properties are sometimes interesting: for example, it has been known for the last twenty years that TTF BDT(Cu) or tetrathiafulvalene-$CuS_4C_4(CF_3)_4$ has a noticeable transition at 1.5 K in a magnetic field and a saturation moment approaching 0.25 A.m^2 per mole in 15 teslas [3]. This molecule contains one 3d atom, that of copper.

For more information on molecular magnetism, the reader is referred to the comprehensive survey published by Olivier Kahn in 1993 [4], and to a more recent review article that authoritatively covers organic molecular magnetic materials and organo-metallics [5].

1.2. LIVING ORGANIC MATERIALS

Even when they are diamagnetic, the organic molecules that make up plants can sometimes present a sufficient magnetic anisotropy to produce orientation phenomena in intense magnetic fields: for example, G. Maret at the High Magnetic Field Laboratory (LCMI) in Grenoble observed in the 1980's that the orientation of the barbs growing on lily pollen grains (at a rate of 0.5 mm per hour) is random in the absence of a magnetic field, or in a field μ_0H less than 3 T. On the contrary, in 14 teslas, these barbs are all aligned with the field. The benzene rings contained in the plant try to rotate so as to minimize their diamagnetic energy, and in doing so drag with them all of the barbs.

1.3. ORGANOMETALLIC MATERIALS AND BIOLOGY

In the following treatment, we will only discuss the magnetism of organometallic molecules that are found in living organisms, or which could present a therapeutic interest: these molecules almost always contain iron atoms, with the exception of the gastropod family (snails) where nickel replaces iron.

Iron is an indispensable element for life due to its presence in certain molecules that accept oxygen or electrons (such as haemoglobin). It is stored in the organism under various organic or mineral forms. Iron sometimes acts as an acceptor of electrons in metabolism, which explains how certain bacteria are capable of catalysing the production of large quantities of iron sulphide or extra-cellular magnetite. The crystallized minerals can become structural elements, magnetite in chiton teeth, goethite in the radula of the patella (a molluscan gastropod), oxyhydroxides in some unicellular organisms, or to form within a cell ferrimagnetic single domain grains whose role will be discussed later. In a normal adult man, one finds 3 to 4 grams of iron, of which 60 to 65% are in haemoglobin and 25 to 30% as ferritin, the remainder being distributed in myoglobin (3 to 5%) and in trace quantities as heme-containing enzymes.

We will first consider the properties of some complex molecules that include iron atoms, because of their fundamental importance in biology.

1.3.1. Haemoglobin

It is a heteroprotein consisting of globin (a protein) and of a prosthetic group called a heme that contains iron and represents only 4% of the total mass of this heteroprotein. This group gives blood its red color in the oxidized state and a bluish color in the reduced state. The principal constituant of red cells, haemoglobin is responsible for the transport of oxygen from the lungs to the cells in the tissues. Iron is in a divalent form, and it is bound by four of its six coordination valences to the four nitrogen atoms of the four pyrrole rings of the protoporphyrine molecule. The low iron concentration explains the paramagnetic behavior of this molecule. For more information the reader should look up the article "blood" in an encyclopaedia.

1.3.2. Ferritin and haemosiderin

Ferritin and haemosiderin are the two main forms of iron storage in living organisms. The molecular weight of ferritin is on the order of 400,000 to 500,000 Da, and this molecule contains about 16% by weight of nitrogen. Its iron content, measured via its ratio to nitrogen (Fe / N), varies considerably according to the physiological conditions and can reach 4,500 atoms, that is 26% of its weight, in the form of trivalent iron (colloidal micelles of ferric hydroxide and iron phosphate). It occurs as 7.5 nm grains of ferri-hydrate ($9 Fe_3O_4 - 9 H_2O$) encapsulated in a spherical envelope of proteins about 2 nm thick. These grains contain about 4,500 spins coupled antiferromagnetically, with a *weak uncompensated moment*: they could therefore act as magnetic field sensors. Haemosiderin, a related molecule, is considered to be a degraded form of iron reserves in the body.

1.3.3. Fibrin

Fibrin is a polymer that forms during the final stage of blood coagulation. Jim Torbet studied its polymerization under an intense magnetic field: as in the case of the previously mentioned lily pollen, the diamagnetic anisotropy of this molecule is large enough to produce perfect orientation of the fibers at the end of the process. This orientation, which was achieved at the Matformag laboratory of the CNRS, has improved our knowledge of the structure, the assembly and the destruction (lysis) of the fibrin under conditions close to those of the physiological state.

1.4. MAGNETIC MINERALS ENCAPSULATED IN ORGANIC MATTER

Systems belonging to this class are particles of magnetite synthesized by certain algae and bacteria. So are those nanoparticles and encapsulated microspheres, natural or synthetic, which all present the characteristic of being magnetic minerals completely surrounded by organic matter. They all provide notably magnetic and possibly biocompatible assemblies. These materials can be used very effectively in imaging (one can localize zones containing magnetic matter), or as separating agents in the purification of biomaterials, or again to identify minute quantities of organisms, of cells or of genomic material, and finally in the treatment of malignant tumors.

1.4.1. Algae and magnetotactic bacteria

The presence in living organisms of ferrimagnetic single domain grains, generally magnetite (Fe_3O_4) but also greigite (Fe_3S_4), suggests that these organisms can use such grains to sense the orientation and perhaps the intensity of the magnetic field due to the torque the field produces on a grain with uniaxial anisotropy. This is undoubtedly the case in magnetotactic bacteria. These were discovered in 1975 by Blakemore [6]. They feature a chain of single domain grains along the length of their bodies, which are in the shape of rods terminated by one or two cilia that propel them through water.

Figure 25.2 shows a bacterium that has a flagella at each of its extremities, while most other bacteria only possess a single flagella at only one extremity. All these bacteria are permanently oriented parallel to the magnetic field lines. They can remain in their preferred environment, fine grained sediments, thanks to their magnetic orientation with respect to their unique direction of motion. In the northern hemisphere, they move toward the north pole, and in the southern hemisphere toward the south pole. In either case, due to the inclination of the earth magnetic field, they are thus led toward the bottom of the water. This assures their survival as these organisms do not withstand an excess of oxygen. It should be noted that a small proportion of these bacteria is born disoriented and rapidly disappear, but these will save the species in case of a reversal of the geomagnetic field! Bacteria containing magnetite and those containing greigite do not have the same genes and developed independently. The decomposition and sedimentation of greigite-based bacteria could be the origin of mineral greigite.

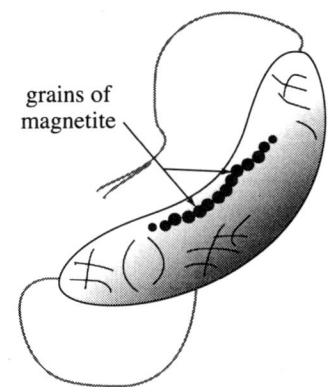

Figure 25.2 - Bacterium
Aquaspirillum magnetotacticum,
enlarged about 20,000 times

grains of
magnetite

1.4.2. Animal magnetism

A sensitivity to magnetic fields has been demonstrated on numerous animal species which habitually travel with few visual references or over large distances: flying insects (bees), migratory birds (pigeon), migratory or deep water fish (salmon, selachians), cetaceans (dolphins, whales), reptiles (sea turtle), amphibians (salamander), mollusks (sea slug), arthropods (lobster)... However, their detection device is not necessarily based on the incorporation in a sensory organ of a magnetic moment as in bacteria. A competing system could be envisaged, based on the current induced in a conducting loop moving through the field. This hypothesis was proved in selachians, which are also sensitive to an electric field. They can refind a direction only after having turned in circle several times, inducing in their electric "circuit" a current which depends on its orientation with respect to the ambient magnetic field. The joint demonstration of sensitivity to a magnetic field, and of the presence of magnetite in the pigeon, bee, salmon or dolphin, is therefore a strong presumption, but not a proof, of the integration of this magnetite in a magnetometric organ. For further information on magnetic orientation in animal species, one should consult reference [7].

1.4.3. Encapsulated Microspheres

Ferritin

This substance, essential to the life of higher organisms, was already described in § 1.3.2. These particles are found in a natural state in all sorts of organisms and in particular in the bacteria *Escherichia coli* and *Azotobacter*, and one now knows how to synthesise them. They are of interest both to physicians, because they are a material of choice for applications, and to physicists who are looking for a macroscopic quantum tunnelling effect [8].

Dextran ferrite

These are microspheres of colloidal magnetic iron oxide, coated in dextran. They have a sizeable specific magnetic moment, from 16 to 48 $A.m^2.kg^{-1}$, and they reach saturation in fields of 20 to 60 mT. The grains, which are 3 to 15 nm in diameter, are single domain, and have a superparamagnetic behavior. This system is biocompatible and presents a very low toxicity. The surface properties of these particles are determinant in their ability to spread in living organisms. We will see in § 3.6 the use that can be made of this type of material in medicine.

1.5. HUMAN SENSITIVITY TO MAGNETIC FIELDS

Yves Rocard conducted a scientific survey on dowsers, and concluded that a magnetic effect exists in certain particularly sensitive individuals [9]. One can estimate the magnetic fields involved in this type of phenomena to be $B_0 = 10^{-9}$ T. Let us evaluate the energy involved by considering only the contribution of the ferritin molecules: a molecule of ferritin has around 4,500 Fe^{3+} cations coupled antiferromagnetically, but with a compensation defect that one can estimate as 1% of all the sites. Under such a field, the absolute value of the potential energy mB is therefore the energy of a magnetic moment m of 0.02×10^{-19} $A.m^2$ subject to B_0, that is 2×10^{-30} J. If one compares this energy to that associated with the threshold sensitivity of the eye (around 10^{-17} J), one sees that a human being will theoretically only be capable of detecting such a signal if a very large number of molecules carrying a magnetic moment ($N > 10^{13}$) react collectively to this magnetic field. However, as there are around 2×10^{18} molecules of ferritin in the human body, and if one considers that all of these molecules take part, then such a magnetic energy becomes easily detectable, but by what mechanism? The terrestrial magnetic field corresponds to a much larger signal but no convincing experiment has yet shown the existence of magnetotropism in a human being. This topic therefore still remains very controversial.

The recent discovery of particular tissues or of lineages of white blood cells enriched in magnetite, has probably more to do with the metabolism of iron, in relation to a possible pathological differentiation. *This presence calls for caution regarding the present claims of the harmlessness of exposure to strong fields or field gradients during MRI scans. Cells rich in magnetite could, in theory, suffer lesions during their exposure.*

On the other hand, the sensitivity of living cells to *variable* magnetic fields has been abundantly proven, but it concerns effects related to the frequency, rather than to the intensity of the field.

2. MAGNETIC EXPLORATION TECHNIQUES FOR LIVING ORGANISMS

Magnetic clinical examination methods are very often non-invasive, that is completely painless. They are at present undergoing a rapid development. For example electro-myography, which consists in inserting a needle into a muscle of the hand, and then measuring the electrical potential associated with the work of this muscle, will be advantageously replaced by magnetomyography. The latter monitors at a distance the magnetic field associated with the same electric signal, but without any pain for the patient! Without being exhaustive, the list of magnetic techniques which will be described below shows the diversity of the approaches used:

◆ resonance methods (MRI),
◆ detection of magnetic fields produced by living tissues,
◆ magnetic marker techniques,
◆ magnetic sensors.

2.1. RESONANCE METHODS

MRI (magnetic resonance imaging), because of its importance, is treated separately in chapter 23. We just note that this technique submits the patient to intense static magnetic fields whose harmlessness at the cell level appeared to be accepted but has again become controversial (see § 1.5).

2.2. DETECTION OF MAGNETIC FIELDS PRODUCED BY LIVING TISSUES

The biological functions associated with muscular work or nerve impulse set off a cascade of chained chemical transformations, electric polarizations and depolarizations, which result in peaks of electric activity detectable using appropriate voltmeters (electromyography, electroencephalography). This electric activity generates varying magnetic fields at the same relatively low frequencies: the electric potentials involved (typically in the range of μV to mV) are closely correlated to the associated magnetic fields, which vary from of 2×10^{-14} to 2×10^{-11} T, as was shown by Williamson and Kaufman in an early but still remarkable review article [10] dedicated to biomagnetism.

The detection and the measurement of such weak magnetic fields requires magnetometers with a very high sensitivity (SQUIDs), which have to be very well protected against all spurious outside sources, liable to produce much larger fields. The present terrestrial magnetic field has a mean value of 45 μT (4.5×10^{-5} T), greater by six orders of magnitude than the most intense biomagnetic field, that associated with the activity of the heart (magnetocardiography or MCG).

The ultimate limit of magnetometer sensitivity depends on the "noise equivalent signal", S_N that is universally expressed in $T^2 . Hz^{-1}$. The magnetic field equivalent to the noise, measured over a band width ΔF, is thus given by $(S_N \Delta F)^{1/2}$. Hence $(S_N)^{1/2}$, which is expressed in tesla $.$ (hertz) $^{-1/2}$, will be the noise equivalent field for $\Delta F = 1$ Hz.

A survey dating from the 1980's [10] compares the sensitivity of different magnetometers at 10 Hz:
- 3×10^{-11} T for a commercial flux gate,
- 2×10^{-12} T for a laboratory flux gate (NASA),
- 3×10^{-13} T for an induction coil with a ferrite core,
- 8×10^{-14} T for a SQUID system in an urban environment,
- 8×10^{-15} T for a SQUID with magnetic shielding, and
- 6×10^{-15} T for a SQUID far away from any urban noise.

These performance figures do not seem to have been surpassed in the recent literature. Experimentally one observes a $1/f$ decrease in the noise at low frequencies, until one meets the Johnson noise. The noise that experimentally limits the best present day magnetometers seems to come from the magnetic fields generated by the thermal agitation of electrons in conductors situated close to the SQUID, principally in the cryostat walls and superinsulation, but also in the laboratory shielding itself. The lower limit in the $(S_N)^{1/2}$ would seem in practice to be a few 10^{-15} T $.$ Hz $^{-1/2}$. Certainly, the magnetic noise of thermal origin produced by the human body is 10 times weaker, but the noise caused by the measurement system will never allow this ultimate sensitivity of 10^{-16} T $.$ Hz $^{-1/2}$ to be reached [11].

However, it can happen that some parts of the human body produce magnetic fields several orders of magnitude greater than this limit: for example the field produced by magnetic dust in the lungs of certain workers (arc welders) and magnetised by external fields. Without magnetic shielding, the sensitivity is poorer, but it can already be sufficient to allow magnetocardiography measurements. In order to cancel external noise to first order, many magnetometers operate as gradiometers: two coils are placed some centimeters apart and connected in opposition. Thus, all noise coming from distant sources is eliminated, while only the differential signal from the source, that is situated closer to one coil than the other, is measured. Closed circuit liquid helium systems allow lighter and even portable equipment to be used. Thus, it is now possible to examine by MCG the cardiac rhythm of a six month old fœtus.

2.3. MAGNETIC MARKING TECHNIQUES

These techniques can be applied either on the macroscopic scale (for example, to measure the gastro-intestinal transit speed of a meal charged with magnetic powder), or on the microscopic scale, for example to follow cellular division or to measure the concentration of a medicine in the blood. In the first case, a portable magnetometer will localise zones presenting a strong magnetism. The second case will require much finer analyses of magnetic susceptibility on a sample taken from the organism (blood sample), or again detection by a SQUID of the magnetism induced by a weak magnetic field in the magnetic microspheres. When a bacterium is marked, it is possible to separate it from the other non-marked cells, by applying a magnetic field gradient: this is the **magnetic separation technique** or **magnetotaxy** that appeared in the middle of the 1970's. Today, the marking and magnetic separation allow a more reliable, simple and fast detection of the tuberculosis bacterium, than by traditional methods [12].

2.4. MAGNETIC SENSORS

The foreseeable development of microsensors will allow the measurement *in situ* of all sorts of physical parameters. Strain or pressure gauges will be able to take advantage of magnetoelastic effects, for example for measuring the pressure in the eye toward detecting a possible risk of glaucoma. This measurement should be carried out during the entire night-day cycle (24 hours in a row), which is impossible at the present time due to the invasive character of the techniques used and the size of the equipment. A miniaturization of these measurements would require a microsystem including a magnetostrictive microactuator acting at regular time intervals on the cornea, a deformation microsensor and a microprocessor capable of transmitting information to a receiver located close to the patient. All the necessary energy would be transmitted by induction, and the signal would also be picked up remotely: the system would be contained in a corneal lens and could therefore be carried by the patient for 24 hours. The interest in magnetic sensors obviously lies in their essential property of being able to be read remotely, without any connecting wires, and important developments can be expected in this promising field.

3. MAGNETIC TECHNIQUES FOR INTERVENTION IN VIVO

This does not concern an examination, but an intervention in order to treat the patient. This area too is in full evolution, and its field of application grows all the time. An excellent review article deals with magnetically aided instrumentation in medical research [13]. We will only describe some typical applications, going from the macroscopic to the more microscopic aspects.

3.1. CARDIAC VALVE

The heart is made up of four cavities (two atria and two ventricles) with four valves opening passages between atrium and ventricle or between ventricle and artery. The essential role of the valves is to ensure a unidirectional blood flow in the heart.

When a natural cardiac valve is deficient and cannot be repaired by a surgeon, it is necessary to replace it by an artificial valve (120,000 implants per year worldwide, 1/3 of these valves are biological and the remaining 2/3 are mechanical). The classic mechanical valvular prostheses consist in general of a circular ring (the seat) inside of which shutters (called leaflets) rotate about hinges. When they are open, these leaflets allow blood to flow; when closed, they prevent the blood from returning. The opening and closing movements of these valves are guided by the thrust exercised respectively by the flow and ebb of the blood. *This mode of operation is fundamentally different from that of natural valves.* In the case of the mitral valve, tendons fixed to the edge of the valve membranes and the ventricule wall pull on them to open the valve, and the combined action of an adverse atrioventricular pressure gradient and post-valvular turbulence closes the valves without any reverse flow. Contrary to natural valves, the energy necessary for the displacements of the mechanical prosthesis shutters is taken from the total energy of the blood flow at each pulse. This results in *an extra workload and therefore an unnecessary strain on the heart* to maintain an identical flow.

To overcome this disadvantage, Professor Carpentier (Broussais Hospital, Paris), invented an active prosthesis whose leaflets are moved partly by magnetic forces. Samarium-cobalt magnets are included in the leaflets and the valve seat in order to produce an opening torque on the closed leaflets, and a closing torque on the open leaflets. Thus, magnetic energy is used to open and to close the leaflets of this prosthesis. The key to the working of the prosthesis lies in the fine balancing of the opening and closing magnetic torques created by the stationary magnets attached to the valve seat on the mobile magnets fixed to the leaflets. The adjustment of the opening and closing magnetic torques is performed experimentally *in vitro* on a cardiovascular simulator, developed by the Laboratory of Cardiovascular Biomechanics, Ecole Supérieure de Mécanique, Marseille. This active mechanical valve prosthesis was developed recently by a French company (SICN, F-38113 Veurey-Voroize).

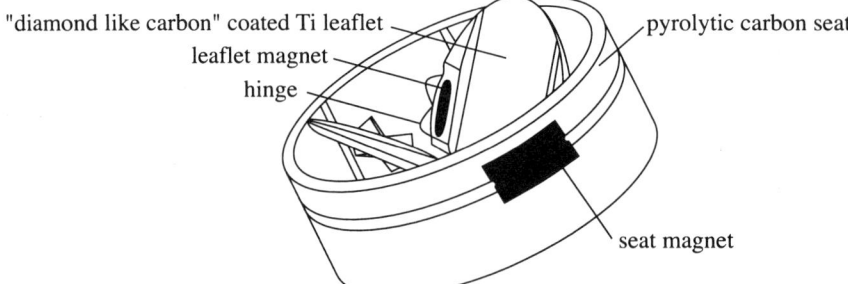

Figure 25.3 - Active cardiac valve (SICN document)

3.2. MAGNETIC MANIPULATION OF CATHETERS

An increasing number of surgical operations call on microsurgery techniques, which in their preliminary phase require a micro-instrument to be directed towards the site of the intervention. Generally, this micro-instrument is carried by a catheter that passes along the blood vessels; but at a bifurcation, it can happen that the catheter obstinately takes the wrong path.

It is then impossible to carry out the operation, and the patient could die: this is why, as from 1951, the technique of magnetic guidance of intravascular catheters was developed under the leadership of Tillander in Sweden [14]. The first applications only concerned those vessels that were sufficiently large to admit these devices, which were at that time quite bulky: aorta, renal and pancreatic arteries.

The technique is still the same: a magnet is fixed to the end of a flexible catheter. During the progression of the catheter along the artery or vein of the patient, should the catheter try to take the incorrect route, then the application of an ad hoc magnetic field gradient will deviate the extremity of the catheter towards the correct direction and allow the micro-instrument to continue its progress towards the target.

This remarkable technique has continued to progress, in particular with the appearance of magnets of much higher energy density so that the same result is achieved for a considerably reduced volume.

Such progress has since allowed this technique to be applied to neurosurgery: for example in the case of the treatment of aneurysms, A. Lacaze (CNRS Grenoble) developed a magnetic guidance system which takes advantage of the remarkable properties of modern samarium-cobalt magnets.

Figure 25.4 illustrates the principles of this technique: the catheter C should reach the aneurysm A, but it has a tendency to pass into the vein B1. A strong mini-magnet M under the influence of a field gradient in the direction of the arrow H attracts the catheter into the vein B2, then by reversing the field gradient, into the aneurysm A.

This magnetic guidance technique is used in a number of different operations, because it is less invasive than classic surgical techniques and avoids the necessity of making large incisions which take longer to heal: for example the treatment of varicose veins can now be carried out from inside the veins, without having to perform multiple incisions along the length of the vein.

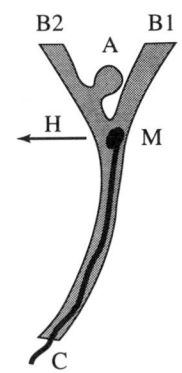

Figure 25.4
Aneurysm therapy

3.3. DENTAL CARE

In odontology, it can be sometimes advantageous to fix a crown on to a base by magnetic means: a Japanese group has developed a dental prosthesis incorporating rare earth magnets, which despite their small size, develop a force of 4.9 N. The magnet is sealed by laser to its stainless steel cover to provide protection against corrosion [15]. The effects of the permanent exposure of brain cells to this magnetic field are unknown.

The filling of the dental channel can also use cements (based on phosphates) containing 50% magnetic particles: their application is carried out under an alternating magnetic field gradient that encourages the penetration of the cement.

3.4. MICRO-ACTUATORS

When we discussed microsensors in § 2.4, we mentioned the need for a microactuator to apply a local pressure, at regular intervals, on the cornea. Such microactuators are due for a considerable development, to assist in many surgical or medical tasks. For example, the use of remote-controlled micro-scalpels is being studied, as well as that of magnetostrictive micro-pumps intended to deliver minute doses of drugs on demand. In cardiology a micro-pump could be inserted during an operation, and it will then function for many years. Should a cardiac problem occur, the starting up will be done using an external magnetic field, which is triggered by an alarm signal produced by a sensor and controlled by a microprocessor. The pump will then deliver into the patient's blood the correct dose of cardiac tonic to overcome the crisis. Such microsystems will use thin film magnetostrictive techniques (see chap. 18 and 20).

3.5. USE OF MAGNETOTACTIC BACTERIA

It is possible to produce magnetic patterns on the surface of very soft transformer sheet metal (coated FeSi) without any damage, by the Bitter method (see § 6.1.1 of chap. 5) using solutions with a high concentration of magnetotactic bacteria to reveal the magnetic domain walls via the gradients in the intensity of the stray field [16]. Apart from this very technical application, the main applications envisaged for these bacteria lie in the medical field. The first of these concerns the treatment of malignant tumors by heat. Under the action of appropriate field gradients, the magnetotactic bacteria are concentrated in the tumor, then heated up to 40°C by the application of an alternating magnetic field. This hyperthermal treatment will selectively destroy only those cells localized in the tumor. Another application concerns fixing poisonous chemicals by magnetotactic bacteria, then guiding these bacteria. When concentrated on the tumor, the bacteria are killed, and the released poisonous substance can act solely on the tumor, without any of the secondary effects that are observed with classical chemotherapy.

3.6. *USE OF OTHER MAGNETIC MATERIALS*

To create a local hyperthermia, without a risk of overheating, Japanese researchers have used amorphous magnetic materials based on Fe, P, C and Cr with a low Curie temperature: as soon as the temperature rises too much, the material loses its magnetism and so warms up less under the action of the variable magnetic field. The temperature is thus stabilised in the neighborhood of the Curie temperature.

Another approach uses synthetic colloidal magnetic iron oxide. Chan *et al.* optimized the manufacturing process of this material in order to ensure, not only a low toxicity and a good biocompatibility, but also an excellent energy yield. The dextran stabilised magnetic iron oxide produced by ultrasonic treatment gives heating rates of the order of 5°C per minute on tissues doped with only 1 mg of iron per cm^3 when they are submitted to a radiation of 1 $W.cm^{-3}$ at 1 MHz, while the heating of unloaded tissue remains negligible [17].

3.7. *MAGNETOTHERAPY*

Here we are faced with a vast topic, and in certain regards controversial. Since antiquity, the Chinese healed otitis by applying magnets to the sick ear. Shortly before the French Revolution, Mesmer claimed he could heal all kinds of illnesses using magnetism, but was convicted of fraud by the Academy of Sciences in Paris.

The famous Japanese company TDK sells gold necklaces inset with a chain of rare earth magnets; these "shoulder TDK" are patented and are claimed to heal pains in the neck, shoulders and the back [18].

In auriculotherapy, one of the techniques used consists of fixing very small needles on the ear and then waving a small magnet close by.

So magnetotherapy is nowadays thriving, with an increased following, but still remains unexplained: the main affections liable to being healed by magnetotherapy seem to be benign rheumatology, minor traumatology and problems concerning tonic posture activity [19].

Some scientific studies attempt to specify the mechanisms that could be involved, for example at the cell level. Thus, the circulation of blood under a static magnetic field would lead to a change of the clotting rate [20].

It is possible to relieve some affections of purely mechanical origin with the help of magnetic fields: for example foreign metallic bodies (copper) have been extracted from the eye using pulsed magnetic field gradients [21].

All these treatments rely on the influence that a magnetic field exercises on a biological substance, via *magnetic dipolar interaction*, which is a well-established scientific fact and that is employed by *magneticians*.

They have nothing in common with the treatment by *hypnotisers* who also claim to use Magnetism, but whose action at a distance has never been analysed quantitatively, nor explained scientifically, and whose results have never been subjected to scientific study and rigorous statistics.

4. CONCLUSIONS

Since ancient times magnetism has fascinated mankind, who has tried to appropriate its virtues, including those for healing. Living organisms generally contain very small quantities of magnetic material, but our knowledge on the interactions between magnetism and biologic functions remains embryonic even to this day, and one cannot yet establish with any certainty the existence of effects of a static magnetic field on the behavior of living organisms.

However, this situation should evolve rapidly since research is very active in this field, as is shown in the already old but excellent review article of Williamson and Kaufman [10], and the more recent book by Wadas [22], both of which are devoted to biomagnetism. In the same way, biotechnologies associated with magnetism are making great strides forward.

This science, situated at the crossroads of magnetism, biology and organic chemistry, calls more and more on the most modern techniques of data processing to interpret a magnetoencephalogram or to refine the structure of a magnetically active macromolecule. With such a variety of investigative tools, one can very shortly expect an abundant harvest of fascinating results.

REFERENCES

[1]   G. CHOUTEAU, CL. VEYRET-JEANDEY, *J. Physique* **42** (1981) 1441.

[2]   R. CHIARELLI, A. RASSAT, P. REY, *J. Chem. Soc. Chem. Commun.* **15** (1992) 1081;
        R. CHIARELLI, M.A. NOVAK, A. RASSAT, J.L. THOLENCE, *Nature* **363** (1993) 147.

[3]   D. BLOCH, J. VOIRON, J.C. BONNER, J.W. BRAY, I.S. JACOBS, L.V. INTERRANTE,
        Phys. Rev. Lett. **44** (1980) 294.

[4]   O. KAHN, Molecular Magnetism (1993) VCH Publishers, New York.

[5]   J.S. MILLER, A.J. EPSTEIN, *Angew. Chem. Int. Ed. Engl.* **33** (1994) 385.

[6]   R.P. BLAKEMORE, R.B. FRANKEL, *Sci. American* **245** (1981) 42.

[7]   R. WILTSCHKO, W. WILTSCHKO, Magnetic Orientation in Animals (1995) Springer Verlag,
        Berlin-New York.

[8]   D.D. AWSCHALOM, J.F. SMYTH, G. GRINSTEIN, D.P. DI VICENZO, D. LOSS, *Phys. Rev. Lett.* **68** (1992) 3092.

[9]   Y. ROCARD, La science et les sourciers (1991) Dunod-Bordas, Paris.

[10]   S.J. WILLIAMSON, L. KAUFMAN, *J. Magn. Magn. Mater.* **22** (1981) 129.

[11]   J. NENONEN, J. MONTONEN, T. KATILA, *Rev. Sci. Instrum.* **67** (1996) 2397.

[12]   M.A. VLADIMIRSKY, A.A. KUZNETSOV, V.I. PHILIPPOV, *J. Magn. Magn. Mater.* **122** (1993) 371.

[13]   G.T. GILLIES, R.C. RITTER, W.C. BROADDUS, M.S. GRADY, M.A. HOWARD III, R.G. MCNEIL, *Rev. Sci. Instrum.* **65** (1994) 533.

[14]   H. TILLANDER, *Acta Radiol.* **35** (1951) 62.

[15]   Y. HONKURA, Y. TANAKA, Y. TWAMA, *IEEE Trans. J. Magn. Japan* **6** (1991) 551.

[16]   G. HARASCO, H. PFUTZNER, K. FUTSCHIK, *IEEE Trans. Magn.* **31** (1995) 938.

[17]   D.C.F. CHAN, D.B. KIRPOTIN, P.A. BUNN Jr, *J. Magn. Magn. Mater.* **122** (1993) 374.

[18]   K. NAKAGAWA, *Japan Medical Journal* (4 décembre 1976) 2745.

[19]   J. TRéMOLIèRES, *Electronique Applications* **63** (déc. 1988-janv. 1989) 61.

[20]   R.P. KIKUT, *Latv. PSR Zinat. Akad. Vestis URSS* **3** (1981) 122.

[21]   H. WEBER, G. LANDWEHR, *IEEE Trans. Magn.* **Mag-17** (1981) 2330.

[22]   R.S. WADAS, Biomagnetism (1991) Ellis Horwood, Chichester.

CHAPTER 26

PRACTICAL MAGNETISM
AND INSTRUMENTATION

The first part of this chapter covers the techniques used for the measurement of soft or hard materials. The second part treats the generation of magnetic fields, whilst their measurement is described in the third and final part.

1. MAGNETIZATION MEASUREMENT TECHNIQUES

The method of measurement will be generally different depending on whether it concerns a magnetically soft or hard material. For a soft material, the demagnetizing field can be 1,000 or 10,000 times greater than the internal field. It must be reduced as much as possible in order to determine the initial part of the magnetization curve.

A device that is used to measure magnetization (*magnetometer*) consists of:

- a source of magnetic field (usually an electromagnet or a superconducting coil),
- a provision for changing the temperature, to enable the study of properties as a function of this parameter,
- a system for measuring the magnetization, the magnetic field (and possibly the temperature).

The techniques described in this chapter are force and flux methods.

The determination of the magnetization by the measurement, through a search coil, of the flux originating from the sample, is a very common direct method, and we first recall the theorem that enables this flux to be calculated.

1.1. RECIPROCITY THEOREM

The flux, originating from a supposedly point-like magnetic moment, across a measuring coil (1) (represented here by a loop), must be calculated. The moment magnetic m is related to the polarization $\mathbf{J}$ ($= \mu_0 \mathbf{M}$), and to the volume V by the relation:

$$m = \mathbf{J} V / \mu_0 \qquad (26.1)$$

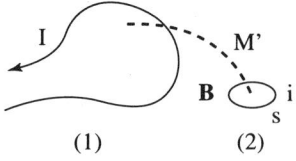

(1)   (2)

**Figure 26.1 - Flux through a coil
from a magnetic moment**

The magnetic moment is equivalent to a current i flowing along a small winding (2) with area s, with $m = i\,s$. Let us assume for the argument's sake that a current I flows in the measurement loop.

This flux ϕ_{12} sent by loop (1) through winding (2) is:

$$\phi_{12} = M'I \tag{26.2}$$

where M' is the coefficient of mutual inductance between (1) and (2), the induction **B** created in (2) being given by: $M'I = \mathbf{B}\,s$.

The flux through loop (1) from winding (2) is: $\phi = M'i = \mathbf{B}\,s\,i\,/\,I$, that is:

$$\phi = (\mathbf{B}\,/\,I)\,m \tag{26.3}$$

The flux from a point magnetic moment m through a measuring coil is equal to the scalar product of this moment and the ratio (B / I), where B is the induction that would be created by a fictitious current I flowing in the measurement coil.

When the magnetic moment can no longer be considered to be a point source, equation (26.3) should be rewritten:

$$\phi = \iiint (\mathbf{B}\,/\,I)\,\mathbf{J}\,dV\,/\,\mu_0 = \iiint (B_x J_x + B_y J_y + B_z J_z)\,dV\,/\,(I\mu_0),$$

where the integral is over the volume of the sample.

If **J** is constant and oriented parallel to the axis of the field and to the Oz axis of the measurement coils, one can take $\mathbf{J} = J_z\hat{\mathbf{U}}$ (unit vector parallel to Oz) out of the integral:

$$\phi = J_z \iiint (B_z\,/\,I)\,dV\,/\,\mu_0 \tag{26.4}$$

This expression will be used in the rest of this chapter. Now consider the case when the component J_x is not zero, and the measuring coils have their axis along Oz. The component B_x is weak, and in general $B_x(x_0) = -B_x(-x_0)$. As a result $\iiint B_x J_x\,dV\,/\,(I\mu_0)$ is zero if the sample lies in a symmetrical position with respect to the axis Ox.

1.2. MEASUREMENT OF SOFT MATERIALS

To study the initial part of the magnetization curve, it is desirable to cancel the demagnetising field by forming as well closed a magnetic circuit as possible.

Figure 26.2 shows the magnetic polarization **J** as a function of the internal field $\mathbf{H_i}$ for soft iron. A saturation magnetic polarization of about 2.18 T is reached in a field of less than 10^{-3} T. The initial part of the curve is not linear: it has an inflexion point (maximum slope) for a field of the order of the coercive field.

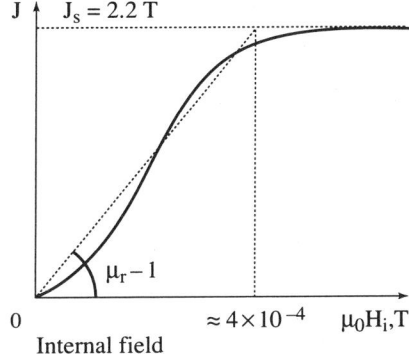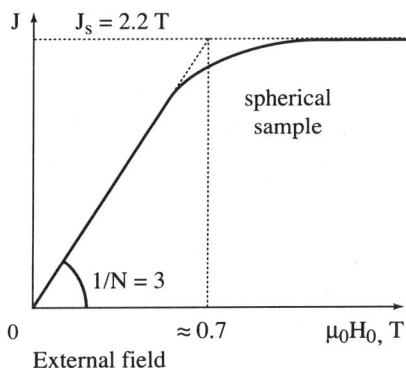

Figure 26.2 - Magnetisation curves of an iron sample
(note the difference in the abscissa scale)

In many cases, the initial part of the curve can be described by a straight line with slope $\chi = (\mu_r - 1)$, μ_r being the relative permeability:

$$\mathbf{J} \approx \mu_0 \, (\mu_r - 1) \mathbf{H_i} \tag{26.5}$$

Since μ_r is greater than 1,000 for a soft material, the slope can be taken simply as: $\chi \approx \mu_r$. One then has: $\mathbf{J} \approx \mu \, \mathbf{H_i}$.

Consider a sample placed in an external field $\mathbf{H_0}$, and suppose that the demagnetizing field is characterized by a constant coefficient N: $\mu_0 \mathbf{H_d} = -N\mathbf{J}$, where N only depends on the shape of the sample.

Therefore: $\mu_0 \, \mathbf{H_i} = \mu_0 \, \mathbf{H_0} - N \, \mathbf{J}$, where $\mathbf{H_0}$ is the external field.

Using the approximation: $\mathbf{J} \approx \mu_0 \, (\mu_r - 1) \, \mathbf{H_i}$, we get: $\mathbf{H_i} \approx \mathbf{H_0} / [1 + N \, (\mu_r - 1)]$, and:

$$\mathbf{J} \approx \mu_0 \, (\mu_r - 1) \, \mathbf{H_0} / [1 + N \, (\mu_r - 1)] \tag{26.6}$$

The apparent permeability, defined by $\mathbf{B} = \mu_a \mu_0 \mathbf{H_0}$, is therefore:

$$\mu_a \approx \mu_r / [1 + N \, (\mu_r - 1)].$$

For large μ_r: $\qquad\qquad \mu_a \approx 1 / (N + 1 / \mu_r) \tag{26.7}$

If $N\mu_r$ is much greater than 1, which is very often the case (except for a closed circuit), one obtains: $\mathbf{J} \approx \mu_0 \, \mathbf{H_0} / N$, and $\mu_a \approx 1 / N$.

The initial part of the magnetization curve is then a straight line with slope $1/N$ up till a field $\mu_0 \, H_{d0} = N \, J_s$ (J_s is the saturation polarization). *Saturating a soft material required an applied field $\mu_0 \, H_0 > N \, J_s$.* For example, to saturate an iron sphere, one must apply a field of 0.7 T, although the internal field is only about 4×10^{-4} T.

1.2.1. Measurement of a toroidal sample

A variable magnetic field is applied to the sample, and the voltage induced in a detection coil is integrated.

The only geometry where the demagnetizing field is zero is a closed circuit such as a toroid.

The magnetic field is created by a current i flowing in an coil consisting of n turns wound directly on the toroid of radius r_0. In general this current i varies sinusoidally at low frequency (for example 1 Hz): $i = I_0 \sin \omega t$. It can also have a saw-tooth waveform. If the material is soft and unsaturated (large μ_r), the magnetic field over the whole toroid has practically the same value, given by Ampere's law, $\int \mathbf{H} \, d\mathbf{l} = n \, i$, hence:

$$H = n \, i / 2\pi \, r_0 \qquad (26.8)$$

Across a second coil (the measurement coil) with n' turns, also wound on the toroid of cross-section S, there is a voltage: $e = -n' \, S \, dB / dt$ corresponding to variations of the flux $\phi = B \, S$ that circulates in the toroid.

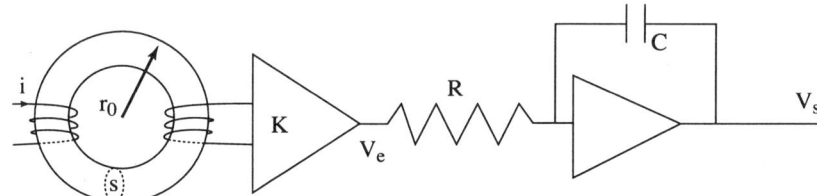

Figure 26.3 - Principle of a fluxmeter

The measuring coil is connected to an integrator (via an impedance matching amplifier with gain K). The output voltage is related to the input voltage by the relationship: $V_0 = -(1/R \, C) \int V_e \, dt$. Since $V_e = -K \, n' \, S \, dB / dt$, we have:

$$V_0 = (K \, n' / R \, C) \, S \, B \qquad (26.9)$$

As $\mathbf{B} = \mu_0 \mathbf{H} + \mathbf{J}$, the measurement of the flux allows the measurement of the polarization to within $\mu_0 \mathbf{H}$.

1.2.2. Epstein permeameter

A closed square-shaped magnetic circuit can be made up from 4p stacked sheets (fig. 26.4). Such a circuit has the advantage that it can be dismantled, but the closure of the circuit is not perfect. It also has the advantage that the field and detection coils do not need to be wound *in situ*.

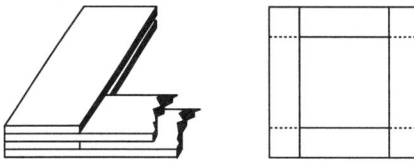

*Figure 26.4
Epstein permeameter*

1.3. MEASUREMENT OF THE MAGNETISATION OF HARD MATERIALS OR OF WEAKLY MAGNETIC MATERIALS

1.3.1. Force methods

The sample is placed in an inhomogeneous field. The force acting on the sample, equal to the product of its magnetic moment and the field gradient, is measured.

The potential energy of a material with volume V, and polarisation $\mathbf{J}$, submitted to a magnetic field $\mathbf{H}$ is: $W = -\mathbf{J}\,\mathbf{H}\,V = -V\,(J_x H_x + J_y H_y + J_z H_z)$.

A sample, placed in a inhomogeneous magnetic field, is subject to a force which is expressed by: $\mathbf{F} = -\mathbf{grad}(W)$, that is:

$$F_x = V\,(J_x\,dH_x/dx + J_y\,dH_y/dx + J_z\,dH_z/dx)$$
$$F_y = V\,(J_x\,dH_x/dy + J_y\,dH_y/dy + J_z\,dH_z/dy) \qquad (26.10)$$
$$F_z = V\,(J_x\,dH_x/dz + J_y\,dH_y/dz + J_z\,dH_z/dz)$$

In this formula, $\mathbf{H}$ is the magnetic field in the sample, that is $\mathbf{H} = \mathbf{H_0} + \mathbf{H_d}$, where $\mathbf{H_0}$ is the applied field and $\mathbf{H_d}$ the demagnetizing field. Derivatives of $\mathbf{H_d}$ are neglected, so that in formula (26.10) $\mathbf{H}$ is taken to be the external field.

One can distinguish two force methods:
♦ the sample is placed in a DC (static) field gradient, producing a DC force,
♦ an AC (alternating) field gradient is applied, leading to an AC force.

Magnetometer with a DC field gradient

Sensitivity: $10^{-8}\,A.m^2$

The sample is placed in a homogeneous field to which a DC magnetic field gradient is superimposed (fig. 26.5), and one measures the magnetic force acting on the sample.

Assume that, in a homogeneous magnetic field along Oz, the sample acquires a polarization J_z, also along Oz (the case of an isotropic sample), and that a field gradient is applied in only one direction, Oy. The sample is subject to a force: $F_z = V\,J_z\,dH_z/dy$.

In practice, one can create the field and the field gradient simultaneously with an electromagnet equipped with asymmetric pole pieces (fig. 26.5-b). Another approach is to create a homogeneous field using an electromagnet, and to produce the field gradient with another coil (fig. 26.5-a). The latter system has the advantage of allowing independent control of the field and the field gradient.

The gradient coil is generally realised using 2 (or 4) coils connected in series-opposition (fig. 26.5-a).

To measure the magnetic force, and therefore the magnetization of the sample, one can simply adapt a commercial precision balance, but this implies that the force is vertical. This technique is quite effective but also restrictive, as the sample is fixed on to a mobile system which also contributes to the measurement signal, and must be free from any friction, air current etc. as one wants to measure very weak forces.

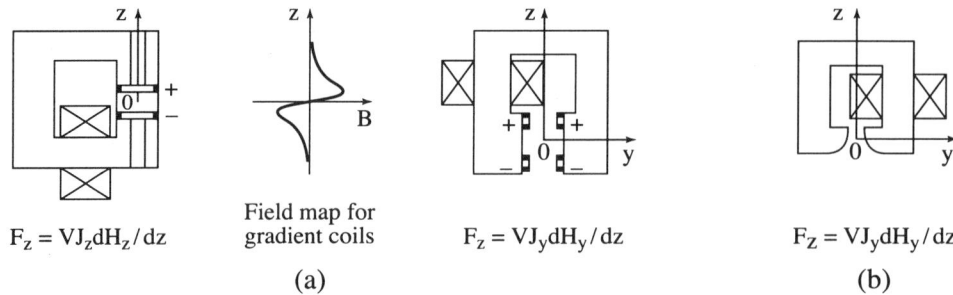

$F_z = VJ_z dH_z / dz$

Field map for gradient coils

$F_z = VJ_y dH_y / dz$

$F_z = VJ_y dH_y / dz$

(a)   (b)

Figure 26.5 - Different ways of creating a field and a superimposed field gradient
(a) gradient independent of the field - (b) gradient proportional to the field

Alternating field gradient magnetometer

The sample, fixed on to a mobile frame connected to a piezo-electric quartz resonator, undergoes at the resonant frequency of the mobile frame an AC force that is detected by the resonator.

Sensitivity: 10^{-11} A . m^2 at room temperature

An AC field gradient is generated for example using the 4 coils shown in figure 26.6. The sample is fixed on to a rod attached to a piezoelectric cantilever. The gradient coil, driven at the resonant frequency of the sample rod (generally of the order of 1 kHz) for better sensitivity, produces an AC force on the sample. This vibration is transmitted to the quartz transducer that produces a voltage proportional to the amplitude of the movement, hence to the magnetization of the sample.

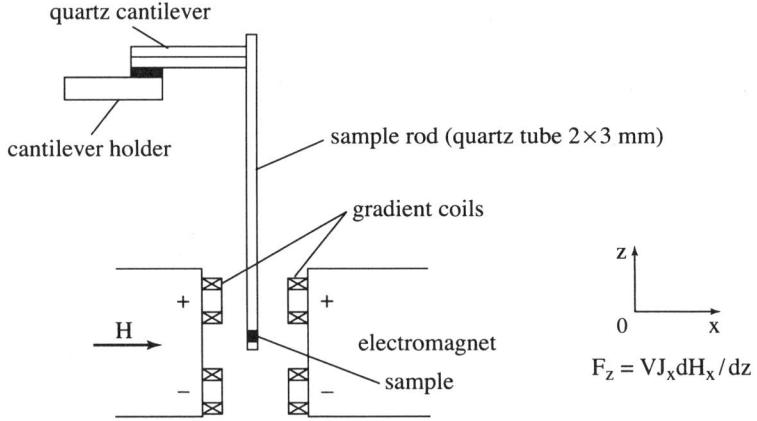

Figure 26.6 - Diagram of an AC field gradient magnetometer
Horizontal magnetic field. 4 gradient coils

Since one measures a force, the instrument is not sensitive to a parasitic flux drift. This allows rapid measurements at the maximum sensitivity when the field varies. The mobile part of the magnetometer needs to be handled with care. For studies at variable temperatures, the resolution decreases because of the piezo-electric quartz resonator. The block diagram in figure 26.7 shows the principle of the measurement using an AC phase lock-in amplifier, where the resolution can reach the nanovolt level.

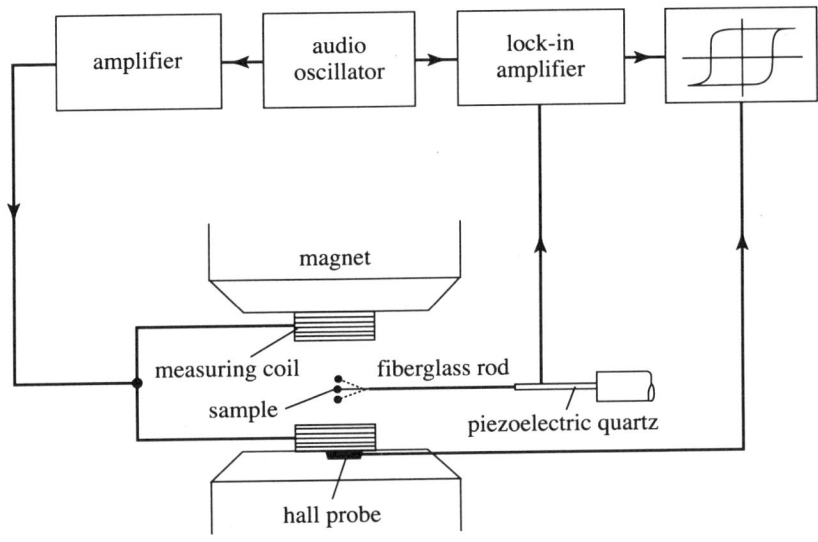

Figure 26.7 - Electronic diagram of an AC gradient magnetometer

1.3.2. Flux methods

The measurement of the flux induced in a coil can be performed by:
- aperiodic displacement of the sample, with measurement of the induced voltage using a digital voltmeter or a SQUID.
- periodic displacement of the sample: this is the vibrating sample magnetometer.
- without any movement of the sample, by applying a variable magnetic field. If the field varies slowly, the apparatus (hysteresigraph) measures the magnetization as a function of the field. Another technique (Hartshorn bridge) uses a weak exciting magnetic field (at frequencies between 10 and 10,000 Hz), and allows the measurement of the effective value of the magnetization in a weak field.

**Aperiodic displacement of the sample and measurement
using a digital voltmeter ("extraction method")**

Sensitivity: $10^{-7} A \cdot m^2$

The coil system generally consists of 4 pick-up coils 1, 2, 3, 4 whose axes are parallel to the field, connected in series opposition, made up of a large number of turns for greater sensitivity (hence made of fine enameled copper wire, see fig. 26.8).

The principal coils 1 and 2 (number of turns n, average cross-section S) are identical, and are therefore axially flux-compensated. Coils 3 and 4 (number of turns n', average cross-section S') provide the radial compensation by respecting the condition:

$$n S = n'S' \qquad (26.11)$$

This double compensation allows a reduction in the induced voltages due to the variations in the parasitic flux coming from the magnetic field source (or of mechanical origin), whereas the flux sent by the sample into the coils is only slightly reduced (typically by 30%) by the presence of coils 3 and 4. The measurement of the magnetisation is obtained by integration of the voltage that appears across the pick-up coils during the displacement of the sample between points A and B (fig. 26.8), the centres of coils 2 and 1, using an integrating digital voltmeter.

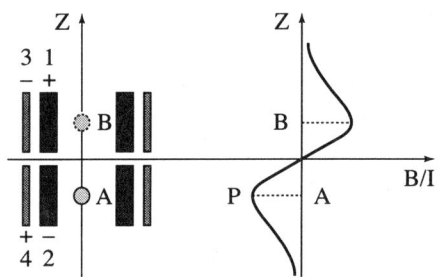

Figure 26.8 - Pick-up system made up of 4 coils connected in series opposition, with the variation of the field in the coils

The variation of the measured flux, as given by the reciprocity theorem (26.3), is proportional to the magnetic moment:

$$\delta\phi = [(B/I)_B - (B/I)_A]\,\mathfrak{m} = 2\,PA.\,\mathfrak{m} \qquad (26.12)$$

The voltage across the pick-up coils is filtered by an RC circuit with a time constant as long as possible, but distinctly smaller than the integration time (for example RC = 0.05 second for an integration time of the order of 1 second).

A sensitivity of 10^{-7} A.m^2 corresponds typically to a noise in the pick-up coils of about 0.2 μV. s.

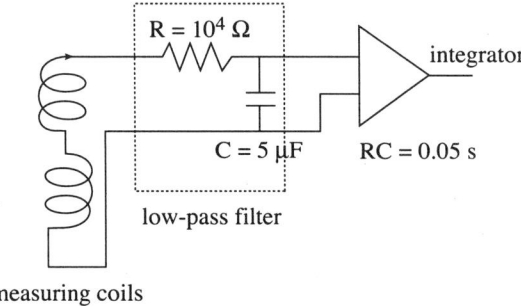

Figure 26.9 - Circuit diagram of an extraction apparatus

Aperiodic displacement of the sample
and measurement by an RF (radiofrequency) SQUID

Sensitivity: 10^{-9} to 10^{-11} A. m^2

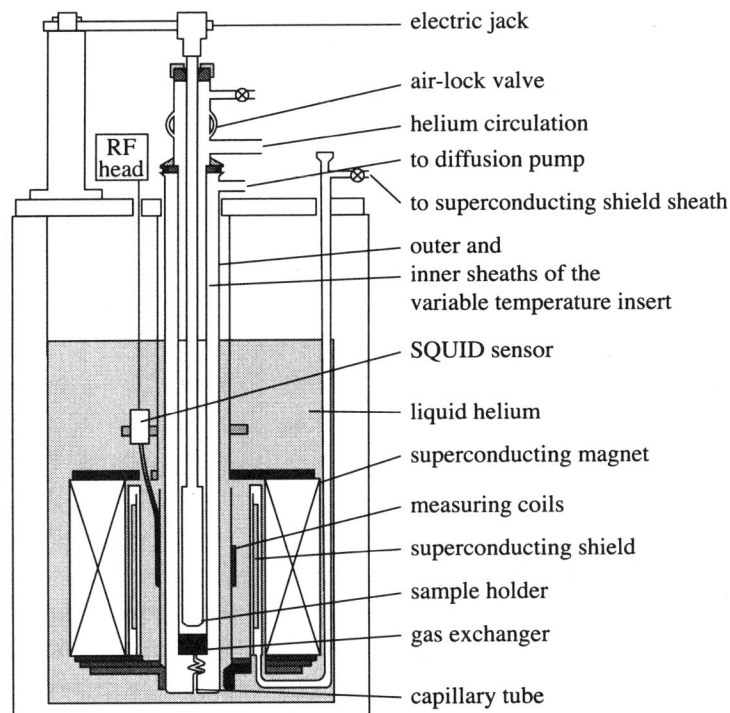

- electric jack
- air-lock valve
- helium circulation
- to diffusion pump
- to superconducting shield sheath
- outer and
- inner sheaths of the variable temperature insert
- SQUID sensor
- liquid helium
- superconducting magnet
- measuring coils
- superconducting shield
- sample holder
- gas exchanger
- capillary tube

RF head

***Figure 26.10 - Diagram of an 8 teslas RF SQUID magnetometer
with variable temperature (1.5 < T < 300 K)***

A parallel resonant R, L, C circuit (fig. 26.11-b) is fed at its resonant frequency $\omega/2\pi$ (chosen between 10 and 300 MHz) by a current $i = I_{RF} \sin \omega t$. The current that flows in the inductance is Qi ($Q = L\omega/R$ is the quality factor, R being the resistance of the inductance), and the peak voltage across this circuit is: $V_t \approx QL\omega I_{RF}$.

A small superconducting ring, interrupted by a Josephson junction, is now placed close to the previous circuit, and coupled to the inductance L by a mutual inductance M'. A Josephson junction can be an insulating gate of the order of a nanometer in thickness. Josephson showed that phase coherence between Cooper pairs (that are at the origin of supraconductivity) persists through such a gate. The interest of the Josephson junction is to lower the critical current i_c of the ring to a few microamperes. The critical flux ϕ_c of the ring $\phi_c = L_s i_c$, where L_s is the inductance of the ring, is then a few flux quanta (a flux quantum is $\phi_0 = 2.07 \times 10^{-15}$ weber).

One sees that the characteristic $V_t(I_{RF})$ is no longer a straight line. Due to the presence of the ring, the current Qi that flows in the inductance induces in the ring an AC flux of amplitude:

$$\phi_1 = M'Q I_{RF} \qquad\qquad (26.13)$$

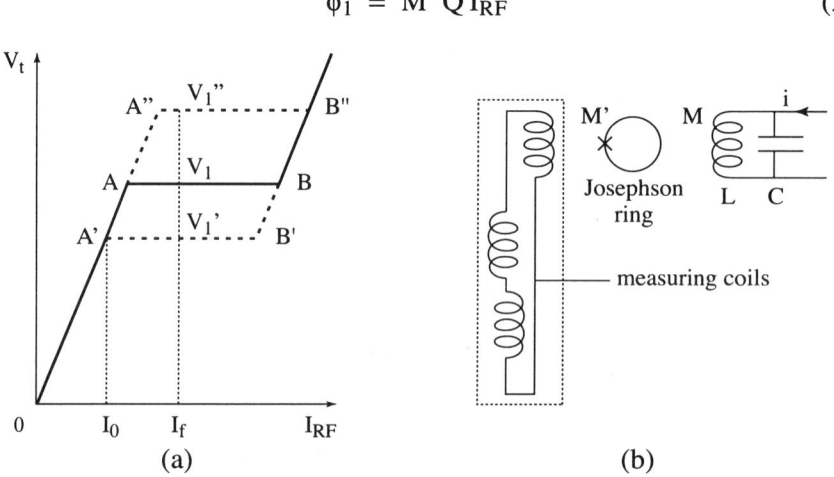

(a)   (b)

Figure 26.11 - (a) Ideal voltage-current characteristic of an RF SQUID
(b) Circuit diagram of a SQUID sensor

When ϕ_1 reaches ϕ_c (for a current $I_0 = \phi_c/M'Q$ –corresponding to point A on the characteristic of figure 26.11-a) or exceeds it, there occurs a periodic admission of a flux quantum flux ϕ_0, followed by the expulsion of this flux quantum $(2n + 1)\,\pi/\omega$ later, where n is an integer which is larger the closer one is to point A.

The voltage across the resonant circuit becomes constant (plateau AB), and equal to:

$$V_T = Q\,L\omega\,I_0 \qquad\qquad (26.14)$$

The end of the plateau, point B on figure 26.11-a, occurs when the admission and expulsion of a flux quantum occur every half-cycle of the current, that is at intervals of π/ω. One chooses an operating current I_f near the middle of the plateau AB. A DC flux $\delta\phi$ is now superimposed on the AC flux: the plateau on the characteristic is reached when the total maximum flux $\phi_1 + \delta\phi$ reaches the value ϕ_c. It occurs earlier (A'B') or later (A"B") for a current $I_0' = (\phi_c - \delta\phi)/M'Q$.

The variation of voltage $\delta V = Q\,L\omega\,(I_0 - I_0')$ corresponds to a variation in flux $\delta\phi = M'Q\,(I_0 - I_0')$ so that:

$$\delta V = (L\omega/M')\,\delta\phi \qquad\qquad (26.15)$$

Thus a *DC flux → peak voltage* converter has been achieved.

The DC flux is applied using a superconducting circuit consisting of a pick-up coil and a small inductance, coupled to the Josephson ring through a mutual inductance M" (fig. 26.11-b). The pick-up coil consists of two coils connected in series opposition, each of 2 or 3 turns, with their axes parallel to the applied field.

When one displaces the sample in the coils, as in the previously described extraction method, to determine its moment $\mathfrak{m}$, the variation of flux given by (26.3) is:

$$\delta\phi = \delta(B/I)\,\mathfrak{m} \qquad\qquad (26.16)$$

The current variation δi in the superconducting measurement circuit with inductance ΣL_i is then given by:

$$\delta\phi = (\Sigma L_i)\delta i \qquad (26.17)$$

The flux variation seen by the superconducting ring:

$$\delta\phi_2 = M''\delta i \qquad (26.18)$$

is compensated by a feedback flux equal and opposite to $\delta\phi_2$, generated by a current δi_{cr} flowing in the inductance of the resonant circuit $\delta\phi_{cr} = M'\delta i_{cr} = -\delta\phi_2$. From equations (26.16) to (26.18), one obtains the magnetic moment of the sample:

$$m = -\frac{M'(\Sigma L_i)\delta i_{cr}}{M''\delta(B/I)} \qquad (26.19)$$

Magnetometers using SQUID sensors are relatively slow because they measure a DC flux. After every magnetic field change, the magnetic field source (almost always a superconducting coil) has a lag, and one must wait until the drift is sufficiently small to be able to make a precise measurement. When measuring in a constant magnetic field, this problem disappears, and the magnetometer has good performance.

DC SQUID detection

A DC SQUID can also be used to convert a change of magnetic flux into a voltage variation.

The voltage across a Josephson junction supplied with a direct current I remains zero as long as $I < I_{c0}$, where I_{c0} is the critical current of the junction. For greater currents one observes a voltage $V \approx R(I - I_{c0})$ (fig. 26.12), R being the equivalent resistance of the junction.

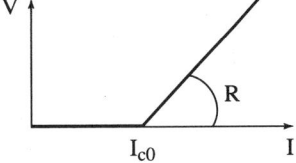

Figure 26.12
Ideal characteristic of a Josephson junction

A DC SQUID is made up of a superconducting loop broken by 2 Josephson junctions each having a critical current I_{c0} (fig. 26.13).

Figure 26.13 - Diagram of a DC SQUID

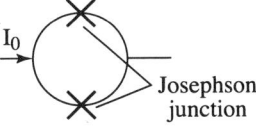

The loop is fed with a direct *bias* current I_0 slightly greater than $2I_{c0}$. The voltage across the SQUID is then simply $V = R(I_0 - 2I_{c0})/2$, corresponding to the fact that the SQUID has a critical current $2I_{c0}$, and an equivalent resistance $R/2$ (fig. 26.14-a). If one applies a flux $\delta\phi$ to the loop, which has an inductance L, an induced shielding current $\delta i = -\delta\phi / L$ will flow so that the total flux in the ring remains zero. In one junction, it adds to the bias current whilst it subtracts in the other (fig. 26.14-b).

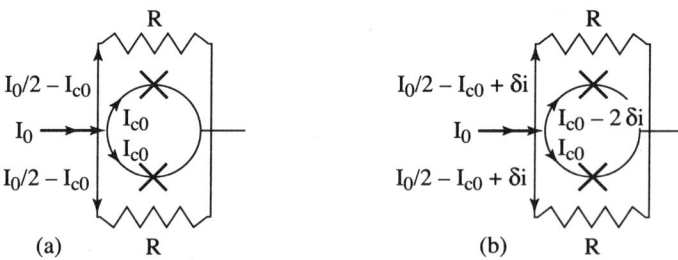

Figure 26.14 - Distribution of currents in a DC SQUID:
(a) without an external flux - (b) with an external flux

The total superconducting current in one junction cannot exceed I_{c0}. In the other junction, it will therefore be equal to $I_{c0} - 2\delta i$. The total critical current is thus: $I_c = 2 I_{c0} - 2\delta i$.

In fact one can show that the critical current features a variation as shown in figure 26.15 with a period ϕ_0.

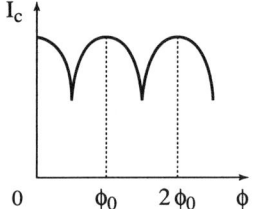

**Figure 26.15 - Critical current in a DC SQUID
as a function of the applied flux**

The voltage across the SQUID therefore depends on the applied flux, and is given by: $V = (R / 2) (I_0 - 2 I_{c0} + 2\delta i)$. The variation of voltage corresponding to a variation of flux: $\delta\phi = L \delta i$ is therefore $\delta V = (R / 2) 2 \delta i = (R / L) \delta\phi$.

With the help of a feedback circuit, one maintains the voltage across the SQUID, and therefore the flux in the SQUID, constant. The feedback current is measured. It is proportional to the flux $\delta\phi$.

Periodic movement of the sample: vibrating sample magnetometer

The induced voltage, produced by the sinusoidal movement of the sample inside a pick-up coil, is measured.

Sensitivity: 10^{-8} to 10^{-10} A . m^2

The sample, placed at the center of a measuring coil, is vibrated along a vertical axis (z), at a frequency f $(= \Omega / 2\pi$, usually between 10 and 100 Hz) with a constant amplitude z_0 which varies from 0.1 to 5 mm depending on the instrument (fig. 26.16).

Therefore: $z = z_0 \sin \Omega t$.

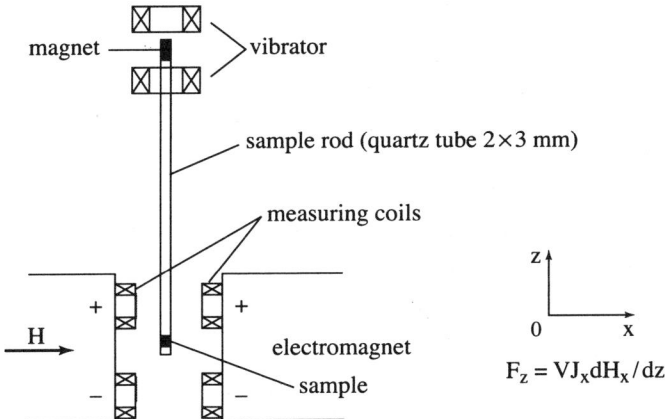

**Figure 26.16 - Principle of a vibrating sample magnetometer
with a horizontal magnetic field**

Let us use the reciprocity theorem to calculate the induced voltage by differentiating (26.3) with respect to the time:

$$e = -\frac{d\phi}{dt} = -m\,\frac{d(B/I)}{dz}\frac{dz}{dt} \qquad (26.20)$$

such that: $e = -E_0 \cos\Omega t$, with $E_0 = z_0\,\Omega\,m\,\dfrac{d(B/I)}{dz}$.

The sensitivity is proportional to $z_0\,\Omega$ which is the amplitude of the velocity of the sample, and to the field gradient in the measurement coils $d(B/I)/dz$.

The applied magnetic field can be along the vertical axis (Oz): one measures a signal proportional to $m_z\,d(B_z/I)/dz$, as the measurement coils have their axis vertical.

The magnetic field can be horizontal (axis Ox): one measures a signal proportional to $m_x\,d(Bx/I)/dz$ with the measurement coils having their axis horizontal.

The measurement or pickup coils consist of 2 (or 4) identical coils, which have the same axis as the direction of the applied field (Ox in fig. 26.16), and are connected in series opposition. Their separation and geometry are calculated so that the sample displacement takes place in a constant field gradient (this is a purely fictitious field gradient, as no current is fed to the coils). The induced voltage is measured with a lock-in amplifier that rejects practically all frequencies other than the measurement frequency. One of the advantages of this measurement method is its rapidity: the disturbing parasitic drift is in the form of pseudo-DC signals that can be easily eliminated by the lock-in amplifier, and measurements can be made even when the applied magnetic field varies.

Hysteresigraph

The field of an electromagnet is varied, with the sample lying between the pole pieces, and the voltage induced in a measurement coil surrounding the sample is integrated.

The sample is placed in the gap on an electromagnet. The separation of the pole pieces can generally be adjusted so that the sample occupies the full width of the pole gap, forming a practically closed magnetic circuit. The measurement coils are placed in the gap, directly around the sample. They consist of two concentric flux compensated coils (including respectively n and n' turns with mean cross-section S and S' satisfying $nS = n'S'$ if the field in the pole gap is homogeneous) so that, when the field varies in the absence of a sample, no voltage is produced across the coils. In the presence of a sample, the inner coil sees a greater flux than the outer coil because it is nearer the sample, and has more turns ($n > n'$ because $S < S'$).

The voltage induced by the sample in the pickup coils is obtained by differentiating (26.4): $e = -(dJ_z/dt) \iiint (B_z/I) \, dv/\mu_0$, where we assumed that J_z is constant over space in order to take it out of the integrand. The measurement of the magnetisation is made, as for a toroidal sample (see § 1.2.1), by integrating the voltage across the pickup coils during the variation of the magnetic field. The latter is measured with a Hall probe placed close to the sample. The field and magnetisation signals are sent to the X and Y inputs of a chart recorder to enable a hysteresis curve to be plotted.

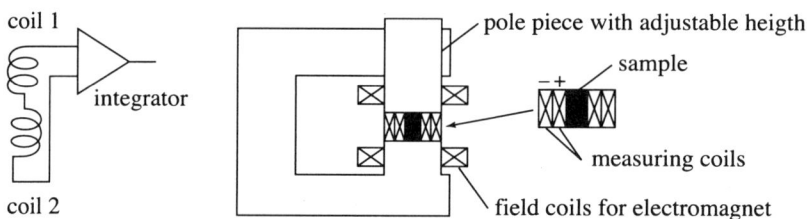

Figure 26.17 - Diagram of a hysteresigraph

It can easily be seen easy that this method is not suitable for the study of soft magnetic materials. Let us suppose that the sample with permeability μ_s occupies the width l_g of the gap; l_y is the length of the yoke of the electromagnet, assumed to have constant permeability μ_y in which the field is H_y. In the neighborhood of the poles the field is weaker due to magnetic leakage, and is equal to H_y/σ ($\sigma > 1$ is the Hopkinson coefficient).

Applying Ampere's law to a closed circuit which includes the sample: $H_y l_y + H_s l_s = n''i$ where n'' is the number of turns of the excitation coil carrying the current i. At the surface separating the sample and the magnet yoke, from the conservation of the normal component of the induction: $B_y = \mu_y H_y /\sigma = \mu_s H_s$,

therefore:
$$H_s = \frac{n''i}{l_y + \sigma l_y \, \mu_s /\mu_y} \tag{26.21}$$

By the same method, one obtains the field in the gap H_g, that is the field measured by a Hall probe:

$$H_g = \frac{n''i}{l_g + \sigma l_y \, \mu_0 /\mu_y} \tag{26.22}$$

From these equations (26.21) and (26.22) it can be seen that H_s and H_g are considerably different if $\mu_s \gg 1$.

One can estimate the error in the field for a hard magnetic material (magnet) whose permeability is equal to 2. One supposes $\mu_y = 2,000\,\mu_0$ and $l_y = 50\,l_g$. One gets $H_0 = 1.05 \times H_s$. For a material with a coercive force of 0.3 T, one measures $H_c = 0.315$ tesla.

1.3.3. Hartshorn bridge

The sample is placed in a weak AC magnetic field. The voltage induced in a measuring coil is proportional to the sample's magnetisation.

A solenoid (primary coil, see fig. 26.18), is supplied with a current $i = I_0 \sin \omega t$, creating a weak alternating magnetic field ($\sim 10^{-4}$ to 10^{-3} T). Inside this coil, in the zone of homogeneous field, one places 2 identical measuring coils (secondary), connected in series opposition. In the absence of a sample, the coils are perfectly compensated, and no voltage is produced across the secondary coil.

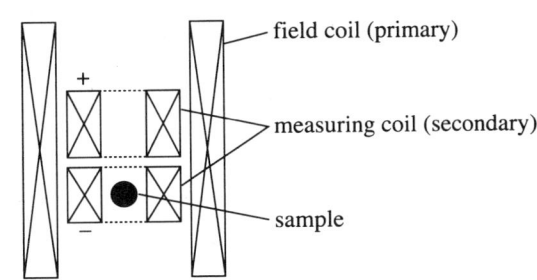

Figure 26.18
Diagram of a Hartshorn bridge

field coil (primary)

measuring coil (secondary)

sample

In the presence of a sample, a voltage appears. It can be measured using a lock-in amplifier. The induced voltage, given by the reciprocity theorem is: $e = -d\phi / dt$ with $\phi = (B / I)\,m$.

If one considers that the magnetization $M = m / V$ varies linearly with the magnetic field $H_0 \sin \omega t$, then: $M = \chi\,H_0 \sin \omega t$, where χ is the volume susceptibility of the material (see chap. 2), hence: $e = -(B / I)\chi\,V\,H_0\,\omega \cos \omega t$.

The amplitude of the voltage is proportional to the magnetisation $\chi V H_0$ and the frequency. One could be tempted to increase the frequency so as to increase the sensitivity, but this results in an increase in the eddy currents, which can produce an important parasitic contribution to the magnetization.

The interest of this method is its simplicity because the sample is stationary. It also allows the determination of transition temperatures.

Another application is to make dynamic measurements as a function of frequency. One can also superimpose a continuous magnetic field in order to move along the magnetization curve, and to measure the slope or differential susceptibility.

1.3.4. Measurement of anisotropy

The material with volume V, having a magnetic moment $\mathbf{J}\,V/\mu_0$, is placed in a field $\mathbf{B}$. If the material is anisotropic, $\mathbf{B}$ and $\mathbf{J}$ are no longer parallel when $\mathbf{B}$ does not coincide with an axis of easy magnetisation. There exists a torque (eq. 2.27) $\boldsymbol{\Gamma} = V\mathbf{J}\times\mathbf{B}/\mu_0$, that tends to rotate $\mathbf{J}$ towards $\mathbf{B}$. One of the techniques for the measurement of anisotropy consists in measuring this torque.

A more direct method consists of simultaneously measuring, by the extraction method (which is quite simple to put in practice), the three components of magnetisation along the axes Ox, Oy, Oz using 3 sets of coils connected to three digital voltmeters.

In figure 26.19, a miniaturized mechanism enables the sample to be turned around a horizontal axis parallel to the axis Ox, the magnetic field being vertical (Oz), which allows the anisotropy to be investigated according to the orientation of the sample with respect to the field.

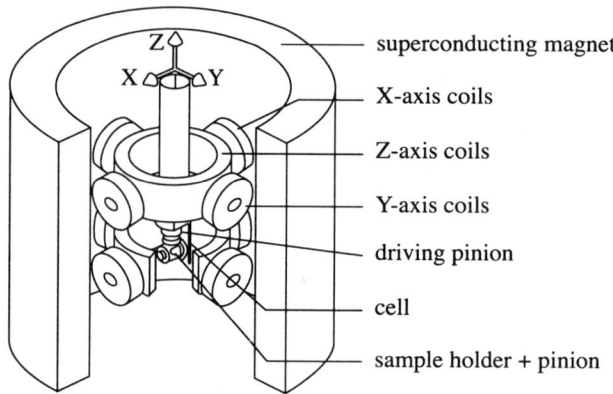

superconducting magnet

X-axis coils

Z-axis coils

Y-axis coils

driving pinion

cell

sample holder + pinion

Figure 26.19
Measuring coils and sample rotation device allowing the study of anisotropy

I.3.5. Calibration of magnetization

The calibration of magnetization is carried out using a standard sample of nickel whose polarization when saturated in a field of 1 T is well known at any temperature.

There exist standards of susceptibility (ratio J/B): palladium at room temperature, gadolinium sulphate above 1.5 kelvin. The combined use of a standard of magnetization and a standard of susceptibility also allows a control of the magnetic field.

2. *GENERATION OF MAGNETIC FIELDS*

To produce a magnetic field, one can distinguish two different techniques:

♦ those where the field is produced by an electric current flowing in a resistive or superconducting coil, without using any magnetic material. This is a linear process, i.e. the magnetic field is proportional to the current that produces it, and can be calculated using formulae such as Biot and Savart's law, Ampere's law, the dipole expression, etc.

♦ and those where the magnetic field source uses magnetic materials, associated (for example an electromagnet) or not (magnet) to electric currents. To solve this non-linear problem one uses approximate solutions.

2.1. MAGNETIC FIELD GENERATION WITHOUT USING MAGNETIC MATERIALS

2.1.1. Calculation of the field produced by a solenoid at any point in space

This problem is easy to solve numerically with a microcomputer. One chooses a system of orthogonal axes Oxyz, Oz being the axis of the solenoid in which a current I flows.

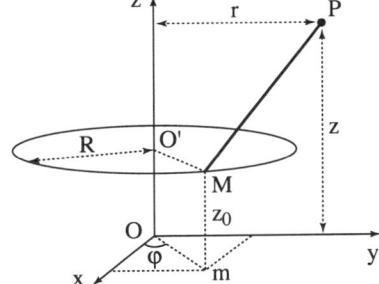

Figure 26.20 - Calculation of the field created by a winding at a point P

One can calculate the field created by a turn at height z_0 along the solenoid at an arbitrary point P in the plane yOz, having coordinates $(0, r, z)$, using Biot and Savart's law (see § 1.1.1 in chapter 2).

$\mathbf{dl}$, $\mathbf{MP}$ and $\mathbf{dl} \times \mathbf{MP}$ have as coordinates:

$$\mathbf{dl} \begin{vmatrix} -R \sin \varphi \, d\varphi \\ R \cos \varphi \, d\varphi \\ 0 \end{vmatrix} \qquad \mathbf{MP} \begin{vmatrix} -R \cos \varphi \\ r - R \sin \varphi \\ z - z_0 \end{vmatrix} \qquad \mathbf{dl} \times \mathbf{MP} \begin{vmatrix} (z - z_0) R \cos \varphi \, d\varphi \\ (z - z_0) R \sin \varphi \, d\varphi \\ (R - r \sin \varphi) R \, d\varphi \end{vmatrix}$$

and $MP^2 = R^2 + r^2 + (z - z_0)^2 - 2 R r \sin \varphi$, so that:

$$B_y = \frac{\mu_0 I}{4\pi} \int_0^{2\pi} (z - z_0) R \sin \varphi \left[R^2 + r^2 + (z - z_0)^2 - 2r R \sin \varphi \right]^{-3/2} d\varphi$$

$$B_z = \frac{\mu_0 I}{4\pi} \int_0^{2\pi} (R - r \sin \varphi) R \left[R^2 + r^2 + (z - z_0)^2 - 2r R \sin \varphi \right]^{-3/2} d\varphi$$

From the symmetry, $B_x = 0$.

To calculate the magnetic field created by the solenoid of length l at any point, it suffices to carry out a triple numerical integration, φ varying between 0 and 2π, then z_0 varying between $-l/2 + d/2$ and $l/2 - d/2$, with a step equal to d, the diameter of the wire making up the solenoid, and finally R varying from $d_1 + d/2$ and $d_2 - d/2$ with a step d, where d_1 and d_2 are the inside and outside diameters of the solenoid.

2.1.2. The solenoid

Maximum field produced without cooling less than 0.1 tesla

A current I flows in a coil, usually wound with enameled copper wire (low resistivity, therefore reduced losses by Joule heating) (fig. 26.21). By cooling the coil through water circulation, one can reach magnetic fields of 0.4 tesla, or even more, but at the price of increasing size.

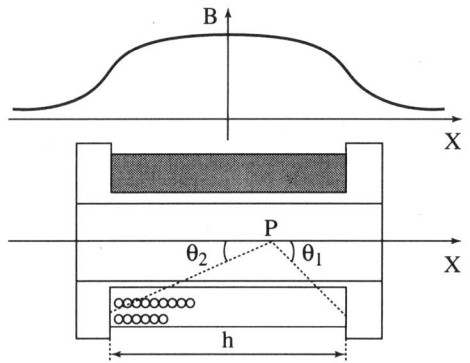

Figure 26.21 - Multi-layer solenoid and its field variation

The field created by a thin solenoid, at a P point along its axis, is given by: $B = \mu_0 \, n \, I \, (\cos\theta_1 + \cos\theta_2)/2h$ (tesla) where n is the number of turns, h the thickness of the windings (meter), θ_1 and θ_2 the angles subtended by the sides of the solenoid from point P.

In the case of a thick solenoid, a far from negligible error can be made by considering that θ_1 and θ_2 are the angles subtended by the average winding at the two ends of the solenoid.

For a coil with a current density j, by using the reduced coordinates $a = a_2/a_1$, $b = 0.5 \, h/a_1$, $c = x/a_1$ (a_1 and a_2 are the inside and outside radii of the coil), the expression of the magnetic field on the axis is:

$$\mu_0 H_x = \frac{\mu_0}{2} j a_1 \left\{ (b+c)\ln\frac{a+\left[a^2+(b+c)^2\right]^{1/2}}{1+\left[1+(b+c)^2\right]^{1/2}} - (c-b)\ln\frac{a+\left[a^2+(c-b)^2\right]^{1/2}}{1+\left[1+(c-b)^2\right]^{1/2}} \right\}$$

For an elongated solenoid, one can see that the field on axis at the end of the solenoid is practically equal to half the field at the center.

2.1.3. Superconducting coil

Magnetic fields presently up to 20 teslas

This is a solenoid constructed with a superconducting wire wound on an aluminium, stainless steel or fiberglass holder (fig. 26.21). The coil is cast in a special resin which must withstand the electrodynamic forces between conductors, perfectly immobilize them, and retain good mechanical properties at low temperatures.

The wire used in the coil is generally made up of a large number of superconducting filaments (typically from 50 to 500,000 depending on to the application, for conductors with a cross-section less than 1 mm^2) set in a resistive matrix, usually copper (fig. 26.22).

Figure 26.22 - Section of a superconducting wire
The small black circles are the filaments

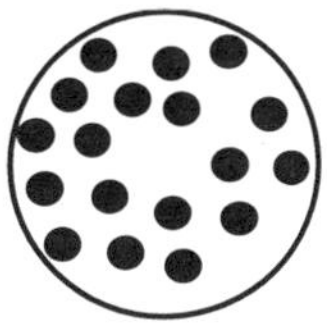

One uses type II superconductors because they have high critical fields: the magnetic flux penetrates progressively, above a certain critical field H_{c1}, through cylindrical zones called vortices, with axes parallel to the applied field, and with radius approximately the penetration depth. The flux in a vortex is equal to the flux quantum ϕ_0 (see chap. 19). Beyond H_{c1}, any growth of the field produces an increase in the flux that penetrates into the superconductor, that is in the number of vortices: they therefore must rearrange themselves in space (they come closer together as their number increases), which gives rise to a dissipation of energy. One greatly reduces the possibilities of vortex displacement, and therefore the energy dissipated, by using very fine filaments, whose size varies between a few tens of nanometres and a few tens of micrometers.

The current densities exceed 500 A.mm^{-2}.

The coil is maintained at a temperature well below the critical temperature of the superconductor used, because the critical field H_c of the superconductor (which must be greater than the field to be produced) decreases with increasing temperature.

The superconductors currently used have critical temperatures T_{crit} lower than 20 kelvin, and are cooled in a liquid helium cryostat (the boiling temperature of helium is 4.2 kelvin):

NbTi: $T_{crit} = 9.5$ K $H_c = 12$ teslas

practical maximum field at T = 4.2 K: $H_c \approx 9$ teslas

Nb_3Sn: $T_{crit} = 18$ K $H_c \approx 22.5$ teslas

practical maximum field at T = 4.2 K: $H_c \approx 19.5$ teslas.

These coils are often equipped with a superconducting wire *short circuit*, connected in parallel with the coil (fig. 26.23), and which can be heated above its critical temperature to make it resistive. If it is cooled down whilst a current flows in the coil, the current thus trapped continues to circulate without expending any energy, and one can even disconnect the electric supply from the coil.

The magnetic field is proportional to the current that creates it, and it can be measured using a shunt in series with the coil (fig. 26.23).

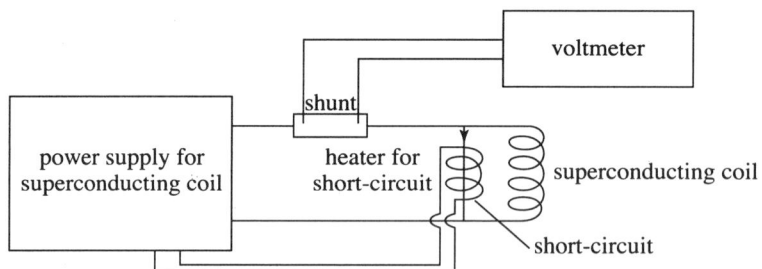

Figure 26.23 - Electric circuit diagram of a superconducting coil

The use of superconducting coils is currently limited to laboratories because the helium needed to cool the coils is expensive, and for this reason, at least in Europe, it circulates in a closed circuit. The liquid helium from a liquefier is transferred into the cryostat containing the superconducting coil. It is rapidly vaporized due to various losses (radiation, conduction), then it is reliquified.

The possibility of an industrial application of superconducting coils appeared recently: a coil of Nb_3Sn is cooled down to a temperature of 10 kelvin using a mechanical closed circuit cryocooler. There is no longer any need for liquid helium, and the device can function all year round with reduced maintenance whilst producing magnetic fields of 8, or even 14 teslas.

2.1.4. Bitter coils

Magnetic field: 25 teslas or more

This technique, based on iron-free coils (no magnetic materials), has been developed in some specialized laboratories. The conductors are copper disks which are stacked, and pierced with small vertically aligned holes to allow cooling by a high pressure water flow. Currents of the order of 10,000 amperes, with very high current densities reaching 300 A.mm^{-2}, can flow in the disks. The installed electric power is often greater than 10 megawatts. It can be shown that the maximum magnetic field, for a given geometry, is a function of the square root of the power P: $H \approx k\,(P)^{1/2}$. This useful approximation applies to any solenoid.

2.1.5. Hybrid techniques

To create a magnetic field, one can use several field sources based on the same or different techniques. The total field created is the sum of the various contributions (superposition theorem). In order to create a DC field of 38 teslas (world record), one uses a large superconducting coil of NbTi, which produces a field of 10 teslas. This coil is placed in a toroidal cryostat with a central hole about 50 cm in diameter, into which one places a Bitter coil which produces a field of 28 teslas.

2.1.6. Pulsed fields

Magnetic fields up to about 200 teslas

This a technique extensively used in the industrial world. For a reasonable cost it allows the generation of high fields for quite short times (from a few microseconds to milliseconds depending on the design of the circuit and the power of the installation). The usual technique consists of discharging a capacitor bank (stored energy for example 50 kJ under 20 kV) into a coil. To create an intense field, the stored capacitive energy $(1/2)\,CV^2$ should be large, and the inductance L of the coil be small: the coil can be made up of only one turn.

To summarize the results: the current created is $I = I_0 \exp(-t/\tau)\sin\Omega t$ with $\Omega = 2\pi/T$. T is close to $T_0 = 2\pi\,(LC)^{1/2}$ and $1/T = (1/T_0)[1 - (T_0/2\pi\tau)^2]^{1/2}$ with $\tau = 2L/R$. So that: $I_0 = V\,(C/L)^{1/2}\,T/T_0$. In practice, in many cases: $I_0 \approx V\,(C/L)^{1/2}$.

To give some orders of magnitude for a 200 teslas system: $C = 130\ \mu F$, $V = 40\ kV$. Stored energy: 100 kJ. Maximum current: 2.5×10^6 A; internal inductance: 20 nH; internal resistance: 0.003 Ω; rise time: 2.4 μs (quarter of a period).

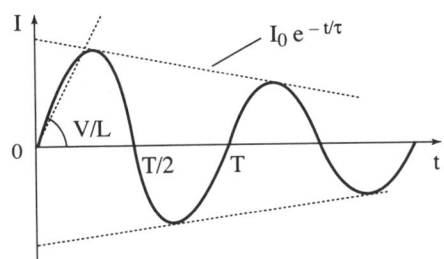

Figure 26.24 - Shape of the discharge current of the capacitor bank

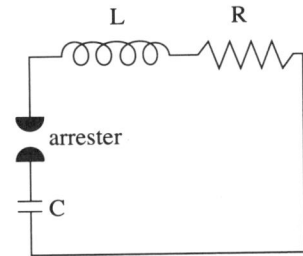

Figure 26.25 - Circuit diagram of pulsed field generator

If the material to be magnetized is metallic, the very fast variation of the magnetic field produces eddy currents that oppose the penetration of the magnetic field.

In this case, the essential parameter is the skin depth δ, the thickness of the metallic material producing an attenuation of the electric or magnetic field by a factor $1/e = 0.36$. In the case of a cylindrical symmetry and a sinusoidal AC field, it is given by: $\delta = (2/\mu\sigma\Omega)^{1/2}$, where $\mu = \mu_0\mu_r$ is the permeability of the material, σ the electric conductivity and $\Omega = 2\pi f$, f being the frequency (or the reciprocal of the characteristic time).

Lately, this technique has been applied to microcoils capable of producing large fields (40 T) in very small volumes (see fig. 1.5 in chap. 1).

2.2. GENERATION OF MAGNETIC FIELDS USING MAGNETIC MATERIALS

2.2.1. Some useful approximations

In chapter 2 it was shown that the field created by matter could be calculated as in electrostatics from magnetic masses with volume density: $\rho = -\mathrm{div}\,(\mathbf{J})/\mu_0$ and surface density $\sigma = \mathbf{J}.\hat{\mathbf{n}}/\mu_0$ where $\mathbf{J}$ is the polarization and $\hat{\mathbf{n}}$ the unit vector perpendicular to the surface of the material, pointing outwards. As we will see, the problem is often simplified when there are no volume distributions of masses, i.e. when $\mathrm{div}\,(\mathbf{J}) \approx 0$ or $\mathrm{div}\,(\mathbf{J}) = 0$. The latter case occurs if the material, placed in a constant magnetic field, is bounded by a second order surface (sphere, ellipsoid of revolution) or if the material has helimagnetic magnetization because, in these two particular cases, the divergence of $\mathbf{J}$ is rigorously zero.

Approximation $\rho = -\mathrm{div}\,(\mathbf{J})/\mu_0 \approx 0$

This leads to a big simplification for two very important types of material:

♦ **Soft unsaturated material**

In this case, one has: $\mathbf{B} = \mu_0\,\mathbf{H_i} + \mathbf{J} = \mu\,\mathbf{H_i} = \mu_0\,\mu_r\,\mathbf{H_i}$ with $\mu_r \gg 1$, where $\mathbf{H_i}$ is the internal field, and $\mathrm{div}\,(\mathbf{B}) = \mathrm{div}\,\{\mathbf{J}[1 + 1/(\mu_r - 1)]\} = 0$. Hence $\mathrm{div}\,(\mathbf{J}) = 0$.
In fact, real soft materials are not quite linear, but as μ_r is often the order of 1,000 or greater, one can write: $\mathrm{div}\,(\mathbf{J}) \approx 0$. Therefore, in a material with a high enough permeability, one can consider that there are no volume masses. One also has: $\mathrm{div}\,(\mathbf{H_i}) = \mathrm{div}\,(\mathbf{H_0} + \mathbf{H_d}) \approx 0$. If the applied magnetic field $\mathbf{H_0}$ is constant, then: $\mathrm{div}\,(\mathbf{H_d}) \approx 0$.

♦ **Rigid magnet**

For a rigid and perfect magnet, $\mathbf{J} \approx \mathbf{J_0}$, and this value is constant, so that also in this case: $\mathrm{div}\,(\mathbf{J}) \approx 0$. For a magnet placed only in its demagnetizing field, then: $\mathrm{div}\,(\mathbf{H_d}) \approx 0$.

For a system with axial symmetry, these results lead to valuable simplifications. If one knows the axial variation H_{dz} of the demagnetizing field, it is possible to deduce the radial field H_{dr}, at least close to the axis, by using Stokes' theorem:

$$\iiint \mathrm{div}\,(\mathbf{H_d})\,\mathrm{dv} = \iint \mathbf{H_d}\,\mathbf{dS}$$

One considers a disk of thickness dz, perpendicular to the z-axis and of radius a. Writing that the flux $\mathbf{H_d}$ coming out of the disk is zero:

$$2\pi\,a\,\mathrm{dz}\,H_{dr}(z) + \pi\,a^2\,[H_{dz}(z + dz) - H_{dz}(z)] = 0,$$

so that $H_{dr} = -(a/2)(dH_{dz}/dz)$.

Similary $\mathrm{div}\,(\mathbf{J}) \approx 0$ leads to an identical formula for $\mathbf{J}$.

Approximation $H_i = 0$ if μ_r is large

In a soft and unsaturated material, the internal field $H_i = H_0 + H_d$ is generally small in comparison to the applied field and to the demagnetizing field (except in the case of a closed magnetic circuit). Two consequences are possible:

♦ if the external field H_0 is constant, the demagnetizing field will be almost constant, and practically equal to $-H_0$,

♦ if H_0 is not constant the demagnetizing field will compensate its variations.

2.2.2. Flux channeling - case of a closed circuit

It is usually accepted that a high permeability magnetic material *channels* the flux, so that it behaves like an induction tube. It is easy to understand the mechanism with the help of an approximate calculation using magnetic masses. Assume a winding carrying a current I lies on a toroid of soft magnetic material: from § 2.2.1 we have div $(\mathbf{J}) \approx 0$ so that one can conclude that there are no volume magnetic masses.

Let H_0 be the field produced by the winding in the absence of any magnetic material, H_d the demagnetizing field. The resultant field (or internal field) is $H_i = H_0 + H_d$. At the center P_1 of the turn, the field $H_0(P_1)$ is greater than $H_0(P_2)$, P_2 being the diametrically opposite point on the toroid (fig. 26.26). To say that the flux is channeled means that the internal field $H_i(P_2)$ is practically the same as $H_i(P_1)$ since it creates the polarization, and therefore the induction. Surface magnetic masses, corresponding to field lines leaving the toroid, create a demagnetizing field that opposes at P_1, and at P_2 adds to, the field H_0 produced by the winding (fig. 26.26).

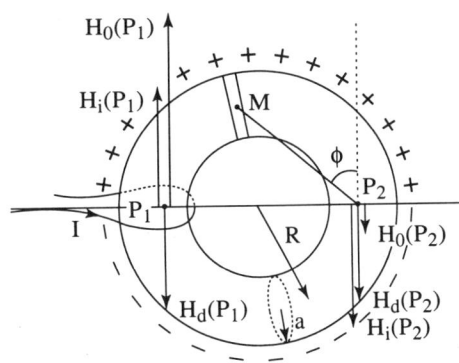

Figure 26.26
Flux channeling in a toroid
of soft magnetic material

The polarization $\mathbf{J}$ is the sum of a tangential component $\mathbf{J}_\ell$ and of a weak radial component $\mathbf{J}_r$: $\mathbf{J} = \mathbf{J}_\ell + \mathbf{J}_r$.

Between P_1 and P_2 we assume that the tangential component of the internal field decreases linearly so that: $dJ_c / d\ell = \mu_0 (\mu_r - 1) [|H_{i\ell}(P_2)| - |H_{i\ell}(P_1)|] / \pi R$, where $2\pi R$ is the mean perimeter.

We divide the toroid into thin disks whose radii are perpendicular to the director of the toroid (fig. 26.26), taking the faces of each disk as parallel, which involves the approximation R >> a. Applying the result of the § 2.2.1, we obtain an average density of surface charges: $\mu_0 \sigma = J_r = -0.5\, a\, dJ_c\, /d\ell$.

We can estimate $H_d(P_2)$ by using the equation:

$$H_d(P_2) = \iint \frac{\sigma \cos \phi\, ds}{4\pi\, MP_2^{\,2}}$$

where M is a point situated on the surface of the toroid, ds a differential surface area around M, and ϕ the angle between MP_2 and the direction of $\mathbf{H_0}$ at P_2. Since, by symmetry, $\mathbf{H_d}(P_2)$ is along this direction too.

Under some approximations, the analytical calculation shows that the relative difference between the fields at P_1 and P_2 is inversely proportional to the relative permeability of the toroid:

$$\frac{\left|H_i(P_2)\right| - \left|H_i(P_1)\right|}{\left|H_i(P_2)\right|} = \frac{1}{k\mu_r} \frac{\left|H_0(P_2)\right| - \left|H_0(P_1)\right|}{\left|H_0(P_2)\right| + \left|H_0(P_1)\right|}$$

with $k \approx (0.5\, a/\pi R)(1 - 0.5\, a/R)$. If μ_r is of the order of 1,000 and $a/R = 0.2$, this approximate calculation shows that the relative difference between $H_i(P_2)$ and $H_i(P_1)$ is close to $-1/k\mu_r$, and equal to -0.035. This corresponds to good (96.5%) flux channeling.

Remark - *We just calculated the demagnetizing field in a toroid. It is zero only if the coil that creates the field is wound regularly on the periphery of the toroid.*

2.2.3. Flux channeling

Case of a circuit with a gap: the electromagnet

Flux channeling is less efficient when there is a gap in the magnetic circuit. This is the case for example in the very widespread electromagnet.

An electromagnet can produce a variable magnetic field exceeding 2 teslas with an air gap of a few centimeters (fields of 3 T are possible with truncated cone-shaped pole pieces made of of iron-cobalt alloy).

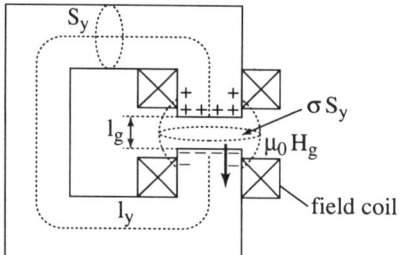

Figure 26.27
Diagram of an electromagnet

A coil of n turns carrying a current I is wound on a soft iron yoke of cylindrical cross-section S_y (diameter d) and of length l_y (fig. 26.27). The air gap is l_g, and has a cross-section equal that of the yoke $S_g = S_y$. Let B_y be the induction in the yoke. B_g is the average field in the air gap. If flux channeling was effective, then: $\phi_y = B_y\, S_y = B_g\, S_y$. Actually, one observes:

$$\phi_y \;=\; B_y\, S_y \;=\; \sigma\, B_g\, S_y \qquad (26.23)$$

where σ (> 1) is the Hopkinson coefficient. The induction in the neighborhood of the pole pieces is therefore B_y/σ. The numerical value of σ for poles of circular cross-section (diameter d) can be found in Hadfield [1]: $\sigma = 1 + 7\, l_g/d$.

Applying Ampere's law to a circuit made up of a yoke and an air gap:

$$H_y\, l_y + H_g\, l_g \;=\; n\, I \qquad (26.24)$$

This equation involves two approximations:
- that in the air gap the field $H_g = B_g/\mu_0$ is more or less constant, which is more correct the smaller the air gap.
- that H_y is constant. Far from the gap, the yoke can be considered as an induction tube such that the field $H_y = B_y/\mu_y$ which creates this induction is itself more or less constant.

One can rewrite equation (26.24) using equation (26.23): $B_g\,(\sigma l_y/\mu_y + l_g/\mu_0) = n\, I$.

If $\mu_y/\mu_0 \approx 500$ to $1{,}000$, and $l_y/l_g \approx 10$ to 20, and therefore $\sigma l_y/\mu_y \ll l_g/\mu_0$, then we have the general result (applicable to many electromagnets):

$$B_g \;\approx\; B_y/\sigma \;\approx\; \mu_0\, n\, I/l_g \qquad (26.25)$$

Therefore $\int B_g\, dl_g \approx$ constant in the region of the air gap: the circulation of the field in the region of the pole pieces is constant, which means that to a good approximation they are equipotential surfaces. The field lines are therefore normal to the surface of the pole pieces (provided they are not saturated, fig. 26.28), which also results from the law of refraction of field lines. This result is useful because it provides access to the stray field from the electromagnet. If a field line between the two poles has a length of twice the gap, then the average field along this field line will be twice as weak. One can approximate the lines of force leaving the sides of the pole pieces by semi-circles. One can therefore understand why, in the region of the gap, a substantial part of the flux $(1 - 1/\sigma)\, B_y\, S_y$ escapes through the side walls of the yoke: about 40% if $l_g = 0.1\, d$.

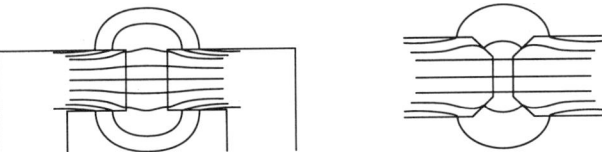

Figure 26.28 - Field lines of electromagnets
with flat or truncated-cone shaped pole pieces

Equation (26.24) can be re-written: $\phi\,(L_y/\mu_y\, S_y + l_g/\mu_0\, S_g') = n\, I$ with $S_g' = \sigma\, S_g$. The quantity $L/\mu\, S$ is called the reluctance $\mathscr{R}$ (see § 1.5 in chap. 2). Making the correspondence with Ohm's law, the quantity $n\, I$ (called the magnetomotive force $\mathscr{E}$) corresponds to a voltage, and the flux ϕ corresponds to the current I. The previous equation can be written: $\mathscr{E} = (\mathscr{R}_f + \mathscr{R}_e) \phi$. One can then apply the laws for resistances in series or in parallel to reluctances. The validity of the assumptions made, principally regarding the permeability of the yoke of the electromagnet, should always be checked.

Example - *Produce an induction $B_g = 1.8\,T$ in a gap $l_g = 0.04\,m$. Using equation (26.25), one obtains: $nI = 57,600$ ampere-turns.*

If one uses a current of 20 amperes, then the magnetizing coil is made of 2,880 turns of wire of section 4 mm², so that the cross-sectional area of the coil will be a little bigger than $4 \times 2,880$ mm², that is about 125 cm² (fig. 26.27). This electromagnet will be quite a bulky device.

One can also roughly estimate the electric power required to produce this field. The average length of a turn $l \approx 60$ cm (fig. 26.27). The electrical resistance of the coil is $R = \rho\,n\,l/S$ where $\rho = 1.8 \times 10^{-8}\ \Omega\cdot m$ is the electrical resistivity of the copper winding at room temperature. This gives $R = 4\ \Omega$. The power dissipated by Joule heating in the electromagnet is 1,600 W. Cooling is therefore necessary.

2.2.4. Permanent magnets

Magnetic fields reached: up to 1 tesla for classical installations (2 T in special cases)

This technique has developed very rapidly due to the recent progress in permanent magnets. The inductions obtained reach 1 T or more for $SmCo_5$ or $Nd_2Fe_{14}B$ magnets, and the coercive forces are well above 1 tesla. The relative permeability of a permanent magnet is not much greater than unity ($1 < \mu_r < 5$).

To calculate the induction created by a cylindrical magnet (length l, cross-section S) with constant polarization $J = J_0$, it can be seen that it is equivalent to a solenoid (n turns with a current I) having the same geometry with the correspondence: $\mu_0 n I/\ell \rightarrow J$ (magnetic polarization), or again, which comes to the same thing: $\mu_0 n I S \rightarrow J V$ ($= \mu_0 \mathfrak{m}$, where $\mathfrak{m}$ is the magnetic moment of the material of volume V).

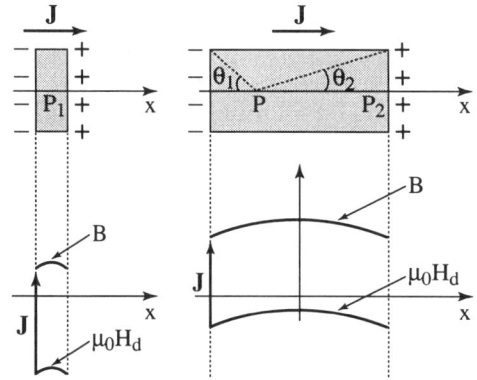

Figure 26.29 - Induction and demagnetizing field in magnets of various shapes

We assume the magnet to be rigid so that there are no volume poles. For a cylindrical magnet with a circular base having its magnetization parallel to the axis of the cylinder, surface magnetic poles $\sigma = \pm J/\mu_0$ appear on its two ends. As in electrostatics, the field created by a uniformly charged disk at a point P on its axis is: $\mu_0 H = 2\pi J (1 - \cos \theta_1)/4\pi$, where θ_1 is the semi-angle of the cone subtended by the disk at P (fig. 26.29).

The total demagnetizing field is therefore:

$$\mu_0 H_d = -J [1 - (\cos \theta_1 + \cos \theta_2)/2] \qquad (26.26)$$

so that the demagnetizing field coefficient: $N = 1 - (\cos \theta_1 + \cos \theta_2)/2$ is not constant.

The induction is: $B = \mu_0 H_d + J = J (\cos \theta_1 + \cos \theta_2)/2$ (tesla) (26.27)

This is similar to the equation for a solenoid (2.1.2). The magnet does not constitute an induction tube, since **B** varies. Using equations (26.26) and (26.27), the demagnetizing field and the induction along the axis are plotted as a function of the abscissa x (fig. 26.29), for two cylindrical magnets of the same material, one flat (in the form of a circular disk), and the other elongated. The induction created at P_1 by the flat magnet is about 3 times weaker than the induction created at P_2 by the other magnet, P_1 and P_2 being on the end faces of the cylinders.

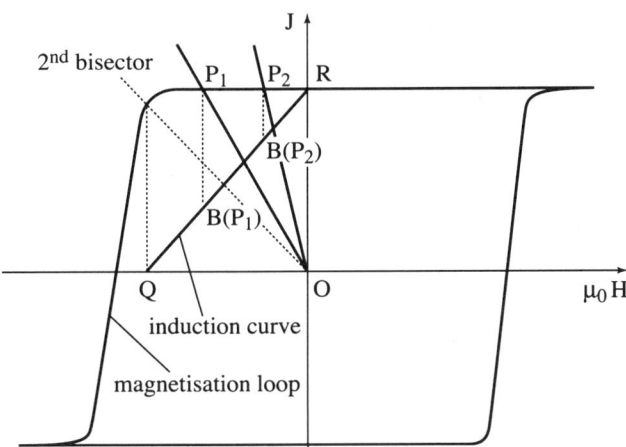

Figure 26.30 - Working points of 2 magnets with different shapes

The slopes of the straight lines OP_1 and OP_2 are $1/N1$ and $1/N2$.

It was assumed that the intensity of magnetization of the cylinder is constant, so that the material has an almost horizontal recoil curve, and its magnetization in its own demagnetizing field is practically constant.

The working point of a magnet is the intersection of the straight line with equation: $J = -\mu_0 H_d/N$ where N is the average demagnetizing field coefficient, with the recoil curve (the portion of the magnetization loop in the 2nd quadrant, see fig. 26.30).

The induction curve is easily obtained from the recoil curve, since, for a magnet placed in its own demagnetizing field, $B = \mu_0 H_d + J = J (1 - N)$.

For $N = 0 (1/N \rightarrow \infty)$, which corresponds to a very elongated sample, point R on the vertical axis is common to the two curves (magnetization and induction).

For $N = 1$ (2nd bisector - very flat sample): $B = 0$ (point Q).

2.2.5. *Almost closed U-shaped magnet*

Let us apply Ampère's law to a magnet of length l_m, of cross-section S_m, inside of which the average demagnetizing field is H_m. H_g is the field in the gap of length l_g. There is no threading current, so that: $H_m l_m + H_g l_g = 0$. It is important to note that H_m and H_g circulate in opposite directions.

On the two opposite faces of the pole pieces, the surface magnetic poles have a density J_m (polarization of the material in the field H_m). The magnetic field created by the magnetic charges on the pole pieces, calculated along the axis, is:

$$B_g = J_m [1 - (\cos \theta_1 + \cos \theta_2) / 2] \text{ (tesla)}.$$

If $l_g{}^2 / S_m << 1$, one has approximately $B_g = \mu_0 H_g \approx J_m$.

Note that in this approximation the average demagnetizing field is: $\mu_0 \mathbf{H_m} = -\mathbf{J_m} l_g / l_m$, so that the average demagnetizing field coefficient is: $N = l_g / l_m$.

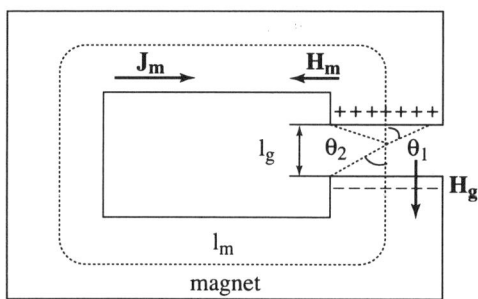

Figure 26.31 - Field produced by a U-shaped magnet

An induction greater than J_m can be produced by using truncated cone shaped pole pieces, if the gap is sufficiently small. The induction is greater than J_m because, in addition to the poles J_m on the plane faces of the pole pieces, there are magnetic poles $J_m \sin \theta$ on the sides of the cone with half-angle θ. Using magnets of $SmCo_5$ or $Nd_2Fe_{14}B$, a few centimeters in size, it is thus possible to create fields reaching or even exceeding 1 tesla.

2.2.6. *The magic cylinder*

Magnetic field created: 2 teslas

This cylinder is made up from magnets in the shape of trapezoidal segments (8 in fig. 26.32): in every segment, the magnetization vector is constant. From one segment to the next, the magnetization turns each time by 90°. The magnetic field thus created in a certain volume inside the cylinder is nearly constant in intensity and in direction, and is relatively high. One can use two concentric cylinders to create an even greater field. By rotating the cylinders about their axes in opposite senses, one creates a field with a given direction but of variable amplitude.

Figure 26.32
Magic cylinder

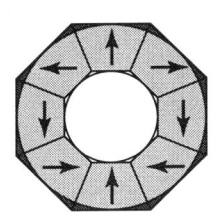
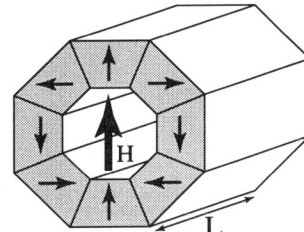

3. MEASUREMENT OF MAGNETIC FIELDS

One generally uses a Hall probe, although other methods, possibly more sensitive but more elaborate such as the "flux gate magnetometer" can be used. One can also rotate at high speed a small coil in the field to be measured: a voltage is induced, which is proportional to the speed and the field to be measured.

3.1. HALL PROBE

Sensitivity: generally from 10^{-5} to 10^{-6} tesla

A Hall probe is a small (surface area a few mm^2), very thin sensor made of a semiconductor material.

Figure 26.33
Circuit diagram
of a Hall effect probe

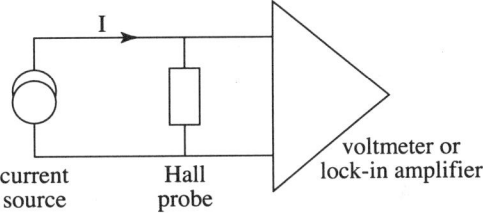

A DC or AC current of a few milliamperes is fed to the sensor. It is connected with four wires so as to measure only the Hall voltage, not the potential drop in the leads. The voltage, which is proportional to the field sensed by the probe, can be measured with a digital voltmeter. If an AC current is fed to the Hall element, one can measure the voltage with a lock-in amplifier for better resolution. A Hall effect probe can measure a DC or AC magnetic field. However, to measure an alternating field, the probe must be fed with a direct current.

Essential qualities of a Hall effect probe

In zero field a perfect probe should give zero voltage. In practice, one measures a voltage called *offset voltage* V_0 which is low for good quality sensors (for example

1 mV for a probe having a sensitivity S of 100 mV. T^{-1} so that the measured voltage is: $V = V_0 + SH$.

For an ordinary probe, S is not rigorously constant, and for example can vary for example by 10% in the range 0 to 2 teslas. The sensitivity for certain probes reaches 2 V. T^{-1}, which corresponds to a resolution of 10^{-6} tesla if the voltage is measured to a microvolt.

Some probes have a low temperature coefficient ($\approx 10^{-4}$ / kelvin) which allows the probe to be used without a calibration as a function of the temperature. The temperature range of ordinary probes is $-55°$ to $+100°C$. For measurements at lower temperatures, one often can use plain probes because *cryogenic* probes are very expensive.

3.2. *FLUX GATE MAGNETOMETER [2]*

This device allows to measure a dc field and also an ac field, in particular at low frequency, by an ac method. Its sensitivity can reach 10^{-11} tesla.

It usually consists of an elongated rod, having a small demagnetizing field coefficient and made of a material with high permeability μ_r. The sensor measures the component H_0 of the dc field parallel to the axis of the rod, which carries two windings:

♦ an excitation winding fed with an ac current: $i = I_0 \sin \omega t$. This current creates an alternating magnetic field : $H_a = H_m \sin \omega t$, whose intensity is sufficient to magnetize the core to saturation in alternate directions (figure 26.34 a and b),

♦ and a secondary coil measuring the output signal. The voltage across this coil is: $V_s = n S \, dB / dt$, where the induction B results from the applied field $H_0 + H_a$. n is the number of turns in the secondary coil and S the cross section of the magnetic rod.

In the absence of external field H_0, dB / dt shows positive and negative peaks at intervals π / ω, corresponding to a signal of frequency $f = \omega / 2\pi$ (dotted lines in figure 26.34 c and d). These peaks correspond to magnetization reversal in the core.

When the external magnetic field H_0 is not zero, the peaks no longer appear at equal intervals. The peaks at times 0, $2\pi / \omega$, $4\pi / \omega$ … are advanced (full lines in figure 26.34 c and d), whereas they are delayed at odd multiples of π / ω. The time shifts between two peaks is now $\pi / \omega + \delta t$ and $\pi / \omega - \delta t$. The signal, which for $H_0 = 0$ had only odd harmonics of frequencies f, 3f …, also features, for $H_0 \neq 0$, even harmonics, in particular that at frequency 2f. This originates from the non linearity of B with respect to H. To show this, let us assume for the sake of simplicity that there is no demagnetizing field effect (case of a very long rod with respect to its cross section). This does not change the principle of the device. The internal field H in the rod is then simply $H = H_0 + H_a$. Let us consider the beginning of saturation.

It corresponds to a downward curvature which can be accounted for by writing:

$$B = \mu H + \mu_3 H^3$$

where μ_3 is a third order permeability. The time derivative of B is then:

$$dB / dt = \mu H_m \omega \cos \omega t + 3\mu_3 H_m \omega \cos \omega t (H_0 + H_m \sin \omega t)^2$$

$$= (\mu + 3\mu_3 H_0^2) H_m \omega \cos \omega t + 3\mu_3 H_m^3 \omega \sin^2 \omega t \cos \omega t + 3\mu_3 H_m^2 H_0 \sin 2\omega t$$

This equation shows that the voltage induced in the secondary has a contribution at frequency 2f which is proportional to H_0, the field that one wants to measure. A more realistic approach of the dependence of B with respect to H leads to the same conclusion.

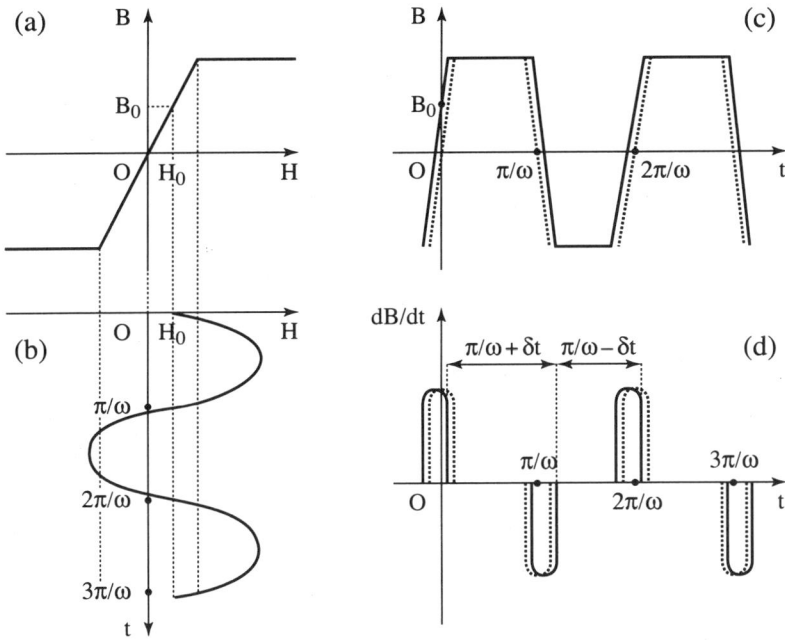

Figure 26.34
Schematic representation of the principle of the flux gate magnetometer

In practice H_0 is determined either by analysing the even harmonics or by measuring δt. Note that a third technique, which consists in measuring the amplitude of the signal, is also used.

The different types of flux gate magnetometer are based on the same principle. Figure 26.35 shows various construction modes of the flux gate probe. As can be seen from this figure the primary and the secondary windings can be merged.

Remarks - *Flux gate magnetometers are vectorial sensors, i.e. they allow to measure the three components of the field.*

- These sensors are much less sensitive to temperature variations than Hall probes.

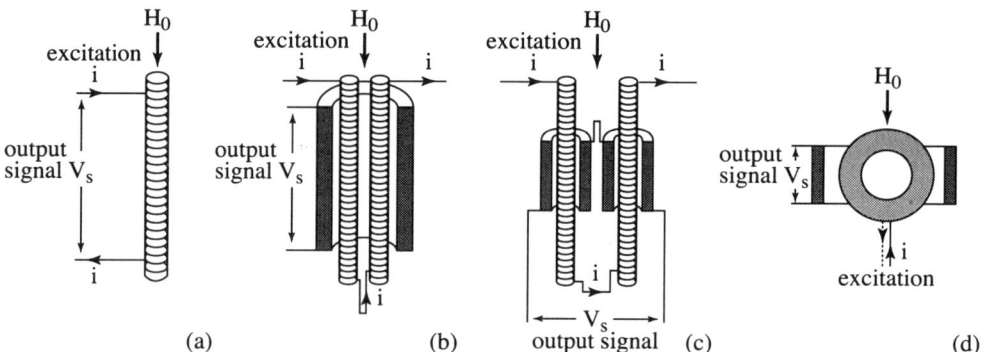

Figure 26.35 - Various configurations of the flux gate magnetometer

(a) device with a unique core rod with a unique winding (used for both the excitation and the signal) or with two windings (one for the excitation and the other for the signal)

(b) Vacquier type device with two core rods in which the ac magnetization directions are opposite. There is a unique winding for the output signal.

(c) Förster type device. It is also based on two core rods in which the ac magnetizations are opposite. The output signal is now measured by two windings connected in series.

(d) sensor with a magnetic core ring

REFERENCES

[1]   D. HADFIELD, Permanent magnets and magnetism (1962) Iliffe Books Ltd, London.

[2]   P. RIPKA, Sensors and Actuators A, *Physical* **33** (1992) 129.
        R.H. KOCH, J.R. ROZEN, *Appl. Phys. Letters* **78** (2001) 1897.
        J.G. DEAK, A.H. MIKLICH, *Appl. Phys. Letters* **69** (1996) 1156.

APPENDICES

APPENDIX 1

SYMBOLS USED IN THE TEXT

a   : Lattice parameter ; ka, Ma: 10^3, 10^6 years (Chap. 24). $\mathcal{A}$: Surface area.

A   : Number of nucleons ; A_{ex}: Exchange constant; **A**: Magnetic vector potential.

B   : Magnetic induction; B_1, B_2, ... $B^{\mu,\ell}$: Magnetoelastic coupling coefficients.

$\mathcal{B}$: Brillouin function.

c   : Speed of light; lattice parameter; c_{ij}, $c_{ij}{}^\mu$: Elastic moduli.

C   : Capacitance of a condensator, specific heat; $\mathcal{C}$: Curie constant.

d   : Density, distance.

e   : emf; – e: Electron charge; $\wp$: Electric potential (voltage).

E   : Energy density; E_F: Fermi energy. **E**: Electric field.

$\mathcal{E}$: Magnetomotive force; energy.

f   : Function, frequency.

F   : Free energy density; **F**: Force.

$\mathcal{F}$: Free energy.

g   : Acceleration of gravity; gap (airgap); g, g_J: Landé factor.

G   : Shear modulus, free enthalpy density; $\mathcal{G}$: Free enthalpy (Gibbs function).

h   : Planck's constant $\{\hbar = h/2\pi\}$.

H   : Magnetic field; H_C: Coercive field; H_{crit}, H_c: Critical field;
      H_A: Anisotropy field; H_d: Demagnetising field; H_0: Applied field (external);
      H_{mat}: Field produced by matter; H_{mol}: Molecular field.

$\mathcal{H}$: Hamiltonian.

i   : Instantaneous electric current; I: Effective electric current.

j   : Electric current density; $j = \sqrt{-1}$.

J   : Total angular momentum operator (in units of $\hbar$); $\mathbf{J}\,(=\mathbf{B}-\mu_0\mathbf{H})$: Magnetic polarisation.

$\mathcal{J}$: Exchange integral; J, j_i: Total angular momentum quantum number.

k   : Wave vector; k_B: Boltzmann's constant.

K_i : Magnetocrystalline anisotropy constants.

L   : Self inductance; L, l_i: Orbital angular momentum quantum number.

L, ℓ : Orbital angular momentum operator (in units of $\hbar$).

$\pounds$: Angular momentum (= $\mathbf{r} \times \mathbf{p}$).

$\mathscr{L}$: Langevin function.

m   : Mass; m: Reduced magnetisation {= M(T, H) / M_0}.

$\mathfrak{m}$: Magnetic moment; $\mathfrak{m}_{eff}$, $\mathfrak{m}_o$, $\mathfrak{m}_s$: Effective, orbital, spin magnetic moment. $\mathfrak{m}$: Magnetic moment (modulus).

M   : Mutual inductance; M_J, M_L, M_S, m_i: Magnetic quantum number.

M   : Magnetisation ($\mathbf{M} = d\mathfrak{m} / dV$); M_s: Spontaneous magnetisation; M_0: M_s (T = 0 K).

n,n_i : Principal quantum number; n_{ij}: Molecular field coefficient.

N   : Demagnetising field coefficient; Avogadro's number; N(ε): density of states.

p   : Pressure.

p   : Momentum (= $m\mathbf{v} + q\mathbf{A}$).

P   : Power.

q   : Electric charge; q_m: magnetic mass.

Q   : Quantity of electricity, quality factor, magneto-optic coefficient.

R   : Electric resistance; R_H: Hall coefficient.

$\mathscr{R}$: Reluctance; $\mathfrak{R}_{ij}$: Components of rotation (i $\neq$ j).

[s_{ij}]   : Elastic compliance tensor (in reduced notation).

S   : Entropy density; Surface area.

S   : Spin operator (in units of $\hbar$); oriented surface area; S, s_i: Spin quantum number.

t   : Time.

T   : Period, temperature; T_N: Néel temperature; T_C : Curie temperature.

T_{crit} : Critical temperature (e.g. superconducting); T_{comp}: Compensation temperature.

u   : Potential difference (voltage); U: Effective potential difference.

u   : Displacement vector; $u_{i,j}$: Distortion.

U   : Internal energy density; U: Magnetic potential difference.

$\mathscr{U}$: Internal energy.

v   : Velocity.

$\hat{\mathbf{v}}$: Unit vector (for instance : $\hat{\mathbf{r}} = \mathbf{OM} / |\mathbf{OM}|$).

V   :   Volume, electric potential.

$\mathscr{V}$:   Magnetic potential.

w   :   Molecular field coefficient.

W   :   Work.

Y   :   Young's modulus.

Z   :   Electrical impedance, atomic number.

α :   Damping coefficient; α_T: Linear thermal expansion coefficient.

α_i :   Direction cosines of magnetisation.

β_i :   Direction cosines of the observation direction.

γ_i :   Direction cosines of direction of application of a stress.

γ :   Gyromagnetic ratio; γ: Wall energy per unit area.

$[\gamma_{ij}]$:   Electric conductivity tensor.

Γ :   Torque, moment of a force.

δ :   Wall thickness. δ_{ij}: Kronecker symbol.

ε :   Permittivity, energy; ε_0: Permittivity of free space; $\boldsymbol{\varepsilon}$, $[\varepsilon_{ij}]$: Permittivity tensor, strain tensor.

η_p :   Viscous friction coefficient.

Θ_p :   Paramagnetic Curie temperature.

κ :   Compressibility.

λ :   Wavelength, spin-orbit coupling parameter; elongation; λ_s, λ^α, $\lambda^{\mu,\ell}$: Magnetostriction coefficient ($\mu = \gamma$, δ, ε, ζ ...; $\ell = 2$, 4, ...).

Λ :   Permeance.

μ :   Magnetic permeability; μ_0: Permeability of free space; μ_r: Relative permeability.

μ_B :   Bohr magneton.

ν :   Poisson's coefficient; frequency.

ξ :   Torsion; coherence length.

ρ :   (Mass) density, resitivity; ρ_e: Charge density; ρ_m: Magnetic mass density.

$[\rho_{ij}]$:   Electrical resistivity tensor.

σ :   Surface electric charge density; specific magnetic moment ($\sigma = \mathfrak{m}/m$); spin quantum number.

σ_m :   Surface magnetic mass density.

$[\sigma_{ij}]$:   Mechanical stress tensor; $\boldsymbol{\sigma}$: Conductivity tensor.

τ :   Relaxation time.

Φ :   Magnetic flux.

χ :   Magnetic susceptibility (χ_p, χ_d: paramagnetic, diamagnetic).

$\psi(\mathbf{r})$:   Wave function for the electron $\{\psi_k(\mathbf{r}) = u_k(\mathbf{r})\exp(i\mathbf{k}.\mathbf{r})$ for a Bloch wave$\}$.

ω :   Angular frequency, relative volume change; ω_L: Larmor (angular) frequency.

$\boldsymbol{\omega}$:   Angular velocity.

Vectors and tensors are printed in bold type. For instance : $\mathbf{r}_{IJ} = $ vector $\overrightarrow{IJ}$.

The modulus of a vector $\mathbf{V}$ is written : $|\mathbf{V}|$.

Unit vectors are printed with a circonflex accent (example : $\hat{\mathbf{r}} = \mathbf{OM}/|\mathbf{OM}|$).

The cross product is written : $\mathbf{A} \times \mathbf{B}$.

UNITS AND UNIVERSAL CONSTANTS

1. CONVERSION OF MKSA UNITS INTO THE CGS SYSTEM, AND OTHER UNIT SYSTEMS OF COMMON USE

Length (metre)   : 1 m $= 10^2$ cm $= 39.37$ " (inch) $= 10^{10}$ angström (Å).

Force (newton)   : 1 N $= 10^5$ dyn $= 0.102$ kgf.

Energy (joule)   : 1 J $= 10^7$ erg $= 0.7243 \times 10^{23}$ K $= 0.6241 \times 10^{19}$ eV.

Energy density   : 1 J.m^{-3} $= 10$ erg.cm^{-3}.

Power (watt)   : 1 W $= 10^7$ erg.s^{-1} $= 1.359 \times 10^{-3}$ CV $= 1.340 \times 10^{-3}$ hp.

Pressure (pascal)   : 1 Pa $= 10$ baryes $= 10^{-5}$ bar $= 1.02 \times 10^{-5}$ kgf.cm^{-2}
$= 7.49 \times 10^{-3}$ torrs $= 1.45 \times 10^{-4}$ psi.

Magnetic induction (tesla): $1\,T = 10^4$ gauss $(= 1\,Wb.m^{-2})$.

B can be called magnetic induction, magnetic induction field or magnetic flux density.

Magnetic field (ampere / metre):   1 A.m^{-1} $= 4\pi \times 10^{-3}$ œrsted.

The magnetic field **H** is often expressed in units of $\mu_0 H$, hence in tesla (T) or its submultiple, the gamma (γ) which is equal to 10^{-9} T. A field of 1 A.m^{-1} corresponds to 1.2566 μT.

Magnetisation (ampere / metre):   1 A.m^{-1} $= 10^{-3}$ emu.cm^{-3}.

Magnetic moment (ampere-square metre or joule per tesla):   1 A.m^2 $= 1$ J.T^{-1} $= 10^3$ emu.

Specific magnetic moment: 1 A.m^2.kg^{-1} $= 1$ emu.g^{-1}.

Note that, sometimes, magnetisation (**M**), magnetic moment ($\mathfrak{m}$) and specific magnetic moment (σ) are expressed in tesla, Weber-meter and Weber-meter per kilogram, respectively : the reason is that we have adopted as definition of magnetisation (d$\mathfrak{m}$/dV): $\mathbf{M} = \mathbf{B}/\mu_0 - \mathbf{H}$, whereas some authors call "magnetisation" the quantity $\mathbf{B}_i = \mathbf{J} = \mathbf{B} - \mu_0\mathbf{H}$ which is usually called "(magnetic) polarisation".

Temperature : to know the temperature in degrees Celsius or centigrades (°C), subtract 273.15 from the value in kelvin (K).

2. *SOME FUNDAMENTAL PHYSICAL CONSTANTS*

Name and symbol	Numerical value
Speed of light in vacuum, c	2.9979×10^8 m.s^{-1}
Permeability of vacuum, μ_0	$4\pi \times 10^{-7}$ H.m^{-1}
Permittivity of vacuum, $\varepsilon_0 = 1/c^2\mu_0$	8.8542×10^{-12} F.m^{-1}
Planck's constant, h	6.6261×10^{-34} J.s
$\hbar = h/2\pi$	1.0546×10^{-34} J.s
Acceleration of gravity, g	9.8066 m.s^{-2}
Electron rest mass, m	9.1094×10^{-31} kg
Electron charge (absolute value), e	1.6022×10^{-19} C
Bohr magneton, $\mu_B = e\hbar/2m$	9.2742×10^{-24} A.m^2
Flux quantum, $h/2e$	2.0678×10^{-15} Wb
Avogadro's number N	6.0221×10^{23} mol^{-1}
Boltzmann's constant, k_B	1.3807×10^{-23} J.K^{-1}

PERIODIC TABLE OF THE ELEMENTS

In the cells of the table below, for each element, the first line indicates its symbol, with its atomic number as superscript, the second one its atomic weight, the third one its spectroscopic ground state in the neutral atomic state, and finally the fourth one gives the electronic configuration of the external shells, again in the neutral atomic state.

Legend:
element — H^1 — atomic number
atomic mass — 1.008
$^2S_{1/2}$ — spectroscopic state
$1s^1$ — configuration of external shells

Main table:

H^1 1.008 $^2S_{1/2}$ $1s^1$																	He^2 4.003 $^1S_0$ $1s^2$
Li^3 6.940 $^2S_{1/2}$ $2s^1$	Be^4 9.012 $^1S_0$ $2s^2$											B^5 10.811 $^2P_{1/2}$ $2p^1$	C^6 12.011 $^3P_0$ $2p^2$	N^7 14.007 $^4S_{3/2}$ $2p^3$	O^8 15.999 $^3P_2$ $2p^4$	F^9 18.998 $^2P_{3/2}$ $2p^5$	Ne^{10} 20.180 $^1S_0$ $2p^6$
Na^{11} 22.990 $^2S_{1/2}$ $3s^1$	Mg^{12} 24.305 $^1S_0$ $3s^2$											Al^{13} 26.982 $^2P_{1/2}$ $3p^1$	Si^{14} 28.086 $^3P_0$ $3p^2$	P^{15} 30.974 $^4S_{3/2}$ $3p^3$	S^{16} 32.060 $^3P_2$ $3p^4$	Cl^{17} 35.453 $^2P_{3/2}$ $3p^5$	Ar^{18} 39.948 $^1S_0$ $3p^6$
K^{19} 39.100 $^2S_{1/2}$ $3d^04s^1$	Ca^{20} 40.08 $^1S_0$ $3d^04s^2$	Sc^{21} 44.95 $^2D_{3/2}$ $3d^14s^2$	Ti^{22} 47.90 $^3F_2$ $3d^24s^2$	V^{23} 50.941 $^4F_{3/2}$ $3d^34s^2$	Cr^{24} 51.996 $^7S_3$ $3d^54s^1$	Mn^{25} 54.938 $^6S_{5/2}$ $3d^54s^2$	Fe^{26} 55.847 $^5D_4$ $3d^64s^2$	Co^{27} 58.933 $^4F_{9/2}$ $3d^74s^2$	Ni^{28} 58.70 $^3F_4$ $3d^84s^2$	Cu^{29} 63.546 $^2S_{1/2}$ $3d^{10}4s^1$	Zn^{30} 65.38 $^1S_0$ $3d^{10}4s^2$	Ga^{31} 69.72 $^2P_{1/2}$ $4p^1$	Ge^{32} 72.59 $^3P_0$ $4p^2$	As^{33} 74.922 $^4S_{3/2}$ $4p^3$	Se^{34} 78.96 $^3P_2$ $4p^4$	Br^{35} 79.904 $^2P_{3/2}$ $4p^5$	Kr^{36} 83.80 $^1S_0$ $4p^6$
Rb^{37} 85.47 $^2S_{1/2}$ $4d^05s^1$	Sr^{38} 87.62 $^1S_0$ $4d^05s^2$	Y^{39} 88.906 $^2D_{3/2}$ $4d^15s^2$	Zr^{40} 91.22 $^3F_2$ $4d^25s^2$	Nb^{41} 92.906 $^6D_{1/2}$ $4d^45s^1$	Mo^{42} 95.94 $^7S_3$ $4d^55s^1$	Tc^{43} ~98 $^6S_{5/2}$ $4d^55s^2$	Ru^{44} 101.07 $^5F_5$ $4d^75s^1$	Rh^{45} 102.91 $^4F_{9/2}$ $4d^85s^1$	Pd^{46} 106.4 $^1S_0$ $4d^{10}5s^0$	Ag^{47} 107.87 $^2S_{1/2}$ $4d^{10}5s^1$	Cd^{48} 112.40 $^1S_0$ $4d^{10}5s^2$	In^{49} 114.82 $^2P_{1/2}$ $5p^1$	Sn^{50} 118.69 $^3P_0$ $5p^2$	Sb^{51} 121.75 $^4S_{3/2}$ $5p^3$	Te^{52} 127.60 $^3P_2$ $5p^4$	I^{53} 126.90 $^2P_{3/2}$ $5p^5$	Xe^{54} 131.30 $^1S_0$ $5p^6$
Cs^{55} 132.91 $^2S_{1/2}$ $5d^06s^1$	Ba^{56} 137.34 $^1S_0$ $5d^06s^2$	La^{57} 138.91 $^2D_{3/2}$ $5d^16s^2$	Hf^{72} 178.49 $^3F_2$ $5d^26s^2$	Ta^{73} 180.95 $^4F_{3/2}$ $5d^36s^2$	W^{74} 183.85 $^5D_0$ $5d^46s^2$	Re^{75} 186.21 $^6S_{5/2}$ $5d^56s^2$	Os^{76} 190.2 $^5D_4$ $5d^66s^2$	Ir^{77} 192.22 $^4F_{9/2}$ $5d^76s^2$	Pt^{78} 195.09 $^3D_3$ $5d^96s^1$	Au^{79} 196.97 $^2S_{1/2}$ $5d^{10}6s^1$	Hg^{80} 200.59 $^1S_0$ $5d^{10}6s^2$	Tl^{81} 204.37 $^2P_{1/2}$ $6p^1$	Pb^{82} 207.2 $^3P_0$ $6p^2$	Bi^{83} 208.98 $^4S_{3/2}$ $6p^3$	Po^{84} ~209 $^3P_2$ $6p^4$	At^{85} ~210 $6p^5$	Rn^{86} ~222 $^1S_0$ $6p^6$
Fr^{87} ~223 $6d^07s^1$	Ra^{88} 226.03 $^1S_0$ $6d^07s^2$	Ac^{89} ~227 $^2D_{3/2}$ $6d^17s^2$															

Lanthanides:

Ce^{58} 140.12 $^3H_4$ $4f^1$ $5d^16s^2$	Pr^{59} 140.91 $^4I_{9/2}$ $4f^3$ $5d^06s^2$	Nd^{60} 144.24 $^5I_4$ $4f^4$ $5d^06s^2$	Pm^{61} ~145 $4f^5$ $5d^06s^2$	Sm^{62} 150.35 $^7F_0$ $4f^6$ $5d^06s^2$	Eu^{63} 151.96 $^8S_{7/2}$ $4f^7$ $5d^06s^2$	Gd^{64} 157.25 $^9D_2$ $4f^7$ $5d^16s^2$	Tb^{65} 158.92 $^8H_{17/2}$ $4f^8$ $5d^16s^2$	Dy^{66} 162.50 $^7J_{10}$ $4f^9$ $5d^16s^2$	Ho^{67} 164.93 $^6K_{10}$ $4f^{10}$ $5d^16s^2$	Er^{68} 167.26 $^5K_{10}$ $4f^{11}$ $5d^16s^2$	Tm^{69} 168.93 $^2F_{7/2}$ $4f^{13}$ $5d^06s^2$	Yb^{70} 173.04 $^1S_0$ $4f^{14}$ $5d^06s^2$	Lu^{71} 174.97 $^2D_{5/2}$ $4f^{14}$ $5d^16s^2$

Actinides:

Th^{90} 232.04 $^3F_2$ $5f^0$ $6d^27s^2$	Pa^{91} 231.04 $^4F_{3/2}$ $5f^2$ $6d^17s^2$	U^{92} 238.03 $^5L_6$ $5f^3$ $6d^17s^2$	Np^{93} 237.05 $5f^4$ $6d^17s^2$	Pu^{94} ...	Am^{95}	Cm^{96}	Bk^{97}	Cf^{98}	Es^{99}	Fm^{100}	Md^{101}	No^{102}	Lw^{103}

MAGNETIC SUSCEPTIBILITIES

We list the magnetic susceptibilities (MKSA) of some so-called "non magnetic" substances: pure elements in the solid and liquid state, and materials of common use: alloys, plastics, glasses, and ceramics. The values usually were published in cgs units of specific susceptibility (χ_{cgs}). We therefore give the value of the density which was used to convert the susceptibility into MKSA units (χ is dimensionless in this system with our convention for the magnetization: it is simply given by: $\chi = 4\pi.d.\chi_{cgs}$ where d is the density expressed in g/cm^3).

We begin with diamagnetic substances, and then move on to paramagnetic substances.

Table A 4.1 - Susceptibility of some diamagnetic pure elements ($\times 10^6$)

Element	d g.cm^{-3}	$-\chi$	Element	d g.cm^{-3}	$-\chi$	Element	d g.cm^{-3}	$-\chi$
Ag	10.492	25	Au	18.88	34	B	2.535	20
Bi	9.78	165	C-diamond	3.52	22	Cd	8.65	19
Cu	8.933	11.8	Ga	5.93	23	Ge	5.46	7.25
H$_2$	0.0763	2.6	Hg	14.2	29	I	4.94	22
In	7.28	51	red P	2.20	20	Pb	11.342	16
S (α)	2.07	12.6	S (β)	1.96	11.4	Sb	6.62	67
Se	4.82	19	Si	2.42	3.4	grey Sn	7.30	29
Te	6.25	24	Tl (α)	11.86	37	Zn	6.92	15

All these elements were studied in their solid phase except for hydrogen and mercury, which are in the liquid state (in grey). The data were obtained at room temperature, except for hydrogen, measured at 13 K. The data are mostly taken from reference [1]. For the case of copper, for instance, one can check [2] that the thermal variations of χ are very weak from 4.2 K to 300 K.

Table A 4.2 - Susceptibility at room temperature
of some diamagnetic materials of common use ($\times 10^6$)

Material	d g.cm^{-3}	$-\chi$	Material	d g.cm^{-3}	$-\chi$	Material	d g.cm^{-3}	$-\chi$
Cu Sn 5	8.9	11	Cu Sn 12	8.9	14	Cu Sn 20	8.8	19
Cu Zn 10	8.9	12	Cu Zn 20	8.9	16	Cu Zn 40	8.9	24
Corning7052	2.27	8	Zerodur	2.52	12	Plexiglas	1.19	9
Polyethylene	0.923	10	Polyimide	1.43	9	PTFE	2.15	10
Alumina	3.87	12	Macor	2.52	11	Silica	2.21	12

The convention for the two first lines is that, for example Cu Sn 5 is a tin-copper alloy with 5 wt% tin. The absolute values of the susceptibility of bronzes (first line), and brasses (second line) increases when the copper content decreases [1]. The third line shows two glasses [3], and Plexiglas [4]. The fourth line presents three plastics (PTFE is often called Teflon), and the fifth line, two ceramics and fused silica (amorphous) [2].

Table A 4.3 - Paramagnetic susceptibility at room temperature
of some pure elements ($\times 10^6$)

Element	d g.cm^{-3}	χ	Element	d g.cm^{-3}	χ	Element	d g.cm^{-3}	χ
Al	2.70	21	Ba	3.5	6.6	Ca	1.55	20
Cs	1.873	5.2	Hf	13.3	70	Ir	22.42	38
K	0.87	5.8	La	6.174	66	Li	0.534	14
Lu	9.842	>0	Mg	1.74	5.5	Mo	9.01	105
Na	0.9712	8.5	Nb	8.4	232	Os	22.5	15
Pd	12.16	815	Pt	21.37	278	Rb	1.53	38
Re	20.53	97	Rh	12.44	170	Ru	12.1	65
Sc	2.992	263	white Sn	7.30	2.4	Sr	2.60	34
Ta	14.6	177	Th	11.0	79	Ti	4.5	180
U-α	18.7	404	V	5.87	370	W	19.3	78
Y	4.478	120	Yb	6.959	126	Zr	6.44	108

All the elements given in this table are solid are room temperature. Magnesium was found diamagnetic by Pascal [5], and paramagnetic by Foëx [1] and then by Thomas, and Mendoza [6]. The reported susceptibility is that given by the latter. One can notice that the weakest susceptibilities are observed for alkali metals.

Table A 4.4 - Paramagnetic susceptibilities of some alloys of common use, at various temperatures ($\times 10^6$) [7]

	304 (N)	304 L	316 (LN)	90Cu 10Ni
T = 300 K	3.1	2.9	3.2	210
T = 77 K	6.6	7.6	8.7	...
T = 4.2 K	9.0	11.3	14.2	280

Note that stainless steels (304 and 316) exhibit paramagnetic susceptibilities comparable to those of alkali metals, i.e. at least 20 times smaller than for other weakly magnetic alloys, even at low temperature. This is why they are widely used in magnetic instrumentation, e.g. for the manufacturing of cryostats for super-conducting coils.

There exist several tables of numerical value where engineers can find complementary information on paramagnetism in solids, e.g. that edited by Landolt et Börnstein [8]. *Important comment: some tables still give numerical values in cgs units, and the reader should thus be careful, and not confuse them with data in SI / m^3, which can be **of the same order of magnitude**!*

Note that a ferromagnetic impurity can strongly disturb the magnetic properties of very weakly magnetic substances, even for concentrations of the order of a ppm: the perturbation will generally be temperature dependent, and be larger at low temperature.

On the other hand, special care must be taken in processing materials expected to exhibit very weak magnetism: brazing or soldering under argon atmosphere is for instance liable to precipitate a strongly magnetic phase in a weakly magnetic alloy. Stainless steels that are well known to be very weakly magnetic, are then seen to become slightly ferromagnetic! Special care is therefore required, and it is important to check the magnetic properties of an allegedly non magnetic alloy before using it, and after thermal or mechanical processing. An extremely simple method to check the magnetism of a metallic part consists in hanging a strong magnet (Nd-Fe-B or $SmCo_5$) at the end of a 1 metre long thin nylon thread, and bringing it near the part to be checked: if the slightest attraction is observed, the substance is already slightly ferromagnetic or at least very strongly paramagnetic.

REFERENCES

[1]   Tables de constantes et Données numériques: 7. Constantes sélectionnées: G. FOëX, Diamagnétisme et paramagnétisme; C.J. GORTER, L.J. SMITS, Relaxation paramagnétique (1957) Masson & Cie, Paris, 317 p.

[2]   C.M. HURD, *Cryogenics* **6** (1966) 264.

[3]   P.T. KEYSER, S.R. KEFFERTS, *Rev. Sci. Instrum.* **60** (1989) 2711.

[4]   PH. LETHUILLIER, private communication.

[5]   P. PASCAL, Nouveau traité de chimie minérale, Tome IV (1958) 143, Masson, Paris.

[6]   J.G. THOMAS, E. MENDOZA, *Phil. Mag.* **43** (1952) 900.

[7]   Handbook on materials for superconducting machinery, metals and ceramics (1974) Battelle Information Center, Columbus (Ohio).

[8]   LANDöLT, BöRNSTEIN, Zahlenwerte und Funktionen aus Physik. Chemie. Astronomie. Geophysik. Technik II - 9 - Magnetische Eigenschaften 1, Section 29.1111 (1962) Springer, Berlin.

FERROMAGNETIC MATERIALS

We summarize the structural physical properties, the linear thermal expansion coefficient, the elastic properties (tab. A 5.1) and the intrinsic magnetic properties (tab. A 5.2) of some common strongly magnetic substances .

Table A 5.1 - Physical properties of some strongly magnetic substances, measured at room temperature

Substance	Molecular weight	Structure	a	c / a	Density	α_T	c_{11}	c_{12}	c_{44}
	g		nm		$g.cm^{-3}$	$10^{-6} K^{-1}$	GPa	GPa	GPa
iron	55.847	bcc	0.287	1	7.85	12.1	237	140	116
cobalt	58.933	hcp	0.251	1.622	8.84	12.4	307*	165*	75.5*
nickel	58.70	fcc	0.352	1	8.90	12.8	250	160	118.5
gadolinium	157.25	hcp	0.363	1.591	7.90	-	67.8*	25.6*	20.8*
Fe_3O_4	231.54	spinel	0.839	1	5.19	-	273	106	97
$CoFe_2O_4$	234.63	spinel	0.838	1	5.29	-	-	-	-
$NiFe_2O_4$	234.39	spinel	0.834	1	5.38	7.5	220	109.4	81.2
$Y_3Fe_5O_{12}$	737.95	garnet	1.238	1	5.17	-	269	107.7	76.4
Fe-80%Ni	(58.10)	fcc	0.354	1	8.65	12.0	-	-	-
Cu_2MnAl	209.01	bcc	0.596	1	6.55	-	135.3	97.3	94
$BaFe_{12}O_{19}$	1111.5	h	0.589	3.937	4.5	-	-	-	-
$SmCo_5$	445.01	h	0.500	0.794	8.58	-	-	-	-
Sm_2Co_{17}	1302.7	h	0.838	0.973	?	-	-	-	-
$Nd_2Fe_{14}B$	1081.1	tetra.	0.879	1.389	7.60	-	-	-	-

*   *For hexagonal substances, two additional elastic constant have to be considered: for cobalt, they are $c_{13} = 103$ GPa and $c_{33} = 358$ GPa; for gadolinium, $c_{13} = 20.7$ GPa and $c_{33} = 71.2$ GPa.*

Table A 5.2 - Magnetic properties of some strongly magnetic substances

Substance	T_C K	M_s $kA.m^{-1}$	$\mu_0 M_s$ T	σ_s $A.m^2.kg^{-1}$	K_1 $kJ.m^{-3}$
iron	1043	1720	2.16	218	48
cobalt	1394	1370	1.72	162	530
nickel	631	485	0.61	56	− 4.5
gadolinium	289	2117	2.66	268	(at O K)
Fe_3O_4	858	477	0.60	91.0	− 13
$CoFe_2O_4$	793	398	0.50	75.2	180
$NiFe_2O_4$	858	271	0.34	50.4	− 6.9
$Y_3Fe_5O_{12}$	553	139	0.17	26.1	− 2.5
Fe-80%Ni	595	828	1.04	95.7	− 2*
Cu_2MnAl	610	560	0.70	85.5	− 0.47
$BaFe_{12}O_{19}$	723	382	0.48	84.9	250
$SmCo_5$	995	836	1.05	97.4	17 000
Sm_2Co_{17}	1190	1030	1.29	101	3 300
$Nd_2Fe_{14}B$	585	1280	1.61		4 900

The values of magnetization (M_s and T_s) and of anisotropy (K_1) are for room temperature (300 K), except in the case of gadolinium, for which the values at O K are given.

* K_1 *is very sensitive to thermal treatments for Permalloy (Fe-80%Ni), and changes from* − 2 kJ.m^{-3} *to a value ten times smaller after quenching.*

Note - *The magnetization M of a substance, the magnetic moment of which is expressed in μ_B per molecule $\sigma(\mu_B)$, is obtained in A/m through the relation:*

$$M = 5.585 \times \frac{\sigma(\mu_B) \times d(kg.m^{-3})}{M_{mol}(kg)} = 5.585 \times 10^6 \times \frac{\sigma(\mu_B) \times d(g.cm^{-3})}{M_{mol}(g)}.$$

ECONOMIC ASPECTS OF MAGNETIC MATERIALS

The aim of this Appendix is to illustrate, through figures, the economic impact of the major applications covered in this book, without in any way claiming to offer an exhaustive account of the global economic facts pertaining to the magnetic material market, the importance of which is generally underestimated. Information comes from diverse sources and, at times, it is difficult to separate out the turnover relevant to raw material from that of processed material, or from a system that has reached the market.

1. INTRODUCTION

In order to grasp the reality behind the figures below, we shall give here some essential facts. The various materials produced represent an *annual turnover* for the firms that produce them: this is this figure which is estimated here. These materials are incorporated into systems, devices, etc., which have a value higher than that of the material itself, frequently an order of magnitude higher. However, without the material, nothing would exist. It is estimated that the component amounts to 5-10% of the total cost.

Magnetic materials are *key components* of our everyday lives, in traditional domestic applications as well as in rapidly growing technologies such as information and telecommunications. Constant research leads both to improvements in existing materials and to the discovery of others, thus raising the performance of current products and enabling the creation of new ones. In the field of energy, the principal market for magnetic materials, raising the efficiency of electric motors with medium power from 90 to 92%, corresponds, on the world scale, to the electricity output of a nuclear power plant.

First we shall consider the three principal markets. These are:
- the hard materials,
- the soft materials for electrical engineering and electronics, and
- the materials for magnetic recording.

We then discuss, more briefly, some other magnetic materials.

2. HARD MATERIALS FOR PERMANENT MAGNETS

2.1. THE APPLICATIONS FOR MAGNETS

Permanent magnets are used in very diverse systems and may be divided into four
different categories of application.

2.1.1. Electromechanical systems

These are the DC brushless motors, the linear or rotating actuators used principally in
the car industry for electrical accessories (e.g. window lift motors), in domestic
electrical appliances, and also in disk or floppy disk drives, in computer printers, in
electro-acoustic applications (loudspeakers, microphones…) and in galvanometric
measuring instruments.

2.1.2. Systems using magnetic force at the macroscopic scale

Applications are varied, they range from magnetic pins, used to affix posters, and the
closure mechanisms of refrigerator doors, to levitating vehicles, and include eddy
current retarders.

2.1.3. Applications to particle physics

Magnets are found primarily in the systems used for guiding electrons or ions:
cathode ray tubes, travelling wave tubes, bending magnets in particle accelerators,
undulators and w*igglers* in synchrotron radiation sources.

2.1.4. Medical applications

These applications, which are now expanding rapidly, require magnets as components
in mini-actuators or micro-motors, for guiding catheters, and also as sources of
DC magnetic field for medical imaging (see chap. 23). This last application has been
first developed in Japan by Hitachi Medicals.

2.2. THE MAIN FAMILIES OF PERMANENT MAGNETS

The world market in permanent magnets was equivalent to six billion euros in 1999
and has been growing by more than 10% per annum for several years. This strong
growth is fed by the discovery of new, more efficient and more stable magnets, which
has allowed the development of new applications which were technically inaccessible
before. The commercialization of neodymium-iron-boron magnets since the end of the
1980's is, for the moment, the principal driving force for this growth.

Four main families of permanent magnets essentially share the world market. They
complement each other and each occupies a dominant position in specific applications.

2.2.1. Hard ferrite magnets

These magnets have a weak specific energy but resist demagnetization well, are chemically very stable and may be used for temperatures up to 200°C. The least expensive of the magnets available on the market, both per unit volume and per unit energy, they make up more than 90% of the world production in weight and 55% in value. The automotive and acoustics industries are the principal users (electrical equipment for vehicles, sensors for ABS, loudspeakers and, recently, starter motors etc.).

2.2.2. Samarium-cobalt magnets

Invented at the end of the 1960's, these magnets gave rise to new applications by virtue of their clearly superior magnetic performance over the magnets of the period (AlNiCo, ferrites). They currently represent 8% of the market, but their relative importance is diminishing with the commercialization of the neodymium-iron-boron magnets. Their use rests principally in applications where the demand for temperature stability is more important than cost: aeronautics, space, telecommunications, large magnetic motors for military equipment and magnetic couplings, in particular in the chemical industry. They show a strong remanent induction, and are resistant both to corrosion and to demagnetization up to 400°C. The high cost of these magnets, linked to the strategic nature of cobalt, has led their users to prefer the neodymium-iron-boron magnets.

2.2.3. Neodymium-iron-boron magnets

The most recent arrivals on the market, they have been the object of continuous and successful research aiming at a reduction of their original disadvantages: their tendency to rust, and the very strong degradation of their magnetic properties above 200°C. At room temperature, they present a remanent magnetization higher than that of the samarium cobalt magnets. A whole range of grades are currently available, which allows these magnets to be very much at the forefront in many applications, particularly in the computer industry. Thanks to their lower cost for the same performance, they have replaced some of the samarium-cobalt magnets. They could be used in the propulsion motors of electric vehicles of the future and serve, especially in Japan, to create the static magnetic fields for medical NMR imaging.

The Nd-Fe-B magnets represented about 33% of the global turnover of the magnet industry in 1998, a figure rising by 13% per year.

2.2.4. AlNiCo magnets

The oldest of all, these magnets have retained about 5% of the world market. They cover almost exclusively the market for measuring devices: voltage, current, electric meters, etc. These are the magnets that demonstrate the best thermal stability around room temperature.

2.3. BONDED MAGNETS

In equal use are the ferrites and Nd-Fe-B in dilute form known as bonded magnets (see chap. 15): we should bear in mind that the base material is a coercive powder mixed with a polymer. This results in a 40% reduction in the maximum induction. The bonded ferrites are obtained by injection, extrusion or compression. This procedure allows the production of complex shapes, unobtainable by casting or sintering. All refrigerator doors are equipped with closure mechanisms of this type. The Nd-Fe-B bonded magnets are manufactured from ribbons which are generally obtained by roller quenching , they are then finely ground before being mixed with a polymer. They perform better than the sintered ferrites but their cost is too high for them to replace the latter in small electric motors, except for very special applications.

This technology has been developing fast since the 1980's. The bonded magnets represent 20% of the turnover for the total production of permanent magnets.

2.4. GEOGRAPHICAL DISTRIBUTION

With regard to the distribution of the world production of permanent magnets in 2000, China and Japan each produce 40% on their own territory (Japan controls more than 50% of the market). The USA and Europe, with about 10% each, are the third largest producers. This estimate by country is evolving very quickly in a climate of increasing globalization, which tends to bring the production of material ever closer to its point of use.

3. SOFT MATERIALS

The soft magnetic material market is essentially that for electrical engineering, dominated by ferromagnetic alloys. In electronics, where the operating frequency lies in the radio and microwave ranges, insulators or semi-conductors such as ferro- or ferrimagnetic oxides are used. In the intermediate frequency range, special alloys are used.

3.1. SOFT IRON

This market is difficult to quantify (about 1,000 tonnes a year): it is primarily concerned with flux guidance in traditional applications (magnetic circuit closure, all kinds of electromagnets, the sorting of minerals or metals, hoisting, small domestic electrical appliances, etc.) where performance becomes secondary to cost. The standard material is based on "normal" iron, sometimes alloyed with carbon or manganese.

3.2. IRON-SILICON ALLOYS

These are used in classical electrical engineering and are primarily iron-silicon sheets (with 3% silicon), either grain oriented (transformers) or non-oriented (rotating machines). The world production of grain oriented sheets is 1.5 million tonnes, and for non grain oriented sheets 6 million tonnes, representing a market of around six billion euros. Japan is the world leader; Europe's contribution is around 20% of the world production.

In addition, about 1,000 tonnes of non grain-oriented iron-silicon sheets with 6.5% silicon are produced per annum. These are used for their high permeability, but their cost is high.

3.3. SPECIAL ALLOYS

Other more elaborate industrial products are used for very high frequency applications, up to 50-100 kHz in telephony, in security systems, in micro-electronics... The most commonly used materials in this field are the iron-nickel alloys (Fe-Ni 50-50, Permalloys, Supermalloys...) and the iron-cobalt alloys. The world production is 12,000 tonnes per annum but their high cost puts a brake on their development. The metallic amorphous materials are produced in the USA, Germany and Japan. In the USA they are used in power distribution transformers of average power. The annual world production is around 20,000 tonnes, with a value of about 180 million euros.

3.4. HIGH FREQUENCY MATERIALS

In the radio frequency range, substituted ferrites, manganese-zinc or nickel-zinc depending on the application, are used (see chap. 17). Here we can mention cellular telephony, a market in full expansion. As regards microwave frequency applications, the hexaferrites and rare earth substituted yttrium iron garnet YIG are suitable. The soft magnetic material market for electronic applications can be assumed to be in excess of one billion euros.

4. MATERIALS FOR MAGNETIC RECORDING

The magnetic and magneto-optical recording industry covers the needs of both the "general public" and the professionals. The most popular applications fall within the *audio-visual* province: audio cassettes, video cassettes, video tapes and video cameras. However *computer technology* uses more than half of the world output of products in the field of recording in the form of hard disks, diskettes, DVD, magnetic tape and reading heads, with a very strong growth in demand.

In 2001 the turnover in *materials* used for magnetic recording represented more than 47 billion euros globally: 40 for hard disk drives, 5 for tape drives, 2 for floppy drives.

This industry, which is in constant technical evolution, has as its primary objective the enhancement of storage capacity per unit area; for example, the demand for computers is rising at 60% per annum, half of which is for hard disks, and this trend is expected to continue for at least ten years. In 2001 the most common density stood at 1.5 to 3.0 $Gb.cm^{-2}$ (see fig. 1.8). The technologies permitting 10 $Gb.cm^{-2}$ to be attained are known and will be implemented rapidly, making them available to the market early in the new millennium. The substitution of longitudinal recording by a perpendicular method should allow surface densities of 15 to 30 $Gb.cm^{-2}$ to be reached, with reading heads made of giant magneto-resistance transducers.

5. OTHER MAGNETIC MATERIALS

Compared to the three giants –the hard and soft materials, and the materials used in recording– the production of other magnetic materials for industrial applications remains very marginal, in tonnage as well as in terms of their market share.

5.1. MAGNETOSTRICTIVE MATERIALS

Subject to competition from piezo-electric ceramics and handicapped by its price, Terfenol-D, which appeared on the market in the 1980's, is developing slowly. One Swedish company has ceased production altogether. Nevertheless the costs for mass production are decreasing considerably and the company Etrema Products has increased its production area three-fold in the USA. TDK is now offering a composite at 2 $euros.g^{-1}$, which is three times less expensive than Terfenol D in the 1990's. This product can be envisaged as remaining competitive in the space and aeronautic industries. Eight producers have already put this product on the market, four in Japan, two in Europe and two in the USA.

The magnetostrictive materials for sensors, in particular the Metglas 2605 SC which also reached the market in the 1980's, have a promising future before them, even though they involve very modest quantities of material by comparison with the tonnages required by transformers.

5.2. FERROFLUIDS

Here too, the quantities of material involved are very modest, and most of the applications envisaged –described in chapter 22– are only at the experimental stage. As yet few companies commercialise this product: to take one example, Ferrofluidics in the USA both produces ferrofluids and sells the applications for these materials; a company such as Advanced Fluid Systems in Great Britain also sells applications.

GENERAL REFERENCES

AHARONI A. - *Introduction to the theory of ferromagnetism*. Clarendon Press, Oxford, 1996.

BRISSONNEAU P. - *Magnétisme et matériaux magnétiques pour l'électrotechnique*. Hermès, Paris, 1997.

BLUNDELL S. - *Magnetism in Condensed Matter*. Oxford University Press, Oxford, 2001.

BUSCHOW K.H. & WOHLFARTH E.P. Eds - *Ferromagnetic Materials*. North Holland, Amsterdam (a multi-volume series).

CHIKAZUMI S. - *Physics of magnetism*. Wiley, New York, 1964; *Physics of ferromagnetism*. Clarendon Press, Oxford, 1997.

COEY J.M.D. Ed - *Rare earth permanent magnets*. Clarendon Press, Oxford, 1996.

CULLITY B.D. - *Introduction to magnetic materials*. Addison-Wesley, Reading, 1972.

DURAND E. - *Magnétostatique*. Masson, Paris, 1968.

HERPIN A. - *Théorie du magnétisme*. Presses Universitaires de France, Paris, 1968.

HUBERT A. & SCHAEFER R. - *Magnetic domains. The Analysis of Magnetic Microstructures*. Springer, Berlin-Heidelberg-New York, 1998.

LANDAU L.D., LIFSHITZ E.M. & PITAEVSKII L.P. - *Electrodynamics of Continuous Media*. Elsevier, 1997.

LANDAU L.D. & LIFSHITZ E.M. - *Classical theory of fields*. Butterworth-Heinemann, 1997.

MATTIS D.C. - *Theory of magnetism*. Harper & Row, New York, 1965.

MORRISH A.H. - *The physical principles of magnetism*. Wiley, New York, 1965.

NEEL L. - *Oeuvres scientifiques*. Editions du CNRS, 1978.

O'HANDLEY R.C. - *Modern magnetic materials: principles and applications*. Wiley, New York, 2000.

SKOMSKI R. & COEY J.M.D. - *Permanent magnetism*. Institute of Physics Publishing, Bristol and Philadelphia, 1999.

INDEX BY MATERIAL

INDEX BY SUBJECT

$$\frac{10}{R10-3A}$$